Complex Analysis

Edition: 2025 Updated Edition

Website: https://www.westmont.edu/people/russell-w-howell-phd

To Kay and Fran

Table of Contents

Preface

Approach After six editions of publishing with Jones and Bartlett Learning we are delighted that we can now make this text available as an open-source document, and have designed it for students in mathematics, physics, and engineering at the undergraduate level. Our goal is to illustrate the theoretical concepts and proofs with practical applications, and to present them in a style that is enjoyable for students to read. We believe both mathematicians and scientists should be exposed to a careful presentation of mathematics. Our use of the term "careful" here means paying attention to such things as ensuring required assumptions are met before using a theorem, checking that algebraic operations are valid, and confirming that formulas have not been blindly applied. We do not mean to equate care with rigor, as we present our proofs in a self-contained manner that is understandable by students having studied multivariable calculus. For example, we include Green's theorem and use it to prove the Cauchy-Goursat theorem, although we also include the proof by Goursat. Depending on the level of rigor desired, students may look at one or the other—or both.

We aim to give sufficient applications to motivate and illustrate how complex analysis is used in applied fields. Computer graphics help show that complex analysis is a computational tool of practical value. The exercise sets offer a wide variety of choices for computational skills, theoretical understanding, and applications that were class tested for six editions of the text when it was available for purchase (*i.e.*, prior to this free open-source edition). We provide answers to most odd-numbered problems. For those problems that require proofs, we attempt to model what a good proof should look like, often guiding students up to a point and then asking them to fill in the details.

The purpose of the first six chapters is to lay the foundation for the study of complex analysis and develop the topics of analytic and harmonic functions, the elementary functions, and contour integration. Chapters 7 and 8, dealing with residue calculus and applications, may be skipped if there is more interest in conformal mapping and applications of harmonic functions, which are the topics of Chapters 9 and 10 respectively. For courses requiring even more applications, Chapter 11 investigates Fourier and Laplace transforms.

Features The answers to most odd-numbered exercises should help instructors as they deliberate on problem assignments, and should help students as they review material. We present conformal mapping in a visual and geometric manner so that compositions and images of curves and regions can be more easily understood. We first solve boundary value problems for harmonic functions in the upper half-plane so that we can use conformal mapping by elementary functions to obtain solutions in other domains. We carefully develop the Schwarz-Christoffel transformation and present applications. We use two-dimensional mathematical models for applications in the areas of ideal fluid flow, steady-state temperatures, and electrostatics. We accurately portray streamlines, isothermals, and equipotential curves with computer-drawn figures.

An early introduction to sequences and series appears in Chapter 4, which facilitates the definition of the exponential function via series. The section on Julia and Mandelbrot sets illustrates how complex analysis connects with contemporary topics in mathematics. Included are computer-generated illustrations such as Riemann surfaces, contour and surface graphics for harmonic functions, the Dirichlet problem, streamlines involving harmonic and analytic functions, and conformal mapping. We also have a section on the Joukowski airfoil.

Acknowledgments A textbook does not make it to the sixth edition without the support of a long list of colleagues from various institutions. Their help has been invaluable, and we owe them much more than the brief acknowledgment we are able to provide here. Alphabetically by institution they are: Edward G. Thurber (Biola University); Robert A. Calabretta (Boeing Corporation); Vencil Skarda (Brigham Young University; Stuart Goldenberg (California Polytechnic State University, San Luis Obispo); Vuryl Klassen, Gerald Marley, and Harris Shultz (California State University, Fullerton); Michael Stob (Calvin College); Al Hibbard(Central College); Paul Martin (Colorado School of Mines); R.E. Williamson (Dartmouth College); William Trench (Drexel University); Arlo Davis (Indiana University of Pennsylvania); Elgin H. Johnston (Iowa State University); Richard A. Alo (Lamar University); Martin Bazant (Massachusetts Institute of Technology); Carroll O. Wilde (Naval Postgraduate School); Holland Filgo (Northeastern University); E. Melvin J. Jacobsen (Rensselaer Polytechnic Institute); Christine Black (Seattle University); Geoffrey Prince and John Trienz (United States Naval Academy); William Yslas Velez (University of Arizona); Charles P. Luehr (University of Florida); Robert D. Brown and T.E. Duncan (University of Kansas); Donald Hadwin (University of New Hampshire); Calvin Wilcox (University of Utah); Robert Heal (Utah State University); and C. Ray Rosentrater (Westmont College).

We also wish to thank the students of California Baptist University, Cal State Fullerton, the United States Air Force Academy, University of Maryland, and Westmont College for their helpful suggestions and words of encouragement. In production matters we are indebted to David Farmer of the American Institute of Mathematics. He generously gave an enormous amount of time in helping the PreTeXt version of this text come to fruition. Finally, we thank in advance those of you who will make suggestions for improvements to the text as it now stands, whether that be the PreTeXt or PDF version. We welcome correspondence via surface or e-mail.

Russell W. Howell
Kathryn Smith Professor of Mathematics
Department of Mathematics and Computer Science
Westmont College
Santa Barbara, CA 93108
howell@westmont.edu

John H. Mathews
Professor of Mathematics, Emeritus
Department of Mathematics
California State University, Fullerton
Fullerton, CA 92634
fmathews394@gmail.com

Chapter 1

Complex Numbers

Get ready for a treat. You are about to begin studying some of the most beautiful ideas in mathematics. They are ideas with surprises. They evolved over several centuries, yet they greatly simplify extremely difficult computations, making some as easy as sliding a hot knife through butter. They also have applications in a variety of areas, ranging from fluid flow, to electric circuits, to the mysterious quantum world. Generally, they belong to the area of mathematics known as complex analysis, which is the subject of this book. This chapter focuses on the development of entities we now call *complex numbers*.

1.1   The Origin of Complex Numbers

Complex analysis can roughly be thought of as the subject that applies the theory of calculus to imaginary numbers. But what exactly are imaginary numbers? Usually, students learn about them in high school with introductory remarks from their teachers along the following lines: "We can't take the square root of a negative number. But let's *pretend* we can and begin by using the symbol $i = \sqrt{-1}$." Rules are then learned for doing arithmetic with these numbers. At some level the rules make sense: if $i = \sqrt{-1}$, it stands to reason that $i^2 = -1$. However, it is not uncommon for students to wonder whether they are really doing magic rather than mathematics.

If you ever felt that way, congratulate yourself!—you are in the company of some of the great mathematicians from the sixteenth through the nineteenth centuries. They also were perplexed by the notion of roots of negative numbers. Our purpose in this section is to highlight some of the episodes in the very colorful history of how thinking about imaginary numbers developed. We intend to show you that, contrary to popular belief, there is really nothing *imaginary* about "imaginary umbers." They are just as real as "real numbers."

Our story begins in 1545. In that year, the Italian mathematician Girolamo Cardano published *Ars Magna* (*The Great Art*), a 40-chapter masterpiece in which he gave for the first time a method for solving the general cubic equation

$$z^3 + a_2 z^2 + a_1 z + a_0 = 0. \tag{1.1}$$

Cardano did not have at his disposal the power of today's algebraic notation, and he tended to think of cubes or squares as geometric objects rather than algebraic quantities. Essentially, however, his solution began by making the substitution $z = x - \frac{a_2}{3}$. This move transformed Equation (1.1) into a cubic equation without a squared term, which is called a **depressed cubic**.

To illustrate, begin with $z^3 + 9z^2 + 24z + 20 = 0$ and substitute $z = x - \frac{a_2}{3} = x - \frac{9}{3} = x - 3$. The equation then becomes $(x-3)^3 + 9(x-3)^2 + 24(x-3) + 20 = 0$, which simplifies to $x^3 - 3x + 2 = 0$.

You need not worry about the computational details here, but in general the substitution $z = x - \frac{a_2}{3}$ transforms Equation (1.1) into

$$x^3 + bx + c = 0, \tag{1.2}$$

where $b = a_1 - \frac{1}{3}a_2^2$, and $c = -\frac{1}{3}a_1 a_2 + \frac{2}{27}a_2^3 + a_0$.

If Cardano could get any value of x that solved a depressed cubic, he could easily get a corresponding solution to Equation (1.1) from the identity $z = x - \frac{a_2}{3}$. Happily, Cardano knew how to solve a depressed cubic. The technique had been communicated to him by Niccolo Fontana who, unfortunately, came to be known as Tartaglia (the *stammerer*) due to a speaking disorder that was caused when he was 12 years old. (Evidently, during the Italian wars, French troops sacked his home in Brescia, Italy in 1512, and struck Tartaglia in the face with a saber.) The procedure was also independently discovered some 30 years earlier by Scipione del Ferro of Bologna. Ferro and Tartaglia showed that one of the solutions to Equation (1.2) is

$$x = \sqrt[3]{-\frac{c}{2} + \sqrt{\frac{c^2}{4} + \frac{b^3}{27}}} + \sqrt[3]{-\frac{c}{2} - \sqrt{\frac{c^2}{4} + \frac{b^3}{27}}}. \tag{1.3}$$

Although Cardano would not have reasoned in the following way, today we can take this value for x and use it to factor the depressed cubic into a linear and quadratic term. The remaining roots can then be found with the quadratic formula. For example, to solve the (full) cubic equation $z^3 + 9z^2 + 24z + 20 = 0$, use the substitution $z = x - 3$ to get $x^3 - 3x + 2 = 0$, which is a depressed cubic in the form of Equation (1.2). Next, apply the "Ferro-Tartaglia" formula with $b = -3$ and $c = 2$ to get

$$x = \sqrt[3]{-\frac{2}{2} + \sqrt{\frac{2^2}{4} + \frac{(-3)^3}{27}}} + \sqrt[3]{-\frac{2}{2} - \sqrt{\frac{2^2}{4} + \frac{(-3)^3}{27}}} = \sqrt[3]{-1} + \sqrt[3]{-1} = -2.$$

Since $x = -2$ is a root, $x + 2$ must be a factor of $x^3 - 3x + 2$. Dividing $x + 2$ into $x^3 - 3x + 2$ gives $x^2 - 2x + 1 = (x-1)^2$, so that the remaining (duplicate) roots are $x = 1, x = 1$. The solutions to $z^3 + 9z^2 + 24z + 20 = 0$ are obtained by recalling $z = x - 3$, which yields the three roots $z_1 = -2 - 3 = -5$, and $z_2 = z_3 = 1 - 3 = -2$.

So, by using Tartaglia's work and a clever transformation technique, Cardano was able to crack what had seemed to be the impossible task of solving the general cubic equation. Surprisingly, this development played a significant role in helping to establish the legitimacy of imaginary numbers. Roots of negative numbers, of course, had come up earlier in the simplest of quadratic equations, such as $x^2 + 1 = 0$. The solutions we know today as $x = \pm\sqrt{-1}$, however, were easy for mathematicians to ignore. In Cardano's time, negative numbers were still being treated with some suspicion, as it was difficult to conceive of any physical reality corresponding to them. Taking square roots of such quantities was surely all the more ludicrous. Nevertheless, Cardano made some genuine attempts to deal with $\sqrt{-1}$. Unfortunately, his geometric thinking made it hard to make much headway. At one point he commented that the process of arithmetic that deals with quantities such as $\sqrt{-1}$ "involves mental tortures and is truly sophisticated." At another point he concluded that the process is "as refined as it is useless." Many mathematicians held this view, but finally there was a breakthrough.

In his 1572 treatise *L'Algebra*, Rafael Bombelli showed that roots of negative numbers have great utility. Consider the depressed cubic $x^3 - 15x - 4 = 0$. Using Formula (1.3), we compute

$$x = \sqrt[3]{2 + \sqrt{-121}} + \sqrt[3]{2 - \sqrt{-121}} = \sqrt[3]{2 + 11\sqrt{-1}} + \sqrt[3]{2 - 11\sqrt{-1}}.$$

Simplifying this expression would have been very difficult if Bombelli had not come up with what he called a "wild thought." He suspected that if the original depressed cubic had real solutions, then the two parts of x in the preceding equation could be written as $u + v\sqrt{-1}$ and $u - v\sqrt{-1}$ for some real numbers u and v. That is, Bombelli believed $u + v\sqrt{-1} = \sqrt[3]{2 + 11\sqrt{-1}}$ and $u - v\sqrt{-1} = \sqrt[3]{2 - 11\sqrt{-1}}$, which would mean

$$(u + v\sqrt{-1})^3 = 2 + 11\sqrt{-1}, \quad \text{and} \quad (u - v\sqrt{-1})^3 = 2 - 11\sqrt{-1}.$$

Then, using the well-known algebraic identity $(a + b)^3 = a^3 + 3a^2 b + 3ab^2 + b^3$, and assuming that roots of negative numbers obey the rules of algebra, he obtained

$$
\begin{aligned}
(u + v\sqrt{-1})^3 &= u^3 + 3(u^2)v\sqrt{-1} + 3(u)(v\sqrt{-1})^2 + (v\sqrt{-1})^3 \\
&= u^3 + 3(u)(v\sqrt{-1})^2 + 3(u^2)v\sqrt{-1} + (v\sqrt{-1})^3 \\
&= (u^3 - 3uv^2) + (3u^2 v - v^3)\sqrt{-1} \\
&= u(u^2 - 3v^2) + v(3u^2 - v^2)\sqrt{-1} \qquad (1.4) \\
&= 2 + 11\sqrt{-1}. \qquad (1.5)
\end{aligned}
$$

By equating like parts of Equations (1.4) and (1.5) Bombelli reasoned that $u(u^2 - 3v^2) = 2$ and $v(3u^2 - v^2) = 11$. Perhaps thinking even more wildly, Bombelli then supposed that u and v were integers. The only integer factors of 2 are 2 and 1, so the equation $u(u^2 - 3v^2) = 2$ led Bombelli to conclude that $u = 2$ and $u^2 - 3v^2 = 1$. From this conclusion it follows that $v^2 = 1$, or $v = \pm 1$. Amazingly, $u = 2$ and $v = 1$ solve the second equation $v(3u^2 - v^2) = 11$, so Bombelli declared the values for u and v to be $u = 2$ and $v = 1$, respectively.

Since $(2 + \sqrt{-1})^3 = 2 + 11\sqrt{-1}$, we clearly have $2 + \sqrt{-1} = \sqrt[3]{2 + 11\sqrt{-1}}$. Similarly, Bombelli showed that $2 - \sqrt{-1} = \sqrt[3]{2 - 11\sqrt{-1}}$, so that

$$\sqrt[3]{2 + 11\sqrt{-1}} + \sqrt[3]{2 - 11\sqrt{-1}} = (2 + \sqrt{-1}) + (2 - \sqrt{-1}) = 4, \qquad (1.6)$$

which was a proverbial bombshell. Prior to Bombelli, mathematicians could easily scoff at imaginary numbers when they arose as solutions to quadratic equations. With cubic equations, they no longer had this luxury. That $x = 4$ was a correct solution to the equation $x^3 - 15x - 4 = 0$ was indisputable, as it could be checked easily. However, to arrive at this very real solution, mathematicians had to take a detour through the uncharted territory of "imaginary numbers." Thus, whatever else might have been said about these numbers (which, today, we call *complex numbers*), their utility could no longer be ignored.

1.1.1 Geometric Progress of John Wallis

As significant as Bombelli's work was his results left many issues unresolved. For example, his technique applied only to a few specialized cases. Could it be extended? Even if it could be extended a larger question remained: What possible physical representation could complex numbers have? That question remained unanswered for more than two centuries. Paul J. Nahin's book *An Imaginary Tale: the Story of $\sqrt{-1}$* describes the progress in answering it as occurring in several stages. A preliminary step came in 1685 when the English mathematician John Wallis published *A Treatise of Algebra, both Historical and Practical.* Among the many contributions in that book two are particularly noteworthy for our purposes. They are displayed in Wallis' analysis of a problem from classical geometry that, at first glance, seems completely unrelated to complex numbers.

Problem 1 (Classical Geometry Problem).
Construct a triangle determined by two sides and an angle not *included between those sides.*

We will get to Wallis' contributions in a moment. First, observe that Figure 1.1 illustrates the standard solution to this classical problem. Given side length a (represented by segment AB), angle α (determined by segments AB and BC), and side length b, draw an arc of radius b whose center is at point A. If the arc intersects segment BC at points E and F, then the resulting triangles ABE and ABF each satisfy the problem requirement.

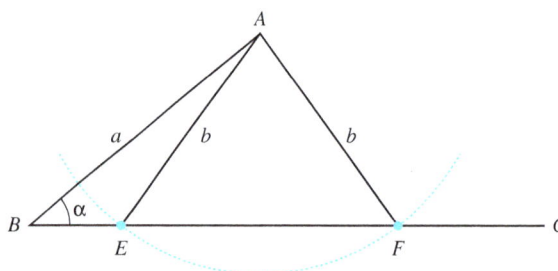

Figure 1.1: The standard solution to Wallis' problem

1.1.2   A Geometric Representation of Real Numbers

Wallis' first contribution allowed him to associate numbers with the points E and F of Figure 1.1. The association came by way of a construct that may sound completely trivial to us, but that is only because we have been raised with Wallis' idea: the number line. By choosing an arbitrary point to represent the number zero on a given line, Wallis declared that positive numbers could be viewed as corresponding distances to the *right* of zero, and negative numbers as corresponding (positive) distances to the *left* of zero.

To complete the association refer to Figure 1.2 and think of segment BC as lying on a portion of the x-axis. Then draw a perpendicular segment AD to BC and designate D to be the origin. If the length of AD is c the Pythagorean theorem gives $\sqrt{b^2 - c^2}$ for the length of segments ED and DF.

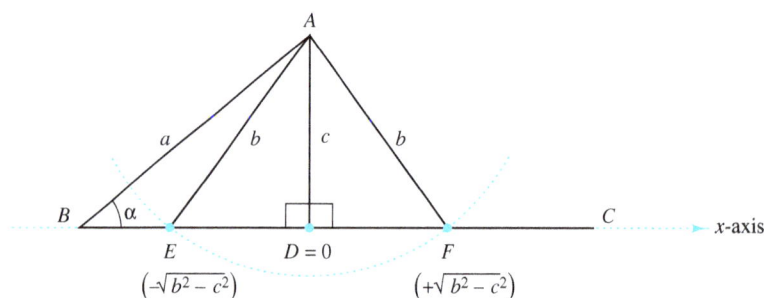

Figure 1.2: Wallis' depiction of real numbers

4

Combining this result with Wallis' number line results in points E and F representing the numbers

$$E = -\sqrt{b^2 - c^2}, \quad \text{and} \quad F = +\sqrt{b^2 - c^2}.$$

Thus, if $b = 5$ and $c = 4$, points E and F would represent -3 and $+3$, respectively, because

$$E = -\sqrt{5^2 - 4^2} = -3, \quad \text{and} \quad F = +\sqrt{5^2 - 4^2} = +3.$$

From both an algebraic and geometric viewpoint this procedure only makes sense if the stipulated length b is greater than or equal to c. If b were less than c then the algebraic expressions for points E and F ($-\sqrt{b^2 - c^2}$ and $+\sqrt{b^2 - c^2}$) would be meaningless, as the quantity $b^2 - c^2$ inside the square root would be negative. Viewed geometrically, if b were less than c then the arc of radius b that is centered at A would not be able to intersect segment BC. In other words, if b were less than c, Problem 1 would appear to have no solution.

1.1.3 A Geometric Representation of Complex Numbers

Appearances, of course, can be deceiving, and Wallis reinforced the truth of that ancient proverb when he came up with his second—and bolder—contribution. It was a solution Problem 1 in the case when b is less than c. Figure 1.3 illustrates how he did it. From the midpoint of AD Wallis drew a circle with diameter AD. Then, with A as a center he drew an arc of radius b. Because b is less than c the arc will intersect the circle at two points, say E and F.

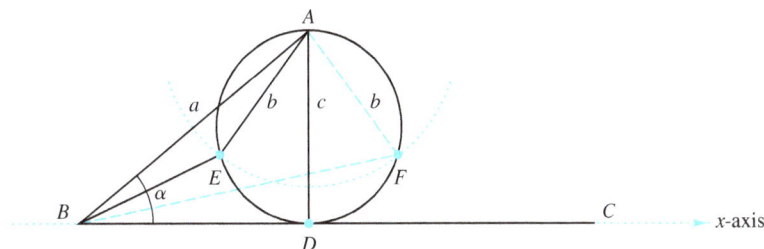

Figure 1.3: Wallis' depiction of complex numbers

Again we get two triangles: ABE and ABF. Wallis claimed that these triangles each satisfy the requirement of Problem 1. You might object to this construction on the grounds that angle α is not part of either triangle. If you read the problem statement carefully, however, you will notice that it never states that the angle α has to be part of any triangle, only that it must play a role in *determining* a triangle. From this perspective Wallis completely satisfied the requirement.

Notice, also, that points E and F are no longer on the x-axis as they were when b was greater than c (and when $\sqrt{b^2 - c^2}$ was a real number). They are now somewhere above the x-axis, and it is not unreasonable to conclude that points E and F give, respectively, geometric representations of the expressions $-\sqrt{b^2 - c^2}$ and $+\sqrt{b^2 - c^2}$ when b is less than c (and when $\sqrt{b^2 - c^2}$ is a complex number).

Although Wallis only hinted at such a conclusion, he nevertheless helped set the stage for thinking of real numbers as being embedded in a larger set of complex numbers, and that these numbers could be represented as "points in the plane." Unfortunately, if we tried to apply Wallis' method to construct complex numbers we would find it had some serious defects. For example, if $b = 0$ and $c = 1$ the expression $\pm\sqrt{b^2 - c^2}$ becomes $\pm\sqrt{-1}$, and points E and F

now coincide at point A. But we surely would not want to say that $-\sqrt{-1}$ and $+\sqrt{-1}$ are the same number. Thus, even with Wallis' work the jigsaw of getting a legitimate picture of complex numbers remained. It would be yet another century before someone put most of the pieces together.

1.1.4 Caspar Wessel Makes a Breakthrough

Points in the plane can also be thought of as vectors, which are directed line segments from the origin to those points. In 1797 Caspar Wessel presented a paper to the Danish Academy of Sciences in which he described how to manipulate vectors geometrically. This description eventually led to the current representation of complex numbers.

To add two vectors, make a copy of the second vector and place its tail on the head of the first vector. The resultant vector is the directed line segment drawn from the tail of the first vector to the head of the second copy vector. Figure 1.4a illustrates the addition of vector b to vector a.

When Wessel gave his paper the procedure for adding vectors was already known. The unique contribution that he made was his description of how to *multiply* two vectors.

To understand Wessel's thinking recall that any non-zero vector can be represented by two quantities: its length, and its angular displacement from the positive x-axis. Figure 1.4b illustrates this idea for vector a: its length is r, and its angular displacement from the positive x-axis is α.

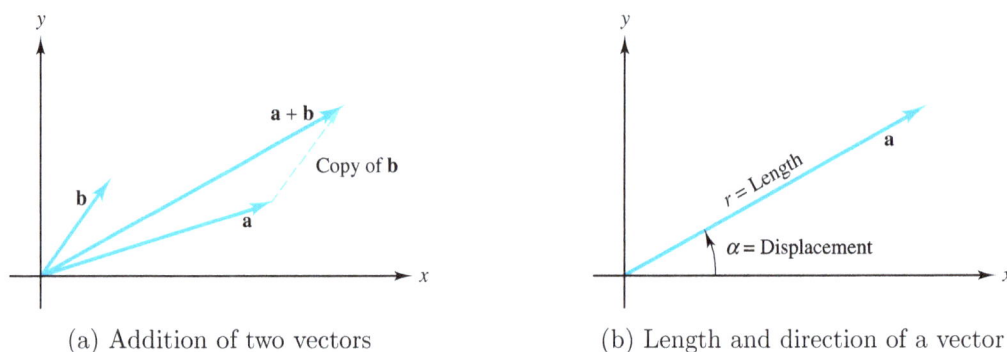

(a) Addition of two vectors   (b) Length and direction of a vector

Figure 1.4: The geometry of vectors

Wessel stated that, to multiply two vectors, the length of the product vector should be the product of the lengths of its factors. Should the angular displacement of the product vector likewise be the product of the angular displacements of its factors? Definitely not, and you will see in the exercises why Wessel knew that such a provision would have been a bad idea. What, then, should be the angular displacement of the product?

In answering this question Wessel drew an analogy from the multiplication of real numbers. He observed that, if $c = ab$, then $\frac{c}{a} = b = \frac{b}{1}$, and $\frac{c}{b} = a = \frac{a}{1}$. In other words, the ratio of the product to any given factor is the same as the ratio of the other factor to the number one.

What vector represents the number one? It seems obvious that, using the number line of Wallis, it should be the directed line segment from the origin to the number one on the positive x-axis. Let's call this vector the *standard unit vector*, as illustrated in Figure 1.5.

With this identification in mind, and using the multiplication analogy just mentioned, Wessel made a brilliant move. He reasoned that the (angular) displacement of the product of two

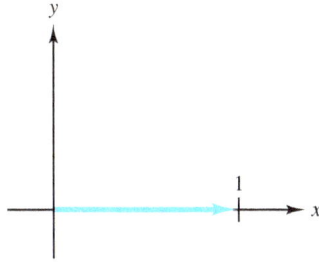

Figure 1.5: The standard unit vector

vectors should differ from the displacement of any given factor by the same amount that the displacement of the other factor differs from the displacement of the standard unit vector. That's quite a mouthful; let's see what it means.

What is the (angular) displacement of the standard unit vector? Clearly, its displacement is zero radians, as it coincides with the positive x-axis. Thus, if vectors a and b have displacements of α and β, respectively, and vector $c = ab$, then the displacement of c should be $\alpha + \beta$, as shown in Figure 1.6a. The reason for this assertion is that, with such an arrangement, Wessel's displacement protocol works out perfectly: the displacement of c (which is $\alpha + \beta$) differs from the displacement of a (which is α) by β. This is the same amount that the displacement of b (which is β) differs from the displacement of the standard unit vector (which is 0). Likewise, the displacement of c differs from the displacement of b by α, which is the same amount that the displacement of a differs from the displacement of the standard unit vector.

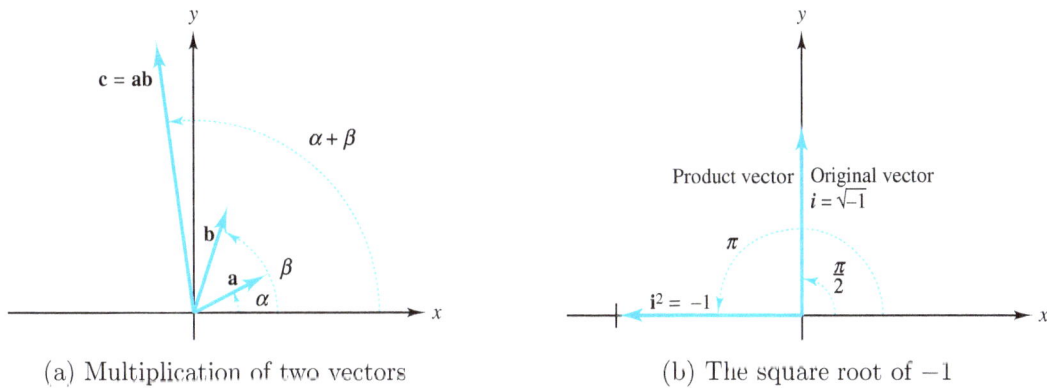

(a) Multiplication of two vectors

(b) The square root of -1

Figure 1.6: Wessel's multiplication scheme for vectors

How does Wessel's procedure lead to a geometric representation of complex numbers? Consider what happens if a unit vector is drawn from the origin straight up the y-axis, and then multiplied by itself. By Wessel's rules the length of the product vector is one unit, as the length of each factor is one unit. What about its direction? The angular displacement of the original vector is $\frac{\pi}{2}$ radians, so by Wessel's rules again the product vector has a displacement of $\frac{\pi}{2} + \frac{\pi}{2} = \pi$ radians. Thus, the product vector is aligned along the x-axis, but is directed from the origin *to the left* by one unit, as shown in Figure 1.6b. Using Wallis' number line we see that the product vector is naturally identified with the number -1. Label the original vector as i. What do you conclude? Obviously, that $i^2 = -1$, which must mean that $i = \sqrt{-1}$. Neat!

Neat, yes, but the material we presented leading up to this result was (if you'll pardon the pun) complex, so you need not worry if you had some difficulty following it. Sections 1.2–1.6

will flesh out these ideas in much more detail.

It should be pointed out that Wessel was not the only mathematician—or even the first— who began thinking of complex numbers as vectors, or, as points in the plane. As early as 1732 the great Swiss mathematician Leonard Euler (pronounced "oiler") adopted this view concerning the n solutions to the equation $x^n - 1 = 0$. You will learn shortly that these solutions can be expressed as $\cos\theta + \sqrt{-1}\sin\theta$ for various values of θ. Euler thought of them as being located at the vertices of a regular polygon in the plane. Euler was also the first to use the symbol i for $\sqrt{-1}$. Today this notation is still the most popular, although some electrical engineers prefer the symbol j instead so that they can use i to represent current.

Two additional mathematicians deserve mention. The Frenchman Augustin-Louis Cauchy (1789–1857) formulated many of the classic theorems that are now part of the corpus of complex analysis. The German Carl Friedrich Gauss (1777–1855) reinforced the utility of complex numbers when he used them in his several proofs of the fundamental theorem of algebra (see Section 6.4). In an 1831 paper, he produced a clear geometric representation of $x + iy$ by identifying it with the point (x, y) in the coordinate plane. He also described how to perform arithmetic operations with these new numbers.

It would be a mistake, however, to conclude that in 1831 complex numbers were transformed into legitimacy. In that same year the prolific logician Augustus De Morgan commented in his book, *On the Study and Difficulties of Mathematics,* "We have shown the symbol $\sqrt{-a}$ to be void of meaning, or rather self-contradictory and absurd. Nevertheless, by means of such symbols, a part of algebra is established which is of great utility."

There were, indeed, serious logical problems associated with complex numbers. For example, with real numbers $\sqrt{ab} = \sqrt{a}\sqrt{b}$ so long as both sides of the equation are defined. Applying this identity to complex numbers leads to $1 = \sqrt{1} = \sqrt{(-1)(-1)} = \sqrt{-1}\sqrt{-1} = -1$. Plausible answers to these problems can be given, however, and you will learn how to resolve this apparent contradiction in Section 2.2. De Morgan's remark illustrates that many factors are needed to persuade mathematicians to adopt new theories. In this case, as always, a firm logical foundation was crucial, but so, too, was a willingness to modify some ideas concerning certain well-established properties of numbers.

As time passed, mathematicians gradually refined their thinking, and by the end of the nineteenth century complex numbers were firmly entrenched. Thus, as it is with many new mathematical or scientific innovations, the theory of complex numbers evolved by way of a very intricate process. But what is the theory that Tartaglia, Ferro, Cardano, Bombelli, Wallis, Euler, Cauchy, Gauss, and so many others helped produce? That is, how do we now think of complex numbers? We explore this question in the remainder of this chapter.

Exercises for Section 1.1 (Selected answers or hints are on page 427.)

1. Show that $2 - \sqrt{-1} = \sqrt[3]{2 - 11\sqrt{-1}}$.

2. Explain why cubic equations, rather than quadratic equations, played a pivotal role in helping to obtain the acceptance of complex numbers.

3. Find all solutions to the following depressed cubics.

   (a) $27x^3 - 9x - 2 = 0$. *Hint:* Get an equivalent monic polynomial.
   (b) $x^3 - 27x + 54 = 0$.

4. Explain why Wallis's view of complex numbers results in $-\sqrt{-1}$ being represented by the same point as is $\sqrt{-1}$.

5. Use Bombelli's technique to get all solutions to the following depressed cubics.

   (a) $x^3 - 30x - 36 = 0$.
   (b) $x^3 - 87x - 130 = 0$.
   (c) $x^3 - 60x - 32 = 0$.

6. Use Cardano's technique (of substituting $z = x - \frac{a_2}{3}$) to solve the following cubics.

   (a) $z^3 - 6z^2 - 3z + 18 = 0$.
   (b) $z^3 + 3z^2 - 24z + 28 = 0$.

7. Is it possible to modify slightly Wallis's picture of complex numbers so that it is consistent with the representation used today? To help you answer this question, refer to the article by Alec Norton and Benjamin Lotto, "Complex Roots Made Visible," *The College Mathematics Journal*, 15(3), June 1984, pp. 248–249.

8. Investigate library or web resources and write up a detailed description explaining why the solution to the depressed cubic, Equation (1.3), is valid.

1.2   The Algebra of Complex Numbers, Part I

We have shown that complex numbers came to be viewed as ordered pairs of real numbers. That is, a complex number z is defined to be

$$z = (x, y), \tag{1.7}$$

where x and y are both real numbers.

The reason we say *ordered* pair is because we are thinking of a point in the plane. The point $(2, 3)$, for example, is not the same as $(3, 2)$. The *order* in which we write x and y in Equation (1.7) makes a difference. Clearly, then, two complex numbers are equal if and only if their x coordinates are equal *and* their y coordinates are equal. In other words,

$$(x, y) = (u, v) \quad \text{iff} \quad x = u \quad and \quad y = v.$$

(Throughout this text, "iff" means *if and only if*.)

A meaningful number system requires a method for combining ordered pairs. The definition of algebraic operations must be consistent so that the sum, difference, product, and quotient of any two ordered pairs will again be an ordered pair. The key to defining how these numbers should be manipulated is to follow Gauss's lead and equate (x, y) with $x + iy$. Then, if $z_1 = (x_1, y_1)$ and $z_2 = (x_2, y_2)$ are arbitrary complex numbers, we have

$$\begin{aligned}
z_1 + z_2 &= (x_1, y_1) + (x_2, y_2) \\
&= (x_1 + iy_1) + (x_2 + iy_2) \\
&= (x_1 + x_2) + i(y_1 + y_2) \\
&= (x_1 + x_2, \ y_1 + y_2).
\end{aligned}$$

Thus, the following definitions should make sense.

Definition 1.1 (Addition).

$$z_1 + z_2 = (x_1, y_1) + (x_2, y_2)$$
$$= (x_1 + x_2, \; y_1 + y_2). \tag{1.8}$$

Definition 1.2 (Subtraction).

$$z_1 - z_2 = (x_1, y_1) - (x_2, y_2)$$
$$= (x_1 - x_2, \; y_1 - y_2). \tag{1.9}$$

Example 1.1. If $z_1 = (3, 7)$ and $z_2 = (5, -6)$, then

$$z_1 + z_2 = (3, 7) + (5, -6) = (8, 1) \quad \text{and}$$
$$z_1 - z_2 = (3, 7) - (5, -6) = (-2, 13).$$

We can also use the notation $z_1 = 3 + 7i$ and $z_2 = 5 - 6i$:

$$z_1 + z_2 = (3 + 7i) + (5 - 6i) = 8 + i \quad \text{and}$$
$$z_1 - z_2 = (3 + 7i) - (5 - 6i) = -2 + 13i.$$

Given the rationale we devised for addition and subtraction, it is tempting to define the product $z_1 z_2$ as $z_1 z_2 = (x_1 x_2, y_1 y_2)$. It turns out, however, that this is not a good definition, and we ask you in the exercises for this section to explain why. How, then, should products be defined? Again, if we equate $(x, \; y)$ with $x + iy$ and assume, for the moment, that $i = \sqrt{-1}$ makes sense (so that $i^2 = -1$), we have

$$z_1 z_2 = (x_1, y_1)(x_2, y_2)$$
$$= (x_1 + iy_1)(x_2 + iy_2)$$
$$= x_1 x_2 + ix_1 y_2 + ix_2 y_1 + i^2 y_1 y_2$$
$$= x_1 x_2 - y_1 y_2 + i(x_1 y_2 + x_2 y_1)$$
$$= (x_1 x_2 - y_1 y_2, \; x_1 y_2 + x_2 y_1).$$

It appears, therefore, that we are forced into the following definition.

Definition 1.3 (Multiplication).

$$z_1 z_2 = (x_1, y_1)(x_2, y_2)$$
$$= (x_1 x_2 - y_1 y_2, \; x_1 y_2 + x_2 y_1). \tag{1.10}$$

Example 1.2. If $z_1 = (3, \; 7)$ and $z_2 = 5 - 6i$, then

$$z_1 z_2 = (3, 7)(5, -6)$$
$$= (3 \cdot 5 - 7 \cdot (-6), \; 3 \cdot (-6) + 5 \cdot 7)$$
$$= (15 + 42, \; -18 + 35)$$
$$= (57, 17).$$

We get the same answer by using the notation $z_1 = 3 + 7i$ and $z_2 = 5 - 6i$:

$$z_1 z_2 = (3, 7)(5, -6)$$
$$= (3 + 7i)(5 - 6i)$$
$$= 15 - 18i + 35i - 42i^2$$
$$= 15 - 42(-1) + (-18 + 35)i$$
$$= 57 + 17i$$
$$= (57, 17).$$

Of course, it makes sense that the answer came out as we expected because we used the notation $x + iy$ as motivation for our definition in the first place. Exercise 14 asks you to show that Wessel's analogy for the norm and angular displacement discussed in Section 1.1.4 leads to the same rule for multiplication as that given in Definition 1.3.

To motivate our definition for division, we proceed along the same lines as we did for multiplication, assuming that $z_2 \neq 0$:

$$
\frac{z_1}{z_2} = \frac{(x_1, y_1)}{(x_2, y_2)}
$$

$$
= \frac{(x_1 + iy_1)}{(x_2 + iy_2)}.
$$

We need to figure out a way to write the preceding quantity in the form $x + iy$. To do so, we use a standard technique and multiply the numerator and denominator by $x_2 - iy_2$, which gives

$$
\frac{z_1}{z_2} = \frac{(x_1 + iy_1)(x_2 - iy_2)}{(x_2 + iy_2)(x_2 - iy_2)}
$$

$$
= \frac{x_1 x_2 + y_1 y_2 + i(-x_1 y_2 + x_2 y_1)}{x_2^2 + y_2^2}
$$

$$
= \left(\frac{x_1 x_2 + y_1 y_2}{x_2^2 + y_2^2} \right) + i \left(\frac{-x_1 y_2 + x_2 y_1}{x_2^2 + y_2^2} \right)
$$

$$
= \left(\frac{x_1 x_2 + y_1 y_2}{x_2^2 + y_2^2}, \frac{-x_1 y_2 + x_2 y_1}{x_2^2 + y_2^2} \right).
$$

Thus we finally arrive at a rather odd definition.

Definition 1.4 (Division).

$$
\frac{z_1}{z_2} = \frac{(x_1, y_1)}{(x_2, y_2)}
$$

$$
= \left(\frac{x_1 x_2 + y_1 y_2}{x_2^2 + y_2^2}, \frac{-x_1 y_2 + x_2 y_1}{x_2^2 + y_2^2} \right) \quad \text{for} \quad z_2 \neq 0. \tag{1.11}
$$

Example 1.3. If $z_1 = (3, 7)$ and $z_2 = (5, -6)$, then

$$
\frac{z_1}{z_2} = \frac{(3, 7)}{(5, -6)} = \left(\frac{15 - 42}{25 + 36}, \frac{18 + 35}{25 + 36} \right) = \left(-\frac{27}{61}, \frac{53}{61} \right).
$$

As with the example for multiplication, we also get this answer if we use the notation $x + iy$:

$$
\frac{z_1}{z_2} = \frac{(3, 7)}{(5, -6)}
$$

$$
= \frac{3 + 7i}{5 - 6i}
$$

$$
= \left(\frac{3 + 7i}{5 - 6i} \right) \left(\frac{5 + 6i}{5 + 6i} \right)
$$

$$
= \frac{15 + 18i + 35i + 42i^2}{25 + 30i - 30i - 36i^2}
$$

$$
= \frac{15 - 42 + (18 + 35)i}{25 + 36}
$$

$$
= -\frac{27}{61} + \frac{53}{61}i = \left(-\frac{27}{61}, \frac{53}{61} \right).
$$

To perform operations on complex numbers, mathematicians use the notation $x + iy$ and engage in algebraic manipulations, as we did here, rather than apply the complicated-looking definitions we gave for those operations on ordered pairs. This procedure is valid because we used the $x + iy$ notation as a guide for defining the operations in the first place. Remember, though, that the $x + iy$ notation is nothing more than a convenient bookkeeping device for keeping track of how to manipulate ordered pairs. It is the ordered pair algebraic definitions that form the real foundation on which the complex number system is based. In fact, if you were to program a computer to do arithmetic on complex numbers, your program would perform calculations on ordered pairs, using exactly the definitions that we gave.

It turns out that our algebraic definitions give complex numbers all the properties we normally ascribe to the real number system. Taken together, they describe what algebraists call a **field**. In formal terms, a field is a set (in this case, the complex numbers) together with two binary operations (in this case, addition and multiplication) having the following properties.

1. **(P1) Commutative law for addition:** $z_1 + z_2 = z_2 + z_1$.

2. **(P2) Associative law for addition:** $z_1 + (z_2 + z_3) = (z_1 + z_2) + z_3$.

3. **(P3) Additive identity:** There is a complex number ω such that $z + \omega = z$ for all complex numbers z. The number ω is obviously the ordered pair $(0, 0)$.

4. **(P4) Additive inverses:** For any complex number z, there is a unique complex number η (depending on z) with the property that $z + \eta = (0, 0)$. Obviously, if $z = (x, y) = x + iy$, the number η will be $(-x, -y) = -x - iy = -z$.

5. **(P5) Commutative law for multiplication:** $z_1 z_2 = z_2 z_1$.

6. **(P6) Associative law for multiplication:** $z_1(z_2 z_3) = (z_1 z_2) z_3$.

7. **(P7) Multiplicative identity:** There is a complex number ζ such that $z\zeta = z$ for all complex numbers z. As you might expect, $(1, 0)$ is the unique complex number ζ having this property. We ask you to verify this identity in the exercises for this section.

8. **(P8) Multiplicative inverses:** For any complex number $z = (x, y)$ other than the number $(0, 0)$, there is a complex number (depending on z), which we denote z^{-1}, having the property that $zz^{-1} = (1, 0) = 1$. Based on our definition for division, it seems reasonable that the number z^{-1} would be

$$z^{-1} = \frac{(1, 0)}{z} = \frac{1}{z} = \frac{1}{x + iy} = \frac{x - iy}{x^2 + y^2} = \frac{x}{x^2 + y^2} + i\left(\frac{-y}{x^2 + y^2}\right)$$
$$= \left(\frac{x}{x^2 + y^2}, \frac{-y}{x^2 + y^2}\right).$$

We ask you to confirm this result in the exercises for this section.

9. **(P9) The distributive law:** $z_1(z_2 + z_3) = z_1 z_2 + z_1 z_3$.

None of these properties is difficult to prove. Most of the proofs make use of corresponding facts in the real number system. To illustrate, we give a proof of property **(P1)**.

Proof of the commutative law for addition:

Let $z_1 = (x_1, \ y_1)$ and $z_2 = (x_2, \ y_2)$ be arbitrary complex numbers. Then

$$
\begin{aligned}
z_1 + z_2 &= (x_1, y_1) + (x_2, y_2) \\
&= (x_1 + x_2, y_1 + y_2) && \text{(by definition of addition of complex numbers)} \\
&= (x_2 + x_1, y_2 + y_1) && \text{(by the commutative law for \textit{real} numbers)} \\
&= (x_2, y_2) + (x_1, y_1) && \text{(by definition of addition of complex numbers)} \\
&= z_2 + z_1.
\end{aligned}
$$

$\square$

Actually, you can think of the real number system as a subset of the complex number system. To see why, let's agree that, as any complex number of the form $(t, \ 0)$ is on the x axis, we can identify it with the real number t. With this correspondence, we can easily verify that our definitions for addition, subtraction, multiplication, and division of complex numbers are consistent with the corresponding operations on real numbers. For example, if x_1 and x_2 are real numbers, then

$$
\begin{aligned}
x_1 x_2 &= (x_1, 0)(x_2, 0) && \left(\text{by our agreed correspondence}\right) \\
&= (x_1 x_2 - 0, \ 0 + 0) && \text{(by definition of multiplication of complex numbers)} \\
&= (x_1 x_2, 0) && \text{(confirming the consistency of our correspondence)}.
\end{aligned}
$$

It is now time to show specifically how the symbol i relates to the quantity $\sqrt{-1}$. Note that

$$
\begin{aligned}
(0, 1)^2 &= (0, 1)(0, 1) \\
&= (0 - 1, 0 + 0) && \text{(by definition of multiplication of complex numbers)} \\
&= (-1, 0) \\
&= -1 && \text{(by our agreed correspondence)}.
\end{aligned}
$$

If we use the symbol i for the point $(0, 1)$, the preceding identity gives

$$
i^2 = (0, 1)^2 = -1,
$$

which means $i = (0, \ 1) = \sqrt{-1}$. So, the next time you are having a discussion with your friends and they scoff when you claim that $\sqrt{-1}$ is not imaginary, calmly put your pencil on the point $(0, 1)$ of the coordinate plane and ask them if there is anything imaginary about it. When they agree there isn't, you can tell them that this point, in fact, represents the mysterious $\sqrt{-1}$ in the same way that $(1, 0)$ represents the number 1.

We can also see more clearly now how the notation $x + iy$ equates to (x, y). Using the preceding conventions (*i.e.*, $x = (x, 0)$, etc.), we have

$$
\begin{aligned}
x + iy &= (x, 0) + (0, 1)(y, 0) && \text{(by our previously discussed conventions)} \\
&= (x, 0) + (0, y) && \text{(by definition of multiplication of complex numbers)} \\
&= (x, y) && \text{(by definition of addition of complex numbers)}.
\end{aligned}
$$

Thus, we may move freely between the notations $x + iy$ and (x, y), depending on which is more convenient for the context in which we are working. Students sometimes wonder whether it matters where the "i" is located in writing a complex number. It does not. Generally, most texts place terms containing an "i" at the end of an expression, and place the "i" before a variable, but after a constant. Thus, we write $x + iy$, $u + iv$, etc., but $3 + 7i$, $5 - 6i$ and so forth. Because letters lower in the alphabet generally denote constants, you will usually (but not always) see the expression $a + bi$ instead of $a + ib$. Many authors write quantities like $1 + i\sqrt{3}$ instead of $1 + \sqrt{3}i$ to make sure the "i" is not mistakenly thought to be inside the square root symbol. Additionally, if there is concern that the "i" might be missed, it is sometimes placed before a lengthy expression, as in $2\cos(-\frac{5\pi}{6} + 2n\pi) + i2\sin(-\frac{5\pi}{6} + 2n\pi)$.

We close this section with three important definitions and a theorem involving them. We ask you for a proof of the theorem in the exercises.

Definition 1.5 (Real Part). *The **real part** of z, denoted by $\operatorname{Re}(z)$, is the real number x.*

Definition 1.6 (Imaginary Part). *The **imaginary part** of z, denoted by $\operatorname{Im}(z)$, is the real number y.*

Definition 1.7 (Conjugate). *The **conjugate** of z, denoted by $\overline{z}$, is the complex number $(x, -y) = x - iy$.*

Example 1.4. $\operatorname{Re}(-3 + 7i) = -3$ and $\operatorname{Re}[(9, 4)] = 9$, $\operatorname{Im}(-3 + 7i) = 7$. Also, $\operatorname{Im}[(9, 4)] = 4$, $\overline{-3 + 7i} = -3 - 7i$ and $\overline{(9, \ 4)} = (9, -4)$.

Theorem 1.1. *Suppose that z, z_1, and z_2 are arbitrary complex numbers. Then the following identities hold true:*

$$\overline{\overline{z}} = z; \tag{1.12}$$

$$\overline{z_1 + z_2} = \overline{z_1} + \overline{z_2}; \tag{1.13}$$

$$\overline{z_1 z_2} = \overline{z_1} \ \overline{z_2}; \tag{1.14}$$

$$\overline{\left(\frac{z_1}{z_2}\right)} = \frac{\overline{z_1}}{\overline{z_2}} \quad (if \ \ z_2 \neq 0); \tag{1.15}$$

$$\operatorname{Re}(z) = \frac{z + \overline{z}}{2}; \tag{1.16}$$

$$\operatorname{Im}(z) = \frac{z - \overline{z}}{2i}; \tag{1.17}$$

$$\operatorname{Re}(iz) = -\operatorname{Im}(z); \tag{1.18}$$

$$\operatorname{Im}(iz) = \operatorname{Re}(z). \tag{1.19}$$

Because of what it erroneously connotes, it is a shame that the term *imaginary* is used in Definition 1.6. It was coined by the brilliant mathematician and philosopher René Descartes (1596–1650) during an era when quantities such as $\sqrt{-1}$ were thought to be just that. Gauss, who was successful in getting mathematicians to adopt the phrase *complex number* rather than *imaginary number*, also suggested that we use *lateral part* of z in place of *imaginary part* of z. Unfortunately, that suggestion never caught on, and it appears we are stuck with what history has handed down to us.

Exercises for Section 1.2 (Selected answers or hints are on page 427.)

1. Perform the required calculations and express your answers in the form $a + bi$.

   (a) i^{275}.

   (b) $\frac{1}{i^5}$.

   (c) $\operatorname{Re}(i)$.

   (d) $\operatorname{Im}(2)$.

   (e) $(i - 1)^3$.

   (f) $(7 - 2i)(3i + 5)$.

   (g) $\operatorname{Re}(7 + 6i) + \operatorname{Im}(5 - 4i)$.

   (h) $\operatorname{Im}(\frac{1+2i}{3-4i})$.

   (i) $\frac{(4-i)(1-3i)}{-1+2i}$.

   (j) $\overline{(1 + i\sqrt{3})(i + \sqrt{3})}$.

2. Evaluate the following quantities.

   (a) $\overline{(1 + i)(2 + i)}(3 + i)$.

   (b) $(3 + i)/(\overline{2 + i})$.

   (c) $\operatorname{Re}(i - 1)^3]$.

   (d) $\operatorname{Im}[(1 + i)^{-}2]$.

   (e) $\frac{1+2i}{3-4i} - \frac{4-3i}{2-i}$.

   (f) $(1 + i)^{-2}$.

   (g) $\operatorname{Re}[(x - iy)^2]$.

   (h) $\operatorname{Im}(\frac{1}{x-iy})$.

   (i) $\operatorname{Re}[(x + iy)(x - iy)]$.

   (j) $\operatorname{Im}[(x + iy)^3]$.

3. Show that $z\bar{z}$ is always a real number.

4. Verify Identities (1.12)–(1.19).

5. Let $P(z) = a_n z^n + a_{n-1} z^{n-1} + \cdots + a_1 z + a_0$ be a polynomial of degree n.

   (a) Suppose that $a_n, a_{n-1}, \ldots, a_1, a_0$ are all real. Show that if z_1 is a root of P, then $\overline{z_1}$ is also a root. In other words, the roots must be complex conjugates, something you likely learned without proof in high school.

   (b) Suppose not all of $a_n, a_{n-1}, \ldots, a_1, a_0$ are real. Show that P has at least one root whose complex conjugate is not a root. Hint: Prove the contrapositive.

   (c) Find an example of a polynomial that has some roots occurring as complex conjugates, and some not.

6. Let $z_1 = (x_1, y_1)$ and $z_2 = (x_2, y_2)$ be arbitrary complex numbers. Prove or disprove the following.

(a) $\operatorname{Re}(z_1 + z_2) = \operatorname{Re}(z_1) + \operatorname{Re}(z_2)$.

(b) $\operatorname{Re}(z_1 z_2) = \operatorname{Re}(z_1)\operatorname{Re}(z_2)$.

(c) $\operatorname{Im}(z_1 + z_2) = \operatorname{Im}(z_1) + \operatorname{Im}(z_2)$.

(d) $\operatorname{Im}(z_1 z_2) = \operatorname{Im}(z_1)\operatorname{Im}(z_2)$.

7. Prove that the complex number $(1,0)$ (which we identify with the real number 1) is the multiplicative identity for complex numbers.

8. Use mathematical induction to show that the binomial theorem is valid for complex numbers. In other words, show that, if z and w are arbitrary complex numbers and n is a positive integer, then

$$(z + w)^n = \sum_{k=0}^{n} \binom{n}{k} z^k w^{n-k}, \text{ where } \binom{n}{k} = \frac{n!}{k!(n-k)!}.$$

9. Let's use the symbol $*$ for a new type of multiplication of complex numbers defined by $z_1 * z_2 = (x_1 x_2, \ y_1 y_2)$. This exercise shows why this is a bad definition.

   (a) Use the definition given in property **(P7)** and state what the multiplicative identity ζ would have to be for this new multiplication.

   (b) Show that, if you use this new multiplication, nonzero complex numbers of the form $(0, a)$ have no inverse. That is, show that, if $z = (0, a)$, there is no complex number w with the property that $z * w = \zeta$, where ζ is the multiplicative identity you found in part (a).

10. Explain why the complex number $(0,0)$ (which, you recall, we identify with the real number 0) has no multiplicative inverse.

11. Prove property **(P9)**, the distributive law for complex numbers.

12. Verify that, if $z = (x, y)$, with x and y not both 0, then $z^{-1} = \frac{(1,0)}{z}$ $\left(i.e., \ z^{-1} = \frac{1}{z} \right)$. *Hint:* Let $z = (x, y)$ and use the (ordered pair) definition for division to compute $z^{-1} = \frac{(1,0)}{(x,y)}$. Then, with the result you obtained, use the (ordered pair) definition for multiplication to confirm that $zz^{-1} = (1,0) = 1$.

13. From Exercise 12 and basic cancellation laws, it follows that $z^{-1} = \frac{1}{z} = \frac{\bar{z}}{z\bar{z}}$. The numerator here, $\bar{z}$, is trivial to calculate and, as the denominator $z\bar{z}$ is a real number (Exercise 3), computing the quotient $\frac{\bar{z}}{z\bar{z}}$ should be rather straightforward. Use this fact to compute z^{-1} if $z = 2 + 3i$ and again if $z = 7 - 5i$.

14. Recall the following trigonometric identities:

    (a) $\cos(\theta_1 + \theta_2) = \cos\theta_1 \cos\theta_2 - \sin\theta_1 \sin\theta_2$;

    (b) $\sin(\theta_1 + \theta_2) = \cos\theta_1 \sin\theta_2 + \cos\theta_2 \sin\theta_1$.

    Use these identities to show (using Wessel's analogy for the norm and angular displacement discussed in Section 1.1.4) that the ordered-pair definition for the product of two vectors must agree with Equation (1.10) of the text. In other words, show that it must be the case that, if $\mathbf{z_1} = (x_1, y_1)$ and $\mathbf{z_2} = (x_2, y_2)$, then $\mathbf{z_1 z_2} = (x_1 x_2 - y_1 y_1, \ x_1 y_2 + x_2 y_1)$.

15. Show, by equating the real numbers x_1 and x_2 with $(x_1, 0)$ and $(x_2, 0)$, respectively, that the complex definition for division is consistent with the real definition for division.

1.3   The Geometry of Complex Numbers, Part I

Complex numbers are ordered pairs of real numbers, so they can be represented by points in the plane. In this section we show the effect that algebraic operations on complex numbers have on their geometric representations.

We can represent the number $z = x + iy = (x, y)$ by a position vector in the xy plane whose tail is at the origin and whose head is at the point (x, y). When the xy plane is used for displaying complex numbers, it is called the **complex plane**, or more simply, the **z plane**. Recall that $\operatorname{Re}(z) = x$ and $\operatorname{Im}(z) = y$. Geometrically, $\operatorname{Re}(z)$ is the projection of $z = (x, y)$ onto the x axis, and $\operatorname{Im}(z)$ is the projection of z onto the y axis. It makes sense, then, to call the x axis the **real axis** and the y axis the **imaginary axis**, as Figure 1.7 illustrates.

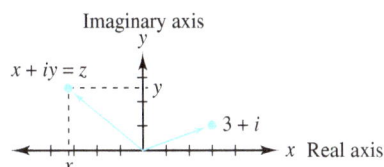

Figure 1.7: The complex plane

Addition of complex numbers is analogous to addition of vectors in the plane. As we saw in Section 1.2, the sum of $z_1 = x_1 + iy_1 = (x_1, y_1)$ and $z_2 = x_2 + iy_2 = (x_2, y_2)$ is $(x_1 + x_2, y_1 + y_2)$. Hence $z_1 + z_2$ can be obtained using the "parallelogram law" for vectors, where the vector sum is represented by the diagonal of the parallelogram formed by the two original vectors, as illustrated by Figure 1.8. The difference $z_1 - z_2$ can be represented by the displacement vector from the point $z_2 = (x_2, y_2)$ to the point $z_1 = (x_1, y_1)$, as Figure 1.9 depicts.

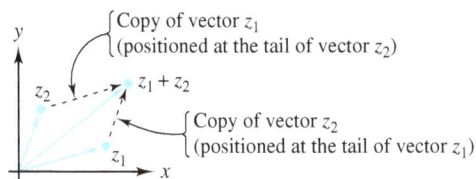

Figure 1.8: The sum $z_1 + z_2$

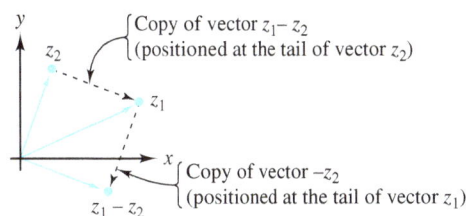

Figure 1.9: The difference $z_1 - z_2$

Definition 1.8 (Modulus). *The **modulus**, or **absolute value**, of the complex number $z = x + iy$ is a nonnegative real number denoted by $|z|$ and defined by the relation*

$$|z| = \sqrt{x^2 + y^2}. \tag{1.20}$$

The number $|z|$ is the distance between the origin and the point $z = (x, y)$. The only complex number with modulus zero is the number 0. The number $z = 4 + 3i$ has modulus $|4 + 3i| = \sqrt{4^2 + 3^2} = \sqrt{25} = 5$, and is depicted in Figure 1.10.

The numbers $|\operatorname{Re}(z)|$, $|\operatorname{Im}(z)|$, and $|z|$ are the lengths of the sides of the right triangle OPQ shown in Figure 1.11. The inequality $|z_1| < |z_2|$ means that the point z_1 is closer to the origin than the point z_2. Although obvious from Figure 1.11, it is still profitable to work out algebraically (which we ask you to do in the exercises) that

$$|x| = |\operatorname{Re}(z)| \leq |z| \quad \text{and} \quad |y| = |\operatorname{Im}(z)| \leq |z|. \tag{1.21}$$

17

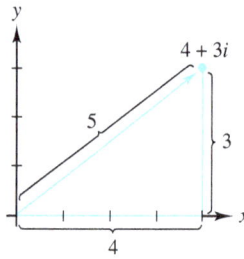

Figure 1.10: The real and imaginary parts of a complex number

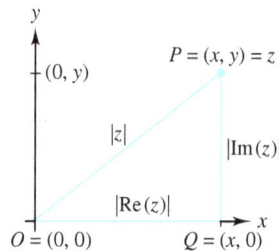

Figure 1.11: The modulus of z and its components

The difference $z_1 - z_2$ represents the displacement vector from z_2 to z_1, so the distance between z_1 and z_2 is given by $|z_1 - z_2|$. We can obtain this distance by using Definitions 1.2 and 1.8 to obtain the familiar formula

$$\text{dist}(z_1, z_2) = |z_1 - z_2| = \sqrt{(x_1 - x_2)^2 + (y_1 - y_2)^2}.$$

If $z = (x, y) = x + iy$, then $-z = (-x, -y) = -x - iy$ is the reflection of z through the origin, and $\overline{z} = (x, -y) = x - iy$ is the reflection of z through the x axis, as illustrated in Figure 1.12

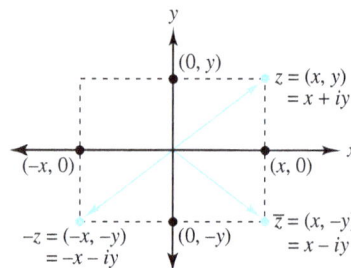

Figure 1.12: The geometry of negation and conjugation

We can use an important algebraic relationship to establish properties of the absolute value that have geometric applications. Its proof is rather straightforward, and we ask you to give it in the exercises for this section.

$$|z|^2 = z\overline{z}. \tag{1.22}$$

An important application of Identity (1.22) is its use in establishing the triangle inequality, which states that the sum of the lengths of two sides of a triangle is greater than or equal to the length of the third side. Figure 1.13 illustrates this inequality.

Theorem 1.2 (The triangle inequality). *If z_1 and z_2 are arbitrary complex numbers, then*

$$|z_1 + z_2| \le |z_1| + |z_2|. \tag{1.23}$$

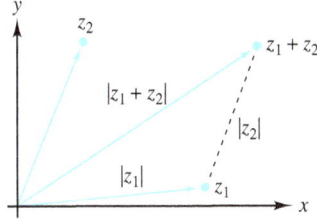

Figure 1.13: The triangle inequality

Proof. We appeal to basic results:

$$
\begin{aligned}
|z_1 + z_2|^2 &= (z_1 + z_2)\overline{(z_1 + z_2)} && \text{(by Identity (1.22))} \\
&= (z_1 + z_2)(\overline{z_1} + \overline{z_2}) && \text{(by Identity (1.13))} \\
&= z_1\overline{z_1} + z_1\overline{z_2} + z_2\overline{z_1} + z_2\overline{z_2} \\
&= |z_1|^2 + z_1\overline{z_2} + \overline{z_1}z_2 + |z_2|^2 && \text{(by Identity (1.22) and the commutative law)} \\
&= |z_1|^2 + z_1\overline{z_2} + \overline{(z_1\overline{z_2})} + |z_2|^2 && \text{(by Identities (1.12) and (1.14))} \\
&= |z_1|^2 + 2\mathrm{Re}(z_1\overline{z_2}) + |z_2|^2 && \text{(by Identity (1.16))} \\
&\leq |z_1|^2 + 2|z_1\overline{z_2}| + |z_2|^2 && \text{(by Identity (1.21))} \\
&= (|z_1| + |z_2|)^2.
\end{aligned}
$$

Taking square roots yields the desired inequality. □

Example 1.5. To produce an example of which Figure 1.13 is a reasonable illustration, let $z_1 = 7 + i$ and $z_2 = 3 + 5i$. Then $|z_1| = \sqrt{49 + 1} = \sqrt{50}$ and $|z_2| = \sqrt{9 + 25} = \sqrt{34}$. Clearly, $z_1 + z_2 = 10 + 6i$; hence $|z_1 + z_2| = \sqrt{100 + 36} = \sqrt{136}$. In this case, we can verify the triangle inequality without appealing to calculator approximations because

$$
|z_1 + z_2| = \sqrt{136} = 2\sqrt{34} = \sqrt{34} + \sqrt{34} < \sqrt{50} + \sqrt{34} = |z_1| + |z_2|.
$$

We can also establish other important identities by means of the triangle inequality. Note that

$$
\begin{aligned}
|z_1| &= |(z_1 + z_2) + (-z_2)| \\
&\leq |z_1 + z_2| + |-z_2| \\
&= |z_1 + z_2| + |z_2|.
\end{aligned}
$$

Subtracting $|z_2|$ from the left and right sides of this string of inequalities gives an important relationship that is used in determining lower bounds of sums of complex numbers:

$$
|z_1 + z_2| \geq |z_1| - |z_2|. \tag{1.24}
$$

From Identity (1.22) and the commutative and associative laws, it follows that

$$
|z_1 z_2|^2 = (z_1 z_2)\overline{(z_1 z_2)} = (z_1\overline{z_1})(z_2\overline{z_2}) = |z_1|^2 |z_2|^2.
$$

Taking square roots of the terms on the left and right establishes another important identity:

$$
|z_1 z_2| = |z_1||z_2|. \tag{1.25}
$$

As an exercise, we ask you to show that

$$
\left|\frac{z_1}{z_2}\right| = \frac{|z_1|}{|z_2|}, \quad \text{provided} \quad z_2 \neq 0. \tag{1.26}
$$

19

Example 1.6. If $z_1 = 1+2i$ and $z_2 = 3+2i$, then $|z_1| = \sqrt{1+4} = \sqrt{5}$ and $|z_2| = \sqrt{9+4} = \sqrt{13}$. Also $z_1 z_2 = -1 + 8i$; hence $|z_1 z_2| = \sqrt{1+64} = \sqrt{65} = \sqrt{5}\sqrt{13} = |z_1||z_2|$.

Figure 1.14 illustrates the multiplication shown in Example 1.6. The length of the $z_1 z_2$ vector apparently equals the product of the lengths of z_1 and z_2, confirming Equation (1.25), but why is it located in the second quadrant when both z_1 and z_2 are in the first quadrant? The answer to this question will become apparent to you as you read Section 1.4.

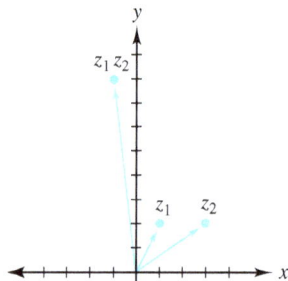

Figure 1.14: The geometry of multiplication

Exercises for Section 1.3 (Selected answers or hints are on page 428.)

1. Evaluate the following quantities. Put your answer in the form $a+ib$ and show your work.

   (a) $|(1+i)(2+i)|$.

   (b) $\left|\frac{4-3i}{2-i}\right|$.

   (c) $|(1+i)^{50}|$.

   (d) $|z\bar{z}|$, where $z = x + iy$.

   (e) $|z-1|^2$, where $z = x + iy$.

2. Plot z_1 and z_2 as vectors, then find and plot $z_1 + z_2$ and $z_1 - z_2$ when

   (a) $z_1 = 2 + 3i$ and $z_2 = 4 + i$..

   (b) $z_1 = -1 + 2i$ and $z_2 = -2 + 3i$..

   (c) $z_1 = 1 + i\sqrt{3}$ and $z_2 = -1 + i\sqrt{3}$.

3. Which of the following points lie inside the circle $|z - i| = 2$? Explain.

   (a) $\frac{1}{2} + i$.

   (b) $\sqrt{2} + i(\sqrt{2} + 1)$.

   (c) $2 + 3i$.

   (d) $\frac{-1}{2} + i\sqrt{3}$.

4. Prove the following Identities for $z = x + iy$:

   (a) Identity (1.21): $|x| = |\text{Re}(z)| \le |z|$ and $|y| = |\text{Im}(z)| \le |z|$.

   (b) Identity (1.22), the triangle inequality: $|z_1 + z_2| \le |z_1| + |z_2|$.

   (c) Identity (1.26): $\left|\frac{z_1}{z_2}\right| = \frac{|z_1|}{|z_2|}$,   provided $z_2 \ne 0$.

5. Show that nonzero vectors z_1 and z_2 are perpendicular iff $\text{Re}(z_1\overline{z_2}) = 0$.

6. Sketch the sets of points determined by the following relations.

   (a) $|z + 1 - 2i| = 2$.

   (b) $\text{Re}(z + 1) = 0$.

   (c) $|z + 2i| \le 1$.

   (d) $\text{Im}(z - 2i) > 6$.

7. Prove that $\sqrt{2}|z| \ge |\text{Re}(z)| + |\text{Im}(z)|$.

8. Show that the point $\frac{z_1 + z_2}{2}$ is the midpoint of the line segment joining z_1 to z_2.

9. Show that $|z_1 - z_2| \le |z_1| + |z_2|$.

10. Prove that $|z| = 0$ iff $z = 0$.

11. Show that, if $z \ne 0$, the four points $z, \overline{z}, -z$, and $-\overline{z}$ are the vertices of a rectangle with its center at the origin.

12. Show that, if $z \ne 0$, the four points $z,\ iz,\ -z$, and $-iz$ are the vertices of a square with its center at the origin.

13. Show that the equation of the line through the points z_1 and z_2 can be expressed in the form $z = z_1 + t(z_2 - z_1)$, where t is a real number.

14. Show that nonzero vectors z_1 and z_2 are parallel iff $\text{Im}(z_1\overline{z_2}) = 0$.

15. Show that $|z_1 z_2 z_3| = |z_1||z_2||z_3|$.

16. Show that $|z^n| = |z|^n$, where n is an integer.

17. Suppose that either $|z| = 1$ or $|w| = 1$. Prove that $|z - w| = |1 - \overline{z}w|$.

18. Prove the Cauchy-Schwarz inequality: $\left|\sum_{k=1}^{n} z_k w_k\right| = \sqrt{\sum_{k=1}^{n} |z_k|^2 \sqrt{\sum_{k=1}^{n} |w_k|^2}}$.

19. Show that $\left||z_1| - |z_2|\right| \le |z_1 - z_2|$.

20. Show that $z_1\overline{z_2} + \overline{z_1}z_2$ is a real number.

21. If you study carefully the proof of the triangle inequality, you will note that the reasons for the *inequality* hinge on $\text{Re}(z_1\overline{z_2}) \le |z_1\overline{z_2}|$. Under what conditions will these two quantities be equal, thus turning the triangle inequality into an equality?

22. Prove that $|z_1 - z_2|^2 = |z_1|^2 - 2\text{Re}(z_1\overline{z_2}) + |z_2|^2$.

23. Prove by induction that $\left|\sum_{k=1}^{n} z_k\right| \leq \sum_{k=1}^{n} |z_k|$ for all natural numbers n.

24. Let z_1 and z_2 be two distinct points in the complex plane, and let K be a positive real constant that is less than the distance between z_1 and z_2.

    (a) Show that the set of points $\{z : |z - z_1| - |z - z_2| = K\}$ is a hyperbola with foci z_1 and z_2.

    (b) Find the equation of the hyperbola with foci ± 2 that goes through the point $2 + 3i$.

    (c) Find the equation of the hyperbola with foci ± 25 that goes through the point $7 + 24i$.

25. Let z_1 and z_2 be two distinct points in the complex plane, and let K be a positive real constant that is greater than the distance between z_1 and z_2.

    (a) Show that the set of points $\{z : |z - z_1| + |z - z_2| = K\}$ is an ellipse with foci z_1 and z_2.

    (b) Find the equation of the ellipse with foci $\pm 3i$ that goes through the point $8 - 3i$.

    (c) Find the equation of the ellipse with foci $\pm 2i$ that goes through the point $3 + 2i$.

1.4   The Geometry of Complex Numbers, Part II

In Section 1.3 we saw that a complex number $z = x + iy$ could be viewed as a vector in the xy plane with its tail at the origin and its head at the point (x, y). A **vector** can be uniquely specified by giving its magnitude (*i.e.*, its length) and direction (*i.e.*, the angle it makes with the positive x axis). In this section, we focus on these two geometric aspects of complex numbers.

Let r be the modulus of z (*i.e.*, $r = |z|$) and let θ be the angle that the line from the origin to the complex number z makes with the positive x axis. (Note: The number θ is undefined if $z = 0$). Then, as Figure 1.15(a) shows,

$$z = (r\cos\theta,\ r\sin\theta) = r(\cos\theta + i\sin\theta). \tag{1.27}$$

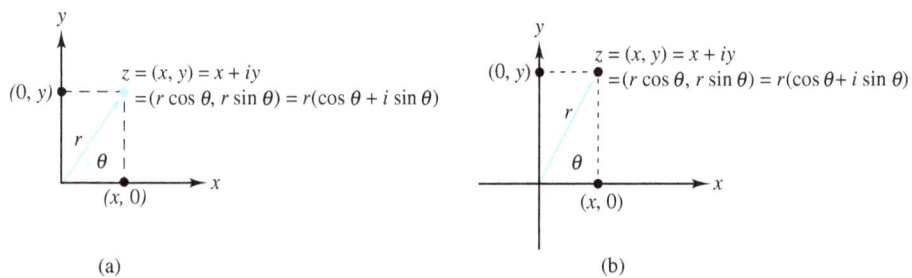

Figure 1.15: Polar representation of complex numbers

Definition 1.9 (Polar Representation). *Identity* (1.27) *is known as a **polar representation** of z, and the values r and θ are called **polar coordinates** of z.*

Example 1.7. If $z = 1 + i$, then $r = \sqrt{2}$ and $z = (\sqrt{2}\cos\frac{\pi}{4}, \sqrt{2}\sin\frac{\pi}{4}) = \sqrt{2}(\cos\frac{\pi}{4} + i\sin\frac{\pi}{4})$ is a polar representation of z. The polar coordinates in this case are $r = \sqrt{2}$, and $\theta = \frac{\pi}{4}$.

As Figure 1.15(b) shows, θ can be *any* value for which the identities $\cos\theta = \frac{x}{r}$ and $\sin\theta = \frac{y}{r}$ hold. For $z \neq 0$, the collection of all values of θ for which $z = r(\cos\theta + i\sin\theta)$ is denoted $\arg z$. Formally, we have the following definitions.

Definition 1.10 (arg z). *If* $z \neq 0$,

$$\arg z = \{\theta : z = r(\cos\theta + i\sin\theta)\}. \tag{1.28}$$

If $\theta \in \arg z$, *we say that* θ *is **an argument** of* z.

Note that we write $\theta \in \arg z$ as opposed to $\theta = \arg z$. We do so because $\arg z$ is a set, and the designation $\theta \in \arg z$ indicates that θ belongs to that set. Note also that, if $\theta_1 \in \arg z$ and $\theta_2 \in \arg z$, then there exists some integer n such that

$$\theta_1 = \theta_2 + 2n\pi. \tag{1.29}$$

Example 1.8. Because $1 + i = \sqrt{2}(\cos\frac{\pi}{4} + i\sin\frac{\pi}{4})$, we have

$$\arg(1+i) = \left\{\frac{\pi}{4} + 2n\pi : n \text{ is an integer}\right\} = \left\{\cdots, -\frac{7\pi}{4}, \frac{\pi}{4}, \frac{9\pi}{4}, \frac{17\pi}{4}, \cdots\right\}.$$

Mathematicians have agreed to single out a special choice of $\theta \in \arg z$. It is that value of θ for which $-\pi < \theta \leq \pi$, as the following definition indicates.

Definition 1.11 (Arg z). *Let* $z \neq 0$ *be a complex number. Then*

$$\mathrm{Arg}\, z = \theta \quad provided \quad z = r(\cos\theta + i\sin\theta) \quad and \quad -\pi < \theta \leq \pi. \tag{1.30}$$

If $\theta = \mathrm{Arg}\, z$, *we call* θ ***the argument*** *of* z.

Example 1.9. $Arg(1+i) = \frac{\pi}{4}$.

Remark 1.1. *Clearly, if* $z = x + iy = r(\cos\theta + i\sin\theta)$, *where* $x \neq 0$, *then*

$$\arg z \subset \arctan\frac{y}{x},$$

where $\arctan\frac{y}{x} = \{\theta : \tan\theta = \frac{y}{x}\}$. *Note that, for any real number* t, $\arctan t$ *is a set (as opposed to* $\mathrm{Arctan}\, t$, *which is a number). We specifically identify* $\arg z$ *as a proper subset of* $\arctan\frac{y}{x}$ *because* $\tan\theta$ *has period* π, *whereas* $\cos\theta$ *and* $\sin\theta$ *have period* 2π. *In selecting the proper values for* $\arg z$, *we must be careful in specifying the choices of* $\arctan\frac{y}{x}$ *so that the point* z *associated with* r *and* θ *lies in the appropriate quadrant.*

Example 1.10. If $z = -\sqrt{3} - i = r(\cos\theta + i\sin\theta)$, then $r = |z| = |-\sqrt{3} - i| = 2$ and $\theta \in \arctan\frac{y}{x} = \arctan\frac{-1}{\sqrt{3}} = \{\frac{\pi}{6} + n\pi : n \text{ is an integer}\}$. It would be a mistake to use $\frac{\pi}{6}$ as an acceptable value for θ, as the point z associated with $r = 2$ and $\theta = \frac{\pi}{6}$ is in the first quadrant, whereas $-\sqrt{3} - i$ is in the third quadrant. A correct choice for θ is $\theta = \frac{\pi}{6} - \pi = -\frac{5\pi}{6}$ because

$$-\sqrt{3} - i = 2\cos\left(-\frac{5\pi}{6}\right) + i2\sin\left(-\frac{5\pi}{6}\right)$$

$$= 2\cos\left(-\frac{5\pi}{6} + 2n\pi\right) + i2\sin\left(-\frac{5\pi}{6} + 2n\pi\right),$$

where n is any integer. Notice also, that

$$\mathrm{Arg}(-\sqrt{3} - i) = -\frac{5\pi}{6}, \quad \text{and}$$

$$\arg(-\sqrt{3} - i) = \left\{-\frac{5\pi}{6} + 2n\pi : n \text{ is an integer}\right\},$$

which illustrates that $\arg(-\sqrt{3} - i)$ is indeed a *proper* subset of $\arctan\frac{-1}{-\sqrt{3}}$.

Example 1.11. If $z = x + iy = 0 + 4i$, it would be a mistake to attempt to find $\operatorname{Arg} z$ by looking at $\arctan \frac{y}{x}$, as $x = 0$, so $\frac{y}{x}$ is undefined. If $z \neq 0$ is on the y axis, then

$$\operatorname{Arg} z = \frac{\pi}{2}, \quad \text{if} \quad \operatorname{Im} z > 0 \text{and}$$

$$\operatorname{Arg} z = -\frac{\pi}{2}, \quad \text{if} \quad \operatorname{Im} z < 0.$$

In this case, $\operatorname{Arg}(4i) = \frac{\pi}{2}$ and $\arg(4i) = \{\frac{\pi}{2} + 2n\pi : n \text{ is an integer}\}$.

As you will see in Chapter 2, $\operatorname{Arg} z$ is a discontinuous function of z because it "jumps" by an amount of 2π as z crosses the negative real axis.

In Chapter 5 we define e^z for any complex number z. You will see that this complex exponential has all the properties of real exponentials that you studied in earlier mathematics courses. That is, $e^{z_1} e^{z_2} = e^{z_1 + z_2}$, and so on. You will also see, amazingly, that if $z = x + iy$, then

$$e^z = e^{x+iy} = e^x(\cos y + i \sin y). \tag{1.31}$$

We will establish this result rigorously in Chapter 5, but there is a plausible explanation we can give now. If e^z has the normal properties of an exponential, it must be that $e^{x+iy} = e^x e^{iy}$. Now, recall from Calculus the values of three infinite series:

$$e^x = \sum_{k=0}^{\infty} \frac{1}{k!} x^k, \quad \cos x = \sum_{n=0}^{\infty} \frac{(-1)^n}{(2n)!} x^{2n}, \text{and } \sin x = \sum_{n=0}^{\infty} \frac{(-1)^n}{(2n+1)!} x^{2n+1}.$$

Substituting iy for x in the infinite series for e^x gives

$$e^{iy} = \sum_{k=0}^{\infty} \frac{1}{k!}(iy)^k = \sum_{k=0}^{\infty} \frac{1}{k!} i^k y^k.$$

At this point our argument loses rigor because we have not talked about infinite series of complex numbers, let alone whether such series converge. Nevertheless, if we mercly take the last series as a formal expression and split it into two series according to whether the index k is even ($k = 2n$) or odd ($k = 2n+1$), we get

$$e^{iy} = \sum_{k \text{ is even}} \frac{1}{k!} i^k y^k \quad + \sum_{k \text{ is odd}} \frac{1}{k!} i^k y^k$$

$$= \sum_{n=0}^{\infty} \frac{1}{(2n)!} i^{2n} y^{2n} \quad + \sum_{n=0}^{\infty} \frac{1}{(2n+1)!} i^{2n+1} y^{2n+1}$$

$$= \sum_{n=0}^{\infty} \frac{1}{(2n)!} (i^2)^n y^{2n} \quad + \sum_{n=0}^{\infty} \frac{1}{(2n+1)!} (i^2)^n i y^{2n+1}$$

$$= \sum_{n=0}^{\infty} \frac{1}{(2n)!} (-1)^n y^{2n} + i \sum_{n=0}^{\infty} \frac{1}{(2n+1)!} (-1)^n y^{2n+1}$$

$$= \cos y \qquad\qquad + i \sin y.$$

Thus, it seems the only possible value for e^z is that given by Equation (1.31). We will use this result freely from now on, and, as stated, supply a rigorous proof in Chapter 5.

If we set $x = 0$ and let θ take the role of y in Equation (1.31), we get a famous result known as **Euler's formula**:

$$e^{i\theta} = (\cos\theta + i\sin\theta) = (\cos\theta, \sin\theta).$$ (1.32)

If θ is a real number, $e^{i\theta}$ will be located somewhere on the circle with radius 1 centered at the origin. This assertion is easy to verify because

$$|e^{i\theta}| = \sqrt{\cos^2\theta + \sin^2\theta} = 1.$$ (1.33)

Figure 1.16 illustrates the location of $e^{i\theta}$ for various values of θ.

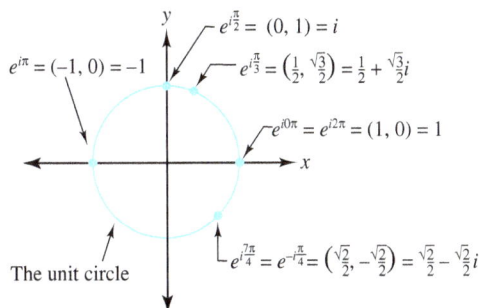

Figure 1.16: The location of $e^{i\theta}$ for various values of θ

Note that, when $\theta = \pi$, we get $e^{i\pi} = (\cos\pi, \sin\pi) = (-1, 0) = -1$, so

$$e^{i\pi} + 1 = 0.$$ (1.34)

Euler was the first to discover this relationship; it is referred to as **Euler's identity**. It has been labeled by many mathematicians as the most amazing relation in analysis—and with good reason. Symbols with a rich history are miraculously woven together—the constant π used by Hippocrates as early as 400 B.C.; e, the base of the natural logarithms; the basic concepts of addition ($+$) and equality ($=$); the foundational whole numbers 0 and 1; and i, the number that is the central focus of this book.

Euler's Formula (1.32) is of tremendous use in establishing important algebraic and geometric properties of complex numbers. You will see shortly that it enables you to multiply complex numbers with great ease. It also allows you to express a polar form of the complex number z in a more compact way. Recall that, if $r = |z|$ and $\theta \in \arg z$, then $z = r(\cos\theta + i\sin\theta)$. Using Euler's Formula we can now write z in its exponential form:

$$z = re^{i\theta}.$$ (1.35)

Example 1.12. With reference to Example 1.10, with $z = -\sqrt{3} - i$, we have $z = 2e^{i(-5\pi/6)}$.

Together with the rules for exponentiation that we will verify in Chapter 5, Equation (1.35) has interesting applications. If $z_1 = r_1 e^{i\theta_1}$ and $z_2 = r_2 e^{i\theta_2}$, then

$$
\begin{aligned}
z_1 z_2 &= r_1 e^{i\theta_1} r_2 e^{i\theta_2} = r_1 r_2 e^{i(\theta_1 + \theta_2)} \\
&= r_1 r_2 [\cos(\theta_1 + \theta_2) + i\sin(\theta_1 + \theta_2)].
\end{aligned}
$$ (1.36)

Figure 1.17 illustrates the geometric significance of this equation.

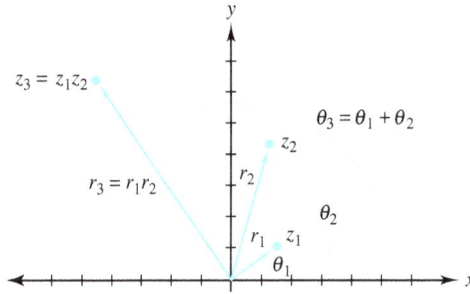

Figure 1.17: The product of two complex numbers $z_3 = z_1 z_2$

We have already shown that the modulus of the product is the product of the moduli; that is, $|z_1 z_2| = |z_1||z_2|$. Identity (1.36) establishes that an argument of $z_1 z_2$ is an argument of z_1 plus an argument of z_2. It also answers the question posed at the end of Section 1.3 regarding why the product $z_1 z_2$ was in a different quadrant than either z_1 or z_2. It further offers an interesting explanation as to why the product of two negative real numbers is a positive real number. The negative numbers, each of which has an angular displacement of π radians, combine to produce a product that is rotated to a point with an argument of $\pi + \pi = 2\pi$ radians, coinciding with the positive real axis.

Using exponential form, if $z \neq 0$, we can write $\arg z$ a bit more compactly as

$$\arg z = \{\theta : z = re^{i\theta}\}. \tag{1.37}$$

Doing so enables us to see a nice relationship between the sets $\arg(z_1 z_2)$, $\arg z_1$, and $\arg z_2$:

Theorem 1.3. If $z_1 = r_1 e^{i\theta_1} \neq 0$ and $z_2 = r_2 e^{i\theta_2} \neq 0$, then as sets,

$$\arg(z_1 z_2) = \arg z_1 + \arg z_2. \tag{1.38}$$

Before proceeding with the proof of this theorem, we recall two important facts about sets. First, to establish the equality of two sets, we must show that each is a subset of the other. Second, the sum of two sets is the sum of all combinations of elements from the first and second sets, respectively. In this case, $\arg z_1 + \arg z_2 = \{\theta_1 + \theta_2 : \theta_1 \in \arg z_1 \text{ and } \theta_2 \in \arg z_2\}$.

Proof. Let $\theta \in \arg(z_1 z_2)$. Because $z_1 z_2 = r_1 r_2 e^{i(\theta_1 + \theta_2)}$, it follows from Formula (1.37) that $\theta_1 + \theta_2 \in \arg(z_1 z_2)$. By Equation (1.29) there is some integer n such that $\theta = \theta_1 + \theta_2 + 2n\pi$. Further, as $z_1 = r_1 e^{i\theta_1}$, $\theta_1 \in \arg z_1$. Likewise, $z_2 = r_2 e^{i\theta_2}$ gives $\theta_2 \in \arg z_2$. But if $\theta_2 \in \arg z_2$, then $\theta_2 + 2n\pi \in \arg z_2$. This result shows that $\theta = \theta_1 + (\theta_2 + 2n\pi) \in \arg z_1 + \arg z_2$. Thus, $\arg(z_1 z_2) \subseteq \arg z_1 + \arg z_2$. The proof that $\arg z_1 + \arg z_2 \subseteq \arg(z_1 z_2)$ is left as an exercise. $\square$

Using Equality (1.35) gives $z^{-1} = \frac{1}{z} = \frac{1}{re^{i\theta}} = \frac{1}{r}e^{-i\theta}$. In other words,

$$z^{-1} = \frac{1}{r}[\cos(-\theta) + i\sin(-\theta)] = \frac{1}{r}e^{-i\theta}.$$

Recalling that $\cos(-\theta) = \cos(\theta)$ and $\sin(-\theta) = -\sin(\theta)$, we also have

$$\bar{z} = r(\cos\theta - i\sin\theta) = r[\cos(-\theta) + i\sin(-\theta)] = re^{-i\theta}, \text{ and}$$
$$\frac{z_1}{z_2} = \frac{r_1}{r_2}[\cos(\theta_1 - \theta_2) + i\sin(\theta_1 - \theta_2)] = \frac{r_1}{r_2}e^{i(\theta_1 - \theta_2)}.$$

If z is in the first quadrant, the positions of the numbers z, $\overline{z}$, and z^{-1} are as shown in Figure 1.18 when $|z| < 1$. Figure 1.19 depicts the situation when $|z| > 1$.

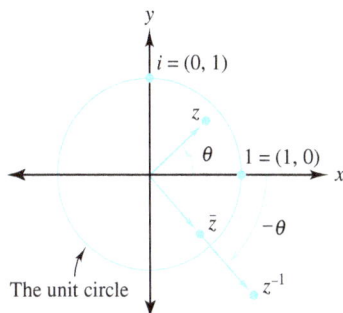

Figure 1.18: Relative positions of z, $\overline{z}$, and z^{-1}, when $|z| < 1$

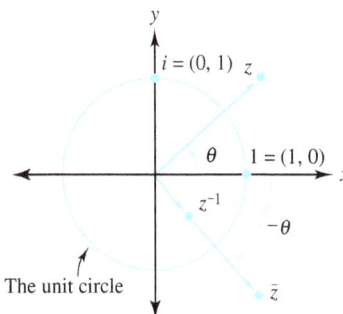

Figure 1.19: Relative positions of z, $\overline{z}$, and z^{-1}, when $|z| > 1$

Example 1.13. If $z = 1 + i$, then $r = |z| = \sqrt{2}$ and $\theta = \text{Arg}\, z = \frac{\pi}{4}$. Therefore $z^{-1} = \frac{1}{\sqrt{2}}[\cos(-\frac{\pi}{4}) + i\sin(-\frac{\pi}{4})] = \frac{1}{\sqrt{2}}\left[\frac{\sqrt{2}}{2} - i\frac{\sqrt{2}}{2}\right]$ and has modulus $\frac{1}{\sqrt{2}} = \frac{\sqrt{2}}{2}$.

Example 1.14. If $z_1 = 8i$ and $z_2 = 1 + i\sqrt{3}$, then representative polar forms for these numbers are $z_1 = 8(\cos\frac{\pi}{2} + i\sin\frac{\pi}{2})$ and $z_2 = 2(\cos\frac{\pi}{3} + i\sin\frac{\pi}{3})$. Hence

$$\frac{z_1}{z_2} = \frac{8}{2}\left[\cos\left(\frac{\pi}{2} - \frac{\pi}{3}\right) + i\sin\left(\frac{\pi}{2} - \frac{\pi}{3}\right)\right] = 4\left(\cos\frac{\pi}{6} + i\sin\frac{\pi}{6}\right) = 2\sqrt{3} + 2i.$$

Exercises for Section 1.4 (Selected answers or hints are on page 429.)

1. Find $\text{Arg}\, z$ for the following values of z.

   (a) $1 - i$.

   (b) $-\sqrt{3} + i$.

   (c) $(-1 - i\sqrt{3})^2$.

   (d) $(1 - i)^3$.

   (e) $\frac{2}{1+i\sqrt{3}}$.

   (f) $\frac{2}{i-1}$.

   (g) $\frac{1+i\sqrt{3}}{(1+i)^2}$.

27

(h) $(1 + i\sqrt{3})(1 + i)$.

2. Use exponential notation to show that

   (a) $(\sqrt{3} - i)(1 + i\sqrt{3}) = 2\sqrt{3} + 2i$.

   (b) $(1 + i)^3 = -2 + 2i$.

   (c) $2i(\sqrt{3} + i)(1 + i\sqrt{3}) = -8$.

   (d) $\frac{8}{1+i} = 4 - 4i$.

3. Represent the following complex numbers in polar form.

   (a) -4.

   (b) $6 - 6i$.

   (c) $-7i$.

   (d) $-2\sqrt{3} - 2i$.

   (e) $\frac{1}{(1-i)^2}$.

   (f) $\frac{6}{i+\sqrt{3}}$.

   (g) $3 + 4i$.

   (h) $(5 + 5i)^3$.

4. Show that $\arg z_1 + \arg z_2 \subseteq \arg z_1 z_2$, thus completing the proof of Theorem 1.3.

5. Express the following in $a + ib$ form.

   (a) $e^{i\frac{\pi}{2}}$.

   (b) $4e^{-i\frac{\pi}{2}}$.

   (c) $8e^{i\frac{7\pi}{3}}$.

   (d) $-2e^{i\frac{5\pi}{6}}$.

   (e) $2ie^{-i\frac{3\pi}{4}}$.

   (f) $6e^{i\frac{2\pi}{3}}e^{i\pi}$.

   (g) $e^2 e^{i\pi}$.

6. Show that $\arg z_1 = \arg z_2$ iff $z_2 = cz_1$, where c is a positive real constant.

7. Let $z_1 = -1 + i\sqrt{3}$ and $z_2 = -\sqrt{3} + i$. Show that the equation $\mathrm{Arg}(z_1 z_2) = \mathrm{Arg}\, z_1 + \mathrm{Arg}\, z_2$ *does not* hold for the specific choice of z_1 and z_2. Why not?

8. Show that the equation $\mathrm{Arg}(z_1 z_2) = \mathrm{Arg}\, z_1 + \mathrm{Arg}\, z_2$ is true provided that the inequalities $-\frac{\pi}{2} < \mathrm{Arg}\, z_1 \leq \frac{\pi}{2}$ and $-\frac{\pi}{2} < \mathrm{Arg}\, z_2 \leq \frac{\pi}{2}$ are satisfied. Describe the set of points that meets this criterion.

9. Describe the set of complex numbers for which $\mathrm{Arg}(\frac{1}{z}) \neq -\mathrm{Arg}(z)$. Prove your assertion.

10. Establish the identity $\arg(\frac{z_1}{z_2}) = \arg z_1 - \arg z_2$.

11. Show that $\arg(\frac{1}{z}) = -\arg z$.

12. Show that $\arg(z_1 \overline{z_2}) = \arg z_1 - \arg z_2$.

13. Show that, if $z \neq 0$, then

    (a) $\mathrm{Arg}(z\overline{z}) = 0$.

    (b) $\mathrm{Arg}(z + \overline{z}) = 0$ when $\mathrm{Re}(z) > 0$.

14. Let z_1, z_2, and z_3 form the vertices of a triangle as indicated in Figure 1.20. Show that $\alpha \in \arg(\frac{z_2 - z_1}{z_3 - z_1}) = \arg(z_2 - z_1) - \arg(z_3 - z_1)$ is the angle at the vertex z_1.

15. Let $z \neq z_0$. Show that the polar representation $z - z_0 = \rho(\cos\phi + i\sin\phi)$ can be used to denote the displacement vector from z_0 to z, as indicated in Figure 1.21

16. Show that $\mathrm{Arg}\,\overline{(z - w)} = -\mathrm{Arg}\,(z - w)$ iff $z - w$ is not a negative real number.

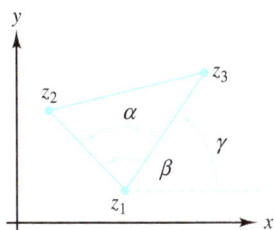

Figure 1.20: For Exercise 14

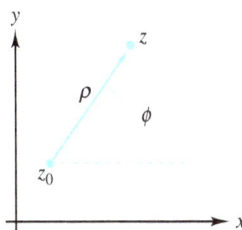

Figure 1.21: For Exercise 15

1.5   The Algebra of Complex Numbers, Part II

The real numbers are deficient in the sense that not all algebraic operations on them produce real numbers. Thus, for $\sqrt{-1}$ to make sense, we must consider the domain of complex numbers. Do complex numbers have this same deficiency? That is, if we are to make sense of expressions such as $\sqrt{1 + i}$, must we appeal to yet another new number system? The answer to this question is *no*. In other words, any reasonable algebraic operation performed on complex numbers gives complex numbers. Later we show how to evaluate intriguing expressions such as i^i. For now we only look at integral powers and roots of complex numbers.

The important players in this regard are the exponential and polar forms of a non-zero complex number $z = re^{i\theta} = r(\cos\theta + i\sin\theta)$. By the laws of exponents (which, you recall, we have promised to prove in Chapter 5) we have

$$z^n = (re^{i\theta})^n = r^n e^{in\theta} = r^n[\cos(n\theta) + i\sin(n\theta)] \quad \text{and} \tag{1.39}$$

$$z^{-n} = (re^{i\theta})^{-n} = r^{-n}e^{-in\theta} = r^{-n}[\cos(-n\theta) + i\sin(-n\theta)].$$

Example 1.15. Show that $(-\sqrt{3} - i)^3 = -8i$ in two ways.

Solution:

(Method 1): The binomial formula (Exercise 8 of Section 1.2) gives

$$(-\sqrt{3} - i)^3 = (-\sqrt{3})^3 + 3(-\sqrt{3})^2(-i) + 3(-\sqrt{3})(-i)^2 + (-i)^3 = -8i.$$

(Method 2): Using Identity (1.39) and Example 1.12 yields

$$(-\sqrt{3} - i)^3 = \left(2e^{i(-\frac{5\pi}{6})}\right)^3 = \left(2^3 e^{i(-\frac{15\pi}{6})}\right) = 8\left[\cos\left(-\frac{15\pi}{6}\right) + i\sin\left(-\frac{15\pi}{6}\right)\right] = -8i.$$

Which method would you use if you were asked to compute $(-\sqrt{3} - i)^{300}$?

Example 1.16. Evaluate $(-\sqrt{3} - i)^{30}$.

Solution:

$$\left(-\sqrt{3} - i\right)^{30} = \left(2e^{i\left(-\frac{5\pi}{6}\right)}\right)^{30} = 2^{30}e^{-i25\pi} = -2^{30}.$$

An interesting application of the laws of exponents comes to light when we put the equation $(e^{i\theta})^n = e^{in\theta}$ in its polar form. Doing so gives

$$(\cos\theta + i\sin\theta)^n = (\cos n\theta + i\sin n\theta), \tag{1.40}$$

which is known as De Moivre's formula, in honor of the French mathematician Abraham De Moivre (1667–1754).

Example 1.17. Use De Moivre's formula $\big($see Equation $(1.40)\big)$ to show that $\cos 5\theta = \cos^5\theta - 10\cos^3\theta\sin^2\theta + 5\cos\theta\sin^4\theta$.

Solution:

If we let $n = 5$ and use the binomial formula the left side of Equation (1.40) becomes

$$\cos^5\theta + i5\cos^4\theta\sin\theta - 10\cos^3\theta\sin^2\theta - 10i\cos^2\theta\sin^3\theta + 5\cos\theta\sin^4\theta + i\sin^5\theta.$$

The real part of this expression is $\cos^5\theta - 10\cos^3\theta\sin^2\theta + 5\cos\theta\sin^4\theta$. Equating this to the real part of $\cos 5\theta + i\sin 5\theta$ on the right side of Equation (1.40) establishes the desired result.

A key aid in determining roots of complex numbers is a corollary to the fundamental theorem of algebra. We prove this theorem in Chapter 6. Our proofs must be independent of the conclusions we derive here because we are going to make use of the corollary now.

Theorem 1.4 (Corollary to the fundamental theorem of algebra). *If $P(z)$ is a polynomial of degree n, $(n > 0)$, with complex coefficients, then the equation $P(z) = 0$ has precisely n (not necessarily distinct) solutions.*

Proof. Refer to Chapter 6. □

Example 1.18. Let $P(z) = z^3 + (2 - 2i)z^2 + (-1 - 4i)z - 2$. This polynomial of degree 3 can be written as $P(z) = (z - i)^2(z + 2)$. Hence the equation $P(z) = 0$ has solutions $z_1 = i$, $z_2 = i$, and $z_3 = -2$. Thus, in accordance with Theorem 1.4, we have three solutions, with z_1 and z_2 being repeated roots.

Theorem 1.4 implies that, if we can find n *distinct* solutions to the equation $z^n = c$ (or $z^n - c = 0$), we will have found *all* the solutions. We begin our search for these solutions by looking at the simpler equation $z^n = 1$. Solving this equation will enable us to handle the more general one quite easily.

To solve $z^n = 1$ we first note that, from Identities (1.29) and (1.37), we can deduce an important condition that determines when two nonzero complex numbers are equal. If we let $z_1 = r_1 e^{i\theta_1}$ and $z_2 = r_2 e^{i\theta_2}$, then

$$z_1 = z_2 \ \left(i.e., r_1 e^{i\theta_1} = r_2 e^{i\theta_2}\right) \quad \text{iff} \quad r_1 = r_2 \quad \text{and} \quad \theta_1 = \theta_2 + 2\pi k, \tag{1.41}$$

where k is an integer. That is, two complex numbers are equal iff their moduli agree and an argument of one equals an argument of the other to within an integral multiple of 2π.

We now find all solutions to $z^n = 1$ in two stages, with each stage corresponding to one direction in the iff part of Relation (1.41) . First, we show that *if* we have a solution to $z^n = 1$, then the solution must have a certain form. Second, we show that any quantity with that form is indeed a solution.

For the first stage, suppose that $z = re^{i\theta}$ is a solution to $z^n = 1$. Putting the latter equation in exponential form gives $r^n e^{in\theta} = 1 \cdot e^{i \cdot 0}$, so Relation (1.41) implies that $r^n = 1$ and $n\theta = 0 + 2\pi k$. In other words,

$$r = 1 \quad \text{and} \quad \theta = \frac{2\pi k}{n}, \tag{1.42}$$

where k is an integer.

So, *if* $z = re^{i\theta}$ is a solution to $z^n = 1$, then Relation (1.42) must be true. This observation completes the first stage of our solution strategy. For the second stage, we note that if $r = 1$, and $\theta = \frac{2\pi k}{n}$, then $z = re^{i\theta} = e^{i\frac{2\pi k}{n}}$ is indeed a solution to $z^n = 1$ because $z^n = (e^{i\frac{2\pi k}{n}})^n = e^{i2\pi k} = 1$. For example, if $n = 7$ and $k = 3$, then $z = e^{i\frac{6\pi}{7}}$ is a solution to $z^7 = 1$ because $(e^{i\frac{6\pi}{7}})^7 = e^{i6\pi} = 1$.

Furthermore, it is easy to verify that we get n *distinct* solutions to $z^n = 1$ (and, therefore, *all* solutions, by Theorem 1.4) by setting $k = 0, 1, 2, \ldots, n-1$. The solutions for $k = n, n+1, \ldots$ merely repeat those for $k = 0, 1, \ldots$, because the arguments so generated agree to within an integral multiple of 2π. The n solutions can be expressed as

$$z_k = e^{i\frac{2k\pi}{n}} = \cos\frac{2\pi k}{n} + i\sin\frac{2\pi k}{n}, \quad \text{for} \quad k = 0, 1, 2, \ldots, n-1. \tag{1.43}$$

They are called the **nth roots of unity**.

When $k = 0$ in Equation (1.43), we get $z_0 = e^{i\frac{2\pi \cdot 0}{n}} = e^0 = 1$, which is a rather trivial result. The first interesting root of unity occurs when $k = 1$, giving $z_1 = e^{i\frac{2\pi}{n}}$. This particular value shows up so often that mathematicians have given it a special symbol.

Definition 1.12 (Primitive nth root). *If n is the smallest natural number for which $z^n = 1$, then z is called a is called a* **primitive nth root of unity**. *Note that, from this definition, it follows that*

$$\omega_n = e^{i\frac{2\pi}{n}} = \cos\frac{2\pi}{n} + i\sin\frac{2\pi}{n}$$

is a primitive nth root of unity for all positive integers n.

By De Moivre's formula—Equation (1.40)—the nth roots of unity can be expressed as

$$1, \omega_n, \omega_n^2, \ldots, \omega_n^{n-1}. \tag{1.44}$$

Geometrically, the nth roots of unity are equally spaced points that lie on the unit circle $C_1(0) = \{z : |z| = 1\}$ and form the vertices of a regular polygon with n sides, which as we mentioned in Section 1.1 is a fact discovered by Leonard Euler.

Example 1.19. The solutions to the equation $z^8 = 1$ are given by the eight values $z_k = e^{i\frac{2\pi k}{8}} = \cos\frac{2\pi k}{8} + i\sin\frac{2\pi k}{8}$, for $k = 0, 1, 2, \ldots, 7$. In Cartesian form, these solutions are ± 1, $\pm i$, $\pm\frac{2+i\sqrt{2}}{2}$, and $\pm\frac{2-i\sqrt{2}}{2}$. The primitive 8th root of unity is $\omega_8 = e^{i\frac{2\pi}{8}} = e^{i\frac{\pi}{4}} = \cos\frac{\pi}{4} + i\sin\frac{\pi}{4} = \frac{\sqrt{2}}{2} + i\frac{\sqrt{2}}{2}$.

From Expressions (1.44) it is clear that $\omega_8 = z_1$ of Equation (1.43), as Figure 1.22 illustrates.

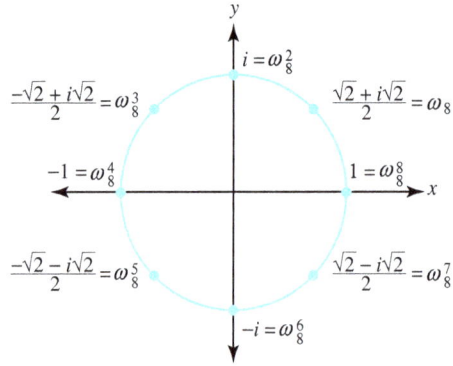

Figure 1.22: The eight eighth roots of unity

The procedure for solving $z^n = 1$ is easy to generalize in solving $z^n = c$ for any nonzero complex number c. If $c = \rho e^{i\phi} = \rho(\cos\phi + i\sin\phi)$ and $z = re^{i\theta}$, then $z^n = c$ iff $r^n e^{in\theta} = \rho e^{i\phi}$. But this last equation is satisfied iff

$$r^n = \rho \quad \text{and}$$
$$n\theta = \phi + 2k\pi, \text{where } k \text{ is an integer.}$$

As before, we get n distinct solutions given by

$$z_k = \rho^{\frac{1}{n}} e^{i\frac{\phi+2\pi k}{n}} = \rho^{\frac{1}{n}}\left(\cos\frac{\phi+2\pi k}{n} + i\sin\frac{\phi+2\pi k}{n}\right), \tag{1.45}$$

for $k = 0, 1, 2, \ldots, n-1$.

Each solution in Equation (1.45) can be considered an nth root of c. Geometrically, the nth roots of c are equally spaced points that lie on the circle $C_{\rho^{\frac{1}{n}}}(0) = \{z : |z| = \rho^{\frac{1}{n}}\}$ and form the vertices of a regular polygon with n sides. Figure 1.23 illustrates the case for $n = 5$.

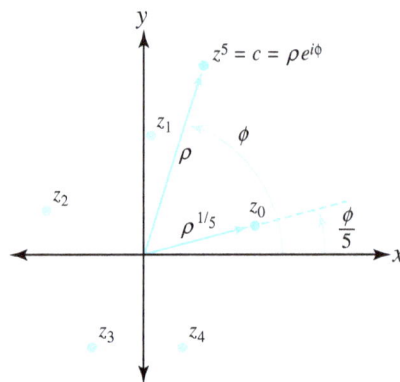

Figure 1.23: The five solutions to the equation $z^5 = c$

It is interesting to note that if ζ is any particular solution to the equation $z^n = c$, then *all* solutions can be generated by multiplying ζ by the various nth roots of unity. That is, the solution set is

$$\zeta, \; \zeta\omega_n, \; \zeta\omega_n^2, \; \ldots, \; \zeta\omega_n^{n-1}. \tag{1.46}$$

The reason for this is that, if $\zeta^n = c$, then for any $j = 0, 1, 2, \ldots, n-1$, $(\zeta \omega_n^j)^n = \zeta^n (\omega_n^j)^n = \zeta^n (\omega_n^n)^j = \zeta^n(1) = c$, and that multiplying a number by $\omega_n = e^{i\frac{2\pi}{n}}$ increases an argument of that number by $\frac{2\pi}{n}$, so that Expressions (1.46) contains n *distinct* values.

Example 1.20. Find all cube roots of $8i = 8(\cos \frac{\pi}{2} + i \sin \frac{\pi}{2})$.

Solution:

Formula (1.45) gives

$$z_k = 2\left[\cos\left(\frac{\frac{\pi}{2} + 2\pi k}{3}\right) + i \sin\left(\frac{\frac{\pi}{2} + 2\pi k}{3}\right)\right] \quad \text{for} \quad k = 0, 1, 2.$$

The Cartesian forms of the solutions are $z_0 = \sqrt{3} + i$, $z_1 = -\sqrt{3} + i$, and $z_2 = -2i$, as shown in Figure 1.24

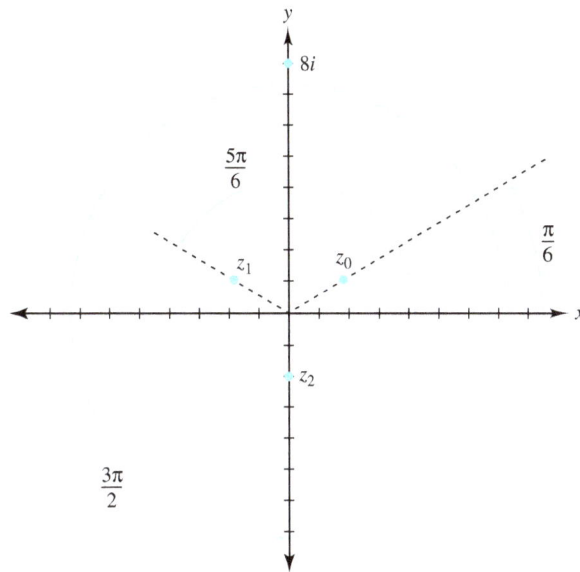

Figure 1.24: The point $z = 8i$ and its three cube roots, z_0, z_1, and z_2

Is the quadratic formula valid in the complex domain? The answer is *yes*, provided we are careful with our terms.

Theorem 1.5 (Quadratic formula). *The equation $az^2 + bz + c = 0$ has $\left\{\frac{-b \mid (b^2 - 4ac)^{\frac{1}{2}}}{2a}\right\}$ as its solution set for z, where by $(b^2 - 4ac)^{\frac{1}{2}}$ we mean all distinct square roots of that expression.*

Proof. The proof is left as an exercise. □

Example 1.21. Find all solutions to the equation $z^2 + (1 + i)z + 5i = 0$.

Solution:

The quadratic formula gives

$$z = \frac{-(1+i) + [(1+i)^2 - 4(1)(5i)]^{\frac{1}{2}}}{2(1)} = \frac{-(1+i) + (-18i)^{\frac{1}{2}}}{2}.$$

As $-18i = 18e^{i(-\frac{\pi}{2})}$, Equations (1.45) give $(-18i)^{\frac{1}{2}} = 18^{\frac{1}{2}}e^{i\frac{(-\frac{\pi}{2} + 2k\pi)}{2}}$, for $k = 0$ and 1. In Cartesian form, this expression reduces to $3 - 3i$ and $-3 + 3i$. Thus, our solution set is $\left\{\frac{-(1+i)+(3-3i)}{2}, \frac{-(1+i)+(-3+3i)}{2}\right\}$, or $\{1 - 2i, -2 + i\}$.

In Exercise 5 of Section 1.2 we asked you to show that a polynomial with non real coefficients must have some roots that do not occur in complex conjugate pairs. This last example gives an illustration of such a phenomenon.

Exercises for Section 1.5 (Selected answers or hints are on page 430.)

1. Calculate the following.

    (a) $(1 - i\sqrt{3})^3(\sqrt{3} + i)^2$

    (b) $\frac{(1+i)^3}{(1-i)^5}$

    (c) $(\sqrt{3} + i)^6$

2. Show that $(\sqrt{3} + i)^4 = -8 + i8\sqrt{3}$ in two different ways:

    (a) by squaring twice using the standard "FOIL" technique;

    (b) by using De Moivre's formula, given in Equation (1.40).

3. Use the method of Example 1.17 to establish trigonometric identities for $\cos 3\theta$ and $\sin 3\theta$.

4. Let z be any nonzero complex number and let n be an integer. Show that $z^n + (\overline{z})^n$ is a real number.

5. Find all the roots in both polar and Cartesian form for each expression.

    (a) $(-2 + 2i)^{\frac{1}{3}}$

    (b) $(-1)^{\frac{1}{5}}$

    (c) $(-64)^{\frac{1}{4}}$

    (d) $(8)^{\frac{1}{6}}$

    (e) $(16i)^{\frac{1}{4}}$

6. Prove Theorem 1.5, the quadratic formula.

7. Find all the roots of the equation $z^4 - 4z^3 + 6z^2 - 4z + 5 = 0$ given that $z_1 = i$ is a root.

8. Solve the equation $(z + 1)^3 = z^3$.

9. Find the three solutions to $z^{\frac{3}{2}} = 4\sqrt{2} + i4\sqrt{2}$.

10. Let m and n be positive integers that have no common factor. Show that there are n distinct solutions to $w^n = z^m$, given by

$$w_k = r^{\frac{m}{n}}\left(\cos\frac{m(\theta + 2\pi k)}{n} + i\sin\frac{m(\theta + 2\pi k)}{n}\right) \quad \text{for } k = 0, 1, \ldots, n - 1.$$

11. Suppose that $z \neq 1$.

    (a) Show that $1 + z + z^2 + \cdots + z^n = \frac{1 - z^{n+1}}{1 - z}$.

    (b) Use part (a) and De Moivre's formula to derive **Lagrange's identity**, which shows that $1 + \cos\theta + \cos 2\theta + \cdots + \cos n\theta = \frac{1}{2} + \frac{\sin\left[(n + \frac{1}{2})\theta\right]}{2\sin\frac{\theta}{2}}$, for $0 < \theta < 2\pi$.

12. If $1 = z_0, z_1, \ldots, z_{n-1}$ are the nth roots of unity, prove that

$$(z - z_1)(z - z_2) \cdots (z - z_{n-1}) = 1 + z + z^2 + \cdots + z^{n-1}.$$

13. Let $z_k \neq 1$ be an nth root of unity. Prove that $1 + z_k + z_k^2 + \cdots + z_k^{n-1} = 0$.

14. Equation (1.40), De Moivre's formula, can be established without recourse to properties of the exponential function. Note that this identity is trivially true for $n = 1$.

    (a) Use basic trigonometric identities to show the identity is valid for $n = 2$.
    (b) Use induction to verify the identity for all positive integers.
    (c) How would you verify this identity for all negative integers?

15. Find all four roots of $z^4 + 4 = 0$, and use them to demonstrate that $z^4 + 4$ can be factored into two quadratics with real coefficients.

16. Verify that Relation (1.41) is valid.

1.6 The Topology of Complex Numbers

In this section we investigate some basic ideas concerning sets of points in the plane. The first concept is that of a curve. Intuitively, we think of a curve as a piece of string placed on a flat surface in some type of meandering pattern. More formally, we define a **curve** to be the range of a continuous complex-valued function $z(t)$ defined on the interval $[a, b]$. That is, a curve C is the range of a function given by $z(t) = \big(x(t), y(t)\big) = x(t) + iy(t)$, for $a \leq t \leq b$, where both $x(t)$ and $y(t)$ are continuous real-valued functions. If both $x(t)$ and $y(t)$ are differentiable and not simultaneously zero (so that the tangent vector is never the zero vector), we say that the curve is **smooth**. A curve for which $x(t)$ and $y(t)$ are differentiable except for a finite number of points is called *piecewise smooth*. We specify a curve C as

$$C : z(t) = x(t) + iy(t) = \big(x(t), y(t)\big) \quad \text{for} \quad a \leq t \leq b, \tag{1.47}$$

and say that $z(t)$ is a **parametrization** for the curve C. Note that, with this parametrization, we are specifying a direction for the curve C, saying that C is a curve that goes from the **initial point** $z(a) = (x(a), y(a)) = x(a) + iy(a)$ to the **terminal point** $z(b) = \big(x(b), y(b)\big) = x(b) + iy(b)$. If we had another function whose range was the same set of points as $z(t)$ but whose initial and final points were reversed, we would indicate the curve that this function defines by $-C$.

Example 1.22. Find parameterizations for C and $-C$, where C is the straight line segment beginning at $z_0 = (x_0, y_0)$ and ending at $z_1 = (x_1, y_1)$.

Solution:

Refer to Figure 1.25. The vector form of a line shows that the direction of C is $z_1 - z_0$. As z_0 is a point on C, its vector equation is

$$C : z(t) = z_0 + (z_1 - z_0)t \quad \text{for} \quad 0 \leq t \leq 1, \quad \text{or} \tag{1.48}$$
$$C : z(t) = [x_0 + (x_1 - x_0)t] + i[y_0 + (y_1 - y_0)t], \quad \text{for} \quad 0 \leq t \leq 1.$$

Using the same technique we see that one parametrization for $-C$ is

$$-C : \gamma(t) = z_1 + (z_0 - z_1)t \quad \text{for} \quad 0 \leq t \leq 1. \tag{1.49}$$

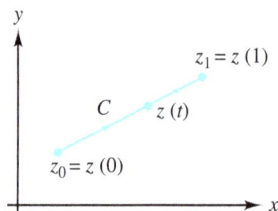

Figure 1.25: The straight-line segment C joining z_0 to z_1

Comparing Equations (1.48) and (1.49) illustrates a general principle: If C is a curve parametrized by $z(t)$ for $0 \leq t \leq 1$, then one parametrization for $-C$ will be $\gamma(t) = z(1-t)$, for $0 \leq t \leq 1$.

A curve C having the property that $z(a) = z(b)$ is said to be a **closed curve**. The line segment of Expression (1.48) is not a closed curve. The range of $z(t) = x(t) + iy(t)$, where $x(t) = \sin 2t \cos t$, and $y(t) = \sin 2t \sin t$ for $0 \leq t \leq 2\pi$ is a closed curve because $z(0) = (0,0) = z(2\pi)$. The range of $z(t)$ is the four-leaved rose shown in Figure 1.26. Note that, as t goes from 0 to $\frac{\pi}{2}$, the point is on leaf 1; from $\frac{\pi}{2}$ to π, it is on leaf 2; between π and $\frac{3\pi}{2}$, it is on leaf 3; and finally, for t between $\frac{3\pi}{2}$ and 2π, it is on leaf 4.

Note further that, at $(0,0)$, the curve has crossed over itself (at points other than those corresponding with $t = 0$ and $t = 2\pi$); we want to be able to distinguish when a curve does not cross over itself in this way. The curve C is called **simple** if it does not cross over itself, except possibly at its initial and terminal points. In other words, the curve $C : z(t)$, for $a \leq t \leq b$, is simple provided that $z(t_1) \neq z(t_2)$ whenever $t_1 \neq t_2$, except possibly when $t_1 = a$ and $t_2 = b$.

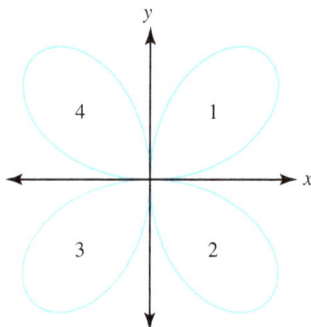

Figure 1.26: The four-leaved rose: $x(t) = \sin 2t \cos t$, $y(t) = \sin 2t \sin t$ for $0 \leq t \leq 2\pi$

Example 1.23. Show that the circle C with center $z_0 = x_0 + iy_0$ and radius r_0 can be parametrized to form a simple closed curve.

Solution:
Note that $C : z(t) = (x_0 + r_0 \cos t) + i(y_0 + r_0 \sin t) = z_0 + r_0 e^{it}$, for $0 \leq t \leq 2\pi$, gives the required parametrization.

Figure 1.27 shows that, as t varies from 0 to 2π, the circle is traversed counterclockwise. If you were traveling around the circle in this manner, its interior would be on your left. When a simple closed curve is parametrized in this fashion, we say that the curve has a **positive orientation**. We will have more to say about this idea shortly.

We need to develop some vocabulary that will help describe sets of points in the plane. One fundamental idea is that of an ε-**neighborhood** of the point z_0. It is the open disk of radius

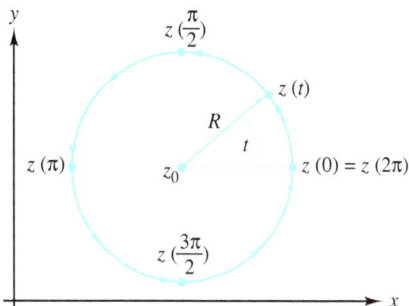

Figure 1.27: The simple closed curve $z(t) = z_0 + r_0 e^{it}$, for $0 \le t \le 2\pi$

$\varepsilon > 0$ about z_0 shown in Figure 1.28. Formally, it is the set of all points satisfying the inequality $\{z : |z - z_0| < \varepsilon\}$ and is denoted by $D_\varepsilon(z_0)$. That is,

$$D_\varepsilon(z_0) = \{z : |z - z_0| < \varepsilon\}. \tag{1.50}$$

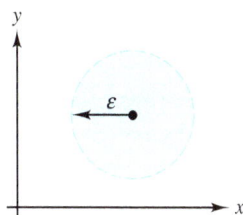

Figure 1.28: An ε-neighborhood of the point z_0

Example 1.24. The solution sets of the inequalities $|z| < 1$, $|z - i| < 2$, and $|z + 1 + 2i| < 3$ are neighborhoods of the points 0, i, and $-1 - 2i$, with radii 1, 2, and 3, respectively. They can also be expressed as $D_1(0)$, $D_2(i)$, and $D_3(-1 - 2i)$.

We also define $\overline{D}_\varepsilon(z_0)$, the **closed disk of radius** ε centered at z_0, and $D_\varepsilon^*(z_0)$, the **punctured disk of radius** ε centered at z_0 as

$$\overline{D}_\varepsilon(z_0) = \{z : |z - z_0| \le \varepsilon\} \quad \text{and} \tag{1.51}$$
$$D_\varepsilon^*(z_0) = \{z : 0 < |z - z_0| < \varepsilon\}. \tag{1.52}$$

The point z_0 is said to be an **interior point** of the set S provided that there exists an ε-neighborhood of z_0 that contains only points of S; z_0 is called an **exterior point** of the set S if there exists an ε-neighborhood of z_0 that contains no points of S. If z_0 is neither an interior point nor an exterior point of S, then it is called a **boundary point** of S and has the property that each ε-neighborhood of z_0 contains both points in S and points not in S. Figure 1.29 illustrates this situation.

The boundary of $D_R(z_0)$ is the circle depicted in Figure 1.27. We denote this circle $C_R(z_0)$ and refer to it as the **circle of radius R centered at** z_0. Thus

$$C_R(z_0) = \{z : |z - z_0| = R\}. \tag{1.53}$$

We use the notation $C_R^+(z_0)$ to indicate that the parametrization we chose for this simple closed curve resulted in a positive orientation; $C_R^-(z_0)$ denotes the same circle, but with a negative orientation. (In both cases, *counterclockwise* denotes the positive direction.) Using notation that we have already introduced, we get $C_R^-(z_0) = -C_R^+(z_0)$.

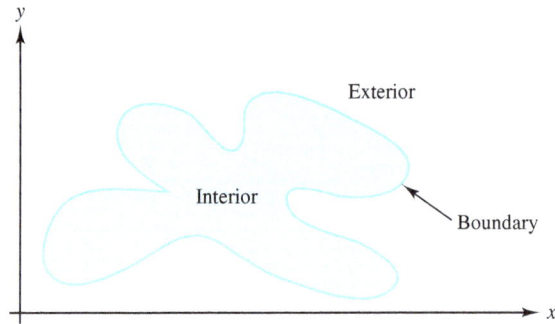

Figure 1.29: The interior, exterior, and boundary of a set

Example 1.25. Find the interior, exterior, and boundary of $S = D_1(0) = \{z : |z| < 1\}$.

Solution:

We show that every point of S is an interior point of S. Let z_0 be a point of S. Then $|z_0| < 1$, and we can choose $\varepsilon = 1 - |z_0| > 0$. We claim that $D_\varepsilon(z_0) \subseteq S$. If $z \in D_\varepsilon(z_0)$, then

$$|z| = |z - z_0 + z_0| \leq |z - z_0| + |z_0| < \varepsilon + |z_0| = 1 - |z_0| + |z_0| = 1.$$

Hence the ε-neighborhood of z_0 is contained in S, which shows that z_0 is an interior point of S. It follows that the interior of S is the set S itself.

Similarly, it can be shown that the exterior of S is $\{z : |z| > 1\}$, and the boundary of S is the unit circle $C_1(0) = \{z : |z| = 1\}$. These claims follow from that fact that, if $z_0 = e^{i\theta_0}$ is any point on the circle, then any ε-neighborhood of z_0 will contain two points: $(1 - \frac{\varepsilon}{2})e^{i\theta_0}$, which belongs to S; and $(1 + \frac{\varepsilon}{2})e^{i\theta_0}$, which does not belong to S. We leave the details of demonstrating this claim as an exercise.

The point z_0 is called an **accumulation point** of the set S if, for each ε, the punctured disk $D_\varepsilon^*(z_0)$ contains at least one point of S. We ask you to show in the exercises that the set of accumulation points of $D_1(0)$ is $\overline{D}_1(0)$, and that there is only one accumulation point of $S = \{\frac{i}{n} : n = 1, 2, \ldots\}$, namely, the point 0. We also ask you to prove that a set is closed if and only if it contains all of its accumulation points.

A set S is called an **open set** if every point of S is an interior point of S. Thus, Example (1.25) shows that $D_1(0)$ is open. A set S is called a **closed set** if it contains all its boundary points. A set S is said to be a **connected set** if every pair of points z_1 and z_2 contained in S can be joined by a curve that lies entirely in S. Roughly speaking, a connected set consists of a "single piece." The unit disk $D_1(0) = \{z : |z| < 1\}$ is a connected open set. We ask you to verify in the exercises that, if z_1 and z_2 lie in $D_1(0)$, then the straight-line segment joining them lies entirely in $D_1(0)$. The annulus $A = \{z : 1 < |z| < 2\}$ is a connected open set because any two points in A can be joined by a curve C that lies entirely in A, as shown in Figure 1.30.

The set $B = \{z : |z + 2| < 1 \text{ or } |z - 2| < 1\}$ consists of two disjoint disks. We leave it as an exercise for you to show that the set is not connected, as shown in Figure 1.31.

We call a connected open set a **domain**. In the exercises we ask you to show that the open unit disk $D_1(0) = \{z : |z| < 1\}$ is a domain and that the closed unit disk $\overline{D}_1(0) = \{z : |z| \leq 1\}$ is not a domain. The term *domain* is a noun and is a type of set. In Chapter 2 we note that it also refers to the set of points on which a function is defined. In the latter context, it does not necessarily mean a connected open set.

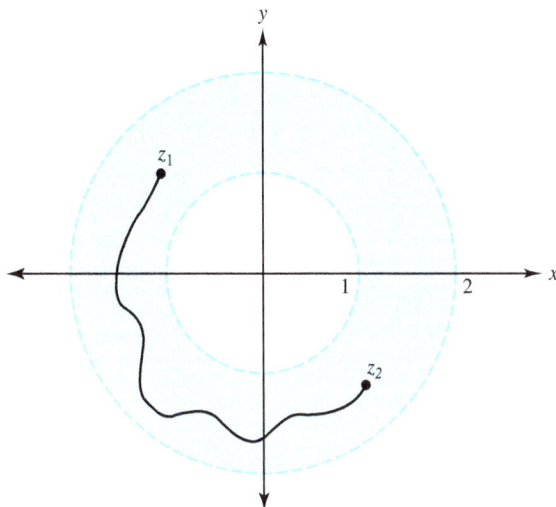

Figure 1.30: The annulus $A = \{z : 1 < |z| < 2\}$ is a connected set

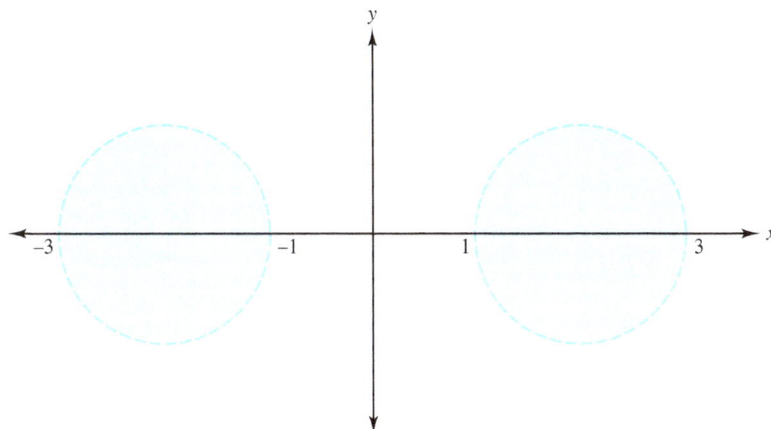

Figure 1.31: The set $B = \{z : |z + 2| < 1 \text{ or } |z - 2| < 1\}$ is not a connected set

Example 1.26. Show that the right half-plane $H = \{z : \operatorname{Re}(z) > 0\}$ is a domain.

Solution:

First we show that H is connected. Let z_0 and z_1 be any two points in H. We claim the obvious: the straight-line segment C given by Equation (1.48) lies entirely within H. To prove this claim, we let $z(t^*) = z_0 + (z_1 - z_0)t^*$, for some $t^* \in [0, 1]$, be an arbitrary point on C. We must show that $\operatorname{Re}\big(z(t^*)\big) > 0$. Now,

$$
\begin{aligned}
\operatorname{Re}\big(z(t^*)\big) &= \operatorname{Re}\big(z_0 + (z_1 - z_0)t^*\big) \\
&= \operatorname{Re}\big(z_0(1 - t^*)\big) + \operatorname{Re}(z_1 t^*) \\
&= (1 - t^*)\operatorname{Re}(z_0) + t^*\operatorname{Re}(z_1).
\end{aligned} \tag{1.54}
$$

If $t = 0$, the last expression becomes $\operatorname{Re}(z_0)$, which is greater than zero because $z_0 \in H$. Likewise, if $t = 1$, then the right side of Equation (1.54) reduces to $\operatorname{Re}(z_1)$, which also is positive. Finally, if $0 < t^* < 1$, then each term in Equation (1.54) is positive, so in this case we also have $\operatorname{Re}\big(z(t^*)\big) > 0$.

To show that H is open, we suppose without loss of generality that $\operatorname{Re}(z_0) \leq \operatorname{Re}(z_1)$. We claim that $D_\varepsilon(z_0) \subseteq H$, where $\varepsilon = \operatorname{Re}(z_0)$. We leave the proof of this claim as an exercise.

39

A domain, together with some, none, or all its boundary points, is called a **region**. For example, the horizontal strip $\{z : 1 < \mathrm{Im}(z) \leq 2\}$ is a region. A set formed by taking the union of a domain and its boundary is called a **closed region**; thus $\{z : 1 \leq \mathrm{Im}(z) \leq 2\}$ is a closed region. A set is said to be a **bounded set** if it can be completely contained in some closed disk, that is, if there exists an $R > 0$ such that for each z in S we have $|z| \leq R$. The rectangle given by $\{z : |x| \leq 4 \text{ and } |y| \leq 3\}$ is bounded because it is contained inside the disk $\overline{D}_5(0)$. A set that cannot be enclosed by any closed disk is called an **unbounded set**.

We mentioned earlier that a simple closed curve is positively oriented if its interior is on the left when the curve is traversed. How do we know, though, that any given simple closed curve will have an interior and exterior? Theorem 1.6 guarantees that this is indeed the case. It is due in part to the work of the French mathematician Camille Jordan (1838–1922).

Theorem 1.6 (The Jordan curve theorem). *The complement of any simple closed curve C can be partitioned into two mutually exclusive domains, I and E, in such a way that I is bounded, E is unbounded, and C is the boundary for both I and E. In addition, $I \cup E \cup C$ is the entire complex plane. The domain I is called the **interior** of C, and the domain E is called the **exterior** of C.*

The Jordan curve theorem is a classic example of a result in mathematics that seems obvious but is very hard to demonstrate, and its proof is beyond the scope of this book. Jordan's original argument, in fact, was inadequate, and not until 1905 was a correct version finally given by the American topologist Oswald Veblen. The difficulty lies in describing the interior and exterior of a simple closed curve analytically, and in showing that they are connected sets. For example, in which domain (interior or exterior) do the two points depicted in Figure 1.32 lie? If they are in the same domain, how, specifically, can they be connected with a curve? If you appreciated the subtleties involved in showing that the right half-plane of Example 1.26 is connected, you can begin to appreciate the obstacles that Veblen had to navigate.

Although an introductory treatment of complex analysis can be given without using this theorem, we think it is important for the well-informed student at least to be aware of it.

Figure 1.32: Are z_1 and z_2 in the interior or exterior of this simple closed curve?

Exercises for Section 1.6 (Selected answers or hints are on page 431.)

1. Find a parametrization of the line that

    (a) joins the origin to the point $1 + i$.
    (b) joins the point 1 to the point $1 + i$.
    (c) joins the point i to the point $1 + i$.
    (d) joins the point 2 to the point $1 + i$.

2. Sketch the curve $z(t) = t^2 + 2t + i(t + 1)$

    (a) for $-1 \leq t \leq 0$.
    (b) for $1 \leq t \leq 2$.

    Hint: Use $x = t^2 + 2t$, $y = t + 1$ and eliminate the parameter t.

3. Find a parametrization of the curve that is a portion of the parabola $y = x^2$ that

    (a) joins the origin to the point $2 + 4i$.
    (b) joins the point $-1 + i$ to the origin.
    (c) joins the point $1 + i$ to the origin.

4. This exercise completes Example (1.26): Suppose that $\operatorname{Re}(z_0) > 0$. Show that $\operatorname{Re}(z) > 0$ for all $z \in D_\varepsilon(z_0)$, where $\varepsilon = \operatorname{Re}(z_0)$.

5. Find a parametrization of the curve that is a portion of the circle $|z| = 1$ that joins the point $-i$ to i if

    (a) the curve is the right semicircle.
    (b) the curve is the left semicircle.

6. Show that $D_1(0)$ is a domain and that $\overline{D}_1(0) = \{z : |z| \leq 1\}$ is not a domain.

7. Find a parametrization of the curve that is a portion of the circle $C_1(0)$ that joins the point 1 to i if

    (a) the parametrization is counterclockwise along the quarter circle.
    (b) the parametrization is clockwise.

8. Fill in the details to complete Example (1.25). That is, show that

    (a) the set $\{z : |z| > 1\}$ is the exterior of the set S.
    (b) the set $C_1(0)$ is the boundary of the set S.

9. Consider the following sets.

    (i) $\{z : \operatorname{Re}(z) > 1\}$.
    (ii) $\{z : -1 < \operatorname{Im}(z) \leq 2\}$.
    (iii) $\{z : |z - 2 - i| \leq 2\}$.
    (iv) $\{z : |z + 3i| > 1\}$.
    (v) $\{re^{i\theta} : 0 < r < 1 \text{ and } -\frac{\pi}{2} < \theta < \frac{\pi}{2}\}$.
    (vi) $\{re^{i\theta} : r > 1 \text{ and } \frac{\pi}{4} < \theta < \frac{\pi}{3}\}$.
    (vii) $\{z : |z| < 1 \text{ or } |z - 4| < 1\}$.

41

(a) Sketch each set.

(b) State, with reasons, which of the following terms apply to the above sets: open; connected; domain; region; closed region; bounded.

10. Show that $D_1(0)$ is connected. *Hint*: Show that if z_1 and z_2 lie in $D_1(0)$, then the straight-line segment joining them lies entirely in $D_1(0)$.

11. Let $S = \{z_1, z_2, \ldots, z_n\}$ be a finite set of points. Show that S is a bounded set.

12. Prove that the boundary of $D_\varepsilon(z_0)$ is the circle $C_\varepsilon(z_0)$.

13. Let S be the open set consisting of all points z such that $|z + 2| < 1$ or $|z - 2| < 1$. Show that S is not connected.

14. Prove 0 is the only accumulation point of $\{\frac{i}{n} : n = 1, 2, \ldots\}$.

15. Regarding the relation between closed sets and accumulation points,

(a) Prove that if a set is closed, then it contains all its accumulations points.

(b) Prove that if a set contains all its accumulation points, then it is closed.

16. Prove that $\overline{D}_1(0)$ is the set of accumulation points of

(a) The set $D_1(0)$.

(b) The set $D_1^*(0)$.

17. Memorize and be prepared to illustrate all the terms in bold in this section.

Chapter 2

Complex Functions

Overview

The last chapter developed a basic theory of complex numbers. For the next few chapters we turn our attention to *functions* of complex numbers. They are defined in a similar way to functions of real numbers that you studied in calculus; the only difference is that they operate on complex numbers rather than real numbers. This chapter focuses primarily on very basic functions, their representations, and properties associated with functions such as limits and continuity. You will learn some interesting applications as well as some exciting new ideas.

2.1   Functions and Linear Mappings

A **complex-valued function** f of the complex variable z is a rule that assigns to each complex number z in a set D one and only one complex number w. We write $w = f(z)$ and call w the **image of** z **under** f. A simple example of a complex-valued function is given by the formula $w = f(z) = z^2$. The set D is called the **domain** of f, and the set of all images $\{w = f(z) : z \in D\}$ is called the **range** of f. When the context is obvious, we omit the phrase *complex-valued*, and simply refer to a function f, or to a complex function f.

We can define the domain to be any set that makes sense for a given rule, so for $w = f(z) = z^2$, we could have the entire complex plane for the domain D, or we might artificially restrict the domain to some set such as $D = D_1(0) = \{z : |z| < 1\}$. Determining the range for a function defined by a formula is not always easy, but we will see plenty of examples later on. In some contexts functions are referred to as *mappings* or *transformations*.

In Section 1.6, we used the term *domain* to indicate a connected open set. When speaking about the domain of a *function*, however, we mean only the set of points on which the function is defined. This distinction is worth noting, and context will make clear the use intended.

Just as z can be expressed by its real and imaginary parts, $z = x + iy$, we write $f(z) = w = u + iv$, where u and v are the real and imaginary parts of w, respectively. Doing so gives us the representation

$$w = f(z) = f(x, y) = f(x + iy) = u + iv.$$

Because u and v depend on x and y, they can be considered to be real-valued functions of the real variables x and y; that is,

$$u = u(x, y) \quad \text{and} \quad v = v(x, y).$$

Combining these ideas, we often write a complex function f in the form

$$f(z) = f(x + iy) = u(x, y) + iv(x, y). \tag{2.1}$$

Figure 2.1 illustrates the notion of a function (mapping) using these symbols.

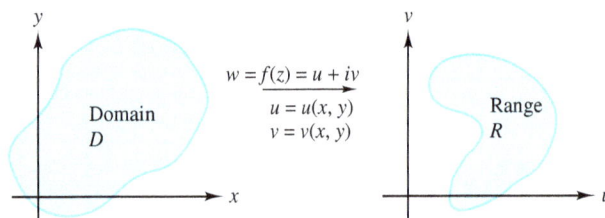

Figure 2.1: The mapping $w = f(z)$

We now give several examples that illustrate how to express a complex function.

Example 2.1. Write $f(z) = z^4$ in the form $f(z) = u(x, y) + iv(x, y)$.

Solution:

Using the binomial formula, we obtain

$$f(z) = (x + iy)^4 = x^4 + 4x^3 iy + 6x^2 (iy)^2 + 4x(iy)^3 + (iy)^4$$
$$= (x^4 - 6x^2 y^2 + y^4) + i(4x^3 y - 4xy^3),$$

so that $u(x, y) = x^4 - 6x^2 y^2 + y^4$ and $v(x, y) = 4x^3 y - 4xy^3$.

Example 2.2. Express the function $f(z) = \overline{z}\,\text{Re}(z) + z^2 + \text{Im}(z)$ in the form $f(z) = u(x, y) + iv(x, y)$.

Solution:

Using the elementary properties of complex numbers, it follows that

$$f(z) = (x - iy)x + (x^2 - y^2 + i2xy) + y = (2x^2 - y^2 + y) + i(xy),$$

so that $u(x, y) = 2x^2 - y^2 + y$ and $v(x, y) = xy$.

Examples 2.1 and 2.2 show how to find $u(x, y)$ and $v(x, y)$ when a rule for computing f is given. Conversely, if $u(x, y)$ and $v(x, y)$ are two real-valued functions of the real variables x and y, they determine a complex-valued function $f(x, y) = u(x, y) + iv(x, y)$, and we can use the formulas

$$x = \frac{z + \overline{z}}{2} \quad \text{and} \quad y = \frac{z - \overline{z}}{2i}$$

to find a formula for f involving the variables z and $\overline{z}$.

Example 2.3. Express $f(z) = 4x^2 + i4y^2$ by a formula involving the variables z and $\overline{z}$.

Solution:

Calculation reveals that

$$f(z) = 4\left(\frac{z + \overline{z}}{2}\right)^2 + i4\left(\frac{z - \overline{z}}{2i}\right)^2$$
$$= z^2 + 2z\overline{z} + \overline{z}^2 - i(z^2 - 2z\overline{z} + \overline{z}^2)$$
$$= (1 - i)z^2 + (2 + 2i)z\overline{z} + (1 - i)\overline{z}^2.$$

Using $z = re^{i\theta}$ in the expression of a complex function f may be convenient. It gives us the polar representation

$$f(z) = f(re^{i\theta}) = u(r, \theta) + iv(r, \theta), \tag{2.2}$$

where u and v are real functions of the real variables r and θ.

Remark 2.1. *For any specific function f, the functions u and v defined here will be different from those of Equation* (2.1) *because Equation* (2.1) *involves Cartesian coordinates and Equation* (2.2) *involves polar coordinates.*

Example 2.4. Express $f(z) = z^2$ in both Cartesian and polar form.

Solution:

For the Cartesian form, a simple calculation gives

$$f(z) = f(x + iy) = (x + iy)^2 = (x^2 - y^2) + i(2xy) = u(x, y) + iv(x, y)$$

so that

$$u(x, y) = x^2 - y^2, \quad \text{and} \quad v(x, y) = 2xy.$$

For the polar form, we refer to Equation (1.39) to get

$$f(re^{i\theta}) = (re^{i\theta})^2 = r^2 e^{i2\theta} = r^2 \cos 2\theta + ir^2 \sin 2\theta = U(r, \theta) + iV(r, \theta),$$

so that

$$U(r, \theta) = r^2 \cos 2\theta \quad \text{and} \quad V(r, \theta) = r^2 \sin 2\theta.$$

Once we have defined u and v for a function f in Cartesian form, we must use different symbols if we want to express f in polar form. As is clear here, the functions u and U are quite different, as are v and V. Of course, if we are working only in one context, we can use any symbols we choose.

Example 2.5. Express $f(z) = z^5 + 4z^2 - 6$ in polar form.

Solution:

Again, using Equation (1.39) we obtain

$$\begin{aligned}
f(z) = f(re^{i\theta}) &= r^5(\cos 5\theta + i \sin 5\theta) + 4r^2(\cos 2\theta + i \sin 2\theta) - 6 \\
&= (r^5 \cos 5\theta + 4r^2 \cos 2\theta - 6) + i(r^5 \sin 5\theta + 4r^2 \sin 2\theta) \\
&= u(r, \theta) + iv(r, \theta).
\end{aligned}$$

We now look at the geometric interpretation of a complex function. If D is the domain of real-valued functions $u(x, y)$ and $v(x, y)$, the equations

$$u = u(x, y) \quad \text{and} \quad v = v(x, y)$$

describe a transformation (or mapping) from D in the xy plane into the uv plane, also called the w plane. Therefore, we can also consider the function

$$w = f(z) = u(x, y) + iv(x, y)$$

to be a transformation (or mapping) from the set D in the z plane onto the range R in the w plane. This idea was illustrated in Figure 2.1. In the following paragraphs we present some additional key ideas. They are staples for any kind of function, and you should memorize all the terms in bold.

If A is a subset of the domain D of f, the set $B = \{f(z) : z \in A\}$ is called the **image** of the set A, and f is said to map A **onto** B. The image of a single point is a single point, and the image of the entire domain, D, is the range, R. The mapping $w = f(z)$ is said to be from A **into** S if the image of A is contained in S. Mathematicians use the notation $f : A \longmapsto S$ to indicate that a function maps A into S.

Figure 2.2 illustrates a function f whose domain is D and whose range is R. The shaded areas depict that the function maps A *onto* B. The function also maps A *into* R, and, of course, it maps D *onto* R.

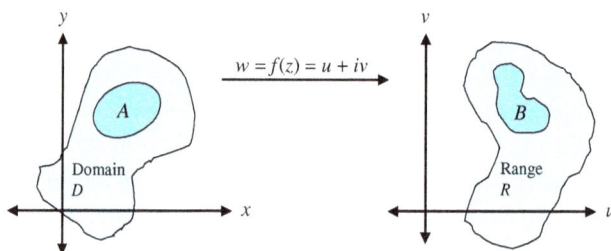

Figure 2.2: f maps A *onto* B; f maps A *into* R

The **inverse image** of a point w is the set of all points z in D such that $w = f(z)$. The inverse image of a point may be one point, several points, or nothing at all. If the last case occurs then the point w is not in the range of f. For example, if $w = f(z) = iz$, the inverse image of the point -1 is the single point i, because $f(i) = i(i) = -1$, and i is the *only* point that maps to -1. In the case of $w = f(z) = z^2$, the inverse image of the point -1 is the set $\{i, -i\}$. You will learn in Chapter 5 that, if $w = f(z) = e^z$, the inverse image of the point 0 is the empty set—there is no complex number z such that $e^z = 0$.

The inverse image of a set of points, S, is the collection of all points in the domain that map into S. If f maps D onto R it is possible for the inverse image of R to be function as well, but the original function must have a special property: a function f is said to be **one-to-one** if it maps distinct points $z_1 \neq z_2$ onto distinct points $f(z_1) \neq f(z_2)$. Many times an easy way to prove that a function f is one-to-one is to suppose $f(z_1) = f(z_2)$, and from this assumption deduce that z_1 must equal z_2. Thus, $f(z) = iz$ is one-to-one because if $f(z_1) = f(z_2)$, then $iz_1 = iz_2$. Dividing both sides of the last equation by i gives $z_1 = z_2$. Figure 2.3 illustrates the idea of a one-to-one function: distinct points get mapped to distinct points.

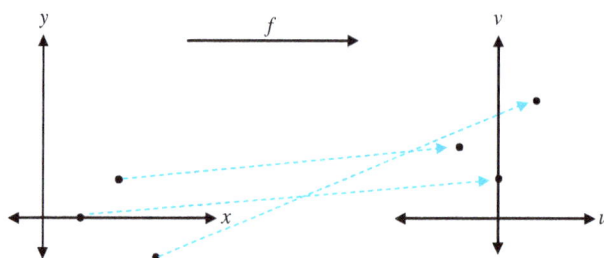

Figure 2.3: A one-to-one function

The function $f(z) = z^2$ is *not* one-to-one. This is because $-i \neq i$, but $f(i) = f(-i) = -1$. Figure 2.4 illustrates this situation: at least two different points get mapped to the same point.

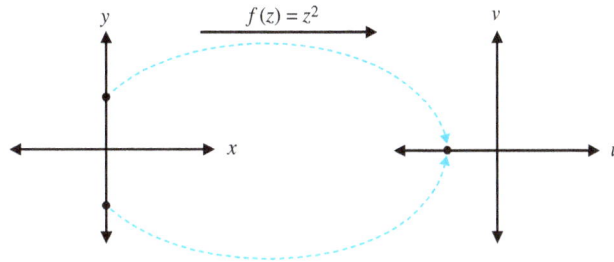

Figure 2.4: A function that is not one-to-one

In the exercises we ask you to demonstrate that one-to-one functions give rise to inverses that are functions. Loosely speaking, if $w = f(z)$ maps the set A one-to-one and onto the set B, then for each w in B there exists exactly one point z in A such that $w = f(z)$. For any such value of z we can take the equation $w = f(z)$ and "solve" for z as a function of w. Doing so produces an inverse function $z = g(w)$ where the following equations hold:

$$g\big(f(z)\big) = z \quad \text{for all} \quad z \in A \quad \text{and}$$
$$f\big(g(w)\big) = w \quad \text{for all} \quad w \in B. \tag{2.3}$$

Conversely, if $w = f(z)$ and $z = g(w)$ are functions that map A into B and B into A, respectively, and Equations (2.3) hold, then f maps the set A one-to-one and onto the set B.

Further, if f is a one-to-one mapping from D into T, and if A is a subset of D, then f is a one-to-one mapping from A onto its image B. We can also show that, if $\zeta = f(z)$ is a one-to-one mapping from A onto B and $w = g(\zeta)$ is a one-to-one mapping from B onto S, then the composite mapping $w = g\big(f(z)\big)$ is a one-to-one mapping from A onto S.

We usually indicate the inverse of f by the symbol f^{-1}. If the domains of f and f^{-1} are A and B, respectively, we can rewrite Equations (2.3) as

$$f^{-1}\big(f(z)\big) = z \quad \text{for all} \quad z \in A \quad \text{and}$$
$$f\big(f^{-1}(w)\big) = w \quad \text{for all} \quad w \in B. \tag{2.4}$$

Also, for $z_0 \in B$ and $w_0 \in A$,

$$w_0 = f(z_0) \quad \text{iff} \quad f^{-1}(w_0) = z_0 \quad \text{and} \quad z_0 = f^{-1}(w_0) \quad \text{iff} \quad f(z_0) = w_0. \tag{2.5}$$

Example 2.6. If $w = f(z) = iz$ for any complex number z, find $f^{-1}(w)$.

Solution:

We can easily show f is one-to-one and onto the entire complex plane. We solve for z, given $w = f(z) = iz$, to get $z = \frac{w}{i} = -iw$. By Equations (2.5), this result implies that $f^{-1}(w) = -iw$ for all complex numbers w.

Remark 2.2. *Once we have specified $f^{-1}(w) = -iw$ for all complex numbers w, we note that there is nothing magical about the symbol w. We could just as easily write $f^{-1}(z) = -iz$ for all complex numbers z.*

We now show how to find the image B of a specified set A under a given mapping $u + iv = w = f(z)$. The set A is usually described with an equation or inequality involving x and y. Using inverse functions, we can construct a chain of equivalent statements leading to a description of the set B in terms of an equation or an inequality involving u and v.

Example 2.7. Show that the function $f(z) = iz$ maps the line $y = x + 1$ in the xy plane onto the line $v = -u - 1$ in the w plane.

Solution:

(Method 1): With $A = \{(x, y) : y = x + 1\}$, we want to describe $B = f(A)$. We let $z = x + iy \in A$ and use Equations (2.5) and Example 2.6 to get

$$u + iv = w = f(z) \in B \iff f^{-1}(w) = z = x + iy \in A$$
$$\iff -iw \in A$$
$$\iff v - iu \in A$$
$$\iff (v, -u) \in A$$
$$\iff -u = v + 1$$
$$\iff v = -u - 1,$$

where $\iff$ means "*if and only if.*"

Note what this result says: $u + iv = w \in B \iff v = -u - 1$. The image of A under f, therefore, is the set $B = \{(u, v) : v = -u - 1\}$.

(Method 2): We write $u + iv = w = f(z) = i(x + iy) = -y + ix$ and note that the transformation can be given by the equations $u = -y$ and $v = x$. Because A is described by $A = \{x + iy : y = x + 1\}$, we can substitute $u = -y$ and $v = x$ into the equation $y = x + 1$ to obtain $-u = v + 1$, which we can rewrite as $v = -u - 1$. If you use this method, be sure to pay careful attention to domains and ranges.

We now look at some elementary mappings. If we let $B = a + ib$ denote a fixed complex constant, the transformation

$$w = T(z) = z + B = x + a + i(y + b)$$

is a one-to-one mapping of the z plane onto the w plane and is called a **translation**. This transformation can be visualized as a rigid translation whereby the point z is displaced through the vector $B = a + ib$ to its new position $w = T(z)$. The inverse mapping is given by

$$z = T^{-1}(w) = w - B = u - a + i(v - b)$$

and shows that T is a one-to-one mapping from the z plane onto the w plane. The effect of a translation is depicted in Figure 2.5.

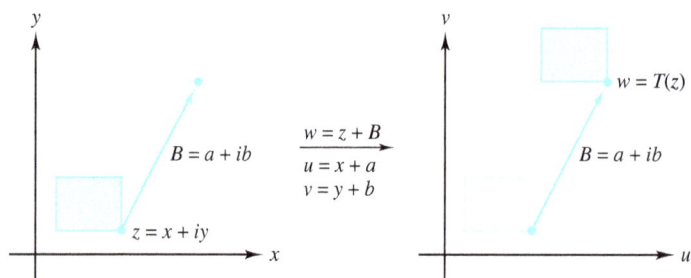

Figure 2.5: The translation $w = T(z) = z + B = x + a + i(y + b)$

If we let α be a fixed real number, then for $z = re^{i\theta}$, the transformation

$$w = R(z) = ze^{i\alpha} = re^{i\theta}e^{i\alpha} = re^{i(\theta + \alpha)}$$

48

is a one-to-one mapping of the z plane onto the w plane and is called a **rotation**. It can be visualized as a rigid rotation whereby the point z is rotated about the origin through an angle α to its new position $w = R(z)$. If we use polar coordinates and designate $w = \rho^{i\phi}$ in the w plane, then the inverse mapping is

$$z = R^{-1}(w) = we^{-i\alpha} = \rho e^{i\phi}e^{-i\alpha} = \rho e^{i(\phi-\alpha)}.$$

This analysis shows that R is a one-to-one mapping of the z plane onto the w plane. The effect of rotation is depicted in Figure 2.6.

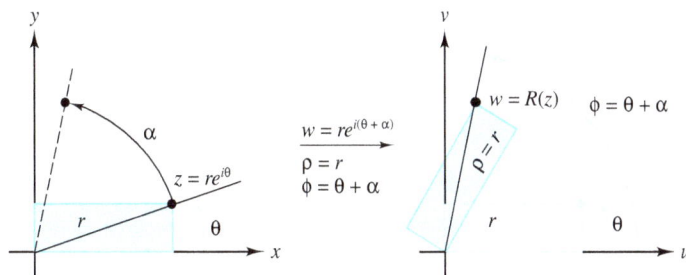

Figure 2.6: The rotation $w = R(z) = re^{i(\theta+\alpha)}$

Example 2.8. The ellipse centered at the origin with a horizontal major axis of four units and vertical minor axis of two units can be represented by the parametric equation

$$s(t) = 2\cos t + i\sin t = (2\cos t, \sin t) \quad \text{for} 0 \le t \le 2\pi.$$

Suppose that we wanted to rotate the ellipse by an angle of $\pi/6$ radians and shift the center of the ellipse 2 units to the right and 1 unit up. Using complex arithmetic, we can easily generate a parametric equation $r(t)$ that does so:

$$\begin{aligned}
r(t) &= s(t)e^{i\frac{\pi}{6}} + (2+i) \\
&= (2\cos t + i\sin t)\left(\cos\frac{\pi}{6} + i\sin\frac{\pi}{6}\right) + (2+i) \\
&= \left(2\cos t\cos\frac{\pi}{6} - \sin t\sin\frac{\pi}{6}\right) + i\left(2\cos t\sin\frac{\pi}{6} + \sin t\cos\frac{\pi}{6}\right) + (2+i) \\
&= \left(\sqrt{3}\cos t - \frac{1}{2}\sin t + 2\right) + i\left(\cos t + \frac{\sqrt{3}}{2}\sin t + 1\right) \\
&= \left(\sqrt{3}\cos t - \frac{1}{2}\sin t + 2, \ \cos t + \frac{\sqrt{3}}{2}\sin t + 1\right) \quad \text{for} \quad 0 \le t \le 2\pi.
\end{aligned}$$

Figure 2.7 shows parametric plots of these ellipses.

If we let $K > 0$ be a fixed positive real number, then the transformation

$$w = S(z) = Kz = Kx + iKy$$

is a one-to-one mapping of the z plane onto the w plane and is called a **magnification**. If $K > 1$, it has the effect of stretching the distance between points by the factor K. If $K < 1$, then it reduces the distance between points by the factor K. The inverse transformation is given by

$$z = S^{-1}(w) = \frac{1}{K}w = \frac{1}{K}u + i\frac{1}{K}v$$

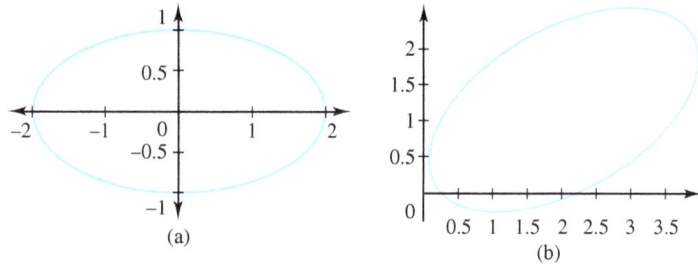

Figure 2.7: Plots of (a) the original ellipse; (b) the rotated ellipse

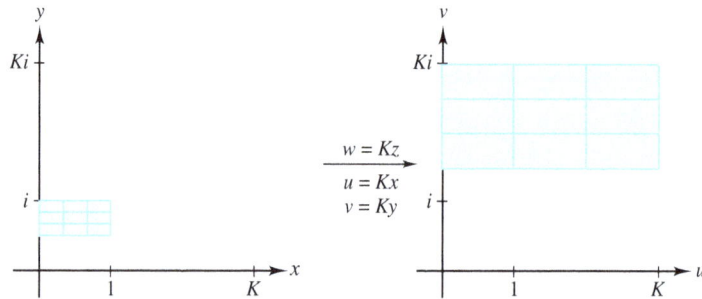

Figure 2.8: The magnification $w = S(z) = Kz = Kx + iKy$

and shows that S is one-to-one mapping from the z plane onto the w plane. The effect of magnification is shown in Figure 2.8.

Finally, if we let $A = Ke^{i\alpha}$ and $B = a + ib$, where $K > 0$ is a positive real number, then the transformation

$$w = L(z) = Az + B$$

is a one-to-one mapping of the z plane onto the w plane and is called a linear transformation. It can be considered as the composition of a rotation, a magnification, and a translation. It has the effect of rotating the plane through an angle given by $\alpha = \text{Arg}(A)$, followed by a magnification by the factor $K = |A|$, followed by a translation by the vector $B = a + ib$. The inverse mapping is given by $z = L^{-1}(w) = \frac{1}{A}w - \frac{B}{A}$ and shows that L is a one-to-one mapping from the z plane onto the w plane.

Example 2.9. Show that the linear transformation $w = iz + i$ maps the right half-plane $\text{Re}(z) \geq 1$ onto the upper half-plane $\text{Im}(w) \geq 2$.

Solution:

(Method 1): Let $A = \{(x, y) : x \geq 1\}$. To describe $B = f(A)$, we solve $w = iz + i$ for z to get $z = \frac{w-i}{i} = -iw - 1 = f^{-1}(w)$. Using Equations (2.5) and the method of Example 2.7 we have

$$\begin{aligned}
u + iv = w = f(z) \in B &\iff f^{-1}(w) = z \in A \\
&\iff -iw - 1 \in A \\
&\iff v - 1 - iu \in A \\
&\iff (v - 1, -u) \in A \\
&\iff v - 1 \geq 1 \\
&\iff v \geq 2.
\end{aligned}$$

Thus $B = \{w = u + iv : v \geq 2\}$, which is the same as saying $\text{Im}(w) \geq 2$.

(Method 2): When we write $w = f(z)$ in Cartesian form as

$$w = u + iv = i(x + iy) + i = -y + i(x + 1),$$

we see that the transformation can be given by the equations $u = -y$ and $v = x+1$. Substituting $x = v-1$ in the inequality $\text{Re}(z) = x \geq 1$ gives $v-1 \geq 1$, or $v \geq 2$, which is the upper half-plane $\text{Im}(w) \geq 2$.

(Method 3): The effect of the transformation $w = f(z)$ is a rotation of the plane through the angle $\alpha = \frac{\pi}{2}$ (when z is multiplied by i) followed by a translation by the vector $B = i$. The first operation yields the set $\text{Im}(w) \geq 1$. The second shifts this set up 1 unit, resulting in the set $\text{Im}(w) \geq 2$.

We illustrate this result in Figure 2.9.

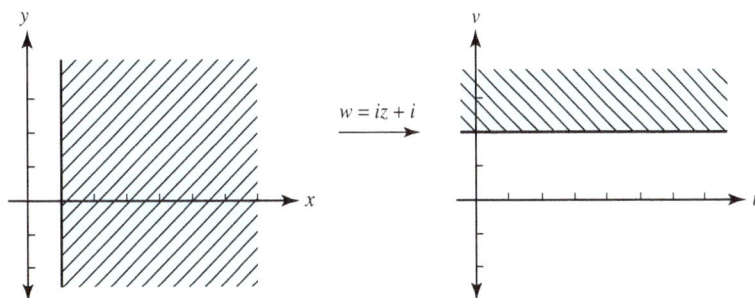

Figure 2.9: The linear transformation $w = f(z) = iz + i$

Translations and rotations preserve angles. First, magnifications rescale distance by a factor K, so it follows that triangles are mapped onto similar triangles, preserving angles. Then, because a linear transformation can be considered to be a composition of a rotation, a magnification, and a translation, it follows that linear transformations preserve angles. Consequently, any geometric object is mapped onto an object that is similar to the original object; hence linear transformations can be called **similarity mappings**.

Example 2.10. Show that the image of $D_1(-1-i) = \{z : |z+1+i| < 1\}$ under the transformation $w = (3-4i)z + 6 + 2i$ is the open disk $D_5(-1+3i) = \{w : |w+1-3i| < 5\}$.

Solution:

The inverse transformation is $z = \frac{w-6-2i}{3-4i}$, so if we designate the range of f as B, then

$$w = f(z) \in B \iff f^{-1}(w) = z \in D_1(-1-i)$$
$$\iff \frac{w-6-2i}{3-4i} \in D_1(-1-i)$$
$$\iff \left| \frac{w-6-2i}{3-4i} + 1 + i \right| < 1.$$
$$\iff \left| \frac{w-6-2i}{3-4i} + 1 + i \right| |3-4i| < 1 \cdot |3-4i|$$
$$\iff |w - 6 - 2i + (1+i)(3-4i)| < 5$$
$$\iff |w + 1 - 3i| < 5.$$

Hence the disk with center $-1 - i$ and radius 1 is mapped one-to-one and onto the disk with center $-1 + 3i$ and radius 5 as shown in Figure 2.10.

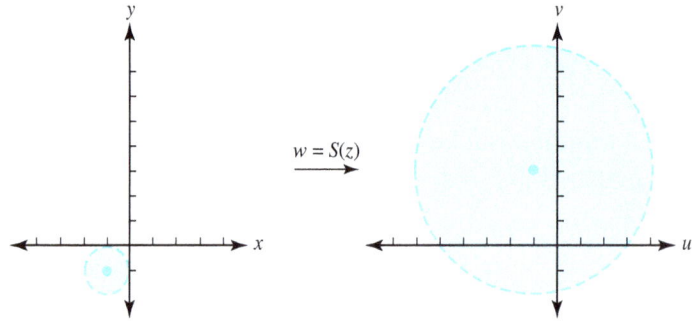

Figure 2.10: The mapping $w = S(z) = (3 - 4i)z + 6 + 2i$

Example 2.11. Show that the image of the right half-plane $\text{Re}(z) \geq 1$ under the linear transformation $w = (-1 + i)z - 2 + 3i$ is the half-plane $v \geq u + 7$.

***Solution*:**

The inverse transformation is given by

$$z = \frac{w + 2 - 3i}{-1 + i} = \frac{u + 2 + i(v - 3)}{-1 + i},$$

which we write as

$$x + iy = \frac{-u + v - 5}{2} + i\frac{-u - v + 1}{2}.$$

Substituting $x = \frac{-u+v-5}{2}$ into $\text{Re}(z) = x \geq 1$ gives $\frac{-u+v-5}{2} \geq 1$, which simplifies to $v \geq u + 7$. Figure 2.11 illustrates the mapping.

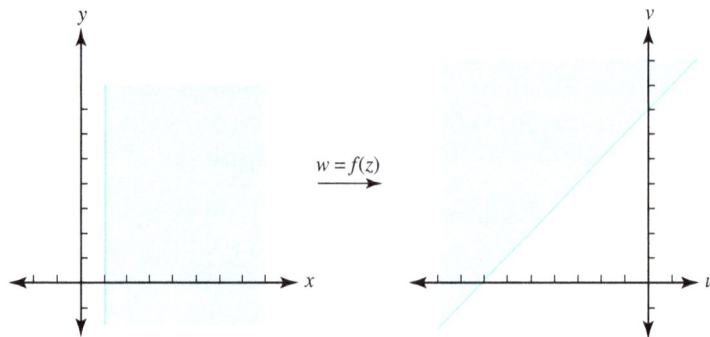

Figure 2.11: The mapping $w = f(z) = (-1 + i)z - 2 + 3i$

Exercises for Section 2.1 (Selected answers or hints are on page 431.)

1. Find $f(1 + i)$ for the following functions.

   (a) $f(z) = z + z^{-2} + 5$
   (b) $f(z) = \frac{1}{z^2+1}$.
   (c) $f(z) = f(x + iy) = x + y + i(x^3y - y^2)$.
   (d) $f(z) = z^2 + 4z\overline{z} - 5\text{Re}(z) + \text{Im}(z)$.

2. Let $f(z) = z^{21} - 5z^7 + 9z^4$. Use polar coordinates to find

   (a) $f(-1 + i)$.
   (b) $f(1 + i\sqrt{3})$.

3. Express the following functions in the form $u(x, y) + iv(x, y)$.

   (a) $f(z) = z^3$.
   (b) $f(z) = \bar{z}^2 + (2 - 3i)z$.
   (c) $f(z) = \frac{1}{z^2}$.

4. Express the following functions in the polar coordinate form $u(r, \theta) + iv(r, \theta)$.

   (a) $f(z) = z^5 + \bar{z}^5$.
   (b) $f(z) = z^5 + \bar{z}^3$.
   (c) For what values of z are the above expressions valid? Why?

5. Let $f(z) = f(x + iy) = e^x \cos y + ie^x \sin y$. Find

   (a) $f(0)$.
   (b) $f(i\pi)$.
   (c) $f(i\frac{2\pi}{3})$.
   (d) $f(2 + i\pi)$.
   (e) $f(3\pi i)$.
   (f) Is f a one-to-one function? Why or why not?

6. For $z \neq 0$, let $f(z) = f(x + iy) = \frac{1}{2} \ln(x^2 + y^2) + i \arctan \frac{y}{x}$. Find

   (a) $f(1)$.
   (b) $f(\sqrt{3} + i)$.
   (c) $f(1 + i\sqrt{3})$.
   (d) $f(3 + 4i)$.
   (e) Is f a one-to-one function? Why or why not?

7. For $z \neq 0$, let $f(z) = \ln r + i\theta$, where $r = |z|$, and $\theta = \text{Arg } z$. Find

   (a) $f(1)$.
   (b) $f(-2)$.
   (c) $f(1 + i)$.
   (d) $f(-\sqrt{3} + i)$.
   (e) Is f a one-to-one function? Why or why not?

8. A line that carries a charge of $\frac{q}{2}$ coulombs per unit length is perpendicular to the z plane and passes through the point z_0. The electric field $\mathbf{E}(z)$ at the point z is a vector, and its intensity is its magnitude $|\mathbf{E}(z)|$. It varies inversely as the distance from z_0, and is directed along the line from z_0 to z. Show that $\mathbf{E}(z) = \frac{k}{\bar{z} - \bar{z}_0}$, where k is some constant.

   Note: in Section 10.11 we show that, in fact, $k = q$, so that actually $\mathbf{E}(z) = \frac{q}{\bar{z} - \bar{z}_0}$.

9. Use the result of Exercise 8 to find the points z where the total charge $\mathbf{E}(z) = 0$ given the following conditions.

   (a) Three positively charged rods carry a charge of $\frac{q}{2}$ coulombs per unit length and pass through the points 0, $1 - i$, and $1 + i$.

   (b) A positively charged rod carrying a charge of $\frac{q}{2}$ coulombs per unit length passes through the point 0 and positively charged rods carrying a charge of q coulombs per unit length pass through the points $2 + i$ and $-2 + i$.

10. Suppose that f maps A into B, g maps B into A, and that Equations (2.3) hold.

    (a) Show that f is one-to-one.

    (b) Show that f maps A *onto* B.

11. Suppose f is a one-to-one mapping from D onto T and that A is a subset of D.

    (a) Show that f is one-to-one from A onto B, where $B = \{f(z) : z \in A\}$.

    (b) Show, additionally, that if g is one-to-one from B onto S, then $h(z)$ is one-to-one from A onto S, where $h(z) = f\big(g(z)\big)$.

12. For each part that follows produce a graphical and mathematical description of the images of the following sets when mapped by the function $w = f(z) = (3 + 4i)z - 2 + i$ (see, for example, the solution to Example 2.11). In each case also indicate graphically the images of $z_1 = 0$, $z_2 = 1 - i$, and $z_3 = 2$.

    (a) The disk $|z - 1| < 1$.

    (b) The line $x = t$, $y = 1 - 2t$ $\quad$ for $-\infty < t < \infty$.

    (c) The half-plane $\mathrm{Im}(z) > 1$.

13. Let $w = (2 + i)z - 2i$. Find the triangle onto which the triangle with vertices $z_1 = -2 + i$, $z_2 = -2 + 2i$, and $z_3 = 2 + i$ is mapped.

14. Let $S(z) = Kz$, where $K > 0$ is a positive real constant. Show that the equation $|S(z_1) - S(z_2)| = K|z_1 - z_2|$ holds and interpret this result geometrically.

15. Find the linear transformations $w = f(z)$ that satisfy the following conditions.

    (a) The points $z_1 = 2$ and $z_2 = -3i$ map onto $w_1 = 1 + i$ and $w_2 = 1$.

    (b) The circle $|z| = 1$ maps onto the circle $|w - 3 + 2i| = 5$, and $f(-i) = 3 + 3i$.

    (c) The triangle with vertices $-4 + 2i$, $-4 + 7i$, and $1 + 2i$ maps onto the triangle with vertices 1, 0, and $1 + i$, respectively.

16. Give a proof that the image of a circle under a linear transformation is a circle. Hint: Let the circle have the parameterization $x = x_0 + R\cos t$, $y = y_0 + R\sin t$.

17. Prove that the composition of two linear transformations is a linear transformation.

18. Show that a linear transformation that maps the circle $|z - z_0| = R_1$ onto the circle $|w - w_0| = R_2$ can be expressed in the form

$$A(w - w_0)R_1 = (z - z_0)R_2, \quad \text{where} \quad |A| = 1.$$

2.2 The Mappings $w = z^n$ and $w = z^{\frac{1}{n}}$

In this section we turn our attention to power functions.

For $z = re^{i\theta} \neq 0$, we can express the function $w = f(z) = z^2$ in polar coordinates as

$$w = f(z) = z^2 = r^2 e^{i2\theta}.$$

If we also use polar coordinates for $w = \rho e^{i\phi}$ in the w plane, we can express this mapping by the system of equations

$$\rho = r^2 \quad \text{and} \quad \phi = 2\theta.$$

Because an argument of the product $(z)(z)$ is twice an argument of z, we say that f doubles angles at the origin. Points that lie on the ray $r > 0$, $\theta = \alpha$ are mapped onto points that lie on the ray $\rho > 0$, $\phi = 2\alpha$. If we now restrict the domain of $w = f(z) = z^2$ to the region

$$A = \{re^{i\theta} : r > 0 \quad \text{and} \quad -\frac{\pi}{2} < \theta \leq \frac{\pi}{2}\}, \tag{2.6}$$

then the image of A under the mapping $w = z^2$ can be described by the set

$$B = \{\rho e^{i\phi} : \rho > 0 \quad \text{and} \quad -\pi < \phi \leq \pi\}, \tag{2.7}$$

which consists of all points in the w plane except the point $w = 0$.

The inverse mapping of f, which we denote g, is then

$$z = g(w) = w^{\frac{1}{2}} = \rho^{\frac{1}{2}} e^{i\frac{\phi}{2}},$$

where $w \in B$. That is,

$$z = g(w) = w^{\frac{1}{2}} = |w|^{\frac{1}{2}} e^{i\frac{\text{Arg}(w)}{2}},$$

where $w \neq 0$. The function g is so important that we call special attention to it with a formal definition.

Definition 2.1 (Principal Square Root). *The function*

$$g(w) = w^{\frac{1}{2}} = |w|^{\frac{1}{2}} e^{i\frac{\text{Arg}(w)}{2}} \quad for \quad w \neq 0, \tag{2.8}$$

*is called the **principal square root function**.*

We leave as an exercise to show that f and g satisfy Equations (2.3) and thus are inverses of each other that map the set A one-to-one and onto the set B and the set B one-to-one and onto the set A, respectively. Figure 2.12 illustrates this relationship.

What are the images of rectangles under the mapping $w = z^2$? To find out, we use the Cartesian form

$$w = u + iv = f(z) = z^2 = x^2 - y^2 + i2xy = (x^2 - y^2, 2xy) = (u, v)$$

and the resulting system of equations

$$u = x^2 - y^2 \quad \text{and} \quad v = 2xy. \tag{2.9}$$

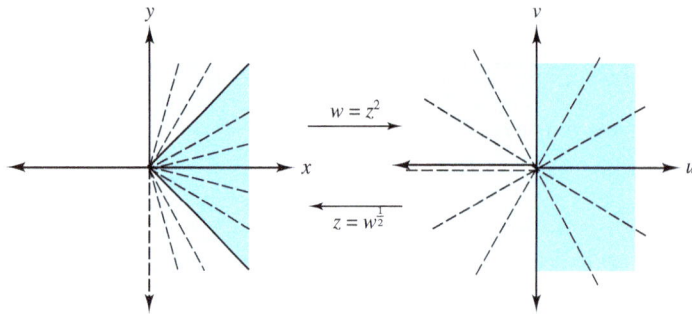

Figure 2.12: The mappings $w = z^2$ and $z = w^{\frac{1}{2}}$

Example 2.12. Show that the transformation $w = f(z) = z^2$, for $z \neq 0$, usually maps vertical and horizontal lines onto parabolas and use this fact to find the image of the rectangle $\{(x, y) : 0 < x < a,\ 0 < y < b\}$.

Solution:

Using Equations (2.9), we determine that the vertical line $x = a$ is mapped onto the set of points given by the equations $u = a^2 - y^2$ and $v = 2ay$. If $a \neq 0$, then $y = \frac{v}{2a}$ and

$$u = a^2 - \frac{v^2}{4a^2}. \tag{2.10}$$

Equation (2.10) represents a parabola with vertex at a^2, oriented horizontally, and opening to the left. If $a > 0$, the set $\{(u, v) : u = a^2 - y^2,\ v = 2ay\}$ has $v > 0$ precisely when $y > 0$, so the part of the line $x = a$ lying above the x axis is mapped to the top half of the parabola.

The horizontal line $y = b$ is mapped onto the parabola given by the equations $u = x^2 - b^2$ and $v = 2xb$. If $b \neq 0$, then as before we get

$$u = -b^2 + \frac{v^2}{4b^2}. \tag{2.11}$$

Equation (2.11) represents a parabola with vertex at $-b^2$, oriented horizontally and opening to the right. If $b > 0$, the part of the line $y = b$ to the right of the y axis is mapped to the top half of the parabola because the set $\{(u, v) : u = x^2 - b^2,\ v = 2bx\}$ has $v > 0$ precisely when $x > 0$.

Quadrant I is mapped onto quadrants I and II by $w = z^2$, so the rectangle $0 < x < a,\ 0 < y < b$ is mapped onto the region bounded by the top halves of the parabolas given by Equations (2.10) and (2.11) and the u axis. The vertices 0, a, $a + ib$, and ib of the rectangle are mapped onto the four points 0, a^2, $a^2 - b^2 + i2ab$, and $-b^2$, respectively, as indicated in Figure 2.13.

Finally, we can verify that the vertical line $x = 0$, $y \neq 0$ is mapped to $\{(-y^2, 0) : y \neq 0\}$. This is simply the set of negative real numbers. Likewise, the horizontal line $y = 0$, $x \neq 0$ is mapped to the set $\{(x^2, 0) : x \neq 0\}$, which is the set of positive real numbers.

What happens to images of regions under the mapping

$$w = f(z) = |z|^{\frac{1}{2}} e^{i\frac{\text{Arg}(z)}{2}} = r^{\frac{1}{2}} e^{i\frac{\theta}{2}} \quad \text{for} \quad z = re^{i\theta} \neq 0,$$

where $-\pi < \theta \leq \pi$? If we use polar coordinates for $w = \rho e^{i\phi}$ in the w plane, we can represent this mapping by the system

$$\rho = r^{\frac{1}{2}} \quad \text{and} \quad \phi = \frac{\theta}{2}. \tag{2.12}$$

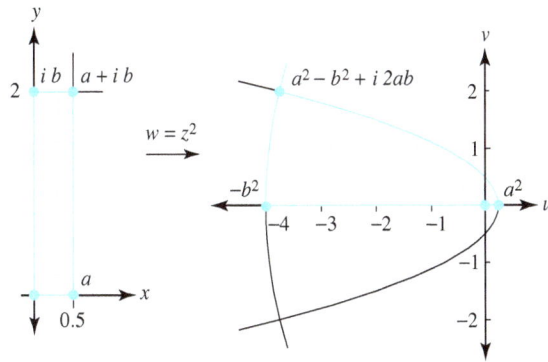

Figure 2.13: The transformation $w = z^2$

Equations (2.12) indicate that the argument of $f(z)$ is half the argument of z and that the modulus of $f(z)$ is the square root of the modulus of z. Points that lie on the ray $r > 0$, $\theta = \alpha$ are mapped onto the ray $\rho > 0$, $\phi = \frac{\alpha}{2}$. The image of the z plane (with the point $z = 0$ deleted) consists of the right half-plane $\text{Re}(w) > 0$ together with the positive v axis. The mapping is shown in Figure 2.14

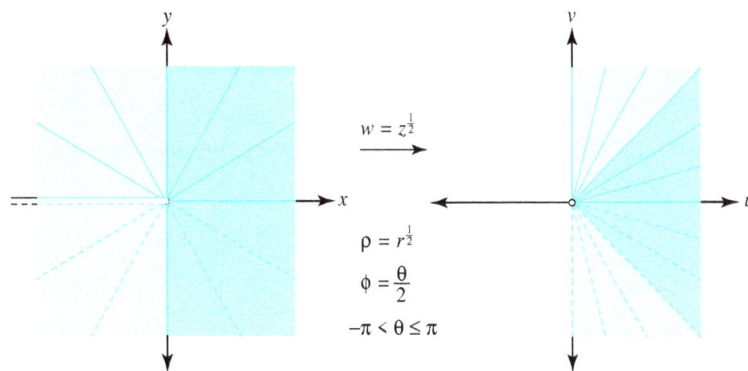

Figure 2.14: The mapping $w = z^{\frac{1}{2}}$

We can use knowledge of the inverse mapping $z = w^2$ to get further insight into how the mapping $w = z^{\frac{1}{2}}$ acts on rectangles. If we let $z = x + iy \neq 0$, then

$$z = w^2 = u^2 - v^2 + i2uv,$$

and we note that the point $z = x + iy$ in the z plane is related to the point $w = u + iv = z^{\frac{1}{2}}$ in the w plane by the system of equations

$$x = u^2 - v^2 \quad \text{and} \quad y = 2uv. \tag{2.13}$$

Example 2.13. Show that the transformation $w = f(z) = z^{\frac{1}{2}}$ usually maps vertical and horizontal lines onto portions of hyperbolas.

Solution:

Let $a > 0$. Equations (2.13) map the right half-plane given by $\text{Re}(z) > a$ (*i.e.*, $x > a$) onto the region in the right half-plane satisfying $u^2 - v^2 > a$ and lying to the right of the hyperbola $u^2 - v^2 = a$. If $b > 0$, Equations (2.13) map the upper half-plane $\text{Im}(z) > b$ (*i.e.*, $y > b$) onto the region in quadrant I satisfying $2uv > b$ and lying above the hyperbola $2uv = b$. This situation

57

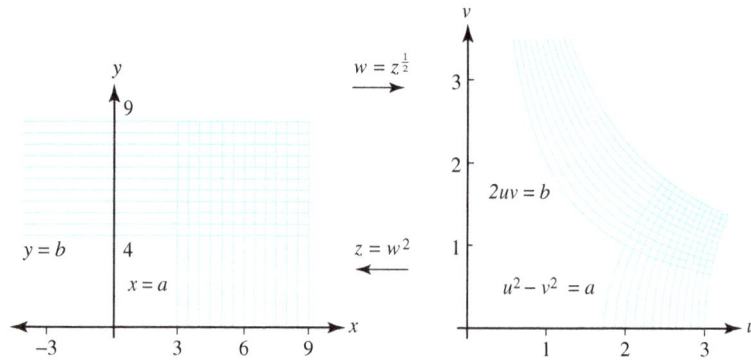

Figure 2.15: The mapping $w = z^{\frac{1}{2}}$

is illustrated in Figure 2.15. We leave as an exercise the investigation of what happens when $a = 0$ or $b = 0$.

We can easily extend what we've done to integer powers greater than 2. We begin by letting n be a positive integer, considering the function $w = f(z) = z^n$, for $z = re^{i\theta} \neq 0$, and then expressing it in the polar coordinate form

$$w = f(z) = z^n = r^n e^{in\theta}. \tag{2.14}$$

If we use polar coordinates $w = \rho e^{i\phi}$ in the w plane, the mapping defined by Equation (2.14) can be given by the system of equations

$$\rho = r^n \quad \text{and} \quad \phi = n\theta.$$

The image of the ray $r > 0$, $\theta = \alpha$ is the ray $\rho > 0$, $\phi = n\alpha$, and the angles at the origin are increased by the factor n. The functions $\cos n\theta$ and $\sin n\theta$ are periodic with period $2\pi/n$, so f is in general an n-to-one function; that is, n points in the z plane are mapped onto each non-zero point in the w plane.

If we now restrict the domain of $w = f(z) = z^n$ to the region

$$E = \left\{ re^{i\theta} : r > 0 \quad \text{and} \quad -\frac{\pi}{n} < \theta \leq \frac{\pi}{n} \right\},$$

then the image of E under the mapping $w = z^n$ can be described by the set

$$F = \{\rho e^{i\phi} : \rho > 0 \quad \text{and} \quad -\pi < \phi \leq \pi\},$$

which consists of all points in the w plane except the point $w = 0$. The inverse mapping of f, which we denote g, is then

$$z = g(w) = w^{\frac{1}{n}} = \rho^{\frac{1}{n}} e^{i\frac{\phi}{n}},$$

where $w \in F$. That is,

$$z = g(w) = w^{\frac{1}{n}} = |w|^{\frac{1}{n}} e^{i\frac{\text{Arg}(w)}{n}},$$

where $w \neq 0$. As with the principal square root function, we make an analogous definition for nth roots.

Definition 2.2 (Principal nth Root). *The function*

$$g(w) = w^{\frac{1}{n}} = |w|^{\frac{1}{n}} e^{i\frac{\text{Arg}(w)}{n}} \quad for \quad w \neq 0$$

*is called the **principal nth root function.***

We leave as an exercise to show that f and g are inverses of each other that map the set E one-to-one and onto the set F and the set F one-to-one and onto the set E, respectively. Figure 2.16 illustrates this relationship.

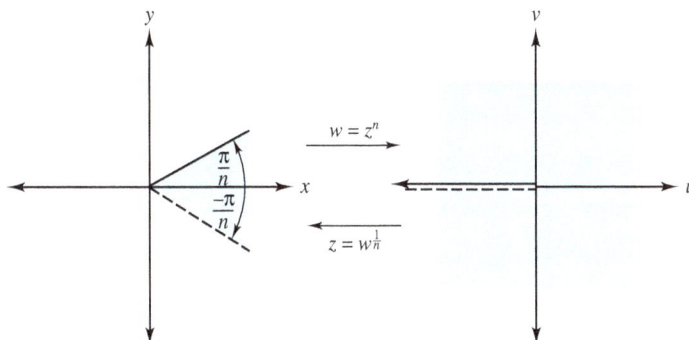

Figure 2.16: The mappings $w = z^n$ and $z = w^{\frac{1}{n}}$

Exercises for section 2.2 (Selected answers or hints are on page 432.)

1. Find the images of the mapping $w = z^2$ in each case, and sketch the mapping.

   (a) The horizontal line $\{(x, y) : y = 1\}$.

   (b) The vertical line $\{(x, y) : x = 2\}$.

   (c) The rectangle $\{(x, y) : 0 < x < 2, \ 0 < y < 1\}$.

   (d) The triangle with vertices 0, 2, and $2 + 2i$.

   (e) The infinite strip $\{(x, y) : 1 < x < 2\}$.

   (f) The right half-plane region to the right of the hyperbola $x^2 - y^2 = 1$.

   (g) The first quadrant region between the hyperbolas $xy = \frac{1}{2}$ and $xy = 4$.

2. For what values of z does $(z^2)^{\frac{1}{2}} = z$ hold if the principal value of the square root is to be used?

3. Sketch the set of points satisfying the following relations.

   (a) $\text{Re}(z^2) > 4$.

   (b) $\text{Im}(z^2) > 6$.

4. Find and illustrate the images of the following sets under the mapping $w = z^{\frac{1}{2}}$.

   (a) $\{re^{i\theta} : r > 1 \text{ and } \frac{\pi}{3} < \theta < \frac{\pi}{2}\}$.

   (b) $\{re^{i\theta} : 1 < r < 9 \text{ and } 0 < \theta < \frac{2\pi}{3}\}$.

   (c) $\{re^{i\theta} : r < 4 \text{ and } -\pi < \theta < \frac{\pi}{2}\}$.

(d) The vertical line $\{(x, y) : x = 4\}$.

(e) The infinite strip $\{(x, y) : 2 < y < 6\}$.

(f) The region to the right of the parabola $x = 4 - \frac{y^2}{16}$.
Hint: Use the inverse mapping $z = w^2$ to show that the answer is the right half-plane $\text{Re}(w) > 2$.

5. Find the image of the right half-plane $\text{Re}(z) > 1$ under the mapping $w = z^2 + 2z + 1$.

6. Find the image of the following sets under the mapping $w = z^3$.

   (a) $\{re^{i\theta} : 1 < r < 2 \text{ and } \frac{\pi}{4} < \theta < \frac{\pi}{3}\}$.

   (b) $\{re^{i\theta} : r > 3 \text{ and } \frac{2\pi}{3} < \theta < \frac{3\pi}{4}\}$.

7. Find the image of $\{re^{i\theta} : r > 2 \text{ and } \frac{\pi}{4} < 0 < \frac{\pi}{3}\}$ under the following mappings.

   (a) $w = z^3$.

   (b) $w = z^4$.

   (c) $w = z^6$.

8. Find the image of the sector $r > 0$, $-\pi < \theta < \frac{2\pi}{3}$ under the following mappings.

   (a) $w = z^{\frac{1}{2}}$.

   (b) $w = z^{\frac{1}{3}}$.

   (c) $w = z^{\frac{1}{4}}$.

9. Use your knowledge of the principal square root function to explain the fallacy in the following logic:
$$1 = \sqrt{(-1)(-1)} = \sqrt{(-1)}\sqrt{(-1)} = (i)(i) = -1.$$

10. Show that the functions $f(z) = z^2$ and $g(w) = w^{\frac{1}{2}} = |w|^{\frac{1}{2}} e^{i\frac{\text{Arg}(w)}{2}}$ with domains given by Equations (2.6) and (2.7), respectively, satisfy Equations (2.3). Thus, f and g are inverses of each other that map the shaded regions in Figure 2.14 one-to-one and onto each other.

11. Show what happens when $a = 0$ and $b = 0$ in Example 2.13.

12. Establish the result referred to in Definition 2.2.

2.3   Limits and Continuity

Let $u = u(x, y)$ be a real-valued function of the two real variables x and y. Recall that u has the limit u_0 as (x, y) approaches (x_0, y_0) provided the value of $u(x, y)$ can be made to get as close as we want to the value u_0 by taking (x, y) to be sufficiently close to (x_0, y_0). When this happens we write
$$\lim_{(x,y) \to (x_0, y_0)} u(x, y) = u_0.$$

In more technical language, u has the limit u_0 as (x, y) approaches (x_0, y_0) iff $|u(x, y) - u_0|$ can be made arbitrarily small by making both $|x - x_0|$ and $|y - y_0|$ small. This condition is like the definition of a limit for functions of one variable. The point (x, y) is in the xy plane, and the distance between (x, y) and $(x_0 y_0)$ is $\sqrt{(x - x_0)^2 + (y - y_0)^2}$. With this perspective we can now give a precise definition of a limit.

Definition 2.3 (Limit of $u(x, y)$). *The expression* $\lim\limits_{(x,y)\to(x_0,y_0)} u(x, y) = u_0$ *means that for each number $\varepsilon > 0$, there is a corresponding number $\delta > 0$ such that*

$$|u(x, y) - u_0| < \varepsilon \quad \text{whenever} \quad 0 < \sqrt{(x - x_0)^2 + (y - y_0)^2} < \delta. \tag{2.15}$$

Example 2.14. Show, if $u(x, y) = \frac{2x^3}{(x^2+y^2)}$, then $\lim\limits_{(x,y)\to(0,0)} u(x, y) = 0$.

***Solution*:**

If $x = r \cos \theta$ and $y = r \sin \theta$, then

$$u(x, y) = \frac{2r^3 \cos^3 \theta}{r^2 \cos^2 \theta + r^2 \sin^2 \theta} = 2r \cos^3 \theta.$$

Because $\sqrt{(x - 0)^2 + (y - 0)^2} = r$ and because $|\cos^3 \theta| < 1$,

$$|u(x, y) - 0| = 2r|\cos^3 \theta| < \varepsilon \quad \text{whenever} \quad 0 < \sqrt{x^2 + y^2} = r < \frac{\varepsilon}{2}.$$

Hence, for any $\varepsilon > 0$, Inequality (2.15) is satisfied for $\delta = \frac{\varepsilon}{2}$; that is, $u(x, y)$ has the limit $u_0 = 0$ as (x, y) approaches $(0, 0)$.

The value u_0 of the limit must not depend on how (x, y) approaches (x_0, y_0), so $u(x, y)$ must approach the value u_0 when (x, y) approaches (x_0, y_0) along any curve that ends at the point (x_0, y_0). Conversely, if we can find two curves C_1 and C_2 that end at (x_0, y_0) along which $u(x, y)$ approaches two distinct values u_1 and u_2, then $u(x, y)$ does not have a limit as (x, y) approaches (x_0, y_0).

Example 2.15. Show that $u(x, y) = \frac{xy}{x^2+y^2}$ does not have a limit as (x, y) approaches $(0, 0)$.

***Solution*:**

If we let (x, y) approach $(0, 0)$ along the x axis, then

$$\lim_{(x,0)\to(0,0)} u(x, 0) = \lim_{(x,0)\to(0,0)} \frac{(x)(0)}{x^2 + 0^2} = 0.$$

But if we let (x, y) approach $(0, 0)$ along the line $y = x$, then

$$\lim_{(x,x)\to(0,0)} u(x, x) = \lim_{(x,x)\to(0,0)} \frac{(x)(x)}{x^2 + x^2} = \frac{1}{2}.$$

Because the value of the limit differs depending on how (x, y) approaches $(0, 0)$, we conclude that $u(x, y)$ does not have a limit as (x, y) approaches $(0, 0)$.

Let $f(z)$ be a complex function of the complex variable z that is defined for all values of z in some neighborhood of z_0, except perhaps at the point z_0. We say that f has the limit w_0 as z approaches z_0, provided the value $f(z)$ can be made as close as we want to the value w_0 by taking z to be sufficiently close to z_0. When this happens we write

$$\lim_{z \to z_0} f(z) = w_0.$$

The distance between the points z and z_0 can be expressed by $|z - z_0|$, so we can give a precise definition similar to the one for a function of two variables.

Definition 2.4 (Limit of $f(z)$). *The expression $\lim\limits_{z \to z_0} f(z) = w_0$ means that for each real number $\varepsilon > 0$, there exists a real number $\delta > 0$ such that*

$$|f(z) - w_0| < \varepsilon \quad whenever \quad 0 < |z - z_0| < \delta.$$

Using the notation of (1.50) and (1.52), we can express the last relationship as

$$f(z) \in D_\varepsilon(w_0) \quad whenever \quad z \in D_\delta^*(z_0).$$

The formulation of limits in terms of open disks provides a good context for looking at this definition. It says that for each disk of radius ε about the point w_0 (represented by $D_\varepsilon(w_0)$) there is a punctured disk of radius δ about the point z_0 (represented by $D_\delta^*(z_0)$) such that the image of each point in the punctured δ disk lies in the ε disk. The image of the δ disk does not have to fill up the entire ε disk; but if z approaches z_0 along a curve that ends at z_0, then $w = f(z)$ approaches w_0. The situation is illustrated in Figure 2.17.

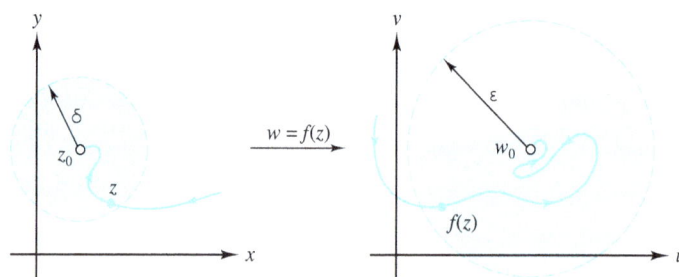

Figure 2.17: As $z \to z_0$ the function values $f(z) \to w_0$

Example 2.16. Show that if $f(z) = \bar{z}$, then $\lim\limits_{z \to z_0} f(z) = \overline{z_0}$, where z_0 is any complex number.

Solution:

As f merely reflects points about the y axis, we suspect that any ε disk about the point $\overline{z_0}$ would contain the image of the punctured δ disk about z_0 if $\delta = \varepsilon$. To confirm this conjecture, we let ε be any positive number and set $\delta = \varepsilon$. Then we suppose that $z \in D_\delta^*(z_0) = D_\varepsilon^*(z_0)$, which means that $0 < |z - z_0| < \varepsilon$. The modulus of a conjugate is the same as the modulus of the number itself, so the last inequality implies that $0 < |\overline{z - z_0}| < \varepsilon$. This inequality is the same as $0 < |\bar{z} - \overline{z_0}| < \varepsilon$. Since $f(z) = \bar{z}$ and $w_0 = \overline{z_0}$, this last inequality becomes $0 < |f(z) - w_0| < \varepsilon$, or $f(z) \in D_\varepsilon(\overline{z_0})$, which is what we needed to show.

If we consider $w = f(z)$ as a mapping from the z plane into the w plane and think about the previous geometric interpretation of a limit, then we are led to conclude that the limit of a function f should be determined by the limits of its real and imaginary parts, u and v. This conclusion also gives us a tool for computing limits.

Theorem 2.1. *Let $f(z) = u(x,y) + iv(x,y)$ be a complex function that is defined in some neighborhood of z_0, except perhaps at $z_0 = x_0 + iy_0$. Then*

$$\lim_{z \to z_0} f(z) = w_0 = u_0 + iv_0 \tag{2.16}$$

iff

$$\lim_{(x,y) \to (x_0,y_0)} u(x,y) = u_0 \quad and \quad \lim_{(x,y) \to (x_0,y_0)} v(x,y) = v_0. \tag{2.17}$$

62

Proof. We first assume that Statement (2.16) is true and show that Statement (2.17) is true. According to the definition of limit, for each $\varepsilon > 0$, there is a corresponding $\delta > 0$ such that

$$f(z) \in D_\varepsilon(w_0) \quad \text{whenever} \quad z \in D_\delta^*(z_0);$$

that is,

$$|f(z) - w_0| < \varepsilon \quad \text{whenever} \quad 0 < |z - z_0| < \delta.$$

Because $f(z) - w_0 = u(x, y) - u_0 + i(v(x, y) - v_0)$, we can use Inequalities (1.21) to conclude that

$$|u(x, y) - u_0| \le |f(z) - w_0| \quad \text{and} \quad |v(x, y) - v_0| \le |f(z) - w_0|.$$

It now follows that $|u(x, y) - u_0| < \varepsilon$ and $|v(x, y) - v_0| < \varepsilon$ whenever $0 < |z - z_0| < \delta$, and so Statement (2.17) is true.

Conversely, assume that Statement (2.17) is true. Then for each $\varepsilon > 0$, there exists $\delta_1 > 0$ and $\delta_2 > 0$ so that

$$|u(x, y) - u_0| < \frac{\varepsilon}{2} \quad \text{whenever} \quad 0 < |z - z_0| < \delta_1, \quad \text{and}$$

$$|v(x, y) - v_0| < \frac{\varepsilon}{2} \quad \text{whenever} \quad 0 < |z - z_0| < \delta_2.$$

We choose δ to be the minimum of the two values δ_1 and δ_2. Then we can use the triangle inequality

$$|f(z) - w_0| \le |u(x, y) - u_0| + |v(x, y) - v_0|$$

to conclude that

$$|f(z) - w_0| < \frac{\varepsilon}{2} + \frac{\varepsilon}{2} = \varepsilon \quad \text{whenever} \quad 0 < |z - z_0| < \delta;$$

that is,

$$f(z) \in D_\varepsilon(w_0) \quad \text{whenever} \quad z \in D_\delta^*(z_0).$$

Hence the truth of Statement (2.17) implies the truth of Statement (2.16), and the proof of the theorem is complete. $\qquad\square$

Example 2.17. Show that $\lim\limits_{z \to 1+i} (z^2 - 2z + 1) = -1$.

Solution:

Let

$$f(z) = z^2 - 2z + 1 = x^2 - y^2 - 2x + 1 + i(2xy - 2y).$$

Computing the limits for u and v, we obtain

$$\lim\limits_{(x,y) \to (1,1)} u(x, y) = 1 - 1 - 2 + 1 = -1 \quad \text{and}$$

$$\lim\limits_{(x,y) \to (1,1)} v(x, y) = 2 - 2 = 0,$$

so our previous theorem implies that $\lim\limits_{z \to 1+i} f(z) = -1$.

Limits of complex functions are formally the same as those of real functions, and the sum, difference, product, and quotient of functions have limits given by the sum, difference, product, and quotient of the respective limits. We state this result as a theorem and leave the proof as an exercise.

Theorem 2.2. *Suppose that* $\lim\limits_{z \to z_0} f(z) = A$ *and* $\lim\limits_{z \to z_0} g(z) = B$. *Then*

$$\lim_{z \to z_0} [f(z) \pm g(z)] = A \pm B, \tag{2.18}$$

$$\lim_{z \to z_0} f(z)g(z) = AB, \quad and \tag{2.19}$$

$$\lim_{z \to z_0} \frac{f(z)}{g(z)} = \frac{A}{B}, \quad where \quad B \neq 0. \tag{2.20}$$

Definition 2.5 (Continuity of $u(x, y)$). *Let $u(x, y)$ be a real-valued function of the two real variables x and y. We say that u is continuous at the point $(x_0,\ y_0)$ if three conditions are satisfied:*

$$\lim_{(x,y) \to (x_0,y_0)} u(x, y) \quad exists, \tag{2.21}$$

$$u(x_0,\ y_0) \quad exists, \quad and \tag{2.22}$$

$$\lim_{(x,y) \to (x_0,y_0)} u(x, y) = u(x_0,\ y_0). \tag{2.23}$$

Condition (2.23) actually implies Conditions (2.21) and (2.22) because the existence of the quantity on each side of Equation (2.23) is implicitly understood to exist. For example, if $u(x, y) = \frac{x^3}{x^2+y^2}$ when $(x, y) \neq (0,\ 0)$ and if $u(0,\ 0) = 0$, then $u(x, y) \to (0,\ 0)$ so that Conditions (2.21), (2.22), and (2.23) are satisfied. Hence $u(x, y)$ is continuous at $(0,\ 0)$.

There is a similar definition for complex valued functions.

Definition 2.6 (Continuity of $f(z)$). *Let $f(z)$ be a complex function of the complex variable z that is defined for all values of z in some neighborhood of z_0. We say that f is continuous at z_0 if three conditions are satisfied:*

$$\lim_{z \to z_0} f(z) \quad exists, \tag{2.24}$$

$$f(z_0) \quad exists, \quad and \tag{2.25}$$

$$\lim_{z \to z_0} f(z) = f(z_0). \tag{2.26}$$

Remark 2.3. *Example 2.16 shows that the function $f(z) = \bar{z}$ is continuous.*

A complex function f is continuous iff its real and imaginary parts, u and v, are continuous. The proof of this fact is an immediate consequence of Theorem 2.1. Continuity of complex functions is formally the same as that of real functions, and sums, differences, and products of continuous functions are continuous; their quotient is continuous at points where the denominator is not zero. These results are summarized by the following theorems. We leave the proofs as exercises.

Theorem 2.3. *Let $f(z) = u(x, y) + iv(x, y)$ be defined in some neighborhood of z_0. Then f is continuous at $z_0 = x_0 + iy_0$ iff u and v are continuous at $(x_0,\ y_0)$.*

Theorem 2.4. *Suppose that f and g are continuous at the point z_0. Then the following functions are continuous at z_0:*

- *The sum $f + g$, where $(f + g)(z) = f(z) + g(z)$;*

- *The difference $f - g$, where $(f - g)(z) = f(z) - g(z)$;*

- *The product fg, where $(fg)(z) = f(z)g(z)$;*

- *The quotient $\frac{f}{g}$, where $\frac{f}{g}(z) = \frac{f(z)}{g(z)}$, provided $g(z_0) \neq 0$; and*

- *The composition $f \circ g$, where $(f \circ g)(z) = f\big(g(z)\big)$, provided f is continuous in a neighborhood of $g(z_0)$.*

Example 2.18. Show that the polynomial function given by

$$w = P(z) = a_0 + a_1 z + a_2 z^2 + \cdots + a_n z^n$$

is continuous at each point z_0 in the complex plane.

Solution:

If a_0 is the constant function, then $\lim\limits_{z \to z_0} a_0 = a_0$; and if $a_1 \neq 0$, then we can use Definition 2.4 with $f(z) = a_1 z$ and the choice $\delta = \frac{\varepsilon}{|a_1|}$ to prove that $\lim\limits_{z \to z_0}(a_1 z) = a_1 z_0$. Using Property (2.19) and mathematical induction, we obtain

$$\lim_{z \to z_0}(a_k z^k) = a_k z_0^k \quad \text{for} \quad k = 0, 1, 2, \ldots, n. \tag{2.27}$$

We can extend Property (2.18) to a finite sum of terms and use the result of Equation (2.27) to get

$$\lim_{z \to z_0} P(z) = \lim_{z \to z_0}\left(\sum_{k=0}^{n} a_k z^k\right) = \sum_{k=0}^{n} a_k z_0^k = P(z_0).$$

Conditions (2.24), (2.25), and (2.26) are satisfied, so we conclude that P is continuous at z_0.

One technique for computing limits is to apply Theorem 2.4 to quotients. If we let P and Q be polynomials and if $Q(z_0) \neq 0$, then

$$\lim_{z \to z_0} \frac{P(z)}{Q(z)} = \frac{P(z_0)}{Q(z_0)}.$$

Another technique involves factoring polynomials. If both $P(z_0) = 0$ and $Q(z_0) = 0$, then P and Q can be factored as $P(z) = (z - z_0)P_1(z)$ and $Q(z) = (z - z_0)Q_1(z)$. If $Q_1(z_0) \neq 0$, then the limit is

$$\lim_{z \to z_0} \frac{P(z)}{Q(z)} = \lim_{z \to z_0} \frac{(z - z_0)P_1(z)}{(z - z_0)Q_1(z)} = \frac{P_1(z_0)}{Q_1(z_0)}.$$

Example 2.19. Show that $\lim\limits_{z \to 1+i} \frac{z^2 - 2i}{z^2 - 2z + 2} = 1 - i$.

Solution:

Here P and Q can be factored: $P(z) = (z - 1 - i)(z + 1 + i)$; $Q(z) = (z - 1 - i)(z - 1 + i)$.

Thus, the limit can be obtained in a straightforward manner:

$$\lim_{z \to 1+i}\left(\frac{z^2 - 2i}{z^2 - 2z + 2}\right) = \lim_{z \to 1+i} \frac{(z-1-i)(z+1+i)}{(z-1-i)(z-1+i)} = \lim_{z \to 1+i}\left(\frac{z+1+i}{z-1+i}\right) = \frac{(1+i)+1+i}{(1+i)-1+i} = \frac{2+2i}{2i} = 1 - i.$$

Exercises for Section 2.3 (Selected answers or hints are on page 433.)

1. Find the following limits.

   (a) $\lim\limits_{z \to 2+i} (z^2 - 4z + 2 + 5i)$.

   (b) $\lim\limits_{z \to i} \frac{z^2+4z+2}{z+1}$.

   (c) $\lim\limits_{z \to i} \frac{z^4-1}{z-i}$.

   (d) $\lim\limits_{z \to 1+i} \frac{z^2+z-2+i}{z^2-2z+1}$.

   (e) $\lim\limits_{z \to 1+i} \frac{z^2+z-1-3i}{z^2-2z+2}$ by factoring.

2. Determine where the following functions are continuous.

   (a) $z^4 - 9z^2 + iz - 2$.

   (b) $\frac{z+1}{z^2+1}$.

   (c) $\frac{z^2+6z+5}{z^2+3z+2}$.

   (d) $\frac{z^4+1}{z^2+2z+2}$.

   (e) $\frac{x+iy}{x-1}$.

   (f) $\frac{x+iy}{|z|-1}$.

3. State why $\lim\limits_{z \to z_0} (e^x \cos y + ix^2 y) = e^{x_0} \cos y_0 + ix_0^2 y_0$.

4. State why $\lim\limits_{z \to z_0} \left[\ln(x^2 + y^2) + iy \right] = \ln(x_0^2 + y_0^2) + iy_0$, provided $|z_0| \neq 0$.

5. Show that

   (a) $\lim\limits_{z \to 0} \frac{|z|^2}{z} = 0$.

   (b) $\lim\limits_{z \to 0} \frac{x^2}{z} = 0$.

6. Let $f(z) = \frac{z \operatorname{Re}(z)}{|z|}$ when $z \neq 0$ and let $f(0) = 0$. Show that $f(z)$ is continuous for all values of z.

7. Let $f(z) = \frac{z^2}{|z|^2} = \frac{x^2 - y^2 + i2xy}{x^2 + y^2}$.

   (a) Find $\lim\limits_{z \to 0} f(z)$ as $z \to 0$ along the line $y = x$.

   (b) Find $\lim\limits_{z \to 0} f(z)$ as $z \to 0$ along the line $y = 2x$.

   (c) Find $\lim\limits_{z \to 0} f(z)$ as $z \to 0$ along the parabola $y = x^2$.

   (d) What can you conclude about the limit of $f(z)$ as $z \to 0$? Why?

8. Let $f(z) = f(x, y) = \frac{xy^3}{x^2+2y^6} + i\frac{x^3 y}{5x^6+y^2}$ when $z \neq 0$, and let $f(0) = 0$.

   (a) Show that $\lim\limits_{z \to 0} f(z) = f(0) = 0$ if z approaches zero along any straight line that passes through the origin.

   (b) Show that f is not continuous at the point 0.

9. For $z \neq 0$, let $f(z) = \frac{\bar{z}}{z}$. Does $f(z)$ have a limit as $z \to 0$?

10. Does $\lim_{z \to -4} \operatorname{Arg} z$ exist? Why?

    Hint: Use polar coordinates and let z approach -4 from the upper and lower half-planes.

11. Let $f(z) = z^{\frac{1}{2}} = r^{\frac{1}{2}}(\cos \frac{\theta}{2} + i \sin \frac{\theta}{2})$, where $z = re^{i\theta}$, $r > 0$, and $-\pi < \theta \leq \pi$. Use the polar form of z and show that

    (a) $f(z) \to i$ as $z \to -1$ along the upper semicircle $r = 1$, $0 < \theta \leq \pi$.

    (b) $f(z) \to -i$ as $z \to -1$ along the lower semicircle $r = 1$, $-\pi < \theta < 0$.

12. Let $f(z) = \frac{x^2 + iy^2}{|z|^2}$ when $z \neq 0$ and let $f(0) = 1$. Show that $f(z)$ is not continuous at $z_0 = 0$.

13. Let $f(z) = xe^y + iy^2 e^{-x}$. Show that $f(z)$ is continuous for all values of z.

14. Use the definition of the limit to show that $\lim_{z \to 3 + 4i} z^2 = -7 + 24i$.

15. Let $f(z) = \frac{\operatorname{Re}(z)}{|z|}$ when $z \neq 0$ and let $f(0) = 1$. Is $f(z)$ continuous at the origin?

16. Let $f(z) = \frac{\left(\operatorname{Re}(z)\right)^2}{|z|}$ when $z \neq 0$ and let $f(0) = 0$. Is $f(z)$ continuous at the origin?

17. Let $f(z) = z^{\frac{1}{2}} = |z|^{\frac{1}{2}} e^{i\frac{\operatorname{Arg}(z)}{2}}$, where $z \neq 0$. Show that $f(z)$ is discontinuous at each point along the negative x axis.

18. Let $f(z) = \ln|z| + i\operatorname{Arg}(z)$, where $-\pi < \operatorname{Arg}(z) \leq \pi$. Show that $f(z)$ is discontinuous at $z_0 = 0$ and at each point along the negative x axis.

19. Let $|g(z)| \leq M$ and $\lim_{z \to z_0} f(z) = 0$. Show that $\lim_{z \to z_0} f(z)g(z) = 0$. Note: Theorem 2.2 is of no use here because you don't know whether $\lim_{z \to z_0} g(z)$ exists. Give an ε, δ argument.

20. Let $\Delta z = z - z_0$. Show that $\lim_{z \to z_0} f(z) = w_0$ iff $\lim_{\Delta z \to 0} f(z_0 + \Delta z) = w_0$.

21. Let $f(z)$ be continuous for all values of z.

    (a) Show that $g(z) = f(\bar{z})$ is continuous for all z.

    (b) Show that $g(z) = \overline{f(z)}$ is continuous for all z.

22. Verify the following identities:

    (a) $\lim_{z \to z_0} [f(z) \pm g(z)] = A \pm B$: Identity (2.18).

    (b) $\lim_{z \to z_0} f(z)g(z) = AB$: Identity (2.19).

    (c) $\lim_{z \to z_0} \frac{f(z)}{g(z)} = \frac{A}{B}$: Identity (2.20).

23. Verify the results of Theorem 2.4.

24. Show that the principal branch of the argument, $\operatorname{Arg} z$, is discontinuous at 0 and all points along the negative real axis.

2.4 Branches of Functions

In Section 2.2 we defined the principal square root function and investigated some of its properties. We left unanswered some questions concerning the choices of square roots. We now look at these questions because they are similar to situations involving other elementary functions.

In our definition of a function in Section 2.1, we specified that each value of the independent variable in the domain is mapped onto one and *only one* value in the range. As a result, we often talk about a single-valued function, which emphasizes the "only one" part of the definition and allows us to distinguish such functions from multiple-valued functions, which we now introduce.

Let $w = f(z)$ denote a function whose domain is the set D and whose range is the set R. If w is a value in the range, then there is an associated inverse relation $z = g(w)$ that assigns to each value w the value (or values) of z in D for which the equation $f(z) = w$ holds. But unless f takes on the value w at most once in D, then the inverse relation g is necessarily many valued, and we say that g is a multivalued function. For example, the inverse of the function $w = f(z) = z^2$ is the square root function $z = g(w) = w^{\frac{1}{2}}$. For each value z other than $z = 0$, then, the two points z and $-z$ are mapped onto the same point $w = f(z)$; hence g is, in general, a two-valued function.

The study of limits, continuity, and derivatives loses all meaning if an arbitrary or ambiguous assignment of function values is made. For this reason we did not allow multivalued functions to be considered when we defined these concepts. When working with inverse functions, you have to specify carefully one of the many possible inverse values when constructing an inverse function, as when you determine implicit functions in calculus. If the values of a function f are determined by an equation that they satisfy rather than by an explicit formula, then we say that the function is defined implicitly or that f is an implicit function. In the theory of complex variables we present a similar concept.

We now let $w = f(z)$ be a multiple-valued function. A branch of f is any single-valued function f_0 that is continuous in some domain (except, perhaps, on the boundary). At each point z in the domain, it assigns one of the values of $f(z)$.

Example 2.20. We consider some branches of the two-valued square root function $f(z) = z^{\frac{1}{2}}$ ($z \neq 0$). Define the principal square root function as

$$f_1(z) = |z|^{\frac{1}{2}} e^{i\frac{\text{Arg}(z)}{2}} = r^{\frac{1}{2}} e^{i\frac{\theta}{2}} = r^{\frac{1}{2}} \cos\frac{\theta}{2} + ir^{\frac{1}{2}} \sin\frac{\theta}{2}, \qquad (2.28)$$

where $r = |z|$ and $\theta = \text{Arg}(z)$ so that $-\pi < \theta \leq \pi$. The function f_1 is a branch of f. Using the same notation, we can find other branches of the square root function. For example, if we let

$$f_2(z) = |z|^{\frac{1}{2}} e^{i\frac{\text{Arg}(z)+2\pi}{2}} = r^{\frac{1}{2}} e^{i\frac{\theta+2\pi}{2}}$$

$$= r^{\frac{1}{2}} \cos\left(\frac{\theta + 2\pi}{2}\right) + ir^{\frac{1}{2}} \sin\left(\frac{\theta + 2\pi}{2}\right), \qquad (2.29)$$

then

$$f_2(z) = r^{\frac{1}{2}} e^{i\frac{\theta+2\pi}{2}} = r^{\frac{1}{2}} e^{i\frac{\theta}{2}} e^{i\pi} = -r^{\frac{1}{2}} e^{i\frac{\theta}{2}} = -f_1(z),$$

so f_1 and f_2 can be thought of as "plus" and "minus" square root functions. The negative real axis is called a branch cut for the functions f_1 and f_2. Each point on the branch cut is a point of discontinuity for both functions f_1 and f_2.

Example 2.21. Show that the function f_1 is discontinuous along the negative real axis.

Solution:

Let $z_0 = r_0 e^{i\pi}$ denote a negative real number. We compute the limit as z approaches z_0 through the upper half-plane $\{z : \operatorname{Im}(z) > 0\}$ and the limit as z approaches z_0 through the lower half-plane $\{z : \operatorname{Im}(z) < 0\}$. In polar coordinates these limits are given by

$$\lim_{(r,\theta)\to(r_0,\pi)} f_1(re^{i\theta}) = \lim_{(r,\theta)\to(r_0,\pi)} r^{\frac{1}{2}} \left(\cos\frac{\theta}{2} + i\sin\frac{\theta}{2}\right) = ir_0^{\frac{1}{2}}, \quad \text{and}$$

$$\lim_{(r,\theta)\to(r_0,-\pi)} f_1(re^{i\theta}) = \lim_{(r,\theta)\to(r_0,-\pi)} r^{\frac{1}{2}} \left(\cos\frac{\theta}{2} + i\sin\frac{\theta}{2}\right) = -ir_0^{\frac{1}{2}}.$$

The two limits are distinct, so the function f_1 is discontinuous at z_0.

Remark 2.4. *Likewise, f_2 is discontinuous at z_0. The mappings $w = f_1(z)$, $w = f_2(z)$, and the branch cut are illustrated in Figure 2.18.*

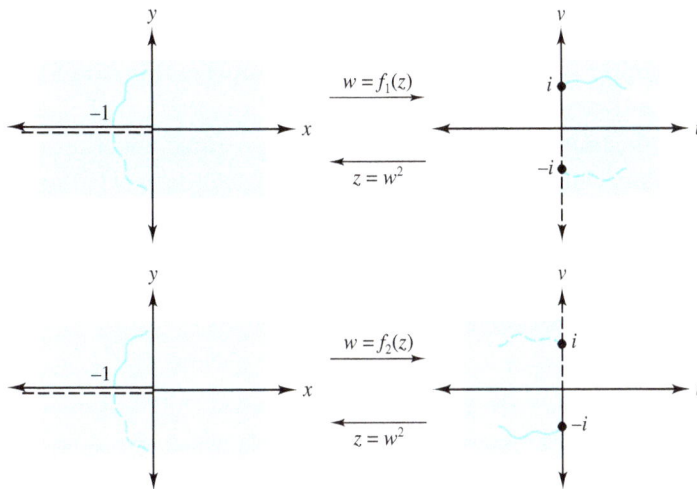

Figure 2.18: The branches f_1 and f_2 of $f(z) = z^{\frac{1}{2}}$

We can construct other branches of the square root function by specifying that an argument of z given by $\theta = \arg z$ is to lie in the interval $\alpha < \theta \leq \alpha + 2\pi$. The corresponding branch is

$$f_\alpha(z) - r^{\frac{1}{2}}\cos\frac{\theta}{2} + ir^{\frac{1}{2}}\sin\frac{\theta}{2}, \quad \text{where} \quad z - re^{i\theta} \neq 0 \quad \text{and} \quad \alpha < \theta \leq \alpha + 2\pi. \tag{2.30}$$

The branch cut for f_α is the ray $r \geq 0$, $\theta = \alpha$, which includes the origin. The point $z = 0$, common to all branch cuts for the multivalued square root function, is called a branch point. The mapping $w = f_\alpha(z)$ and its branch cut are illustrated in Figure 2.19.

2.4.1 The Riemann Surface for $w = z^{\frac{1}{2}}$

A Riemann surface is a construct useful for visualizing a multivalued function. It was introduced by G. F. B. Riemann (1826–1866) in 1851. The idea is ingenious—a geometric construction that permits surfaces to be the domain or range of a multivalued function. Riemann surfaces depend on the function being investigated. We now give a nontechnical formulation of the Riemann surface for the multivalued square root function.

Consider $w = f(z) = z^{\frac{1}{2}}$, which has two values for any $z \neq 0$. Each function f_1 and f_2 in Figure 2.18 is single-valued on the domain formed by cutting the z plane along the negative

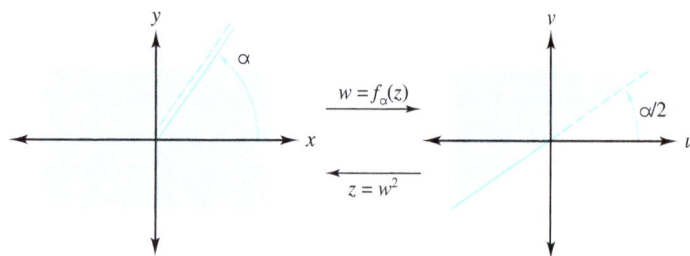

Figure 2.19: The branch f_α of $f(z) = z^{\frac{1}{2}}$

x axis. Let D_1 and D_2 be the domains of f_1 and f_2, respectively. The range set for f_1 is the set H_1 consisting of the right half-plane, and the positive v axis; the range set for f_2 is the set H_2 consisting of the left half-plane and the negative v axis. The sets H_1 and H_2 are "glued together" along the positive v axis and the negative v axis to form the w plane with the origin deleted.

We stack D_1 directly above D_2. The edge of D_1 in the upper half-plane is joined to the edge of D_2 in the lower half-plane, and the edge of D_1 in the lower half-plane is joined to the edge of D_2 in the upper half-plane. When these domains are glued together in this manner, they form R, which is a Riemann surface domain for the mapping $w = f(z) = z^{\frac{1}{2}}$. The portions of D_1, D_2, and R that lie in $\{z : |z| < 1\}$ are shown in Figure 2.20.

Formation of the Riemann surface for $w = z^{\frac{1}{2}}$: (a) a portion of D_1 and its image under $w = z^{\frac{1}{2}}$; (b) a portion of D_2 and its image under $w = z^{\frac{1}{2}}$; (c) a portion of R and its image under $w = z^{\frac{1}{2}}$.

The beauty of this structure is that it makes this "full square root function" continuous for all $z \neq 0$. Normally, the principal square root function would be discontinuous along the negative real axis, as points near -1 but above that axis would get mapped to points close to i, and points near -1 but below the axis would get mapped to points close to $-i$. As Figure 2.20(c) indicates, however, between the point A and the point B, the domain switches from the edge of D_1 in the upper half-plane to the edge of D_2 in the lower half-plane. The corresponding mapped points A' and B' are exactly where they should be. The surface works in such a way that going directly between the edges of D_1 in the upper and lower half-planes is impossible (likewise for D_2). Going counter-clockwise, the only way to get from the point A to the point C, for example, is to follow the path indicated by the arrows in Figure 2.20(c).

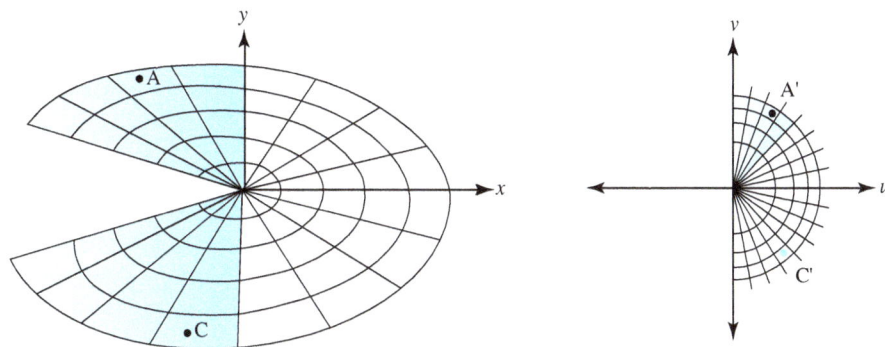

(a) A portion of D_1 and its image under $w = f_1$.

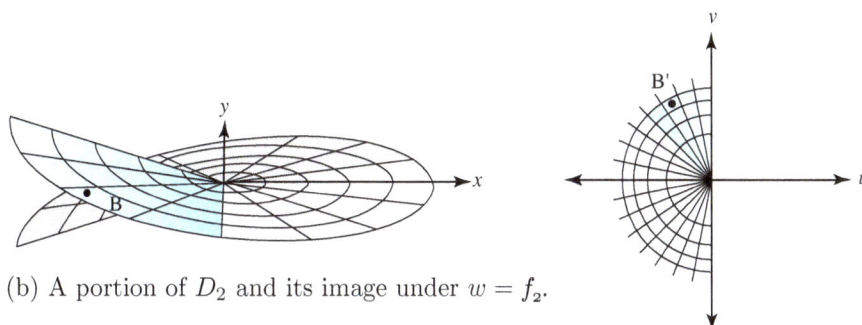

(b) A portion of D_2 and its image under $w = f_2$.

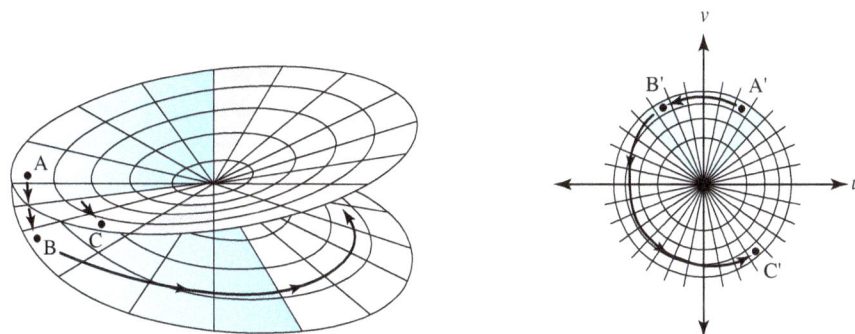

(c) A portion of R and its image under $w = z^{\frac{1}{2}}$.

Figure 2.20: Formation of the Riemann surface for $w = z^{\frac{1}{2}}$: (a) a portion of D_1 and its image under $w = z^{\frac{1}{2}}$; (b) a portion of D_2 and its image under $w = z^{\frac{1}{2}}$; (c) a portion of R and its image under $w = z^{\frac{1}{2}}$

Exercises for Section 2.4 (Selected answers or hints are on page 434.)

1. Let $f_1(z)$ and $f_2(z)$ be the two branches of the square root function given by Equations (2.28) and (2.29), respectively. Use the polar coordinate formulas in Section 2.2 to find the image of

    (a) quadrant II, $x < 0$ and $y > 0$, under the mapping $w = f_1(z)$.

    (b) quadrant II, $x < 0$ and $y > 0$, under the mapping $w = f_2(z)$.

    (c) the right half-plane $\operatorname{Re}(z) > 0$ under the mapping $w = f_1(z)$.

(d) the right half-plane $\text{Re}(z) > 0$ under the mapping $w = f_2(z)$.

2. Let $\alpha = 0$ in Equation (2.30). Find the range of the function $w = f_\alpha(z)$.

3. Let $\alpha = 2\pi$ in Equation (2.30). Find the range of the function $w = f_\alpha(z)$.

4. Find a branch of the square root that is continuous along the negative x axis.

5. Let $f_1(z) = |z|^{\frac{1}{3}} e^{i\frac{\text{Arg}(z)}{3}} = r^{\frac{1}{3}} \cos\frac{\theta}{3} + ir^{\frac{1}{3}} \sin\frac{\theta}{3}$, where $|z| = r \neq 0$, and $\theta = \text{Arg}(z)$. f_1 denotes the principal cube root function.

   (a) Show that f_1 is a branch of the multivalued cube root $f(z) = z^{\frac{1}{3}}$.
   (b) What is the range of f_1?
   (c) Where is f_1 continuous?

6. Let $f_2(z) = r^{\frac{1}{3}} \cos(\frac{\theta+2\pi}{3}) + ir^{\frac{1}{3}} \sin(\frac{\theta+2\pi}{3})$, where $r > 0$ and $-\pi < \theta \leq \pi$.

   (a) Show that f_2 is a branch of the multivalued cube root $f(z) = z^{\frac{1}{3}}$.
   (b) What is the range of f_2?
   (c) Where is f_2 continuous?
   (d) What is the branch point associated with f?

7. Find a branch of the multivalued cube root function that is different from those in Exercises 5 and 6. State the domain and range of the branch you find.

8. Let $f(z) = z^{\frac{1}{n}}$ denote the multivalued nth root, where n is a positive integer.

   (a) Show that f is, in general, an n-valued function.
   (b) Write the principal nth root function.
   (c) Produce a different branch of the multivalued nth root function.

9. Describe a Riemann surface for the domain of definition of

   (a) $w = f(z) = z^{\frac{1}{3}}$.
   (b) $w = f(z) = z^{\frac{1}{4}}$.

10. Discuss how Riemann surfaces should be used for both the domain and range to help describe the multivalued function $w = f(z) = z^{\frac{2}{3}}$.

2.5   The Reciprocal Transformation $w = \frac{1}{z}$

The mapping $w = f(z) = \frac{1}{z}$ is called the *reciprocal transformation*. It maps the z plane one-to-one and onto the w plane except for the point $z = 0$, which has no image. The point $w = 0$ has no preimage or inverse image. Using exponential notation $w = \rho e^{i\phi}$, if $z = re^{i\theta} \neq 0$, we have

$$w = \rho e^{i\phi} = \frac{1}{z} = \frac{1}{r} e^{-i\theta}. \tag{2.31}$$

The geometric description of the reciprocal transformation is now evident. It is an inversion (that is, the modulus of $\frac{1}{z}$ is the reciprocal of the modulus of z) followed by a reflection through

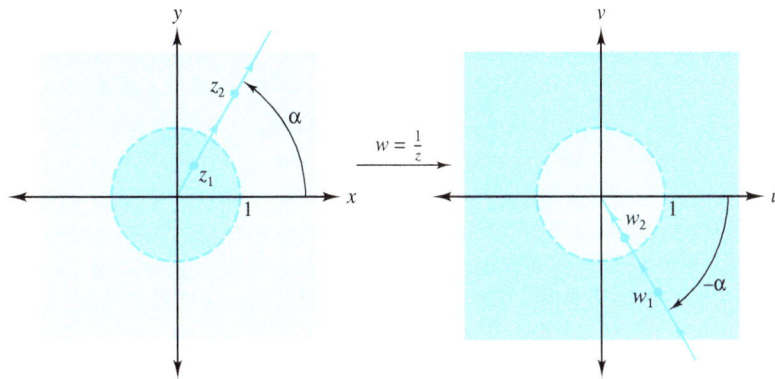

Figure 2.21: The reciprocal transformation $w = \frac{1}{z}$

the x axis. The ray $r > 0$, $\theta = \alpha$, is mapped one-to-one and onto the ray $\rho > 0$, $\phi = -\alpha$. Points that lie inside the unit circle $C_1(0) = \{z : |z| = 1\}$ are mapped onto points that lie outside the unit circle and vice versa, as Figure 2.21 illustrates.

We can extend the system of complex numbers by joining to it an "ideal" point denoted by ∞ and called the point at infinity. This new set is called the extended complex plane. You will see shortly that the point ∞ has the property, loosely speaking, that $z \to \infty$ iff $|z| \to \infty$.

An ε-neighborhood of the point at infinity is the set $\{z : |z| > \frac{1}{\varepsilon}\}$. The usual way to visualize the point at infinity is by using what we call the stereographic projection, which is attributed to Riemann. Let Ω be a sphere of diameter 1 that is centered at the point $(0, 0, \frac{1}{2})$ in three-dimensional space where coordinates are specified by the triple of real numbers (x, y, ξ). Here the complex number $z = x + iy$ is associated with the point $z = (x, y, 0)$.

The point $\mathbb{N} = (0, 0, 1)$ on Ω is called the north pole of Ω. If we let z be a complex number and consider the line segment L in three-dimensional space that joins z to the north pole $\mathbb{N} = (0, 0, 1)$, then L intersects Ω in exactly one point L. The correspondence $z \longleftrightarrow \mathbb{Z}$ is called the stereographic projection of the complex z plane onto the Riemann sphere Ω. A point $z = x + iy$ of unit modulus will correspond with $\mathbb{Z} = (\frac{x}{2}, \frac{y}{2}, \frac{1}{2})$. If z has modulus greater than 1, then L will lie in the upper hemisphere where for points $\mathbb{Z} = (x, y, \xi)$ we have $\xi > \frac{1}{2}$. If z has modulus less than 1, then L will lie in the lower hemisphere where for points $\mathbb{Z} = (x, y, \xi)$ we have $\xi < \frac{1}{2}$. The complex number $z = 0 = 0 + 0i$ corresponds with the south pole, $S = (0, 0, 0)$. Now you can see that indeed $z \to \infty$ iff $|z| \to \infty$ iff $L \to N$. Hence N corresponds with the "ideal" point at infinity. The situation is shown in Figure 2.22.

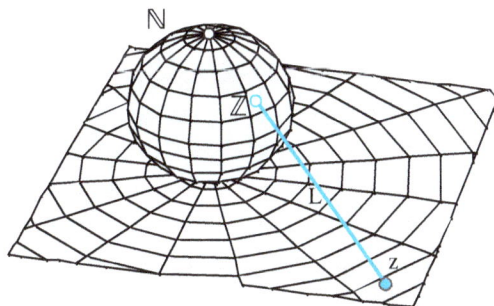

Figure 2.22: The Riemann sphere

Let's reconsider the mapping $w = \frac{1}{z}$ by assigning the images $w = \infty$ and $w = 0$ to the points $z = 0$ and $z = \infty$, respectively. We now write the reciprocal transformation as

$$w = f(z) = \begin{cases} \frac{1}{z} & \text{when} \quad z \neq 0 \quad \text{and} \quad z \neq \infty; \\ 0 & \text{when} \quad z = \infty; \\ \infty & \text{when} \quad z = 0. \end{cases} \tag{2.32}$$

Note that the transformation $w = f(z)$ is a one-to-one mapping of the extended complex z plane onto the extended complex w plane. Further, f is a continuous mapping from the extended z plane onto the extended w plane. We leave the details to you.

Example 2.22. Show that the image of the half-plane $A = \{z : \operatorname{Re}(z) \geq \frac{1}{2}\}$ under the mapping $w = \frac{1}{z}$ is the closed disk $\overline{D}_1(1) = \{w : |w - 1| \leq 1\}$.

Solution:

Proceeding as we did in Example 2.6, we get the inverse mapping of $u + iv = w = f(z) = \frac{1}{z}$ as $z = f^{-1}(w) = \frac{1}{w}$. Then

$$u + iv = w \in B \iff f^{-1}(w) = z = x + iy \in A$$

$$\iff \frac{1}{u + iv} = x + iy \in A$$

$$\iff \frac{u}{u^2 + v^2} + i\frac{-v}{u^2 + v^2} = x + iy \in A$$

$$\iff \frac{u}{u^2 + v^2} = x \geq \frac{1}{2}$$

$$\iff \frac{u}{u^2 + v^2} \geq \frac{1}{2} \tag{2.33}$$

$$\implies u^2 - 2u + 1 + v^2 \leq 1 \tag{2.34}$$

$$\iff (u - 1)^2 + (v - 0)^2 \leq 1,$$

which describes the disk $\overline{D}_1(0)$. As the reciprocal transformation is one-to-one, preimages of the points in the disk $\overline{D}_1(0)$ will lie in the right half-plane $\operatorname{Re}(z) \geq \frac{1}{2}$. Figure 2.23 illustrates this result.

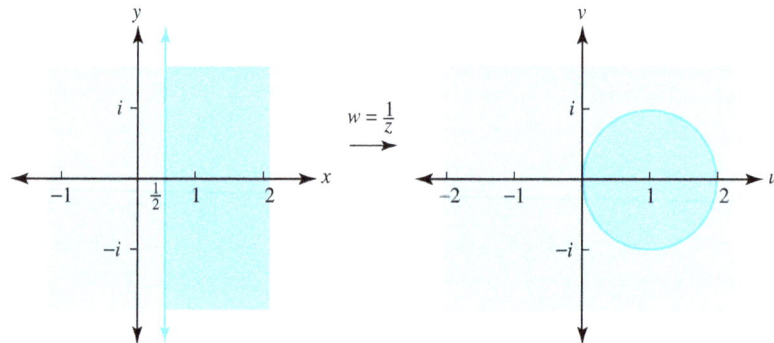

Figure 2.23: The image of $\operatorname{Re}(z) \geq \frac{1}{2}$ under the mapping $w = \frac{1}{z}$

Remark 2.5. *Alas, there is a fly in the ointment here. As our notation indicates, Equations (2.33) and (2.34) are not equivalent. The former implies the latter, but not conversely. That is, Equation (2.34) makes sense when $(u, v) = (0, 0)$, whereas Equation (2.33) does not. Yet*

Figure 2.23 seems to indicate that f maps $\mathrm{Re}(z) \geq \frac{1}{2}$ onto the entire disk $\overline{D}_1(0)$, including the point $(0, 0)$. Actually, it does not, because $(0, 0)$ has no preimage in the complex plane. The way out of this dilemma is to use the complex point at infinity. It is that quantity that gets mapped to the point $(u, v) = (0, 0)$, for as we have already indicated in Equation (2.32), the preimage of 0 under the mapping $\frac{1}{z}$ is indeed ∞.

Example 2.23. For the transformation $\frac{1}{z}$, find the image of the portion of the half plane $\mathrm{Re}(z) \geq \frac{1}{2}$ that is inside the closed disk $\overline{D}_1(\frac{1}{2}) = \{z : |z - \frac{1}{2}| \leq 1\}$.

Solution:

Using the result of Example 2.22, we need only find the image of the disk $\overline{D}_1(\frac{1}{2})$ and intersect it with the closed disk $\overline{D}_1(1)$. To begin, we note that

$$\overline{D}_1\left(\frac{1}{2}\right) = \left\{(x, y) : x^2 + y^2 - x \leq \frac{3}{4}\right\}.$$

Because $z = f^{-1}(w) = \frac{1}{w}$, we have, as before,

$$u + iv = w \in f\left(\overline{D}_1\left(\frac{1}{2}\right)\right) \iff f^{-1}(w) \in \overline{D}_1\left(\frac{1}{2}\right)$$

$$\iff \frac{1}{w} \in \overline{D}_1\left(\frac{1}{2}\right)$$

$$\iff \frac{u}{u^2 + v^2} + i\frac{-v}{u^2 + v^2} \in \overline{D}_1\left(\frac{1}{2}\right)$$

$$\iff \left(\frac{u}{u^2 + v^2}\right)^2 + \left(\frac{-v}{u^2 + v^2}\right)^2 - \frac{u}{u^2 + v^2} \leq \frac{3}{4}$$

$$\iff \frac{1}{u^2 + v^2} - \frac{u}{u^2 + v^2} \leq \frac{3}{4}$$

$$\iff \left(u + \frac{2}{3}\right)^2 + v^2 \geq \left(\frac{4}{3}\right)^2,$$

which is an inequality that determines the set of points in the w plane that lie on and outside the circle $C_{\frac{4}{3}}(-\frac{2}{3}) = \{w : |w + \frac{2}{3}| = \frac{4}{3}\}$. Note that we do not have to deal with the point at infinity this time, as the last inequality is not satisfied when $(u, v) = (0, 0)$.

When we intersect this set with $\overline{D}_1(1)$, we get the crescent-shaped region shown in Figure 2.24.

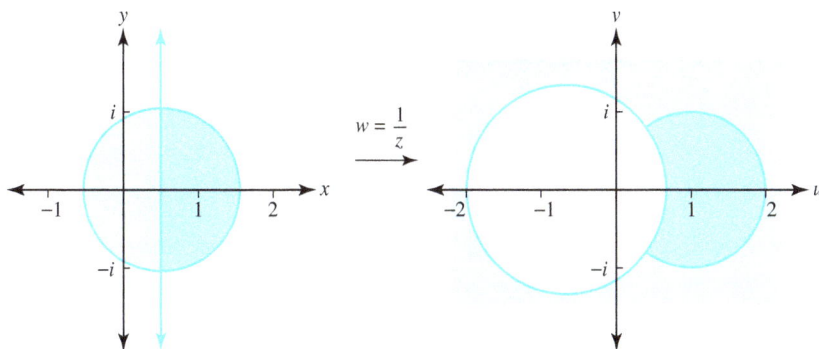

Figure 2.24: The mapping $w = \frac{1}{z}$ discussed in Example 2.23

To study images of "generalized circles," we consider the equation

$$A(x^2 + y^2) + Bx + Cy + D = 0$$

where A, B, C, and D are real numbers. This equation represents either a circle or a line, depending on whether $A \neq 0$ or $A = 0$, respectively. Transforming the equation to polar coordinates gives

$$Ar^2 + r(B\cos\theta + C\sin\theta) + D = 0.$$

Using the polar coordinate form of the reciprocal transformation given in Equation (2.31), we can express the image of the curve in the preceding equation as

$$A + \rho(B\cos\phi - C\sin\phi) + D\rho^2 = 0,$$

which represents either a circle or a line, depending on whether $D \neq 0$ or $D = 0$, respectively. Thus, the reciprocal transformation $w = \frac{1}{z}$ carries the class of lines and circles onto itself.

Example 2.24. Find the images of the vertical lines $x = a$ and the horizontal lines $y = b$ under the mapping $w = \frac{1}{z}$.

Solution:

Considering the point at infinity, the image of the line $x = 0$ is the line $u = 0$; that is, the y-axis is mapped onto the v-axis. Similarly, the x-axis is mapped onto the u-axis. The inverse mapping is $z = \frac{1}{w} = \frac{u}{u^2+v^2} + i\frac{-v}{u^2+v^2}$, so if $a \neq 0$, the vertical line $x = a$ is mapped onto the set of (u, v) points satisfying $\frac{u}{u^2+v^2} = a$. For $(u, v) \neq (0, 0)$, this outcome is equivalent to

$$u^2 - \frac{1}{a}u + \frac{1}{4a^2} + v^2 = \left(u - \frac{1}{2a}\right)^2 + v^2 = \left(\frac{1}{2a}\right)^2.$$

which is the equation of a circle in the w plane with center $w_0 = \frac{1}{2a}$ and radius $\left|\frac{1}{2a}\right|$. The point at infinity is mapped to $(u, v) = (0, 0)$. Similarly, $y = b$ is mapped onto the circle

$$u^2 + v^2 + \frac{1}{b}v + \frac{1}{4b^2} = u^2 + \left(v + \frac{1}{2b}\right)^2 = \left(\frac{1}{2b}\right)^2.$$

which has center $w_0 = -\frac{i}{2b}$ and radius $\left|\frac{1}{2b}\right|$. Figure 2.25 illustrates the images of several lines.

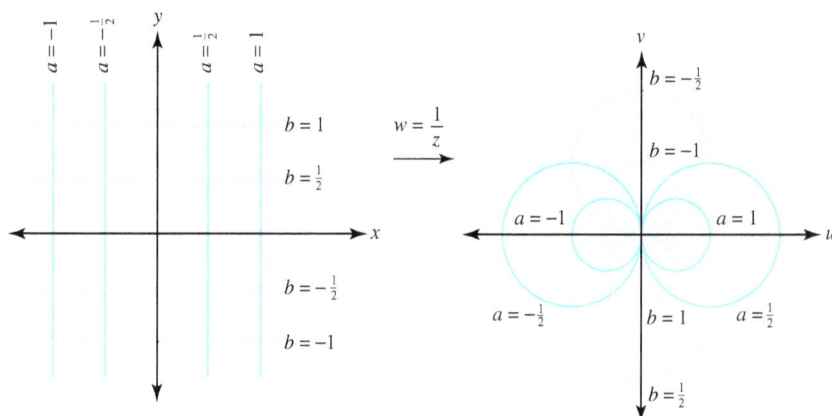

Figure 2.25: The images of horizontal and vertical lines under the reciprocal transformation

Exercises for Section 2.5 (Selected answers or hints are on page 434.)

For Exercises 1–8, find the image of the given circle or line under the reciprocal transformation $w = \frac{1}{z}$.

1. The horizontal line $\text{Im}(z) = \frac{1}{5}$.

2. The circle $C_{\frac{1}{2}}\left(-\frac{i}{2}\right) = \{z : |z + \frac{i}{2}| = \frac{1}{2}\}$.

3. The vertical line $\text{Re}(z) = -3$.

4. The circle $C_1(-2) = \{z : |z + 2| = 1\}$.

5. The line $2x + 2y = 1$.

6. The circle $C_1(\frac{i}{2}) = \{z : |z - \frac{i}{2}| = 1\}$.

7. The circle $C_1(\frac{3}{2}) = \{z : |z - \frac{3}{2}| = 1\}$.

8. The circle $C_2(-1 + i) = \{z : |z + 1 - i| = 2\}$.

9. **Limits involving ∞.** The function $f(z)$ is said to have the limit L as z approaches ∞, and we write $\lim\limits_{z \to \infty} f(z) = L$ iff for every $\varepsilon > 0$ there exists an $R > 0$ such that $f(z) \in D_\varepsilon(L)$ (i.e., $|f(z) - L| < \varepsilon$) whenever $|z| > R$. Likewise, $\lim\limits_{z \to z_0} f(z) = \infty$ iff for every $R > 0$ there exists $\delta > 0$ such that $|f(z)| > R$ whenever $z \in D_\delta^*(z_0)$ (i.e., $0 < |z - z_0| < \delta$). Use this definition to

   (a) show that $\lim\limits_{z \to \infty} \frac{1}{z} = 0$.

   (b) show that $\lim\limits_{z \to 0} \frac{1}{z} = \infty$.

10. Show that the reciprocal transformation $w = \frac{1}{z}$ maps the vertical strip $0 < x < \frac{1}{2}$ onto the region in the right half-plane $\text{Re}(w) > 0$ that is outside the disk $D_1(1) = \{w : |w - 1| < 1\}$.

11. Find the image of the disk $D_{\frac{4}{3}}\left(-\frac{2i}{3}\right) = \{z : |z + \frac{2i}{3}| < \frac{4}{3}\}$ under $f(z) = \frac{1}{z}$.

12. Show that the reciprocal transformation maps the disk $|z - 1| < 2$ onto the region that lies exterior to the circle $\{w : |w + \frac{1}{3}| = \frac{2}{3}\}$.

13. Find the image of the half-plane $y > \frac{1}{2} - x$ under the mapping $w = \frac{1}{z}$.

14. Show that the half-plane $y < x - \frac{1}{2}$ is mapped onto the disk $|w - 1 - i| < \sqrt{2}$ by the reciprocal transformation.

15. Find the image of the quadrant $x > 1$, $y > 1$ under the mapping $w = \frac{1}{z}$.

16. Show that the transformation $w = \frac{2}{z}$ maps the disk $|z - i| < 1$ onto the lower half-plane $\text{Im}(w) < -1$.

17. Show that the transformation $w = \frac{2-z}{z} = -1 + \frac{2}{z}$ maps the disk $|z - 1| < 1$ onto the right half-plane $\text{Re}(w) > 0$.

18. Show that the parabola $2x = 1 - y^2$ is mapped onto the cardioid $\rho = 1 + \cos\phi$ by the reciprocal transformation.

19. Use the definition in Exercise 9 to prove that $\lim\limits_{z \to \infty} \frac{z+1}{z-1} = 1$.

20. Show that $z = x + iy$, when mapped onto the Riemann sphere, has coordinates
$$\left(\frac{x}{x^2 + y^2 + 1}, \frac{y}{x^2 + y^2 + 1}, \frac{x^2 + y^2}{x^2 + y^2 + 1} \right).$$

21. Explain how the quantities $+\infty$, $-\infty$, and ∞ differ. How are they similar?

Chapter 3

Analytic and Harmonic Functions

Overview

Does the notion of a derivative of a complex function make sense? If so, how should it be defined and what does it represent? These and similar questions are the focus of this chapter. As you might guess, complex derivatives have a meaningful definition, and many of the standard derivative theorems from calculus (such as the product rule and chain rule) carry over. There are also some interesting applications. But not everything is symmetric. You will learn in this chapter that the mean value theorem for derivatives does not extend to complex functions. In later chapters you will see that differentiable complex functions are, in some sense, much more "differentiable" than differentiable real functions.

3.1   Differentiable and Analytic Functions

Using our imagination, we take our lead from elementary calculus and define the derivative of f at z_0, written $f'(z_0)$, by

$$f'(z_0) = \lim_{z \to z_0} \frac{f(z) - f(z_0)}{z - z_0}, \tag{3.1}$$

provided the limit exists. If it does, we say that the function f is **differentiable at** z_0. If we write $\Delta z = z - z_0$, then we can express Equation (3.1) in the form

$$f'(z_0) = \lim_{\Delta z \to 0} \frac{f(z_0 + \Delta z) - f(z_0)}{\Delta z}. \tag{3.2}$$

Letting $w = f(z)$ and $\Delta w = f(z) - f(z_0)$ and using the Leibniz notation $\frac{dw}{dz}$ for the derivatives gives

$$f'(z_0) = \frac{dw}{dz} = \lim_{\Delta z \to 0} \frac{\Delta w}{\Delta z}. \tag{3.3}$$

Example 3.1. If $f(z) = z^3$, show that $f'(z) = 3z^2$.

Solution:

Using Equation (3.1), we have

$$f'(z_0) = \lim_{z \to z_0} \frac{z^3 - z_0^3}{z - z_0}$$

$$= \lim_{z \to z_0} \frac{(z - z_0)(z^2 + z_0 z + z_0^2)}{z - z_0}$$

$$= \lim_{z \to z_0} (z^2 + z_0 z + z_0^2)$$

$$= 3z_0^2.$$

We can drop the subscript on z_0 to obtain $f'(z) = 3z^2$ as a general formula.

Pay careful attention to the value Δz in Equation (3.3); the limit must be independent of the manner in which $\Delta z \to 0$. If we can find two curves that end at z_0 along which $\frac{\Delta w}{\Delta z}$ approaches distinct values, then $\frac{\Delta w}{\Delta z}$ does *not* have a limit as $\Delta z \to 0$, so f does *not* have a derivative at z_0. The same observation applies to the limits in Equations (3.1) and (3.2).

Example 3.2. Show that the function $w = f(z) = \bar{z} = x - iy$ is nowhere differentiable.

Solution:

We choose two approaches to the point $z_0 = x_0 + iy_0$ and compute limits of the difference quotients. First, we approach $z_0 = x_0 + iy_0$ along a line parallel to the x-axis by forcing z to be of the form $z = x + iy_0$.

$$\lim_{z \to z_0} \frac{f(z) - f(z_0)}{z - z_0} = \lim_{(x+iy_0) \to (x_0+iy_0)} \frac{f(x + iy_0) - f(x_0 + iy_0)}{(x + iy_0) - (x_0 + iy_0)}$$

$$= \lim_{(x+iy_0) \to (x_0+iy_0)} \frac{(x - iy_0) - (x_0 - iy_0)}{(x - x_0) + i(y_0 - y_0)}$$

$$= \lim_{(x+iy_0) \to (x_0+iy_0)} \frac{x - x_0}{x - x_0}$$

$$= 1.$$

Next, we approach z_0 along a line parallel to the y axis by forcing z to be of the form $z = x_0 + iy$.

$$\lim_{z \to z_0} \frac{f(z) - f(z_0)}{z - z_0} = \lim_{(x_0+iy) \to (x_0+iy_0)} \frac{f(x_0 + iy) - f(x_0 + iy_0)}{(x_0 + iy) - (x_0 + iy_0)}$$

$$= \lim_{(x_0+iy) \to (x_0+iy_0)} \frac{(x_0 - iy) - (x_0 - iy_0)}{(x_0 - x_0) + i(y - y_0)}$$

$$= \lim_{(x_0+iy) \to (x_0+iy_0)} \frac{-i(y - y_0)}{i(y - y_0)}$$

$$= -1.$$

The limits along the two paths are different, so there is no possible value for the right side of Equation (3.1). Therefore $f(z) = \bar{z}$ is not differentiable at the point z_0, and since z_0 was arbitrary, $f(z)$ is nowhere differentiable.

Remark 3.1. *In Section 2.3 we showed that $f(z) = \bar{z}$ is continuous for all z. Thus, we have a simple example of a function that is continuous everywhere but differentiable nowhere. Such functions are hard to construct in real variables. In some sense, the complex case has made pathological constructions simpler!*

We seldom are interested in studying functions that aren't differentiable, or are differentiable at only a single point. Complex functions that have a derivative at all points in a neighborhood of z_0 deserve further study. Indeed, functions that are differentiable in neighborhoods of points are pillars of the complex analysis edifice. We give them a special name, as indicated in the following definition.

Definition 3.1 (Analytic). *The complex function f is **analytic at the point** z_0 provided there is some $\varepsilon > 0$ such that $f'(z)$ exists for all $z \in D_\varepsilon(z_0)$. In other words, f must be differentiable not only at z_0, but also at all points in some ε-neighborhood of z_0.*

If f is analytic at each point in the region R, then we say that f is **analytic on R**. Again, we have a special term if f is analytic on the whole complex plane.

Definition 3.2 (Entire). *If f is analytic on the whole complex plane then f is said to be **entire**.*

Points of non analyticity for a function are called **singular points**. They are important for certain applications in physics and engineering.

Our definition of the derivative for complex functions is formally the same as for real functions and is the natural extension from real variables to complex variables. The basic differentiation formulas are identical to those for real functions, and we obtain the same rules for differentiating powers, sums, products, quotients, and compositions of functions. We can easily establish the proof of the differentiation formulas by using the limit theorems.

Suppose that f and g are differentiable. From Equation (3.1) and the technique exhibited in the solution to Example 3.1, we can establish the following rules, which are virtually identical to those for real-valued functions.

$$\frac{d}{dz} C = 0, \quad \text{where} \quad C \quad \text{is a constant, and} \tag{3.4}$$

$$\frac{d}{dz} z^n = nz^{n-1}, \quad \text{where} \quad n \quad \text{is a positive integer.} \tag{3.5}$$

$$\frac{d}{dz}[Cf(z)] = Cf'(z) \tag{3.6}$$

$$\frac{d}{dz}[f(z) + g(z)] = f'(z) + g'(z) \tag{3.7}$$

$$\frac{d}{dz}[f(z)g(z)] = f(z)g'(z) + g(z)f'(z) \tag{3.8}$$

$$\frac{d}{dz}\frac{f(z)}{g(z)} = \frac{g(z)f'(z) - f(z)g'(z)}{[g(z)]^2} \quad \text{if} \quad g(z) \neq 0, \quad \text{and} \tag{3.9}$$

$$\frac{d}{dz} f\big(g(z)\big) = f'\big(g(z)\big)g'(z). \tag{3.10}$$

Important particular cases of Equations (3.9) and (3.10), respectively, are

$$\frac{d}{dz}\frac{1}{z^n} = \frac{-n}{z^{n+1}}, \quad \text{for } z \neq 0, \quad n \quad \text{a positive integer, and} \tag{3.11}$$

$$\frac{d}{dz}[f(z)]^n = n[f(z)]^{n-1}f'(z), \quad n \quad \text{a positive integer.} \tag{3.12}$$

Example 3.3. If we use Equation (3.12) with $f(z) = z^2 + i2z + 3$ and $f'(z) = 2z + 2i$, then we get

$$\frac{d}{dz}(z^2 + i2z + 3)^4 = 8(z^2 + i2z + 3)^3(z + i).$$

81

The proofs of the rules given in Equations (3.4) through (3.10) depend on the validity of extending theorems for real functions to their complex companions. Equation (3.8), for example, relies on Theorem 3.1.

Theorem 3.1. *If f is differentiable at z_0, then f is continuous at z_0.*

Proof. From Equation (3.1), we obtain

$$\lim_{z \to z_0} \frac{f(z) - f(z_0)}{z - z_0} = f'(z_0).$$

Using the multiplicative property of limits given by Formula (2.19), we get

$$\lim_{z \to z_0} [f(z) - f(z_0)] = \lim_{z \to z_0} \frac{f(z) - f(z_0)}{z - z_0} (z - z_0)$$
$$= \lim_{z \to z_0} \frac{f(z) - f(z_0)}{z - z_0} \lim_{z \to z_0} (z - z_0)$$
$$= f'(z_0) \cdot 0 = 0.$$

This result implies that $\lim_{z \to z_0} f(z) = f(z_0)$, which is equivalent to showing that f is continuous at z_0. $\qquad\square$

We can establish Equation (3.8) from Theorem 3.1. Letting $h(z) = f(z)g(z)$ and using Definition 3.1, we write

$$h'(z_0) = \lim_{z \to z_0} \frac{h(z) - h(z_0)}{z - z_0} = \lim_{z \to z_0} \frac{f(z)g(z) - f(z_0)g(z_0)}{z - z_0}.$$

If we subtract and add the term $f(z_0)g(z)$ in the numerator, we get

$$h'(z_0) = \lim_{z \to z_0} \frac{f(z)g(z) - f(z_0)g(z) + f(z_0)g(z) - f(z_0)g(z_0)}{z - z_0}$$
$$= \lim_{z \to z_0} \frac{f(z)g(z) - f(z_0)g(z)}{z - z_0} + \lim_{z \to z_0} \frac{f(z_0)g(z) - f(z_0)g(z_0)}{z - z_0}$$
$$= \lim_{z \to z_0} \frac{f(z) - f(z_0)}{z - z_0} \lim_{z \to z_0} g(z) + f(z_0) \lim_{z \to z_0} \frac{g(z) - g(z_0)}{z - z_0}.$$

Using the definition of the derivative given by Equation (3.1) and the continuity of g, we obtain $h'(z_0) = f'(z_0)g(z_0) + f(z_0)g'(z_0)$, which is what we wanted to establish. We leave the proofs of the other rules as exercises.

The rule for differentiating polynomials carries over to the complex case as well. If we let $P(z)$ be a polynomial of degree n, so

$$P(z) = a_0 + a_1 z + a_2 z^2 + \cdots + a_n z^n.$$

then mathematical induction, along with Equations (3.5) and (3.7), gives

$$P'(z) = a_1 + 2a_2 z + 3a_3 z^2 + \cdots + n a_n z^{n-1}.$$

Again, we leave the proof of this result as an exercise.

We can use the differentiation rules as aids in determining when functions are analytic. For example, Equation (3.9) tells us that if $P(z)$ and $Q(z)$ are polynomials, then their quotient $\frac{P(z)}{Q(z)}$

is analytic at all points where $Q(z) \neq 0$. This condition implies that the function $f(z) = \frac{1}{z}$ is analytic for all $z \neq 0$. The square root function is more complicated. If $f(z) = z^{\frac{1}{2}} = |z|^{\frac{1}{2}} e^{i \frac{\text{Arg}(z)}{2}}$, then f is analytic at all points except $z = 0$ (because $\text{Arg}(0)$ is undefined) and at points that lie along the negative x axis. The argument function, and therefore the function f itself, are not continuous at points that lie along the negative x axis.

We close this section with a complex extension of a famous theorem, the proof of which will be given in Chapter 7.

Theorem 3.2 (L'Hôpital's rule). *Assume that f and g are analytic at z_0. If $f(z_0) = 0$, $g(z_0) = 0$, and $g'(z_0) \neq 0$, then*

$$\lim_{z \to z_0} \frac{f(z)}{g(z)} = \lim_{z \to z_0} \frac{f'(z)}{g'(z)}.$$

Proof. See Corollary 7.11. $\qquad\qquad\qquad\qquad\qquad\qquad\qquad\qquad\qquad\qquad\qquad\qquad$ $\square$

Exercises for Section 3.1 (Selected answers or hints are on page 435.)

1. Find the derivatives of the following functions.

   (a) $g(z) = (z^2 - iz + 9)^5$.

   (b) $h(z) = \frac{2z+1}{z+2}$ for $z \neq -2$.

   (c) $F(z) = (z^2 + (1 - 3i)z + 1)(z^4 + 3z^2 + 5i)$.

2. Show that the following functions are differentiable nowhere.

   (a) $f(z) = \text{Re}(z)$.

   (b) $f(z) = \text{Im}(z)$.

3. If f and g are entire functions, which of the following are necessarily entire?

   (a) $[f(z)]^3$.

   (b) $f(z)g(z)$.

   (c) $\frac{f(z)}{g(z)}$.

   (d) $f(\frac{1}{z})$.

   (e) $f(z - 1)$.

   (f) $f(g(z))$.

4. Use Equation (3.1) to verify Rule (3.5).

5. Let $P(z) = a_0 + a_1 z + \cdots + a_n z^n$ be a polynomial of degree $n \geq 1$.

   (a) Show that $P'(z) = a_1 + 2a_2 z + \cdots + na_n z^{n-1}$.

   (b) Show that, for $k = 0, 1, \ldots, n$, $a_k = \frac{P^{(k)}(0)}{k!}$, where $P^{(k)}$ denotes the kth derivative of P. (By convention, $P^{(0)}(z) = P(z)$.)

83

6. Let P be a polynomial of degree 2, given by

$$P(z) = (z - z_1)(z - z_2).$$

where $z_1 \neq z_2$. Show that

$$\frac{P'(z)}{P(z)} = \frac{1}{z - z_1} + \frac{1}{z - z_2}.$$

Note: The quotient $\frac{P'(z)}{P(z)}$ is known as the *logarithmic derivative* of P.

7. Use L'Hôpital's rule to find the following limits.

(a) $\lim\limits_{z \to i} \frac{z^4 - 1}{z - i}$.

(b) $\lim\limits_{z \to 1+i} \frac{z^2 - iz - 1 - i}{z^2 - 2z + 2}$.

(c) $\lim\limits_{z \to -i} \frac{z^6 + 1}{z^2 + 1}$.

(d) $\lim\limits_{z \to 1+i} \frac{z^4 + 4}{z^2 - 2z + 2}$.

(e) $\lim\limits_{z \to 1+i\sqrt{3}} \frac{z^6 - 64}{z^3 + 8}$.

(f) $\lim\limits_{z \to -1+i\sqrt{3}} \frac{z^9 - 512}{z^3 - 8}$.

8. Use Equation (3.1) to show that $\frac{d}{dz} \frac{1}{z} = -\frac{1}{z^2}$.

9. Show that $\frac{d}{dz} z^{-n} = -nz^{-n-1}$, where n is a positive integer.

10. Verify the identity.

$$\frac{d}{dz} f(z)g(z)h(z) = f'(z)g(z)h(z) + f(z)g'(z)h(z) + f(z)g(z)h'(z).$$

11. Show that the function $f(z) = |z|^2$ is differentiable only at the point $z_0 = 0$. *Hint*: To show that f is *not* differentiable at $z_0 \neq 0$, choose horizontal and vertical lines through the point z_0 and show that $\frac{\Delta w}{\Delta z}$ approaches two distinct values as $\Delta z \to 0$ along those two lines.

12. Verify the following identities:

(a) $\frac{d}{dz} C = 0$, where C is a constant: Identity (3.4).

(b) $\frac{d}{dz}[f(z) + g(z)] = f'(z) + g'(z)$: Identity (3.7).

(c) $\frac{d}{dz} \frac{f(z)}{g(z)} = \frac{g(z)f'(z) - f(z)g'(z)}{[g(z)]^2}$ if $g(z) \neq 0$: Identity (3.9).

(d) $\frac{d}{dz} f(g(z)) = f'(g(z))g'(z)$: Identity (3.10).

(e) $\frac{d}{dz}[f(z)]^n = n[f(z)]^{n-1} f'(z)$, where n is a positive integer: Identity (3.12).

13. Consider the differentiable function $f(z) = z^3$ and the two points $z_1 = 1$ and $z_2 = i$. Show that there does not exist a point c on the line $y = 1 - x$ between 1 and i such that $\frac{f(z_2) - f(z_1)}{z_2 - z_1} = f'(c)$. This result shows that the mean value theorem for derivatives does not extend to complex functions.

14. Let $f(z) = z^{\frac{1}{n}}$ denote the multivalued "nth root function," where n is a positive integer. Use the chain rule to show that, if $g(z)$ is any branch of the nth root function, then

$$g'(z) = \frac{1}{n}\frac{g(z)}{z}$$

in some suitably chosen domain (which you should specify).

15. Explain why the composition of two entire functions is an entire function.

16. Let f be differentiable at z_0. Show that there exists a function $\eta(z)$ such that

$$f(z) = f(z_0) + f'(z_0)(z - z_0) + \eta(z)(z - z_0),$$

where $\eta(z) \to 0$ as $z \to z_0$.

3.2 The Cauchy-Riemann Equations

In Section 3.1 we showed that computing the derivative of complex functions written in a form such as $f(z) = z^2$ is a rather simple task. But life isn't always so easy. Many times we encounter complex functions written in the form of $f(z) = f(x, y) = u(x, y) + iv(x, y)$. For example, suppose we had

$$f(z) = f(x, y) = u(x, y) + iv(x, y) = (x^3 - 3xy^2) + i(3x^2y - y^3). \tag{3.13}$$

Is there some criterion that we can use to determine whether f is differentiable, and if so, to find the value of $f'(z)$?

The answer to this question is *yes*, thanks to the independent discovery of two important equations by the French mathematician Augustin-Louis Cauchy[1] and the German mathematician Georg Friedrich Bernhard Riemann.

First, let's reconsider the derivative of $f(z) = z^2$. As we have stated, the limit given in Equation (3.1) must *not* depend on how z approaches z_0. We investigate two such approaches: a horizontal approach and a vertical approach to z_0. Recall from our graphical analysis of $w = z^2$ that the image of a square is a "curvilinear quadrilateral." For convenience, we let the square have vertices $z_0 = 2 + i$, $z_1 = 2.01 + i$, $z_2 = 2 + 1.01i$, and $z_3 = 2.01 + 1.01i$. Then the image points are $w_0 = 3 + 4i$, $w_1 = 3.0401 + 4.02i$, $w_2 = 2.9799 + 4.04i$, and $w_3 = 3.02 + 4.0602i$, as shown in Figure 3.1.

We know that f is differentiable, so the limit of the difference quotient $\frac{f(z)-f(z_0)}{z-z_0}$ exists no matter how we approach $z_0 = 2 + i$. Thus we can *approximate* $f'(2 + i)$ by using horizontal or vertical increments in z:

$$f'(2 + i) \approx \frac{f(2.01 + i) - f(2 + i)}{(2.01 + i) - (2 + i)} = \frac{0.0401 + 0.02i}{0.01} = 4.01 + 2i$$

and

$$f'(2 + i) \approx \frac{f(2 + 1.01i) - f(2 + i)}{(2 + 1.01i) - (2 + i)} = \frac{-0.0201 + 0.04i}{0.01i} = 4 + 2.01i.$$

---

[1] A.L. Cauchy (1789–1857) played a prominent role in the development of complex analysis, and his name appears often throughout this text. The last name is *not* pronounced as "kaushee." The first syllable has a long "o" sound, like the word kosher, but with the second syllable having a long "e" instead of "er" at the end. Thus, we pronounce Cauchy as "kōshē." In mathematical circles it is not kosher to mispronounce Cauchy!

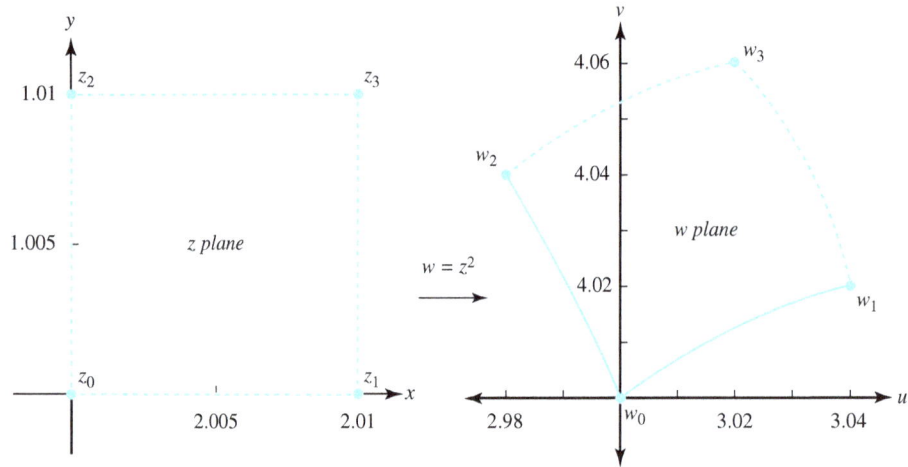

Figure 3.1: The image of a small square with vertex $z_0 = 2 + i$, using $w = z^2$

These computations lead to the idea of taking limits along the horizontal and vertical directions. When we do so, we get

$$f'(2+i) = \lim_{h \to 0} \frac{f(2+h+i) - f(2+i)}{h} = \lim_{h \to 0} \frac{4h + h^2 + i2h}{h} = 4 + 2i$$

and

$$f'(2+i) = \lim_{h \to 0} \frac{f(2+i+ih) - f(2+i)}{ih} = \lim_{h \to 0} \frac{-2h - h^2 + i4h}{ih} = 4 + 2i.$$

We now generalize this idea by taking limits of an arbitrary differentiable complex function and obtain an important result.

Theorem 3.3 (Cauchy-Riemann equations). *Suppose that*

$$f(z) = f(x + iy) = u(x, y) + iv(x, y)$$

is differentiable at the point $z_0 = x_0 + iy_0$. Then the partial derivatives of u and v exist at the point (x_0, y_0), and

$$f'(z_0) = u_x(x_0, y_0) + iv_x(x_0, y_0) \quad \text{and also} \tag{3.14}$$
$$f'(z_0) = v_y(x_0, y_0) - iu_y(x_0, y_0). \tag{3.15}$$

Equating the real and imaginary parts of Equations (3.14) *and* (3.15) *gives*

$$u_x(x_0, y_0) = v_y(x_0, y_0) \quad and \quad u_y(x_0, y_0) = -v_x(x_0, y_0). \tag{3.16}$$

Proof. Because f is differentiable, we know that $\lim_{z \to z_0} \left(\frac{f(z) - f(z_0)}{z - z_0} \right)$ exists regardless of the path we take as $z \to z_0$. We will choose horizontal and vertical lines that pass through the point $z_0 = (x_0, y_0)$ and compute the limiting values of $\frac{f(z) - f(z_0)}{(z - z_0)}$ along these lines. Equating the two resulting limits will yield Equations (3.16). For the horizontal approach to z_0, we set $z = x + iy_0$

and obtain

$$f'(z_0) = \lim_{(x,y_0)\to(x_0,y_0)} \frac{f(x+iy_0) - f(x_0+iy_0)}{x+iy_0 - (x_0+iy_0)}$$

$$= \lim_{x\to x_0} \frac{u(x,y_0) - u(x_0,y_0) + i[v(x,y_0) - v(x_0,y_0)]}{x-x_0}$$

$$= \lim_{x\to x_0} \frac{u(x,y_0) - u(x_0,y_0)}{x-x_0} + i\lim_{x\to x_0} \frac{v(x,y_0) - v(x_0,y_0)}{x-x_0}.$$

The last two limits are the partial derivatives of u and v with respect to x, so

$$f'(z_0) = u_x(x_0,y_0) + iv_x(x_0,y_0).$$

giving us Equation (3.14).

Along the vertical approach to z_0, we have $z = x_0 + iy$, so

$$f'(z_0) = \lim_{(x_0,y)\to(x_0,y_0)} \frac{f(x_0+iy) - f(x_0+iy_0)}{x_0+iy - (x_0+iy_0)}$$

$$= \lim_{y\to y_0} \frac{u(x_0,y) - u(x_0,y_0) + i[v(x_0,y) - v(x_0,y_0)]}{i(y-y_0)}$$

$$= \lim_{y\to y_0} \frac{v(x_0,y) - v(x_0,y_0)}{y-y_0} - i\lim_{y\to y_0} \frac{u(x_0,y) - u(x_0,y_0)}{y-y_0}.$$

The last two limits are the partial derivatives of u and v with respect to y, so

$$f'(z_0) = v_y(x_0,y_0) - iu_y(x_0,y_0).$$

giving us Equation (3.15).

Since f is differentiable at z_0, the limits given by Equations (3.14) and (3.15) must be equal. If we equate the real and imaginary parts in those equations, the result is Equations (3.16), and the proof is complete. □

Note some of the important implications of this theorem.

- If f is differentiable at z_0, then the Cauchy-Riemann Equations (3.16) will be satisfied at z_0, and we can use either Equation (3.14) or (3.15) to evaluate $f'(z_0)$.

- Taking the contrapositive, if Equations (3.16) are *not* satisfied at z_0, then we know automatically that f *is not* differentiable at z_0.

- If Equations (3.16) *are* satisfied at z_0, however, we cannot *necessarily* conclude that f is differentiable at z_0.

We now illustrate each of these points.

Example 3.4. We know that $f(z) = z^2$ is differentiable and that $f'(z) = 2z$. We also have

$$f(z) = z^2 = (x+iy)^2 = (x^2 - y^2) + i(2xy) = u(x,y) + iv(x,y).$$

It is easy to verify that Equations (3.16) are indeed satisfied:

$$u_x(x,y) = 2x = v_y(x,y) \quad \text{and} \quad u_y(x,y) = -2y = -v_x(x,y).$$

Using Equations (3.14) and (3.15), respectively, to compute $f'(z)$ gives

$$f'(z) = u_x(x, y) + iv_x(x, y) = 2x + i2y = 2z, \quad \text{and}$$
$$f'(z) = v_y(x, y) - iu_y(x, y) = 2x - i(-2y) = 2x + i2y = 2z,$$

as expected.

Example 3.5. Show that $f(z) = \bar{z}$ is nowhere differentiable.

Solution:

We have $f(z) = f(x + iy) = x - iy = u(x, y) + iv(x, y)$, where $u(x, y) = x$ and $v(x, y) = -y$. Thus, for any point (x, y), $u_x(x, y) = 1$ and $v_y(x, y) = -1$. The Cauchy-Riemann equations are not satisfied at any point $z = (x, y)$, so we conclude that f is nowhere differentiable.

Example 3.6. Show that the function defined by

$$f(x) = \begin{cases} \frac{(\bar{z})^2}{z} = \frac{x^3 - 3xy^2}{x^2 + y^2} + i\frac{y^3 - 3x^2y}{x^2 + y^2} & \text{when} \quad z \neq 0, \quad \text{and} \\ 0 \quad \text{when} \quad z = 0 \end{cases}$$

is *not* differentiable at the point $z_0 = 0$ even though the Cauchy-Riemann equations are satisfied at $(0, 0)$.

Solution:

We must use limits to calculate the partial derivatives at $(0, 0)$.

$$u_x(0, 0) = \lim_{x \to 0} \frac{u(x, 0) - u(0, 0)}{x - 0} = \lim_{x \to 0} \frac{\frac{x^3 - 0}{x^2 + 0}}{x} = 1.$$

Similarly, we can show that

$$u_y(0, 0) = v_x(0, 0) = 0 \quad \text{and} \quad v_y(0, 0) = 1.$$

Hence the Cauchy-Riemann equations hold at the point $(0, 0)$.

We now show that f is *not* differentiable at $z_0 = 0$. Letting z approach 0 along the x axis gives

$$\lim_{(x,0) \to (0,0)} \frac{f(x + 0i) - f(0)}{x + 0i - 0} = \lim_{x \to 0} \frac{\frac{x^2}{x} - 0}{x - 0} = \lim_{x \to 0} \frac{x - 0}{x - 0} = 1.$$

But if we let z approach 0 along the line $y = x$ given by the parametric equations $x = t$ and $y = t$, then

$$\lim_{(t,t) \to (0,0)} \frac{f(t + it) - f(0)}{t + it - 0} = \lim_{t \to 0} \frac{-\frac{2t^3}{2t^2} + i\left(-\frac{2t^3}{2t^2}\right)}{t + it} = \lim_{t \to 0} \frac{-t - it}{t + it} = -1.$$

The two limits are distinct, so f is not differentiable at the origin.

Example 3.6 reiterates that the mere satisfaction of the Cauchy-Riemann equations is not sufficient to guarantee the differentiability of a function. The following theorem, however, gives conditions that guarantee the differentiability of f at z_0, so that which we can use either Equation (3.14) or (3.15) to compute $f'(z_0)$. They are referred to as the *Cauchy-Riemann conditions* for differentiability.

Theorem 3.4 (Cauchy-Riemann conditions for differentiability). *Let $f(z) = u(x, y) + iv(x, y)$ be a continuous function that is defined in some neighborhood of the point $z_0 = x_0 + iy_0$. If all the partial derivatives u_x, u_y, v_x and v_y are continuous at the point (x_0, y_0), and if the Cauchy-Riemann equations $u_x(x_0, y_0) = v_y(x_0, y_0)$ and $u_y(x_0, y_0) = -v_x(x_0, y_0)$ hold at (x_0, y_0), then f is differentiable at z_0, and $f'(z_0)$ can be computed with either Equation* (3.14) *or* (3.15).

Proof. Let $\Delta z = \Delta x + i\Delta y$ and $\Delta w = \Delta u + i\Delta v$, and let Δz be small enough so that z lies in the ε-neighborhood of z_0 in which the hypotheses hold. We need to show that $\frac{\Delta w}{\Delta z}$ approaches the limit given in Equation (3.15) as Δz approaches zero. We write the difference, Δu, as

$$\Delta u = u(x_0 + \Delta x, y_0 + \Delta y) - u(x_0, y_0).$$

If we subtract and add the term $u(x_0, y_0 + \Delta y)$, then we get

$$\begin{aligned} \Delta u = &[u(x_0 + \Delta x, y_0 + \Delta y) - u(x_0, y_0 + \Delta y)] \\ &+ [u(x_0, y_0 + \Delta y) - u(x_0, y_0)]. \end{aligned} \tag{3.17}$$

The partial derivatives u_x and u_y exist, so the mean value theorem for real functions of two variables implies that a value x^* exists between x_0 and $x_0 + \Delta x$ such that we can write the first term in brackets on the right side of Equation (3.17) as

$$u(x_0 + \Delta x, y_0 + \Delta y) - u(x_0, y_0 + \Delta y) = u_x(x^*, y_0 + \Delta y)\Delta x.$$

Furthermore, as u_x and u_y are continuous at (x_0, y_0), there exists a quantity ε_1 such that

$$u_x(x^*, y_0 + \Delta y) = u_x(x_0, y_0) + \varepsilon_1.$$

where $\varepsilon_1 \to 0$ as $x^* \to x_0$ and $\Delta y \to 0$. Because $\Delta x \to 0$ forces $x^* \to x_0$, we can use the equation

$$u(x_0 + \Delta x, y_0 + \Delta y) - u(x_0, y_0 + \Delta y) = [u_x(x_0, y_0) + \varepsilon_1]\Delta x. \tag{3.18}$$

the second term in brackets on the right side of Equation (3.17) satisfies the equation

$$u(x_0, y_0 + \Delta y) - u(x_0, y_0) = [u_y(x_0, y_0) + \varepsilon_2]\Delta y. \tag{3.19}$$

where $\varepsilon_2 \to 0$ as $\Delta x \to 0$ and $\Delta y \to 0$. Combining Equations (3.18) and (3.19) gives

$$\Delta u = (u_x + \varepsilon_1)\Delta x + (u_y + \varepsilon_2)\Delta y.$$

where partial derivatives u_x and u_y are evaluated at the point (x_0, y_0) and ε_1 and ε_2 tend to zero as Δx and Δy both tend to zero. Similarly, the change Δv is related to the changes Δx and Δy by the equation

$$\Delta v = (v_x + \varepsilon_3)\Delta x + (v_y + \varepsilon_4)\Delta y,$$

where the partial derivatives v_x and v_y are evaluated at the point (x_0, y_0) and ε_3 and ε_4 tend to zero as Δx and Δy both tend to zero. Combining these last two equations gives

$$\Delta w = u_x\Delta x + u_y\Delta y + i(v_x\Delta x + v_y\Delta y) + \varepsilon_1\Delta x + \varepsilon_2\Delta y + i(\varepsilon_3\Delta x + \varepsilon_4\Delta y). \tag{3.20}$$

We can use the Cauchy-Riemann equations in Equation (3.20) to obtain

$$\Delta w = u_x\Delta x - v_x\Delta y + i(v_x\Delta x + u_x\Delta y) + \varepsilon_1\Delta x + \varepsilon_2\Delta y + i(\varepsilon_3\Delta x + \varepsilon_4\Delta y).$$

Now we rearrange the terms and get

$$\Delta w = u_x\left[\Delta x + i\Delta y\right] + iv_x[\Delta x + i\Delta y] + \varepsilon_1\Delta x + \varepsilon_2\Delta y + i(\varepsilon_3\Delta x + \varepsilon_4\Delta y).$$

Since $\Delta z = \Delta x + i\Delta y$, we can divide both sides of this equation by Δz and take the limit as $\Delta z \to 0$:

$$\lim_{\Delta z \to 0}\frac{\Delta w}{\Delta z} = u_x + iv_x + \lim_{\Delta z \to 0}\left[\frac{\varepsilon_1\Delta x}{\Delta z} + \frac{\varepsilon_2\Delta y}{\Delta z} + i\frac{\varepsilon_3\Delta x}{\Delta z} + i\frac{\varepsilon_4\Delta y}{\Delta z}\right]. \tag{3.21}$$

Because ε_1 tends to zero as Δx and Δy both tend to zero, we have

$$\lim_{\Delta z \to 0}\left|\frac{\varepsilon_1\Delta x}{\Delta z}\right| = \lim_{\Delta z \to 0}|\varepsilon_1|\left|\frac{\Delta x}{\Delta z}\right| \leq \lim_{\Delta z \to 0}|\varepsilon_1| = 0.$$

Similarly, the limits of the other quantities in Equation (3.21) involving ε_2, ε_3, ε_4 are zero. Therefore the limit in Equation (3.21) becomes

$$\lim_{\Delta z \to 0}\frac{\Delta w}{\Delta z} = f'(z_0) = u_x(x_0, y_0) + iv_x(x_0, y_0),$$

which completes the proof of the theorem. $\qquad\square$

Example 3.7. At the beginning of this section (Equation (3.13)) we defined the function $f(z) = u(x, y) + iv(x, y) = x^3 - 3xy^2 + i(3x^2y - y^3)$. Show that this function is differentiable for all z, and find its derivative.

Solution:

We compute $u_x(x, y) = v_y(x, y) = 3x^2 - 3y^2$ and $u_y(x, y) = -6xy = -v_x(x, y)$, so the Cauchy-Riemann Equations (3.16) are satisfied. Moreover, u, v, u_x, u_y, v_x, and v_y are continuous everywhere. By Theorem 3.4, f is differentiable everywhere, and Equation (3.14) gives

$$f'(z) = u_x(x, y) + iv_x(x, y) = (3x^2 - 3y^2) + i6xy = 3(x^2 - y^2 + i2xy) = 3z^2.$$

Alternatively, from Equation (3.15),

$$f'(z) = v_y(x, y) - iu_y(x, y) = (3x^2 - 3y^2) - i(-6xy) = 3(x^2 - y^2 + i2xy) = 3z^2.$$

This result isn't surprising because $(x + iy)^3 = x^3 - 3xy^2 + i(3x^2y - y^3)$ and so the function f is really our old friend $f(z) = z^3$.

Example 3.8. Show that the function $f(z) = e^{-y}\cos x + ie^{-y}\sin x$ is differentiable for all z and find its derivative.

Solution:

We first write $u(x, y) = e^{-y}\cos x$ and $v(x, y) = e^{-y}\sin x$ and then compute the partial derivatives.

$$u_x(x, y) = v_y(x, y) = -e^{-y}\sin x \quad \text{and}$$
$$v_x(x, y) = -u_y(x, y) = e^{-y}\cos x.$$

We note that u, v, u_x, u_v, v_x, and v_y are continuous functions and that the Cauchy-Riemann equations hold for all values of (x, y). Hence, using Equation (3.14), we write

$$f'(z) = f'(x, y) = u_x(x, y) + iv_x(x, y) = -e^{-y}\sin x + ie^{-y}\cos x.$$

The Cauchy-Riemann conditions are particularly useful in determining the set of points for which a function f is differentiable.

Example 3.9. Show that the function $f(z) = x^3 + 3xy^2 + i(y^3 + 3x^2y)$ is differentiable on the x and y axes, but analytic nowhere.

Solution:

Recall (Definition 3.1) that when we say a function is analytic at a point z_0 we mean that the function is differentiable not only at z_0, but also at every point in some ε-neighborhood of z_0. With this in mind, we proceed to determine where the Cauchy-Riemann equations are satisfied. We write $u(x, y) = x^3 + 3xy^2$ and $v(x, y) = y^3 + 3x^2y$ and compute the partial derivatives:

$$u_x(x, y) = 3x^2 + 3y^2, \; v_y(x, y) = 3x^2 + 3y^2, \quad \text{and}$$
$$u_y(x, y) = 6xy, \; v_x(x, y) = 6xy.$$

Here u, v, u_x, u_y, and v are continuous, and $u_x(x, y) = v_y(x, y)$ holds for all (x, y). But $u_y(x, y) = -v_x(x, y)$ iff $6xy = -6xy$, which is equivalent to $12xy = 0$. The Cauchy-Riemann equations hold only when $x = 0$ or $y = 0$, and according to Theorem 3.4, f is differentiable only at points that lie on the coordinate axes. But this means that f is nowhere analytic because any ε-neighborhood about a point on either axis contains points that are not on those axes.

When polar coordinates (r, θ) are used to locate points in the plane, we use Expression 2.2 for a complex function for convenience; that is,

$$f(z) = u(x, y) + iv(x, y)$$
$$f(re^{i\theta}) = u(re^{i\theta}) + iv(re^{i\theta})$$
$$= U(r, \theta) + iV(r, \theta).$$

where U and V are real functions of the real variables r and θ. The polar form of the Cauchy-Riemann equations and a formula for finding $f'(z)$ in terms of the partial derivatives of $U(r, \theta)$ and $V(r, \theta)$ are given in Theorem 3.5, which we ask you to prove in Exercise 10. This theorem makes use of the validity of the Cauchy-Riemann equations for the functions u and v, so the relation between them and the functions U and V—namely, $u(x, y) = u(re^{i\theta}) = U(r, \theta)$ and $v(x, y) = v(re^{i\theta}) = V(r, \theta)$—is important.

Theorem 3.5 (Polar form). *Let $f(z) = f(re^{i\theta}) = U(r, \theta) + iV(r, \theta)$ be a continuous function that is defined in some neighborhood of the point $z_0 = r_0e^{i\theta_0}$. If all the partial derivatives U_r, U_θ, V_r and V_θ are continuous at the point (r_0, θ_0) and if the polar form of the Cauchy-Riemann equations,*

$$U_r(r_0, \theta_0) = \frac{1}{r_0}V_\theta(r_0, \theta_0) \quad \text{and} \quad V_r(r_0, \theta_0) = -\frac{1}{r_0}U_\theta(r_0, \theta_0), \tag{3.22}$$

holds, then f is differentiable at z_0 and we can compute the derivative $f'(z_0)$ by using either of the following formulas:

$$f'(z_0) = f'(re^{i\theta_0}) = e^{-i\theta_0}[U_r(r_0, \theta_0) + iV_r(r_0, \theta_0)], \quad \text{or} \tag{3.23}$$

$$f'(z_0) = f'(re^{i\theta_0}) = \frac{1}{r_0}e^{-i\theta_0}[V_\theta(r_0, \theta_0) - iU_\theta(r_0, \theta_0)]. \tag{3.24}$$

Example 3.10. Show that, if f is given by

$$f(re^{i\theta}) = f(z) = z^{\frac{1}{2}} = r^{\frac{1}{2}}\cos\frac{\theta}{2} + ir^{\frac{1}{2}}\sin\frac{\theta}{2}.$$

where the domain is restricted to be $\{re^{i\theta} : r > 0 \text{ and } -\pi < \theta < \pi\}$, then the derivative is given by

$$f'(z) = \frac{1}{2z^{\frac{1}{2}}} = \frac{1}{2}r^{-\frac{1}{2}}\cos\frac{\theta}{2} - i\frac{1}{2}r^{-\frac{1}{2}}\sin\frac{\theta}{2},$$

for every point in the domain.

Solution:

Write

$$U(r,\theta) = r^{\frac{1}{2}}\cos\frac{\theta}{2} \quad \text{and} \quad V(r,\theta) = r^{\frac{1}{2}}\sin\frac{\theta}{2}.$$

Then

$$U_r(r,\theta) = \frac{1}{r}V_\theta(r,\theta) = \frac{1}{2}r^{-\frac{1}{2}}\cos\frac{\theta}{2}, \quad \text{and}$$

$$V_r(r,\theta) = -\frac{1}{r}U_\theta(r,\theta) = \frac{1}{2}r^{-\frac{1}{2}}\sin\frac{\theta}{2}.$$

Since U, V, U_r, U_θ, V_r, and V_θ are continuous at every point in the domain (note the strict inequality in $-\pi < \theta < \pi$), we use Theorem 3.5 and Equation (3.23) to get

$$f'(z) = e^{-i\theta}\left(\frac{1}{2}r^{-\frac{1}{2}}\cos\frac{\theta}{2} + i\frac{1}{2}r^{-\frac{1}{2}}\sin\frac{\theta}{2}\right)$$

$$= e^{-i\theta}\left(\frac{1}{2}r^{-\frac{1}{2}}e^{i\frac{\theta}{2}}\right) = \frac{1}{2}r^{-\frac{1}{2}}e^{-i\frac{\theta}{2}} = \frac{1}{2z^{\frac{1}{2}}}.$$

Note that $f(z)$ is discontinuous on the negative real axis and is undefined at the origin. Using the terminology of Section 2.4, the negative real axis is a branch cut, and the origin is a branch point for this function.

Two important consequences of the Cauchy-Riemann equations close this section.

Theorem 3.6. *Let $f = u + iv$ be an analytic function on the domain D. Suppose for all $z \in D$ that $|f(z)| = K$, where K is a constant. Then f is constant on D.*

Proof. The equation $|f(z)| = K$ implies that, for all $z = (x, y) \in D$,

$$u(x,y)^2 + v(x,y)^2 = K^2. \tag{3.25}$$

If $K = 0$, then it must be that $u(x,y)^2 = 0$ and $v(x,y)^2 = 0$ for all $(x,y) \in D$, so f is identically zero on D. If $K \neq 0$, then we take the partial derivative of both sides of Equation (3.25) with respect to both x and y, resulting in

$$2uu_x + 2vv_x = 0 \quad \text{and} \quad 2uu_y + 2vv_y = 0.$$

where for brevity we write u in place of $u(x,y)$, and so on. We can now use the Cauchy-Riemann equations to rewrite this system as

$$uu_x - vu_y = 0 \quad \text{and} \quad vu_x + uu_y = 0.$$

Treating u and v as coefficients, we have two equations with two unknowns, u_x and u_y. Solving for u_x and u_y gives

$$u_x = \frac{0}{u^2 + v^2} = 0 \quad \text{and} \quad u_y = \frac{0}{u^2 + v^2} = 0.$$

92

Note that it is important here for $K \neq 0$ in Equation (3.25).

A theorem from the calculus of real functions states that, if for all $(x, y) \in D$ we have both $u_x(x, y) = 0$ and $u_y(x, y) = 0$, then for all $(x, y) \in D$, $u(x, y) = c_1$, where c_1 is a constant. Using a similar argument, we find that $v(x, y) = c_2$, for all $(x, y) \in D$, and therefore $f(z) = f(x, y) = c_1 + ic_2$, for all $(x, y) \in D$. In other words, f is constant on D. $\qquad \square$

Theorem 3.7. *Let f be an analytic function in the domain D. If $f'(z) = 0$ for all z in D, then f is constant in D.*

Proof. By the Cauchy-Riemann equations, $f'(z) = u_x(z) + iv_x(z) = v_y(z) - iu_y(z)$ for all $z \in D$. By hypothesis $f'(z) = 0$ for all $z \in D$, so for all $z \in D$ the functions u_x, u_y, v_x, and v_y are identically zero. As with the conclusion to the proof of Theorem 3.6, this situation means both u and v are constant functions, from whence the result follows. $\qquad \square$

Exercises for Section 3.2 (Selected answers or hints are on page 436.)

1. Use the Cauchy Riemann conditions to determine where the following functions are differentiable, and evaluate the derivatives at those points where they exist.

   (a) $f(z) = iz + 4i$.

   (b) $f(z) = f(x, y) = \frac{y + ix}{x^2 + y^2}$.

   (c) $f(z) = -2(xy + x) + i(x^2 - 2y - y^2)$.

   (d) $f(z) = x^3 - 3x^2 - 3xy^2 + 3y^2 + i(3x^2y - 6xy - y^3)$.

   (e) $f(z) = x^3 + i(1 - y)^3$.

   (f) $f(z) = z^2 + z$.

   (g) $f(z) = x^2 + y^2 + i2xy$.

   (h) $f(z) = |z - (2 + i)|^2$.

2. Let f be a differentiable function. Verify the identity $|f'(z)|^2 = u_x^2 + v_x^2 = u_y^2 + v_y^2$.

3. Find the constants a and b so that $f(z) = (2x - y) + i(ax + by)$ is differentiable for all z.

4. Let f be differentiable at $z_0 = r_0 e^{i\theta_0}$. Let z approach z_0 along the ray $r > 0$, $\theta = \theta_0$ and use Equation (3.1) to show that Equation (3.14) holds.

5. Let $f(z) = e^x \cos y + ie^x \sin y$. Show that both $f(z)$ and $f'(z)$ are differentiable for all z.

6. A vector field $\mathbf{F}(z) = U(x, y) + iV(x, y)$ is said to be *irrotational* if $U_y(x, y) = V_x(x, y)$. It is said to be *solenoidal* if $U_x(x, y) = -V_y(x, y)$. If $f(z)$ is an analytic function, show that $\mathbf{F}(z) = \overline{f(z)}$ is both irrotational and solenoidal.

7. Use any method to show that the following functions are nowhere differentiable.

   (a) $h(z) = e^y \cos x + ie^y \sin x$.

   (b) $g(z) = z + \overline{z}$.

8. Use Theorem 3.5 with regard to the following.

(a) Let $f(z) = f(re^{i\theta}) = \ln r + i\theta$, where $r > 0$ and $-\pi < \theta < \pi$. Show that f is analytic in the domain indicated and that $f'(z) = \frac{1}{z}$.

(b) Let $f(z) = (\ln r)^2 - \theta^2 + i2\theta \ln r$ where, $r > 0$ and $-\pi < \theta \leq \pi$. Show that f is analytic for $r > 0$, $-\pi < \theta < \pi$, and find $f'(z)$.

9. Show that the following functions are entire (see Definition 3.2).

(a) $f(z) = \cosh x \sin y - i \sinh x \cos y.p$

(b) $g(z) = \cosh x \cos y + i \sinh x \sin y$.

10. To prove Theorem 3.5, the polar form of the Cauchy-Riemann equations,

(a) Let $f(z) = f(x, y) = f(re^{i\theta}) = u(re^{i\theta}) + iv(re^{i\theta}) = U(r, \theta) + iV(r, \theta)$. Use the transformation $x = r \cos \theta$ and $y = r \sin \theta$ $(i.e., (x, y) = re^{i\theta})$ and the chain rules

$$U_r = u_x \frac{\partial x}{\partial r} + u_y \frac{\partial y}{\partial r} \quad \text{and} \quad U_\theta = u_x \frac{\partial x}{\partial \theta} + u_y \frac{\partial y}{\partial \theta} \quad \text{(similarly for } V).$$

to prove that

$$U_r = u_x \cos \theta + u_y \sin \theta, \qquad U_\theta = -u_x r \sin \theta + u_y r \cos \theta; \quad \text{and}$$
$$V_r = v_x \cos \theta + v_y \sin \theta, \qquad V_\theta = -v_x r \sin \theta + v_y r \cos \theta.$$

(b) Use the original Cauchy-Riemann equations for u and v and the results of part (a) to prove that $rU_r = V_\theta$ and $rV_r = -U_\theta$, thus verifying Equation (3.22)

(c) Use part (a) and Equations (3.14) and (3.15) to show that the right sides of Equations (3.23) and (3.24) simplify to $f'(z_0)$.

11. Determine where the following functions are differentiable and where they are analytic. Explain!

(a) $f(z) = x^3 + 3xy^2 + i(y^3 + 3x^2y)$.

(b) $f(z) = 8x - x^3 - xy^2 + i(x^2y + y^3 - 8y)$.

(c) $f(z) = x^2 - y^2 + i2|xy|$.

12. Let f and g be analytic functions in the domain D. If $f'(z) = g'(z)$ for all z in D, then show that $f(z) = g(z) + C$, where C is a complex constant.

13. Explain how the limit definition for the derivative in complex analysis and the limit definition for the derivative in calculus are different. How are they similar?

14. Let f be an analytic function in the domain D. Show that if $\text{Re}[f(z)] = 0$ at all points in D, then f is constant in D.

15. Let f be a nonconstant analytic function in the domain D. Show that the function $g(z) = \overline{f(z)}$ is *not* analytic in D.

16. Recall that, for $z = x + iy$, $x = \frac{z + \overline{z}}{2}$ and $y = \frac{z - \overline{z}}{2i}$.

(a) Temporarily, think of z and $\overline{z}$ as dummy symbols for real variables. With this perspective, x and y can be viewed as functions of z and $\overline{z}$. Use the chain rule for a function h of two variables to show that

$$\frac{\partial h}{\partial \overline{z}} = \frac{\partial h}{\partial x} \frac{\partial x}{\partial \overline{z}} + \frac{\partial h}{\partial y} \frac{\partial y}{\partial \overline{z}} = \frac{1}{2} \left(\frac{\partial h}{\partial x} + i \frac{\partial h}{\partial y} \right).$$

(b) Now define the operator $\frac{\partial}{\partial \bar{z}} = \frac{1}{2}(\frac{\partial}{\partial x} + i\frac{\partial}{\partial y})$ that is suggested by the previous equation. With this construct, show that if $f = u + iv$ is differentiable at $z = (x, y)$, then, at the point (x, y), $\frac{\partial f}{\partial \bar{z}} = \frac{1}{2}[u_x - v_y + i(y_x + u_y)] = 0$. Equating real and imaginary parts thus gives the complex form of the Cauchy-Riemann equations: $\frac{\partial f}{\partial \bar{z}} = 0$.

3.3 Harmonic Functions

Let $\phi(x, y)$ be a real-valued function of the two real variables x and y defined on a domain D. (Recall that a domain is a connected open set.) The partial differential equation

$$\phi_{xx}(x, y) + \phi_{yy}(x, y) = 0 \tag{3.26}$$

is known as **Laplace's equation** (sometimes referred to as the **potential equation**). If ϕ, ϕ_x, ϕ_y, ϕ_{xx}, ϕ_{xy}, ϕ_{yx}, and ϕ_{yy} are all continuous, and if $\phi(x, y)$ satisfies Laplace's equation, then $\phi(x, y)$ is **harmonic** on D. Harmonic functions are important in applied mathematics, engineering, and mathematical physics. They are used to solve problems involving steady state temperatures, two-dimensional electrostatics, and ideal fluid flow. In Chapter 10 we describe how complex analysis techniques can be used to solve some problems involving harmonic functions. We begin with an important theorem relating analytic and harmonic functions.

Theorem 3.8. *Let $f(z) = u(x, y) + iv(x, y)$ be an analytic function on a domain D. Then both u and v are harmonic functions on D. In other words, the real and imaginary parts of an analytic function are harmonic.*

Proof. In Corollary 6.3 we will show that, if $f(z)$ is analytic, then all partial derivatives of u and v are continuous. Using that result here, we see that, as f is analytic, u and v satisfy the Cauchy-Riemann equations

$$u_x = v_y \quad \text{and} \quad u_y = -v_x.$$

Taking the partial derivative with respect to x of each side of these equations gives

$$u_{xx} = v_{yx} \quad \text{and} \quad u_{yx} = -v_{xx}.$$

Similarly, taking the partial derivative of each side with respect to y yields

$$u_{xy} = v_{yy} \quad \text{and} \quad u_{yy} - -v_{xy}.$$

The partial derivatives u_{xy}, u_{yx}, v_{xy}, and v_{yx} are all continuous, so we use a theorem from the calculus of real functions that states that the mixed partial derivatives are equal; that is,

$$u_{xy} = u_{yx} \quad \text{and} \quad v_{xy} = v_{yx}.$$

Combining all these results finally gives $u_{xx} + u_{yy} = v_{yx} - v_{xy} = 0$, and $v_{xx} + v_{yy} = -u_{yx} + u_{xy} = 0$. Therefore both u and v are harmonic functions on D. $\square$

If we have a function $u(x, y)$ that is harmonic on the domain D and if we can find another harmonic function $v(x, y)$ such that the partial derivatives for u and v satisfy the Cauchy-Riemann equations throughout D, then we say that $v(x, y)$ is a **harmonic conjugate** of $u(x, y)$. It then follows that the function $f(z) = u(x, y) + iv(x, y)$ is analytic on D.

Example 3.11. If $u(x, y) = x^2 - y^2$, then $u_{xx}(x, y) + u_{yy}(x, y) = 2 - 2 = 0$; hence u is a harmonic function for all z. We find that $v(x, y) = 2xy$ is also a harmonic function and that

$$u_x = v_y = 2x \quad \text{and} \quad u_y = -v_x = -2y.$$

Therefore v is a harmonic conjugate of u, and the function f given by

$$f(z) = x^2 - y^2 + i2xy = z^2$$

is an analytic function.

Theorem 3.8 makes the construction of harmonic functions from known analytic functions an easy task.

Example 3.12. The function $f(z) = z^3 = x^3 - 3xy^2 + i(3x^2y - y^3)$ is analytic for all values of z. Hence it follows that

$$u(x, y) = \text{Re}\,[f(z)] = x^3 - 3xy^2$$

is harmonic for all z, and that

$$v(x, y) = \text{Im}\,[f(z)] = 3x^2y - y^3$$

is a harmonic conjugate of $u(x, y)$.

Figures 3.2 and 3.3 show the graphs of these two functions. The partial derivatives are $u_x(x, y) = 3x^2 - 3y^2$, $u_y(x, y) = -6xy$, $v_x(x, y) = 6xy$, and $v_y(x, y) = 3x^2 - 3y^2$. They satisfy the Cauchy-Riemann equations because they are the real and imaginary parts of an analytic function. At the point $(x, y) = (2, -1)$, we have $u_x(2, -1) = v_y(2, -1) = 9$, and these partial derivatives appear along the edges of the surfaces for u and v where $x = 2$ and $y = -1$. Similarly, $u_y(2, -1) = 12$ and $v_x(2, -1) = -12$ also appear along the edges of the surfaces for u and v where $x = 2$ and $y = -1$.

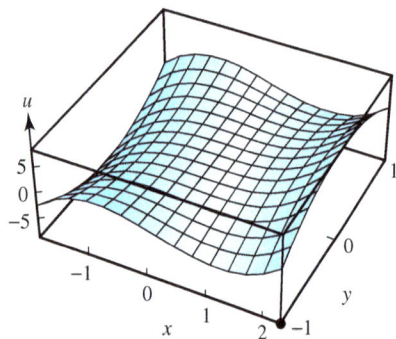

Figure 3.2: $u(x, y) = x^3 - 3xy^2$ Figure 3.3: $v(x, y) = 3x^2y - y^3$

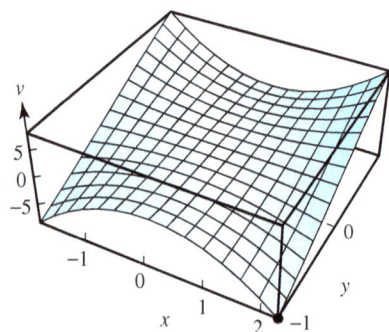

We can use complex analysis to show easily that certain combinations of harmonic functions are harmonic. For example, if v is a harmonic conjugate of u, then their product $\phi(x, y) = u(x, y)v(x, y)$ is a harmonic function. This condition can be verified directly by computing the partial derivatives and showing that Equation (3.26) holds, but the details are tedious. If we use complex variable techniques instead, we can start with the fact that $f(z) = u(x, y) + iv(x, y)$ is an analytic function. Then we observe that the square of f is also an analytic function:

$$[f(z)]^2 = [u(x, y)]^2 - [v(x, y)]^2 + i2u(x, y)v(x, y).$$

We then know immediately that the imaginary part, $2u(x,y)v(x,y)$, is a harmonic function by Theorem 3.8. A constant multiple of a harmonic function is harmonic, so it follows that ϕ is harmonic. We leave as an exercise to show that, if u_1 and u_2 are two harmonic functions that are not related in the preceding fashion, then their product need not be harmonic.

Theorem 3.9 (Construction of a harmonic conjugate). *Let $u(x,y)$ be harmonic in an ε-neighborhood of the point (x_0, y_0). Then there exists a conjugate harmonic function $v(x,y)$ defined in this neighborhood such that $f(z) = u(x,y) + iv(x,y)$ is an analytic function.*

Proof. A conjugate harmonic function v will satisfy the Cauchy-Riemann equations $u_x = v_y$ and $u_y = -v_x$. Assuming that such a function exists, we determine what it would have to look like by using a two-step process. First, we integrate v_y (which should equal u_x) with respect to y and get

$$v(x,y) = \int u_x(x,y)\,dy + C(x). \tag{3.27}$$

where $C(x)$ is a function of x alone that is yet to be determined. Second, we compute $C'(x)$ by differentiating both sides of this equation with respect to x and replacing v_x with $-u_y$ on the left side, which gives

$$-u_y(x,y) = \frac{d}{dx}\int u_x(x,y)\,dy + C'(x).$$

It can be shown (we omit the details) that because u is harmonic, all terms except those involving x in the last equation will cancel, revealing a formula for $C'(x)$ involving x alone. Elementary integration of the single-variable function $C'(x)$ can then be used to discover $C(x)$. We finally observe that the function v so created indeed has the properties we seek. $\square$

Technically we should always specify the domain of function when defining it. When no such specification is given, it is assumed that the domain is the entire complex plane, or the largest set for which the expression defining the function makes sense.

Example 3.13. Show that $u(x,y) = xy^3 - x^3y$ is a harmonic function and find a conjugate harmonic function $v(x,y)$.

***Solution*:**
We follow the construction process of Theorem 3.9. The first partial derivatives are

$$u_x(x,y) = y^3 - 3x^2y \quad \text{and} \quad u_y(x,y) = 3xy^2 - x^3. \tag{3.28}$$

To verify that u is harmonic, we compute the second partial derivatives and note that $u_{xx}(x,y) + u_{yy}(x,y) = -6xy + 6xy = 0$, so u satisfies Laplace's Equation (3.26). To construct v, we start with Equation (3.27) and the first of Equations (3.28) to get

$$v(x,y) = \int (y^3 - 3x^2y)\,dy + C(x) = \frac{1}{4}y^4 - \frac{3}{2}x^2y^2 + C(x).$$

Differentiating the left and right sides of this equation with respect to x and using $-u_y(x,y) = v_x(x,y)$ and Equations (3.28) on the left side yields

$$-3xy^2 + x^3 = 0 - 3xy^2 + C'(x),$$

which implies that

$$C'(x) = x^3.$$

Integrating to get $C(x)$ and using the prior expression for $v(x,y)$ gives

$$v(x,y) = \frac{1}{4}y^4 - \frac{3}{2}x^2y^2 + \frac{1}{4}x^4 + K, \text{ where } K \text{ is some constant.}$$

Harmonic functions arise as solutions to many physical problems. Applications include two-dimensional models of heat flow, electrostatics, and fluid flow. We now give an example of the latter.

We assume that an incompressible and frictionless fluid flows over the complex plane and that all cross sections in planes parallel to the complex plane are the same. Situations such as this occur when fluid is flowing in a deep channel. The velocity vector at the point (x, y) is

$$\mathbf{V}(x, y) = p(x, y) + iq(x, y), \tag{3.29}$$

which we illustrate in Figure 3.4.

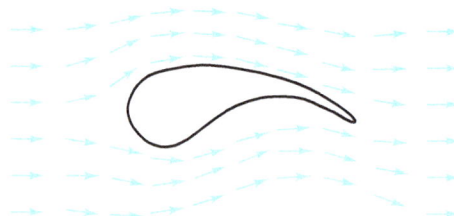

Figure 3.4: The vector field $\mathbf{V}(x, y) = p(x, y) + iq(x, y)$, which can be considered as a fluid flow

The assumption that the flow is irrotational and has no sources or sinks implies that both the curl and divergence vanish; that is, $q_x - p_y = 0$ and $p_x + q_y = 0$. Hence p and q obey the equations

$$p_x(x, y) = -q_y(x, y) \quad \text{and} \quad p_y(x, y) = q_x(x, y). \tag{3.30}$$

Equations (3.30) are similar to the Cauchy-Riemann equations and permit us to define a special complex function:

$$f(z) = u(x, y) + iv(x, y) = p(x, y) - iq(x, y). \tag{3.31}$$

Here we have $u_x = p_x$, $u_y = p_y$, $v_x = -q_x$, and $v_y = -q_y$. We can use Equations (3.30) to verify that the Cauchy-Riemann equations hold for f:

$$u_x(x, y) = p_x(x, y) = -q_y(x, y) = v_y(x, y) \quad \text{and}$$
$$u_y(x, y) = p_y(x, y) = q_x(x, y) = -v_x(x, y).$$

Assuming that the functions p and q have continuous partials, Theorem 3.4 guarantees that function f defined in Equation (3.31) is analytic and that the fluid flow of Equation (3.29) is the conjugate of an analytic function; that is,

$$\mathbf{V}(x, y) = \overline{f(z)}.$$

In Chapter 6 we prove that every analytic function f has an analytic antiderivative F; assuming this to be the case, we can write

$$F(z) = \phi(x, y) + i\psi(x, y). \tag{3.32}$$

where $F'(z) = f(z)$.

Theorem 3.8 implies that $\phi(x, y)$ is a harmonic function. Using the vector interpretation of a complex number, the gradient of ϕ can be written as

$$\operatorname{grad}\phi(x, y) = \phi_x(x, y) + i\phi_y(x, y).$$

The Cauchy-Riemann equations applied to $F(z)$ give $\phi_y(x,y) = -\psi_x(x,y)$. Making this substitution in the preceding equation yields

$$\text{grad}\phi(x,y) = \phi_x(x,y) - i\psi_x(x,y) = \overline{\phi_x(x,y) + i\psi_x(x,y)}.$$

Equation (3.14) says that $\phi_x(x,y) + i\psi_x(x,y) = F'(z)$, which by the preceding equation and Equation (3.32) implies that

$$\text{grad}\phi(x,y) = \overline{F'(z)} = \overline{f(z)}.$$

Finally, from Equation (3.29) ϕ is the scalar potential function for the fluid flow, so

$$\mathbf{V}(x,y) = \text{grad}\phi(x,y).$$

The curves given by $\{(x,y) : \phi(x,y) = \text{constant}\}$ are called *equipotentials*. The curves $\{(x,y) : \psi(x,y) = \text{constant}\}$ are called *streamlines* and describe the path of fluid flow. In Section 10.4 we show that the family of equipotentials is orthogonal to the family of streamlines, as depicted in Figure 3.5.

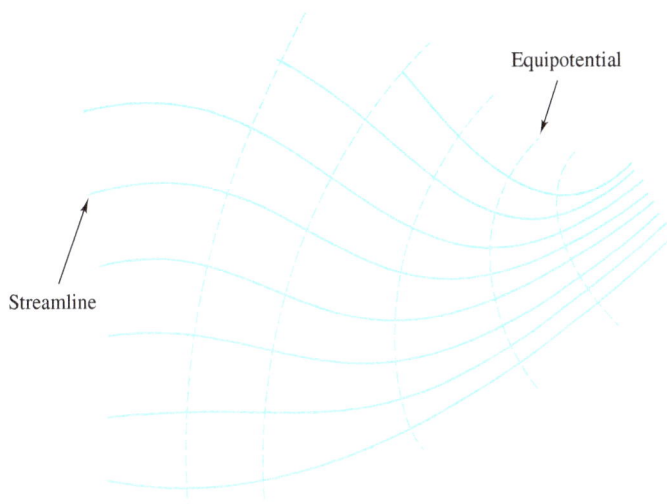

Figure 3.5: The families of orthogonal curves $\{(x,y) : \phi(x,y) = \text{constant}\}$ and $\{(x,y) : \psi(x,y) = \text{constant}\}$ for the function $F(z) = \phi(x,y) + i\psi(x,y)$

Example 3.14. Show that the harmonic function $\phi(x,y) = x^2 - y^2$ is the scalar potential function for the fluid flow expression $\mathbf{V}(x,y) = 2x - i2y$.

Solution:

We can write the fluid flow expression as

$$\mathbf{V}(x,y) = \overline{f(z)} = \overline{2x + i2y} = \overline{2z}.$$

An antiderivative of $f(z) = 2z$ is $F(z) = z^2$, and the real part of $F(z)$ is the desired harmonic function:

$$\phi(x,y) = \text{Re}\,[F(z)] = \text{Im}\,[x^2 - y^2 + i2xy] = x^2 - y^2.$$

Note that the hyperbolas $\phi(x,y) = x^2 - y^2 = C$ are the equipotential curves and that the hyperbolas $\psi(x,y) = 2xy = C$ are the streamline curves; these curves are orthogonal, as shown in Figure 3.6.

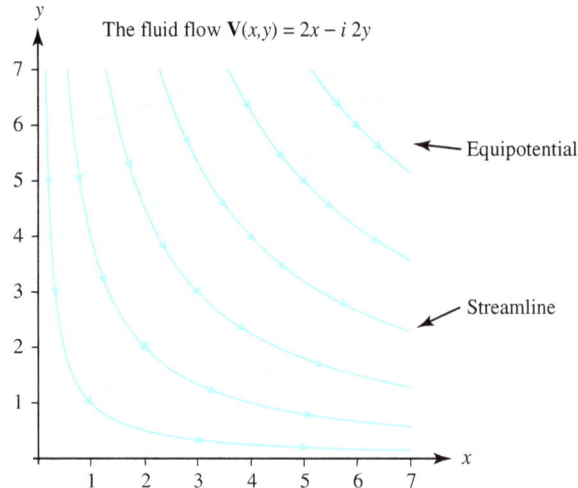

The fluid flow $\mathbf{V}(x,y) = 2x - i\,2y$

Figure 3.6: The equipotential curves $x^2 - y^2 = C$ and streamline curves $2xy = C$ for the function $F(z) = z^2$.

Exercises for Section 3.3 (Selected answers or hints are on page 436.)

1. Determine where the following functions are harmonic.

   (a) $u(x,y) = e^x \cos y$ and $v(x,y) = e^x \sin y$.

   (b) $u(x,y) = \ln(x^2 + y^2)$ for $(x,y) \neq (0,0)$.

2. Does an analytic function $f(z) = u(x,y) + iv(x,y)$ exist for which $v(x,y) = x^3 + y^3$? Why or why not?

3. Let a, b, and c be real constants. Determine a relation among the coefficients that will guarantee that the function $\phi(x,y) = ax^2 + bxy + cy^2$ is harmonic.

4. Let $v(x,y) = \arctan(\frac{y}{x})$ for $x \neq 0$. Compute the partial derivatives of v and verify that v satisfies Laplace's equation.

5. Find an analytic function $f(z) = u(x,y) + iv(x,y)$ given the following information.

   (a) $u(x,y) = y^3 - 3x^2y$.

   (b) $u(x,y) = \sin y \sinh x$.

   (c) $v(x,y) = e^y \sin x$.

   (d) $v(x,y) = \sin x \cosh y$.

6. Let $u_1(x,y) = x^2 - y^2$ and $u_2(x,y) = x^3 - 3xy^2$. Show that u_1 and u_2 are harmonic functions but that their product $u_1(x,y)u_2(x,y)$ is not a harmonic function.

7. Let $u(x,y)$ be harmonic on a region D that is symmetric about the line $y = 0$. Show that $U(x,y) = u(x,-y)$ is harmonic on D. *Hint*: Use the chain rule for differentiation of real functions and note that $u(x,-y)$ is really the function $u\big(g(x,y)\big)$, where $g(x,y) = (x,-y)$.

8. Let v be a harmonic conjugate of u. Show that $-u$ is a harmonic conjugate of v.

9. Let v be a harmonic conjugate of u. Show that $h = u^2 - v^2$ is a harmonic function.

10. Suppose that v is a harmonic conjugate of u and that u is a harmonic conjugate of v. Show that u and v must be constant functions.

11. Let $f(z) = f(re^{i\theta}) = u(r, \theta) + iv(r, \theta)$ be analytic on a domain D that does not contain the origin. Use the polar form of the Cauchy-Riemann equations $u_\theta = -rv_r$, and $v_\theta = ru_r$. Differentiate them first with respect to θ and then with respect to r. Use the results to establish the polar form of Laplace's equation:

$$r^2 u_{rr}(r, \theta) + r u_r(r, \theta) + u_{\theta\theta}(r, \theta) = 0.$$

12. Use the polar form of Laplace's equation given in Exercise 11 to show that the following functions are harmonic.

    (a) $u(r, \theta) = (r + \frac{1}{r}) \cos \theta$ and $v(r, \theta) = (r - \frac{1}{r}) \sin \theta$.

    (b) $u(r, \theta) = r^n \cos n\theta$ and $v(r, \theta) = r^n \sin n\theta$.

13. The function $F(z) = \frac{1}{z}$ is used to determine a field known as a dipole.

    (a) Express $F(z)$ in the form $F(z) = \phi(x, y) + i\psi(x, y)$.

    (b) Sketch the equipotentials $\phi = 1, \frac{1}{2}, \frac{1}{4}$ and streamlines $\psi = 1, \frac{1}{2}, \frac{1}{4}$.

14. Assume that $F(z) = \phi(x, y) + i\psi(x, y)$ is analytic on the domain D and that $F'(z) \neq 0$ on D. Consider the families of level curves $\{\phi(x, y) = \text{constant}\}$ and $\{\psi(x, y) = \text{constant}\}$, which are the equipotentials and streamlines for the fluid flow $\mathbf{V}(x, y) = \overline{F'(z)}$. Prove that the two families of curves are orthogonal.
    Hint: Suppose that (x_0, y_0) is a point common to the two curves $\phi(x, y) = c_1$ and $\psi(x, y) = c_2$. Use the gradients of ϕ and ψ to show that the normals to the curves are perpendicular.

15. We introduce the logarithmic function in Chapter 5. For now, let $F(z) = \text{Log}\, z = \ln |z| + i \text{Arg}\, z$. Here we have $\phi(x, y) = \ln |z|$ and $\psi(x, y) = \text{Arg}\, z$. Sketch the equipotentials $\phi = 0, \ln 2, \ln 3, \ln 4$ and streamlines $\psi = \frac{k\pi}{8}$ for $k = 0, 1, \ldots, 7$.

16. Theorem 3.9 claims that it is possible to prove that $C'(x)$ is a function of x alone. Prove this assertion.

17. Discuss and compare the statements "$u(x, y)$ is harmonic" and "$u(x, y)$ is the imaginary part of an analytic function."

Chapter 4

Sequences, Series, and Fractals

Overview

In 1980 Benoit Mandelbrot led a team of mathematicians in producing some stunning computer graphics from very simple rules for manipulating complex numbers. This event marked the beginning of a new branch of mathematics known as *fractal geometry*. Many of the tools needed to appreciate Mandelbrot's work are contained in this chapter. We look at extensions to the complex domain of sequences and series, ideas that are part of a standard calculus course.

4.1 Sequences and Series

In formal terms, a **complex sequence** is a function whose domain is the positive integers and whose range is a subset of the complex numbers. The following are examples of sequences:

$$f(n) = \left(2 - \frac{1}{n}\right) + \left(5 + \frac{1}{n}\right) i \quad \text{for} \quad n = 1,\, 2,\, 3,\, \dots; \tag{4.1}$$

$$g(n) = e^{i\frac{\pi n}{4}} \quad \text{for} \quad n = 1,\, 2,\, 3,\, \dots; \tag{4.2}$$

$$h(k) = 5 + 3i + \left(\frac{1}{1+i}\right)^k \quad \text{for} \quad k = 1,\, 2,\, 3,\, \dots \quad \text{and} \tag{4.3}$$

$$r(n) = \left(\frac{1}{4} + \frac{i}{2}\right)^n \quad \text{for} \quad n = 1,\, 2,\, 3,\, \dots. \tag{4.4}$$

For convenience, at times we use the term *sequence* rather than *complex sequence*. If we want a function s to represent an arbitrary sequence, we can specify it by writing $s(1) = z_1$, $s(3) = z_3$, and so on. The values z_1, z_2, $z_3, \dots$, are called the **terms** of a sequence, and mathematicians, being generally lazy when it comes to such things, often refer to z_1, z_2, $z_3 \dots$ as the sequence itself, even though they are really speaking of the range of the sequence when they do so. You will usually see a sequence written as $\{z_n\}_{n=1}^{\infty}$, $\{z_n\}_1^{\infty}$, or, when the indices are understood, as $\{z_n\}$. Mathematicians are also not so fussy about starting a sequence at z_1 so that $\{z_n\}_{n=-1}^{\infty}$, $\{z_k\}_{k=0}^{\infty}$, $\dots$ would also be acceptable notation provided all terms were defined. For example, the sequence r given by Equation (4.4) could be written in a variety of ways:

$$\left\{\left(\frac{1}{4} + \frac{i}{2}\right)^n\right\}_{n=1}^{\infty}, \quad \left\{\left(\frac{1}{4} + \frac{i}{2}\right)^n\right\}_1^{\infty}, \quad \left\{\left(\frac{1}{4} + \frac{i}{2}\right)^n\right\},$$

$$\left\{\left(\frac{1}{4} + \frac{i}{2}\right)^{n+3}\right\}_{n=-2}^{\infty}, \quad \left\{\left(\frac{1}{4} + \frac{i}{2}\right)^k\right\}_{k=1}^{\infty}, \quad \dots$$

The sequences f and g given by Equations (4.1) and (4.2) behave differently as n gets larger. The terms in Equation (4.1) approach $2 + 5i = (2, 5)$, but those in Equation (4.2) do not approach any particular number, as they oscillate around the eight eighth roots of unity on the unit circle. Informally, the sequence $\{z_n\}_1^\infty$ has ζ as its limit as n approaches infinity, provided the terms z_n can be made as close as we want to ζ by making n large enough. When this happens, we write

$$\lim_{n\to\infty} z_n = \zeta \quad \text{or} \quad z_n \to \zeta \quad \text{as} \quad n \to \infty. \tag{4.5}$$

If $\lim_{n\to\infty} z_n = \zeta$, we say that the sequence $\{z_n\}_1^\infty$ **converges to** ζ.

We need a rigorous definition for Statement (4.5), however, if we are to do honest mathematics.

Definition 4.1 (Limit of a Sequence). $\lim_{n\to\infty} z_n = \zeta$ *means that for any real number $\varepsilon > 0$ there corresponds a positive integer N_ε (which depends on ε) such that $z_n \in D_\varepsilon(\zeta)$ whenever $n > N_\varepsilon$. That is, $|z_n - \zeta| < \varepsilon$ whenever $n > N_\varepsilon$.*

Remark 4.1. *The reason that we use the notation N_ε is to emphasize the fact that this number depends on our choice of ε. Sometimes, for convenience, we drop the subscript.*

Figure 4.1 illustrates a convergent sequence.

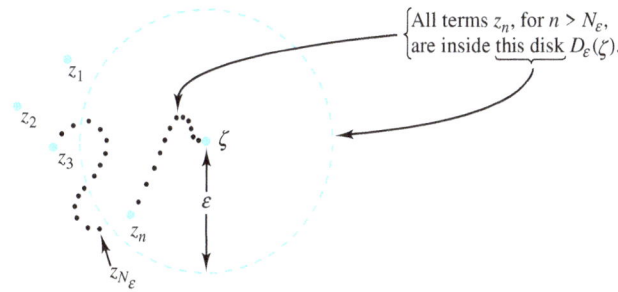

Figure 4.1: A sequence that converges to ζ

In form, Definition (4.1) is exactly the same as the corresponding definition for limits of real sequences. In fact, a simple criterion casts the convergence of complex sequences in terms of the convergence of real sequences.

Theorem 4.1. *Let $z_n = x_n + iy_n$ and $\zeta = u + iv$. Then,*

$$\lim_{n\to\infty} z_n = \zeta \quad \text{iff} \tag{4.6}$$

$$\lim_{n\to\infty} x_n = u \quad \text{and} \quad \lim_{n\to\infty} y_n = v. \tag{4.7}$$

Proof. First we assume that Statement (4.6) is true and then deduce the truth of Statement (4.7). Let ε be an arbitrary positive real number. To establish Statement (4.7), we must show (1) that there is a positive integer N_ε such that the inequality $|x_n - u| < \varepsilon$ holds whenever $n > N_\varepsilon$ and (2) that there is a positive integer M_ε such that the inequality $|y_n - v| < \varepsilon$ holds whenever $n > M_\varepsilon$. Because we are assuming Statement (4.6) to be true, we know (according to Definition 4.1) that there is a positive integer N_ε such that $z_n \in D_\varepsilon(\zeta)$ if $n > N_\varepsilon$. Recall that $z_n \in D_\varepsilon(\zeta)$ is equivalent to the inequality $|z_n - \zeta| < \varepsilon$. Thus, whenever $n > N_\varepsilon$, we have

$$|x_n - u| = |\text{Re}(z_n - \zeta)|$$
$$\leq |z_n - \zeta| \quad \text{(by Inequality (1.21))}$$
$$< \varepsilon.$$

Similarly, we can show that there is a number M_ε such that $|y_n - v| < \varepsilon$ whenever $n > M_\varepsilon$, which proves Statement (4.7).

To complete the proof of this theorem, we must show that the conditions of Statement (4.7) imply Statement (4.6). Let $\varepsilon > 0$ be an arbitrary real number. By Statement (4.7), there exist positive integers N_ε and M_ε such that

$$|x_n - u| < \frac{\varepsilon}{2} \quad \text{whenever} \quad n > N_\varepsilon, \quad \text{and} \tag{4.8}$$

$$|y_n - v| < \frac{\varepsilon}{2} \quad \text{whenever} \quad n > M_\varepsilon. \tag{4.9}$$

Let $L_\varepsilon = \max\{N_\varepsilon, M_\varepsilon\}$; then, if $n > L_\varepsilon$,

$$
\begin{aligned}
|z_n - \zeta| &= |(x_n + iy_n) - (u + iv)| \\
&= |(x_n - u) + i(y_n - v)| \\
&\le |x_n - u| + |i(y_n - v)| &&\text{(What is the reason for this step?)} \\
&= |x_n - u| + |i|\,|y_n - v| &&\text{(by properties of absolute value)} \\
&= |x_n - u|\,|y_n - v| &&\text{(because } |i| = 1) \\
&< \frac{\varepsilon}{2} + \frac{\varepsilon}{2} &&\text{(by Statements (4.8) and (4.9))} \\
&= \varepsilon.
\end{aligned}
$$

$\square$

We needed to show the strict inequality $|z_n - \zeta| < \varepsilon$, and the next-to-last line in the proof gives us precisely that. Note also that we have been speaking of *the* limit of a sequence. Strictly speaking, we are not entitled to use this terminology because we haven't proved that a complex sequence can have only one limit. The proof, however, is almost identical to the corresponding result for real sequences, and we leave it as an exercise.

Example 4.1. Find $\lim\limits_{n\to\infty} z_n$ if $z_n = \frac{\sqrt{n} + i(n+1)}{n}$.

Solution:

We write $z_n = x_n + iy_n = \frac{1}{\sqrt{n}} + i\frac{n+1}{n}$. Using results concerning sequences of real numbers, we find that $\lim\limits_{n\to\infty} x_n = \lim\limits_{n\to\infty} \frac{1}{\sqrt{n}} = 0$ and $\lim\limits_{n\to\infty} y_n = \lim\limits_{n\to\infty} \frac{n+1}{n} = 1$. Therefore, $\lim\limits_{n\to\infty} z_n = \lim\limits_{n\to\infty} \frac{\sqrt{n} + i(n+1)}{n} = i$.

Example 4.2. Show that $\{(1+i)^n\}$ diverges.

Solution:

We have

$$z_n = (1+i)^n = (\sqrt{2})^n \cos\frac{n\pi}{4} + i(\sqrt{2})^n \sin\frac{n\pi}{4}.$$

The real sequences $\{(\sqrt{2})^n \cos\frac{n\pi}{4}\}$ and $\{(\sqrt{2})^n \sin\frac{n\pi}{4}\}$ both diverge, so we conclude that the sequence $\{(1+i)^n\}$ diverges.

Definition 4.2 (Bounded Sequence). *A complex sequence $\{z_n\}$ is **bounded** provided that there exists a positive real number R and an integer N such that $|z_n| < R$ for all $n > N$. In other words, for $n > N$, the sequence $\{z_n\}$ is contained in the disk $D_R(0)$.*

Bounded sequences play an important role in some newer developments in complex analysis that are discussed in Section 4.2. A theorem from real analysis stipulates that convergent sequences are bounded. The same result holds for complex sequences.

Theorem 4.2. *If $\{z_n\}$ is a convergent sequence, then $\{z_n\}$ is bounded.*

Proof. The proof is left as an exercise. $\qquad\square$

As with the real numbers, we also have the following definition.

Definition 4.3 (Cauchy Sequence). *The sequence $\{z_n\}$ is a **Cauchy sequence** if for every $\varepsilon > 0$ there is a positive integer N_ε such that if $n, m > N_\varepsilon$, then $|z_n - z_m| < \varepsilon$, or, equivalently, $z_n - z_m \in D_\varepsilon(0)$.*

The following theorem should now come as no surprise.

Theorem 4.3. *If $\{z_n\}$ is a Cauchy sequence, $\{z_n\}$ converges.*

Proof. Let $z_n = x_n + iy_n$. Using the techniques of Theorem 4.1, we can easily show that both $\{x_n\}$ and $\{y_n\}$ are Cauchy sequences of real numbers. Since Cauchy sequences of real numbers are convergent, we know that

$$\lim_{n\to\infty} x_n = x_0 \quad \text{and} \quad \lim_{n\to\infty} y_n = y_0$$

for some real numbers x_0 and y_0. By Theorem 4.1, then, $\lim_{n\to\infty} z_n = z_0$, where $z_0 = x_0 + iy_0$. In other words, the sequence $\{z_n\}$ converges to z_0. $\qquad\square$

One of the most important notions in analysis (real or complex) is a theory that allows us to add up infinitely many terms. To make sense of such an idea we begin with a sequence $\{z_n\}$, and form a new sequence $\{S_n\}$, called the **sequence of partial sums**, as follows.

$$S_1 = z_1,$$
$$S_2 = z_1 + z_2,$$
$$S_3 = z_1 + z_2 + z_3,$$
$$\vdots$$
$$S_n = z_1 + z_2 + \cdots + z_n = \sum_{k=1}^{n} z_k,$$
$$\vdots$$

Definition 4.4 (Infinite Series). *The formal expression $\sum_{k=1}^{\infty} z_k = z_1 + z_2 + \cdots + z_n + \cdots$ is called an **infinite series**, and $z_1, z_2, \ldots$ are called the **terms** of the series.*

Definition 4.5 (Convergent Series). *If there is a complex number S for which*

$$S = \lim_{n\to\infty} S_n = \lim_{n\to\infty} \sum_{k=1}^{n} z_k,$$

*we say that the infinite series $\sum_{k=1}^{\infty} z_k$ **converges** to S, and that S is the **sum** of the infinite series. When convergence occurs, we write $S = \sum_{k=1}^{\infty} z_k$.*

Definition 4.6 (Absolutely Convergent Series). *The series* $\sum\limits_{k=1}^{\infty} z_k$ *is said to be **absolutely convergent** provided that the (real) series of magnitudes* $\sum\limits_{k=1}^{\infty} |z_k|$ *converges.*

Definition 4.7 (Divergent Series). *If a series does not converge, we say that it **diverges**.*

Remark 4.2. *The first finitely many terms of a series do not affect its convergence or divergence and, in this respect, the beginning index of a series is irrelevant. Thus, we will without comment conclude that if a series* $\sum\limits_{k=N+1}^{\infty} z_k$ *converges, then so does* $\sum\limits_{k=0}^{\infty} z_k$, *where* $z_0, z_1, \ldots, z_N$ *is any finite collection of terms. A similar remark applies to determining divergence of a series.*

As you might expect, many of the results concerning real series carry over to complex series. We now give several of the more standard theorems for complex series, along with examples of how they are used.

Theorem 4.4. *Let* $z_n = x_n + iy_n$ *and* $S = U + iV$. *Then*

$$S = \sum_{n=1}^{\infty} z_n = \sum_{n=1}^{\infty} (x_n + iy_n) \quad \text{iff}$$

$$U = \sum_{n=1}^{\infty} x_n \quad \text{and} \quad V = \sum_{n=1}^{\infty} y_n.$$

Proof. Let $U_n = \sum\limits_{k=1}^{n} x_k$, $V_n = \sum\limits_{k=1}^{n} y_k$, and $S_n = U_n + iV_n$. We use Theorem 4.1 to conclude that $\lim\limits_{n\to\infty} S_n = \lim\limits_{n\to\infty} (U_n + iV_n) = U + iV = S$ iff both $\lim\limits_{n\to\infty} U_n = U$ and $\lim\limits_{n\to\infty} V_n = V$. The completion of the proof now follows from Definition 4.4. $\qquad\square$

Theorem 4.5. *If* $\sum\limits_{n=1}^{\infty} z_n$ *is a convergent complex series, then* $\lim\limits_{n\to\infty} z_n = 0$.

Proof. The proof is left as an exercise. $\qquad\square$

Example 4.3. Show that the series $\sum\limits_{n=1}^{\infty} \frac{1+in(-1)^n}{n^2} = \sum\limits_{n=1}^{\infty} \left[\frac{1}{n^2} + i\frac{(-1)^n}{n} \right]$ is convergent.

Solution:
Recall that the real series $\sum\limits_{n=1}^{\infty} \frac{1}{n^2}$ and $\sum\limits_{n=1}^{\infty} \frac{(-1)^n}{n}$ are convergent. Hence, Theorem 4.4 implies that the given complex series is convergent.

Example 4.4. Show that the series $\sum\limits_{n=1}^{\infty} \frac{(-1)^n+i}{n} = \sum\limits_{n=1}^{\infty} \left[\frac{(-1)^n}{n} + i\frac{1}{n} \right]$ is divergent.

Solution:
We know that the real series $\sum\limits_{n=1}^{\infty} \frac{1}{n}$ is divergent. Hence, Theorem 4.4 implies that the given complex series is divergent.

Example 4.5. Show that the series $\sum\limits_{n=1}^{\infty} (1+i)^n$ is divergent.

Solution:

Here we set $z_n = (1+i)^n$ and observe that $\lim\limits_{n\to\infty} |z_n| = \lim\limits_{n\to\infty} (\sqrt{2})^n = \infty$. Thus $\lim\limits_{n\to\infty} z_n \neq 0$, and Theorem 4.5 implies that the series is not convergent; hence it is divergent.

Theorem 4.6. *Let $\sum\limits_{n=1}^{\infty} z_n$ and $\sum\limits_{n=1}^{\infty} w_n$ be convergent series and let c be a complex number. Then*

$$\sum_{n=1}^{\infty} cz_n = c \sum_{n=1}^{\infty} z_n \quad and \qquad\qquad (4.10)$$

$$\sum_{n=1}^{\infty} (z_n + w_n) = \sum_{n=1}^{\infty} z_n + \sum_{n=1}^{\infty} w_n.$$

Proof. The proof is left as an exercise. □

Definition 4.8 (Cauchy Product). *Let $\sum\limits_{n=0}^{\infty} a_n$ and $\sum\limits_{n=0}^{\infty} b_n$ be convergent series, where a_n and b_n are complex numbers. The **Cauchy product** of the two series is defined to be the series $\sum\limits_{n=0}^{\infty} c_n$, where $c_n = \sum\limits_{k=0}^{n} a_k b_{n-k}$.*

Theorem 4.7. *If the Cauchy product converges, then*

$$\sum_{n=0}^{\infty} c_n = \left(\sum_{n=0}^{\infty} a_n \right) \left(\sum_{n=0}^{\infty} b_n \right).$$

Proof. The proof can be found in a number of texts—for example, *Infinite Sequences and Series,* by Konrad Knopp (translated by Frederick Bagemihl; New York: Dover, 1956). □

Theorem 4.8 (Comparison test). *Let $\sum\limits_{n=1}^{\infty} M_n$ be a convergent series of real nonnegative terms. If $\{z_n\}$ is a sequence of complex numbers and $|z_n| \leq M_n$ holds for all n, then the infinite series $\sum\limits_{n=1}^{\infty} z_n = \sum\limits_{n=1}^{\infty} (x_n + iy_n)$ converges.*

Proof. Using Equations (1.21), we determine that $|x_n| \leq |z_n| \leq M_n$ and $|y_n| \leq |z_n| \leq M_n$ holds for all n. By the comparison test for real series, we conclude that $\sum\limits_{n=1}^{\infty} |x_n|$ and $\sum\limits_{n=1}^{\infty} |y_n|$ are convergent. An absolutely convergent real series is convergent, so $\sum\limits_{n=1}^{\infty} x_n$ and $\sum\limits_{n=1}^{\infty} y_n$ are convergent. With these results, together with Theorem 4.4, we conclude that the infinite series $\sum\limits_{n=1}^{\infty} z_n = \sum\limits_{n=1}^{\infty} x_n + i \sum\limits_{n=1}^{\infty} y_n$ is convergent. □

Corollary 4.1. *If $\sum\limits_{n=1}^{\infty} |z_n|$ converges, then $\sum\limits_{n=0}^{\infty} z_n$ converges. In other words, absolute convergence implies convergence for complex series as well as for real series.*

Proof. The proof is left as an exercise. □

Example 4.6. Show that $\sum_{n=1}^{\infty} \frac{(3+4i)^n}{5^n n^2}$ converges.

Solution:

We calculate $|z_n| = \left| \frac{(3+4i)^n}{5^n n^2} \right| = \frac{1}{n^2} = M_n$. Using the comparison test and the fact that $\sum_{n=1}^{\infty} \frac{1}{n^2}$

converges, we determine that $\sum_{n=1}^{\infty} \left| \frac{(3+4i)^n}{5^n n^2} \right|$ converges, and hence so does $\sum_{n=1}^{\infty} \frac{(3+4i)^n}{5^n n^2}$.

Exercises for Section 4.1 (Selected answers or hints are on page 437.)

1. Find the following limits.

   (a) $\lim_{n\to\infty} (\frac{1}{2} + \frac{i}{4})^n$.

   (b) $\lim_{n\to\infty} \frac{n+(i)^n}{n}$.

   (c) $\lim_{n\to\infty} \frac{n^2+i2^n}{2^n}$.

   (d) $\lim_{n\to\infty} \frac{(n+i)(1+ni)}{n^2}$.

2. Show that $\lim_{n\to\infty} (i)^{\frac{1}{n}} = 1$, where $(i)^{\frac{1}{n}}$ is the principal value of the nth root of i.

3. Suppose that $\lim_{n\to\infty} z_n = z_0$. Show that $\lim_{n\to\infty} \overline{z_n} = \overline{z_0}$.

4. Suppose that the complex series $\{z_n\}$ converges to ζ. Show that $\{z_n\}$ is bounded in two ways.

   (a) Write $z_n = x_n + iy_n$ and use the fact that convergent series of real numbers are bounded.

   (b) For $\varepsilon = 1$, use Definitions 4.1 and 4.2 to show that there is some integer N such that, for $n > N$, $|z_n| = |\zeta + (z_n - \zeta)| \le |\zeta| + 1$. Then set $R = \max\{|z_1|, |z_2|, \ldots, |z_N|, \zeta+1\}$.

5. Show that $\sum_{n=0}^{\infty} (\frac{1}{n+1+i} - \frac{1}{n+i}) = i$.

6. Suppose that $\sum_{n=1}^{\infty} z_n = S$. Show that $\sum_{n=1}^{\infty} \overline{z_n} = \overline{S}$.

7. Does $\lim_{n\to\infty} (\frac{1+i}{\sqrt{2}})^n$ exist? Why or why not?

8. Let $z_n = r_n e^{i\theta_n} \ne 0$, where $\theta_n = \text{Arg}(z_n)$.

   (a) Suppose $\lim_{n\to\infty} r_n = r_0$ and $\lim_{n\to\infty} \theta_n = \theta_0$. Show that $\lim_{n\to\infty} r_n e^{i\theta_n} = r_0 e^{i\theta_0}$.

   (b) Find an example of a sequence $\{z_n\} = \{r_n e^{i\theta_n}\}$ where $\lim_{n\to\infty} z_n = z_0 = r_0 e^{i\theta_0}$ and $\lim_{n\to\infty} r_n = r_0$, but $\lim_{n\to\infty} \theta_n$ does not exist.

   (c) If $\{z_n\} = \{r_n e^{i\theta_n}\}$, is it possible to have $\lim_{n\to\infty} z_n = z_0 = r_0 e^{i\theta_0}$, but $\lim_{n\to\infty} r_n$ does not exist?

9. Show that, if $\sum_{n=1}^{\infty} z_n$ converges, then $\lim_{n\to\infty} z_n = 0$.
   Hint: $z_n = S_n - S_{n-1}$.

10. State (with justification) whether the following series converge.

   (a) $\sum\limits_{n=1}^{\infty} \frac{i^n}{n}$.

   (b) $\sum\limits_{n=1}^{\infty} \left(\frac{1}{n} + \frac{i}{2^n}\right)$.

11. Let $\sum\limits_{n=1}^{\infty} (x_n + iy_n) = u + iv$. If $c = a + ib$ is a complex constant, show that

   $$\sum_{n=1}^{\infty} (a + ib)(x_n + iy_n) = (a + ib)(u + iv).$$

12. If $\sum\limits_{n=0}^{\infty} z_n$ converges, show that $\left|\sum\limits_{n=0}^{\infty} z_n\right| \leq \sum\limits_{n=0}^{\infty} |z_n|$.

13. Complete the proof of Theorem 4.1. In other words, suppose that $\lim\limits_{n\to\infty} z_n = \zeta$, where $z_n = x_n + iy_n$ and $\zeta = u + iv$. Prove that $\lim\limits_{n\to\infty} y_n = v$.

14. A side comment asked you to justify the first inequality in the proof of Theorem 4.1. Give a justification.

15. Prove that a sequence can have only one limit. *Hint*: Suppose that there is a sequence $\{z_n\}$ such that $z_n \to \zeta_1$ and $z_n \to \zeta_2$. Show this assumption implies $\zeta_1 = \zeta_2$ by proving that for all $\varepsilon > 0$, $|\zeta_1 - \zeta_2| < \varepsilon$.

16. Prove Corollary 4.1.

17. Prove that $\lim\limits_{n\to\infty} z_n = 0$ iff $\lim\limits_{n\to\infty} |z_n| = 0$.

4.2 Julia and Mandelbrot Sets

An impetus for studying complex analysis is the comparison of properties of real numbers and functions with their complex counterparts. In this section we take a look at Newton's method for finding solutions to the equation $f(z) = 0$. Then, by examining the more general topic of iteration, we will plunge into a breathtaking world of color and imagination. The mathematics surrounding this topic has generated a great deal of popular attention in the past few years.

Recall from calculus that Newton's method proceeds by starting with a function $f(x)$ and an initial "guess" of x_0 as a solution to $f(x) = 0$. We then generate a new guess x_1 by the computation $x_1 = x_0 - \frac{f(x_0)}{f'(x_0)}$. Using x_1 in place of x_0, this process is repeated, giving $x_2 = x_1 - \frac{f(x_1)}{f'(x_1)}$. We thus obtain a sequence of points $\{x_k\}$, where $x_{k+1} = x_k - \frac{f(x_k)}{f'(x_k)}$. The points $\{x_k\}_{k=0}^{\infty}$ are called the **iterates** of x_0. For functions defined on the real numbers, this method gives remarkably good results, and the sequence $\{x_k\}$ often converges to a solution of $f(x) = 0$ rather quickly. In the late 1800s, the British mathematician Arthur Cayley investigated the question of whether Newton's method can be applied to complex functions. He wrote a paper giving an analysis for how this method works for quadratic polynomials and indicated his intention to publish a subsequent paper for cubic polynomials. Unfortunately, Cayley died before producing this paper. As you will see, the extension of Newton's method to the complex domain and the general question of iteration are quite (if you'll pardon the pun) *complex*.

Example 4.7. Trace the next five iterates of Newton's method for an initial guess of $z_0 = \frac{1}{4} + \frac{1}{4}i$ as a solution to the equation $f(z) = 0$, where $f(z) = z^2 + 1$.

Solution:

For any guess z for a solution, Newton's method gives as the next guess the number $z - \frac{f(z)}{f'(z)} = \frac{z^2 - 1}{2z}$. Table 4.1 gives the iterates, rounded to five decimal places.

k	z_k	$f(z_k)$
0	$0.25000 + 0.25000i$	$1.00000 + 0.12500i$
1	$-0.87500 + 1.12500i$	$0.50000 - 1.96875i$
2	$-0.22212 + 0.83942i$	$0.34470 - 0.37290i$
3	$0.03624 + 0.97638i$	$0.04799 + 0.07077i$
4	$-0.00086 + 0.99958i$	$0.00084 - 0.00172i$
5	$0.00000 + 1.00000i$	$0.00000 + 0.00000i$

Table 4.1: The iterates of $z_0 = \frac{1}{4} + \frac{1}{4}i$ for Newton's method applied to $f(z) = z^2 + 1$.

Figure 4.2 shows the relative positions of these points on the z plane. Note that the points z_4 and z_5 are so close together that they appear to coincide, and that the value for z_5 agrees to five decimal places with the actual solution $z = i$.

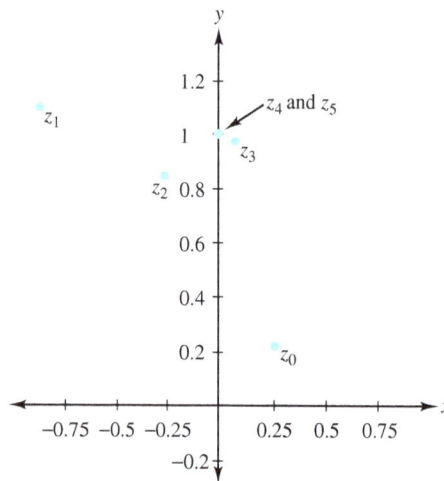

Figure 4.2: The iterates of $z_0 = \frac{1}{4} + \frac{1}{4}i$ for Newton's method applied to $f(z) = z^2 + 1$

The complex version of Newton's method also appears to work quite well. Recall, however, that with functions defined on the reals, not every initial guess produces a sequence that converges to a solution. Example 4.8 shows that the same is true in the complex case.

Example 4.8. Show that Newton's method fails for the function $f(z) = z^2 + 1$ if the initial guess is a real number.

Solution:

From Example 4.7 we know that, for any guess z as a solution of $z^2 + 1 = 0$, the next guess at a solution is $N(z) = z - \frac{f(z)}{f'(z)} = \frac{z^2 - 1}{2z}$. We let z_0 be any real number and $\{z_k\}$ be the sequence of iterations produced by the initial seed z_0. If for any k, $z_k = 0$, the procedure terminates, as z_{k+1} will be undefined. If all the terms of the sequence $\{z_k\}$ are defined, an easy induction argument shows that all the terms of the sequence are real. Because the solutions of $z^2 + 1 = 0$ are $\pm i$, the sequence $\{z_k\}$ cannot possibly converge to either solution. In the exercises we ask you to explore in detail what happens when z_0 is in the upper or lower half-plane.

The case for cubic polynomials is more complicated than that for quadratics. Fortunately, we can get an idea of what's going on by doing some experimentation with computer graphics. We begin with the cubic polynomial $f(z) = z^3 + 1$. (Recall that the roots of this polynomial are at -1, $\frac{1}{2} + \frac{\sqrt{3}}{2}i$, and $\frac{1}{2} - \frac{\sqrt{3}}{2}i$.) We associate a color with each root (blue, red, and green, respectively). We form a rectangular region R, which contains the three roots of $f(z)$, and partition this region into equal rectangles R_{ij}. We then choose a point z_{ij} at the center of each rectangle, and for each of these points apply the following algorithm:

1. With $N(z) = z - \frac{f(z)}{f'(z)}$, compute $N(z_{ij})$. Continue computing successive iterates of this initial point until we either are within a certain preassigned tolerance (say, ε) of one of the roots of $f(z) = 0$, or until the number of iterations has exceeded a preassigned maximum.

2. If Step 1 leaves us within ε of one of the roots of $f(z)$, we color the entire rectangle R_{ij} with the color associated with that root. Otherwise, we assume that the initial point z_{ij} does not converge to any root, and we color the entire rectangle yellow.

Note that this algorithm doesn't prove anything. In Step 2, there is no a priori reason to justify the assumption mentioned, nor is there any necessity for an initial point z_{ij} to have its sequence of iterates converging to one of the roots of $f(z) = 0$ just because a particular iteration is within ε of that root. Finally, the fact that one point in a rectangle behaves in a certain way does not imply that all the points in that rectangle behave in a like manner. Nevertheless, we can use this algorithm as a basis for mathematical explorations. Indeed, computer experiments such as the one described have contributed to a lot of exciting mathematics during the past 30 years. Figure 4.3a shows the results for the cubic polynomial $f(z) = z^3 + 1$.

The points in the blue, red, and green regions are those "initial guesses" that will converge to the roots -1, $\frac{1}{2} + \frac{\sqrt{3}}{2}i$, and $\frac{1}{2} - \frac{\sqrt{3}}{2}i$, respectively. (The roots themselves are located in the middle of the three largest colored regions.) The complexity of this picture becomes apparent when you observe that, wherever two colors appear to meet, the third color emerges between them. But then, a closer inspection of the area where this third color meets one of the other colors reveals again a different color between them. This process continues with an infinite complexity.

There appear to be no yellow regions with any area in Figure 4.3a, indicating that at least most initial guesses z_0 at a solution to $z^3 + 1 = 0$ will produce a sequence $\{z_k\}$ that converges to one of the three roots. Figure 4.3b demonstrates that this outcome does not always occur. It shows the results of applying the preceding algorithm to the polynomial $f(z) = z^3 + (-0.26 + 0.02i)z + (-0.74 + 0.02i)$.

The yellow area shown is often referred to as the *rabbit*. It consists of a main body and two ears, and is an example of a *fractal image*. Mathematicians use the term **fractal** to indicate an object that is self-similar and infinitely replicating. Figure 4.4 illustrates this phenomenon by zooming in on a portion of Figure 4.3b. You can see that each of the ears consists of a main body and two ears, and so on.

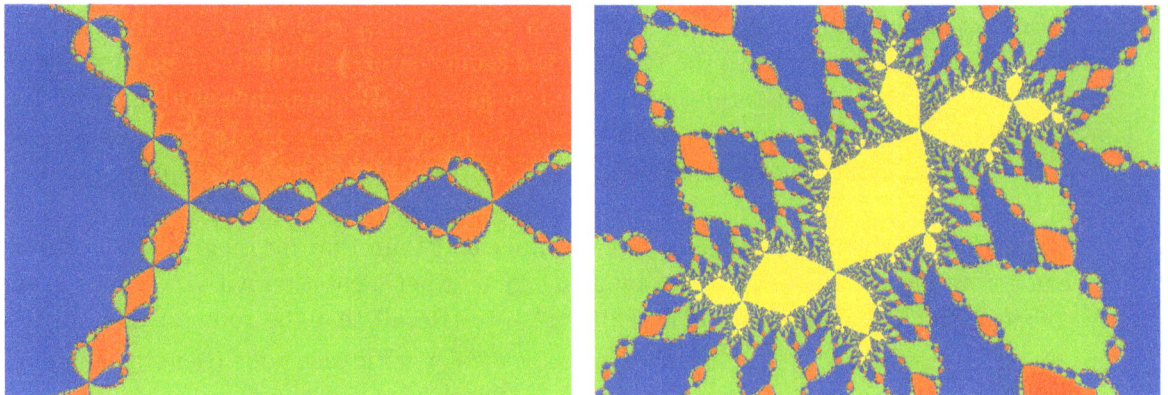

(a) $f(z) = z^3 + 1$ (b) $f(z) = z^3 + (-0.26 + 0.02i)z + (-0.74 + 0.02i)$

Figure 4.3: Newton's method applied to two different cubics

Figure 4.4: A zoom of the rabbit

In 1918, the French mathematicians Gaston Julia and Pierre Fatou noticed this fractal phenomenon when exploring iterations of functions not necessarily connected with Newton's method. Beginning with a function $f(z)$ and a point z_0, they computed the iterates

$$z_1 = f(z_0), \ z_2 = f(z_1), \ \ldots, \ z_{k+1} = f(z_k), \ldots$$

and investigated properties of the sequence $\{z_k\}$. Their findings did not receive a great deal of attention, largely because computer graphics were not available at that time. With the recent proliferation of computers, it is not surprising that these investigations were revived in the 1980s. Detailed studies of Newton's method and the more general topic of iteration were undertaken by a host of mathematicians including Curry, Devaney, Douady, Garnett, Hubbard, Mandelbrot, Milnor, and Sullivan. We now turn our attention to some of their results by focusing on the iterations produced by quadratics of the form $f_c(z) = z^2 + c$. You will be surprised at the startling pictures that graphical iterates of such a simple function produce.

Example 4.9. For $f_c(z) = z^2 + c$, analyze all possible iterations when $c = 0$, that is, for the function f_0 defined by $f_0(z) = z^2 + 0$.

Solution:

We leave as an exercise the claim that, if $|z_0| < 1$, the sequence will converge to 0; if $|z_0| > 1$, the sequence will be unbounded; and if $|z_0| = 1$, the sequence will either oscillate around the unit circle or converge to 1.

For the function f_c, defined by $f_c(z) = z^2 + c$, and an initial seed z_0, the set of iterates given by $z_1 = f_c(z_0)$, $z_2 = f_c(z_1)$, ... is also called the **orbit** of z_0 generated by f_c. We let K_c denote the set of points with a bounded orbit for f_c. Example 4.9 shows that K_0 is the closed unit disk $\overline{D}_1(0)$. The boundary of K_c is known as the **Julia set** for the function f_c. Thus the Julia set for f_0 is the unit circle $C_1(0)$. It turns out that K_c is a nice simple set only when $c = 0$ or $c = -2$; otherwise, K_c is fractal. Figure 4.5a shows $K_{-1.25}$. Its reflective nature has reminded some of St. Mark's square in Venice when flooding occurs. The variation in colors in that figure indicate the length of time it takes for points to become "sufficiently unbounded" according to the following algorithm, which uses the same notation as our algorithm for iterations via Newton's method:

1. Compute $f_c(z_{ij})$. Continue computing successive iterates of this initial point until the absolute value of one of the iterations exceeds a certain bound (say, L), or until the number of iterations has exceeded a preassigned maximum.

2. If Step 1 leaves us with an iteration whose absolute value exceeds L, we color the entire rectangle R_{ij} with a color indicating the number of iterations needed before this value was attained (the more iterations required, the darker the color). Otherwise, we assume that the orbit of the initial point z_{ij} do not diverge to infinity, and we color the entire rectangle black.

Note, again, that this algorithm doesn't prove anything. It merely guides the direction of our efforts to do rigorous mathematics.

Figure 4.5b shows the Julia set for the function f_c, where $c = -0.11 - 0.67i$. The boundary of this set is different from the boundaries of the other sets we have seen, in that it is disconnected. Julia and Fatou independently discovered a simple criterion that can be used to tell when the Julia set for f_c is connected or disconnected. We state their result, but omit the proof, as it is beyond the scope of this text.

Theorem 4.9. *The boundary of K_c is connected if and only if $0 \in K_c$. In other words, the Julia set for f_c is connected if and only if the orbit of 0 is bounded.*

Example 4.10. Show that the Julia set for f_i is connected.

Solution:

We apply Theorem 4.9 and compute the orbit of 0 for $f_i(z) = z^2 + i$. We have $f_i(0) = i$, $f_i(i) = -1 + i$, $f_i(-1 + i) = -i$, and $f_i(-i) = -1 + i$. Thus the orbit of 0 are the sequence $\{0, -1 + i, -i, -1 + i, -i, -1 + i, -i, \ldots\}$, which is clearly a bounded sequence. Thus, by Theorem 4.9, the Julia set for f_i is connected.

(a) $f_{-1.25}(z)$

(b) $f_{-0.11-0.67i}(z)$

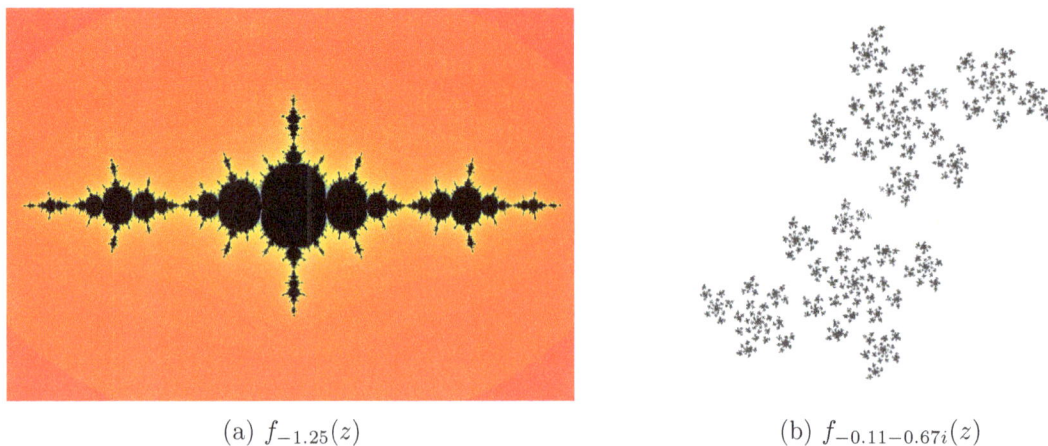

Figure 4.5: Results of iterating $f_c(z)$ for two different values of c

In 1980, the Polish-born mathematician Benoit Mandelbrot used computer graphics to study the set

$$M = \{c : \text{the Julia set for } f_c \text{ is connected}\}$$
$$= \{c : \text{the orbit of 0 determined by } f_c \text{ is bounded}\}.$$

The set M is known as the **Mandelbrot set**. Figure 4.6 shows its intricate nature.

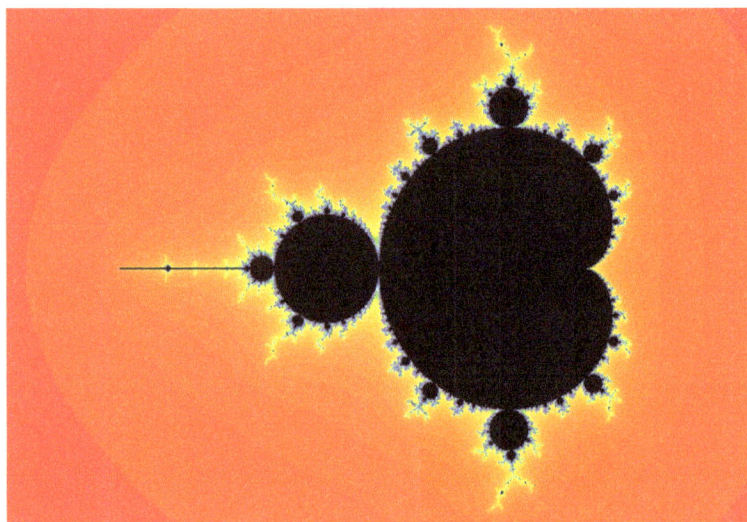

Figure 4.6: The Mandelbrot set

Technically, the Mandelbrot set is not fractal because it is not self-similar (although it may look that way). However, it is infinitely complex. Figure 4.7a shows a zoom over the upper portion of the set shown in Figure 4.6. Likewise, Figure 4.7b zooms in on the upper portion of Figure 4.7a. In Figure 4.7b you can see the emergence of another structure very similar to the Mandelbrot set that we began with. Although it isn't an exact replica, if you zoomed in on this set at almost any spot, you would eventually see yet another "Mandelbrot clone" and so on ad infinitum! In the remainder of this section we look at some of the properties of this amazing set.

(a) A zoom of the upper-portion of Figure 4.6   (b) A zoom of the upper-portion of Figure 4.7a

Figure 4.7: Zooming in on the Mandelbrot set

Example 4.11. Show that $\{c : |c| \leq \frac{1}{4}\} \subseteq M$.

***Solution*:**

Let $\{a_n\}_{n=0}^{\infty}$ be the orbit of 0 generated by $f_c(z) = z^2 + c$, where $|c| \leq \frac{1}{4}$. Then

$$a_0 = 0,$$
$$a_1 = f_c(a_0) = a_0^2 + c = c,$$
$$a_2 = f_c(a_1) = a_1^2 + c, \text{ and in general,}$$
$$a_{n+1} = f_c(a_n) = a_n^2 + c.$$

We show that $\{a_n\}$ is bounded, and, in particular, we show that $|a_n| \leq \frac{1}{2}$ for all n by mathematical induction. Clearly $|a_n| \leq \frac{1}{2}$ if $n = 0$ or 1. We assume that $|a_n| \leq \frac{1}{2}$ for some value of $n \geq 1$ (our goal is to show $|a_{n+1}| \leq \frac{1}{2}$). Now,

$$\begin{aligned}
|a_{n+1}| &= |a_n^2 + c| \\
&\leq |a_n^2| + |c| \quad \text{(by the triangle inequality)} \\
&\leq \frac{1}{4} + \frac{1}{4} = \frac{1}{2} \quad \text{(by our induction assumption and the fact that } |c| \leq \frac{1}{4}.
\end{aligned}$$

In the exercises, we ask you to show that, if $|c| > 2$, then $c \notin M$. Thus the Mandelbrot set depicted in Figure 4.6 contains the disk $\overline{D}_{\frac{1}{4}}(0)$ and is contained in the disk $\overline{D}_2(0)$.

We can use other methods to determine which points belong to M. To do so, we need some additional vocabulary.

Definition 4.9 (Fixed Point). *The point z_0 is a **fixed point** for the function f if $f(z_0) = z_0$.*

Definition 4.10 (Attracting Point). *The point z_0 is an **attracting point** for the function f if $|f'(z_0)| < 1$.*

Theorem 4.10 explains the significance of these terms.

Theorem 4.10. *Suppose that z_0 is an attracting fixed point for the function f. Then there is a disk $D_r(z_0)$ about z_0 such that the iterates of all the points in $D_r^*(z_0)$ are drawn toward z_0 in the sense that, if $z \in D_r^*(z_0)$, then $|f(z) - z_0| < |z - z_0|$. In fact, if z_k is the kth iterate of $z \in D_r^*(z_0)$, then $\lim_{k \to \infty} z_k = z_0$.*

Proof. Because z_0 is an attracting point for f, we know that $|f'(z_0)| < 1$. And because f is differentiable at z_0, we know that for any $\varepsilon > 0$ there exists some $r > 0$ such that if $z \in D_r^*(z_0)$, then $\left|\frac{f(z)-f(z_0)}{z-z_0} - f'(z_0)\right| < \varepsilon$. If we set $\varepsilon = 1 - |f'(z_0)|$, then we have for all z in $D_r^*(z_0)$ that

$$\left|\left|\frac{f(z) - f(z_0)}{z - z_0}\right| - |f'(z_0)|\right| \le \left|\frac{f(z) - f(z_0)}{z - z_0} - f'(z_0)\right| < 1 - |f'(z_0)|,$$

which gives $\left|\frac{f(z)-f(z_0)}{z-z_0}\right| < 1$. Thus $|f(z) - f(z_0)| < |z - z_0|$. Because z_0 is a fixed point for f, this last inequality implies that $|f(z) - z_0| < |z - z_0|$, which is the first part of our theorem.

The proof that $\lim_{k\to\infty} z_k = z_0$ is left as an exercise. $\qquad\square$

In 1905, Fatou showed that, if the function f_c defined by $f_c(z) = z^2 + c$ has attracting fixed points, then the orbit of 0 determined by f_c must converge to one of them. Because a convergent sequence is bounded, this condition implies that c must belong to M. In the exercises we ask you to show that the main cardioid-shaped body of M in Figure 4.6 is composed of those points c for which f_c has attracting fixed points. You will find Theorem 4.11 to be a useful characterization of these points.

Theorem 4.11. *The function $f_c(z) = z^2 + c$ has attracting fixed points iff $|1 + \sqrt{1 - 4c}| < 1$ or $|1 - \sqrt{1 - 4c}| < 1$, where the square root designates the principal square root function.*

Proof. The point z_0 is a fixed point for f_c iff $f_c(z_0) = z_0$. In other words, iff $z_0^2 - z_0 + c = 0$. By Theorem 1.5, the solutions to this equation are

$$z_0 = \frac{1 + \sqrt{1 - 4c}}{2} \quad \text{or} \quad z_0 = \frac{1 - \sqrt{1 - 4c}}{2},$$

where again the square root designates the principal square root function. Now, z_0 is an attracting point iff $|f_c'(z_0)| = |2z_0| < 1$. Combining this result with the solutions for z_0 gives our desired result. $\qquad\square$

Definition 4.11 (n-cycle). *An n-cycle for a function f is a set*

$$\{z_0,\, z_1, \ldots, z_{n-1}\}$$

of n complex numbers such that $z_k = f(z_{k-1})$, for $1 \le k \le n-1$ and $f(z_{n-1}) = z_0$.

Definition 4.12 (Attracting). *An n-cycle $\{z_0, z_1, \ldots, z_{n-1}\}$ for a function f is said to be* ***attracting*** *if $|g_n'(z_0)| < 1$, where g_n is the composition of f with itself n times. For example, if $n = 2$, then $g_2(z) = (f \circ f)(z) = f(f(z))$.*

Example 4.12. Example 4.10 shows that $\{-1 + i, -i\}$ is a 2-cycle for the function f_i. It is not an attracting 2-cycle because $g_2(z) = z^4 + 2iz^2 + i - 1$ and $g_2'(z) = 4z^3 + 4iz$. Hence $|g_2'(-1 + i)| = |4 + 4i|$, so $|g_2'(-1 + i)| > 1$.

In the exercises, we ask you to show that, if $\{z_0, z_1, \ldots, z_{n-1}\}$ is an attracting n-cycle for a function f, then not only does z_0 satisfy $|g_n'(z_0)| < 1$, but also that $|g_n'(z_k)| < 1$, for $k = 1, 2, \ldots, n - 1$.

One can prove that the large disk to the left of the cardioid in Figure 4.6 consists of those points c for which $f_c(z)$ has a 2-cycle. The large disks above and below the main cardioid disk are the points c for which $f_c(z)$ has a 3-cycle.

Continuing with this scheme, we see that the idea of n-cycles explains the appearance of the "buds" that you see on Figure 4.6 . It does not, however, begin to do justice to the enormous complexity of the entire set. Even Figures 4.7a and 4.7b are mere glimpses into its awesome beauty.

Exercises for Section 4.2 (Selected answers or hints are on page 437.)

1. Consider the function $f(z) = z^2 + 1$, where

$$N(z) = z - \frac{f(z)}{f'(z)} = \frac{z^2 - 1}{2z} = \frac{1}{2}\left(z - \frac{1}{z}\right).$$

   (a) Show that, if $\text{Im}(z_0) > 0$, the sequence $\{z_k\}$ formed by successive iterations of z_0 via $N(z)$ lies entirely within the upper half-plane.

   (b) Show that a similar result holds if $\text{Im}(z_0) < 0$.

   (c) Use induction to show that, if all the terms of the sequence $\{z_k\}$ are defined, then the sequence $\{z_k\}$ is real, provided z_0 is real.

   (d) Discuss whether $\{z_k\}$ converges to i if $\text{Im}(z_0) > 0$ and to $-i$ if $\text{Im}(z_0) < 0$.

2. Formulate and solve problems analogous to those in Exercise 1 for the function $f(z) = z^2 - 1$.

3. Prove that Newton's method always works for polynomials having degree 1 (functions of the form $f(z) = az + b$, where $a \neq 0$). How many iterations are necessary before Newton's method produces the solution $z = -\frac{b}{a}$ to $f(z) = 0$?

4. Consider the function $f_0(z) = z^2$ and an initial point z_0. Let $\{z_k\}$ be the sequence of iterates of z_0 generated by f_0. That is, $z_1 = f_0(z_0)$, $z_2 = f_0(z_1)$, and so on.

   (a) Show that, if $|z_0| < 1$, the sequence $\{z_k\}$ converges to 0.

   (b) Show that, if $|z_0| > 1$, the sequence $\{z_k\}$ is unbounded.

   (c) Show that, if $|z_0| = 1$, the sequence $\{z_k\}$ either converges to 1 or oscillates around the unit circle. Give a simple criterion that you can apply to z_0 that will reveal which of these two paths $\{z_k\}$ takes.

5. Show that the Julia set for $f_{-2}(z)$ is connected.

6. Determine the precise structure of the set K_{-2}.

7. Prove that if a complex number c is in the Mandelbrot set, then its conjugate $\bar{c}$ is also in the Mandelbrot set. Thus, the Mandelbrot set is symmetric about the x axis. *Hint*: Use mathematical induction.

8. Show that, if c is any real number greater than $\frac{1}{4}$, then c is not in the Mandelbrot set. *Note*: Combining this condition with Example 4.11 shows that the cusp in the cardioid section of the Mandelbrot set occurs precisely at $c = \frac{1}{4}$.

9. Find a value for c that is in the Mandelbrot set such that its negative, $-c$, is not in the Mandelbrot set.

10. Show that the points c that solve the inequalities of Theorem 4.11 form a cardioid. This cardioid is the main body of the Mandelbrot set shown in Figure 4.6. *Hint*: It may be helpful to write the inequalities of Theorem 4.11 as

$$\left| \frac{1}{2} + \sqrt{\frac{1}{4} - c} \right| < \frac{1}{2} \quad \text{or} \quad \left| \frac{1}{2} - \sqrt{\frac{1}{4} - c} \right| < \frac{1}{2}.$$

11. Use Theorem 4.11 and the paragraph immediately before it to show that the point $-\frac{1}{4}\sqrt{3}i$ belongs to the Mandelbrot set.

12. Suppose that $\{z_0, z_1\}$ is a 2-cycle for f.

    (a) Show that, if z_0 is attracting for $g_2(z)$, then so is the point z_1. *Hint*: Differentiate $g_2(z) = f(f(z))$ using the chain rule, and show that $g_2'(z_0) = g_2'(z_1)$.
    (b) Generalize part (a) to n-cycles.

13. Prove that $\lim_{k \to \infty} z_k = z_0$ in Theorem 4.10.

4.3 Geometric Series and Convergence Theorems

We begin this section by presenting a series of the form $\sum_{n=0}^{\infty} z^n$, which is called a **geometric series** and is one of the most important series in mathematics.

Theorem 4.12 (Geometric series). *If $|z| < 1$ the series $\sum_{n=0}^{\infty} z^n$ converges to $f(z) = \frac{1}{1-z}$ That is, if $|z| < 1$, then*

$$\sum_{n=0}^{\infty} z^n = 1 + z + z^2 + \cdots + z^k + \cdots = \frac{1}{1-z}. \tag{4.11}$$

If $|z| \geq 1$ the series diverges.

Proof. Suppose that $|z| < 1$. By Definition 4.4, we must show $\lim_{n \to \infty} S_n = \frac{1}{1-z}$, where

$$S_n = 1 + z + z^2 + \cdots + z^{n-1}. \tag{4.12}$$

Multiplying both sides of Equation (4.12) by z gives

$$zS_n = z + z^2 + z^3 + \cdots + z^{n-1} + z^n. \tag{4.13}$$

Subtracting Equation (4.13) from Equation (4.12) yields

$$(1 - z)S_n = 1 - z^n$$

so that

$$S_n = \frac{1}{1-z} - \frac{z^n}{1-z}. \tag{4.14}$$

Since $|z| < 1$, $\lim_{n \to \infty} z^n = 0$. (Can you *prove* this assertion? We ask you to do so in the exercises!) Hence $\lim_{n \to \infty} S_n = \frac{1}{1-z}$.

Now suppose $|z| \geq 1$. Clearly, $\lim_{n \to \infty} |z^n| \neq 0$, so $\lim_{n \to \infty} z^n \neq 0$ (see Exercise 17, Section 4.1). Thus, by the contrapositive of Theorem 4.5, $\sum_{n=0}^{\infty} z^n$ must diverge. $\square$

Corollary 4.2. *If $|z| > 1$ the series $\sum\limits_{n=1}^{\infty} z^{-n}$ converges to $f(z) = \frac{1}{z-1}$ That is, if $|z| > 1$ then*

$$\sum_{n=1}^{\infty} z^{-n} = z^{-1} + z^{-2} + \cdots + z^{-n} + \cdots = \frac{1}{z-1}, \quad \text{or equivalently,}$$

$$-\sum_{n=1}^{\infty} z^{-n} = -z^{-1} - z^{-2} - \cdots - z^{-n} - \cdots = \frac{1}{1-z}.$$

If $|z| \le 1$, the series diverges.

Proof. If we let $\frac{1}{z}$ take the role of z in Equation (4.11), we get

$$\sum_{n=0}^{\infty} \left(\frac{1}{z}\right)^n = \frac{1}{1 - \frac{1}{z}}, \quad \text{if} \quad \left|\frac{1}{z}\right| < 1.$$

Multiplying both sides of this equation by $\frac{1}{z}$ gives

$$\frac{1}{z} \sum_{n=0}^{\infty} \left(\frac{1}{z}\right)^n = \frac{1}{z-1}, \quad \text{if} \quad \left|\frac{1}{z}\right| < 1.$$

which, by Equation (4.10), is the same as

$$\sum_{n=0}^{\infty} \left(\frac{1}{z}\right)^{n+1} = \frac{1}{z-1} \quad \text{if} \quad \left|\frac{1}{z}\right| < 1.$$

But this expression is equivalent to saying that $\sum\limits_{n=1}^{\infty} (\frac{1}{z})^n = \frac{1}{z-1}$, if $1 < |z|$, which is what the corollary claims.

It is left as an exercise to show that the series diverges if $|z| \le 1$. $\square$

Corollary 4.3. *If $z \ne 1$ then for all n,*

$$\frac{1}{1-z} = 1 + z + z^2 + \cdots + z^{n-1} + \frac{z^n}{1-z}.$$

Proof. This result follows immediately from Equation (4.14). $\square$

Example 4.13. Show that $\sum\limits_{n=0}^{\infty} \frac{(1-i)^n}{2^n} = 1 - i$.

Solution:
If we set $z = \frac{1-i}{2}$, then $|z| = \frac{\sqrt{2}}{2} < 1$. By Theorem 4.12, the sum is

$$\frac{1}{1 - \frac{1-i}{2}} = \frac{2}{2 - 1 + i} = \frac{2}{1+i} = 1 - i.$$

Example 4.14. Evaluate $\sum\limits_{n=3}^{\infty} (\frac{i}{2})^n$.

Solution:

We can put this expression in the form of a geometric series:

$$\sum_{n=3}^{\infty} \left(\frac{i}{2}\right)^n = \sum_{n=3}^{\infty} \left(\frac{i}{2}\right)^3 \left(\frac{i}{2}\right)^{n-3}$$

$$= \left(\frac{i}{2}\right)^3 \sum_{n=3}^{\infty} \left(\frac{i}{2}\right)^{n-3} \quad \text{(by Equation (4.10) in Theorem 4.5)}$$

$$= \left(\frac{i}{2}\right)^3 \sum_{n=0}^{\infty} \left(\frac{i}{2}\right)^n \quad \text{(by reindexing)}$$

$$= \left(\frac{i}{2}\right)^3 \left(\frac{1}{1 - \frac{i}{2}}\right) \quad \text{(by Theorem 4.12 because } \left|\frac{i}{2}\right| = \frac{1}{2} < 1\text{)}$$

$$= \frac{1}{20} - \frac{i}{10} \quad \text{(by standard simplification procedures)}.$$

Remark 4.3. *The equalities given in Example 4.14 collectively illustrate an important point with regard to evaluating a geometric series whose beginning index is other than zero. The value of $\sum_{n=r}^{\infty} z^n$ equals $\frac{z^r}{1-z}$. If we think of z as the "ratio" by which any term of the series is multiplied to generate successive terms, we note that the sum of a geometric series equals $\frac{first\ term}{1-ratio}$, provided $|ratio| < 1$.*

The geometric series is used in the proof of Theorem 4.12, which is known as the **ratio test**. It is one of the most commonly used tests for determining the convergence or divergence of series. The proof is similar to the one used for real series, and we leave it for you to do.

Theorem 4.13 (d'Alembert's ratio test). *If $\sum_{n=0}^{\infty} \zeta_n$ is a complex series with the property that*

$$\lim_{n \to \infty} \frac{|\zeta_{n+1}|}{|\zeta_n|} = L,$$

then the series is absolutely convergent if $L < 1$ and divergent if $L > 1$.

Example 4.15. Show that $\sum_{n=0}^{\infty} \frac{(1-i)^n}{n!}$ converges.

Solution:

Using the ratio test, we find that

$$\lim_{n \to \infty} \frac{|(1-i)^{n+1}/(n+1)!|}{|(1-i)^n/n!|} = \lim_{n \to \infty} \frac{n!|1-i|}{(n+1)!} = \lim_{n \to \infty} \frac{|1-i|}{n+1}$$

$$= \lim_{n \to \infty} \frac{\sqrt{2}}{n+1} = 0 = L.$$

Because $L < 1$, the series converges.

Example 4.16. Show that the series $\sum_{n=0}^{\infty} \frac{(z-i)^n}{2^n}$ converges for all values of z in the disk $|z-i| < 2$ and diverges if $|z - i| > 2$.

Solution:

Using the ratio test, we see that

$$\lim_{n\to\infty} \frac{|(z-i)^{n+1}/2^{n+1}|}{|(z-i)^n/2^n|} = \lim_{n\to\infty} \frac{|z-i|}{2} = \frac{|z-i|}{2} = L.$$

If $|z-i| < 2$, then $L < 1$, and the series converges. If $|z-i| > 2$, then $L > 1$, and the series diverges.

Our next result, known as the **root test**, is slightly more powerful than the ratio test. Before we present this test, we need to discuss a rather sophisticated idea used with it—the *limit supremum*.

Definition 4.13 (Limit Supremum). *Let $\{t_n\}$ be a sequence of positive real numbers. The* **limit supremum** *of the sequence (denoted $\lim\limits_{n\to\infty} \sup t_n$) is the smallest real number L having the property that for any $\varepsilon > 0$, there are at most finitely many terms in the sequence that are larger than $L + \varepsilon$. If there is no such number L, then $\lim\limits_{n\to\infty} \sup t_n = \infty$.*

Example 4.17. The limit supremum of the sequence

$$\{t_n\} = \{4.1,\ 5.1,\ 4.01,\ 5.01,\ 4.001,\ 5.001, \ldots\} \quad \text{is} \quad \lim_{n\to\infty} \sup t_n = 5,$$

because if we set $L = 5$, then for any $\varepsilon > 0$, there are only finitely many terms in the sequence larger than $L + \varepsilon = 5 + \varepsilon$. Additionally, if L is smaller than 5, then by setting $\varepsilon = 5 - L$, we can find infinitely many terms in the sequence larger than $L + \varepsilon$ (because $L + \varepsilon = 5$).

Example 4.18. The limit supremum of the sequence

$$\{t_n\} = \{1,\ 2,\ 3,\ 1,\ 2,\ 3,\ 1,\ 2,\ 3,\ 1,\ 2,\ 3, \ldots\} \quad \text{is} \quad \lim_{n\to\infty} \sup t_n = 3,$$

because if we set $L = 3$, then for any $\varepsilon > 0$, there are only finitely many terms (actually, there are none) in the sequence larger than $L + \varepsilon = 3 + \varepsilon$. Additionally, if L is smaller than 3, then by setting $\varepsilon = \frac{3-L}{2}$ we can find infinitely many terms in the sequence larger than $L + \varepsilon$ (because $L + \varepsilon < 3$), as the following calculation shows:

$$L + \varepsilon = L + \frac{3-L}{2} = \frac{3+L}{2} = \frac{3}{2} + \frac{L}{2} < \frac{3}{2} + \frac{3}{2} = 3.$$

Example 4.19. The limit supremum of the Fibonacci sequence

$$\{t_n\} = \{1,\ 1,\ 2,\ 3,\ 5,\ 8,\ 13,\ 21,\ 34, \ldots\} \quad \text{is} \quad \lim_{n\to\infty} \sup t_n = \infty.$$

(The Fibonacci sequence satisfies the relation $t_n = t_{n-1} + t_{n-2}$ for $n > 2$.)

The limit supremum is a powerful idea because the limit supremum of a sequence always exists, which is not true for the ordinary limit. However, Example 4.20 illustrates the fact that, if the limit of a sequence does exist, then it will be the same as the limit supremum.

Example 4.20. The sequence

$$\{t_n\} = \{1 + \frac{1}{n}\}$$
$$= \{2,\ 1.5,\ 1.3\overline{3},\ 1.25,\ 1.2,\ \ldots\} \quad \text{has} \quad \lim_{n\to\infty} \sup t_n = 1.$$

We leave verification of this as an exercise.

Theorem 4.14 (Root test). *Suppose the series* $\sum\limits_{n=0}^{\infty} \zeta_n$ *has* $\lim\limits_{n\to\infty} \sup |\zeta_n|^{\frac{1}{n}} = L$. *Then the series is absolutely convergent if* $L < 1$ *and divergent if* $L > 1$.

Proof. Suppose first that $L < 1$. We can select a number r such that $L < r < 1$. By definition of the limit supremum, only finitely many terms in the sequence $\{|\zeta_n|^{\frac{1}{n}}\}$ exceed r, so there exists a positive integer N such that for all $n > N$ we have $|\zeta_n|^{\frac{1}{n}} < r$. That is, $|\zeta_n| < r^n$ for all $n > N$. For $r < 1$, Theorem 4.12 implies that $\sum\limits_{n=N+1}^{\infty} r^n$ converges. But then Theorem 4.8 implies that $\sum\limits_{n=N+1}^{\infty} |\zeta_n|$ converges, and hence so does $\sum\limits_{n=0}^{\infty} |\zeta_n|$. Corollary 4.1 then guarantees that $\sum\limits_{n=0}^{\infty} \zeta_n$ converges.

Now suppose that $L > 1$. We can select a number r such that $1 < r < L$. Again, by definition of the limit supremum we conclude that $|\zeta_n|^{\frac{1}{n}} > r$ for infinitely many n. But this condition means that $|\zeta_n| > r^n$ for infinitely many n, and as $r > 1$, this implies that ζ_n does not converge to 0. By Theorem 4.5, $\sum\limits_{n=0}^{\infty} \zeta_n$ does not converge. $\square$

Note that, in applying either Theorem 4.13 or 4.14, if $L = 1$ the convergence or divergence of the series is unknown, and further analysis is required to determine the true state of affairs.

Exercises for Section 4.3 (Selected answers or hints are on page 438.)

1. Evaluate

   (a) $\sum\limits_{n=0}^{\infty} \frac{(1+i)^n}{2^n}$.

   (b) $\sum\limits_{n=0}^{\infty} \left(\frac{1}{2+i}\right)^n$.

2. Show that $\sum\limits_{n=0}^{\infty} \frac{(z+i)^n}{2^n}$ converges for all values of z in the disk $D_2(-i) = \{z : |z+i| < 2\}$ and diverges if $|z+i| > 2$.

3. Is the series $\sum\limits_{n=0}^{\infty} \frac{(4i)^n}{n!}$ convergent? Why or why not?

4. Use the ratio test to show that the following series converge.

   (a) $\sum\limits_{n=0}^{\infty} \left(\frac{1+i}{2}\right)^n$.

   (b) $\sum\limits_{n=1}^{\infty} \frac{(1+i)^n}{n2^n}$.

   (c) $\sum\limits_{n=1}^{\infty} \frac{(1+i)^n}{n!}$.

   (d) $\sum\limits_{n=0}^{\infty} \frac{(1+i)^{2n}}{(2n+1)!}$.

5. Use the ratio test to find a disk in which the following series converge.

   (a) $\sum_{n=0}^{\infty} (1+i)^n z^n$.

   (b) $\sum_{n=0}^{\infty} \frac{z^n}{(3+4i)^n}$.

   (c) $\sum_{n=0}^{\infty} \frac{(z-i)^n}{(3+4i)^n}$.

   (d) $\sum_{n=0}^{\infty} \frac{(z-3-4i)^n}{2^n}$.

6. Establish the claim in the proof of Theorem 4.12 that, if $|z| < 1$, then $\lim_{n \to \infty} z^n = 0$.

7. In the geometric series, show that if $|z| > 1$, then $\lim_{n \to \infty} |S_n| = \infty$.

8. Prove that the series in Corollary 4.2 diverges if $|z| \leq 1$.

9. Prove Theorem 4.13.

10. Give a rigorous argument to show that $\limsup_{n \to \infty} t_n = 1$ in Example 4.20.

11. For $|z| < 1$, let $f(z) = \sum_{n=0}^{\infty} z^{(2^n)} = z + z^2 + z^4 + \cdots + z^{2^n} + \cdots$. Show that $f(z) = z + f(z^2)$.

12. This exercise makes interesting use of the geometric series.

    (a) Use the formula for geometric series with $z = re^{i\theta}$, where $r < 1$, to show that

    $$\sum_{n=0}^{\infty} z^n = \sum_{n=0}^{\infty} r^n e^{in\theta} = \frac{1 - r\cos\theta + ir\sin\theta}{1 + r^2 - 2r\cos\theta}.$$

    (b) Use part (a) to obtain

    $$\sum_{n=0}^{\infty} r^n \cos n\theta = \frac{1 - r\cos\theta}{1 + r^2 - 2r\cos\theta}, \quad \text{and}$$

    $$\sum_{n=0}^{\infty} r^n \sin n\theta = \frac{r\sin\theta}{1 + r^2 - 2r\cos\theta}.$$

4.4 Power Series Functions

Suppose that we have a series $\sum_{n=0}^{\infty} \zeta_n$, where $\zeta_n = c_n(z-\alpha)^n$. If α and the collection of c_n are fixed complex numbers, we get different series by selecting different values for z. For example, if $\alpha = 2$ and $c_n = \frac{1}{n!}$ for all n, we get the series $\sum_{n=0}^{\infty} \frac{1}{n!}(\frac{i}{2}-2)^n$ if $z = \frac{i}{2}$ and $\sum_{n=0}^{\infty} \frac{1}{n!}(2+i)^n$ if $z = 4+i$. Note that, when $\alpha = 0$ and $c_n = 1$ for all n, we get the geometric series. The collection of points for which the series $\sum_{n=0}^{\infty} c_n(z-\alpha)^n$ converges is the domain of a function $f(z) = \sum_{n=0}^{\infty} c_n(z-\alpha)^n$, which we call a **power series function**. Technically, this series is undefined if $z = \alpha$ and $n = 0$ because 0^0 is undefined. We get around this difficulty by stipulating that the series $\sum_{n=0}^{\infty} c_n(z-\alpha)^n$ is really compact notation for $c_0 + \sum_{n=1}^{\infty} c_n(z-\alpha)^n$. In this section we present some results that are useful in helping establish properties of functions defined by power series.

Theorem 4.15. *Suppose that $f(z) = \sum_{n=0}^{\infty} c_n(z - \alpha)^n$. Then the set of points z for which the series converges is one of the following:*

 i. the single point $z = \alpha$;

 ii. the disk $D_\rho(\alpha) = \{z : |z - \alpha| < \rho\}$, along with part (either none, some, or all) of the circle $C_\rho(\alpha) = \{z : |z - \alpha| = \rho\}$;

 iii. the entire complex plane.

Proof. By Theorem 4.14, the series converges absolutely at those values of z for which $\limsup_{n \to \infty} |c_n(z - \alpha)^n|^{\frac{1}{n}} < 1$. This condition is the same as requiring

$$|z - \alpha|\left(\limsup_{n \to \infty} |c_n|^{\frac{1}{n}}\right) < 1. \tag{4.15}$$

There are three possibilities to consider for the value of $\limsup_{n \to \infty} |c_n|^{\frac{1}{n}}$. If the limit supremum equals ∞, Inequality (4.15) holds iff $z = \alpha$, which is case (i). If $0 < \limsup_{n \to \infty} |c_n|^{\frac{1}{n}} < \infty$, Inequality (4.15) holds iff $|z - \alpha| < \frac{1}{\limsup_{n \to \infty} |c_n|^{\frac{1}{n}}}$ $\left(\text{i.e., iff } z \in D_\rho(\alpha), \text{ where } \rho = \frac{1}{\limsup_{n \to \infty} |c_n|^{\frac{1}{n}}}\right)$ which is case (ii). Finally, if the limit supremum equals 0, the left side of Inequality (4.15) will be 0 for any value of z, which is case (iii). We are unable to say for sure what happens with respect to convergence on $C_\rho(\alpha) = \{z : |z - \alpha| = \rho\}$. You will see in the exercises that there are various possibilities. $\square$

Another way to phrase case (ii) of Theorem 4.15 is to say that the power series $f(z) = \sum_{n=0}^{\infty} c_n(z - \alpha)^n$ converges if $|z - \alpha| < \rho$ and diverges if $|z - \alpha| > \rho$. We call the number ρ the **radius of convergence** of the power series (see Figure 4.8). For case (i) of Theorem 4.15, we say that the radius of convergence is zero and that the radius of convergence is infinity for case (iii).

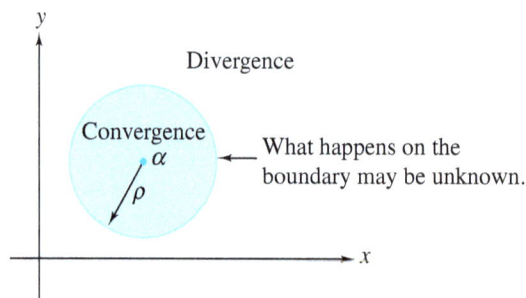

Figure 4.8: The radius of convergence of a power series

Theorem 4.16. *For the power series function $f(z) = \sum_{n=0}^{\infty} c_n(z - \alpha)^n$ we can find ρ, its radius of convergence, by any of the following methods:*

 i. Cauchy's root test: $\rho = \dfrac{1}{\lim_{n \to \infty} |c_n|^{\frac{1}{n}}}$ (provided the limit exists).

ii. Cauchy-Hadamard formula: $\rho = \dfrac{1}{\lim\limits_{n\to\infty} \sup |c_n|^{\frac{1}{n}}}$ *(this limit always exists).*

iii. d'Alembert's ratio test: $\rho = \dfrac{1}{\lim\limits_{n\to\infty} |\frac{c_{n+1}}{c_n}|}$ *(provided the limit exists).*

We set $\rho = \infty$ if the limit equals 0 and $\rho = 0$ if the limit equals ∞.

Proof. If you examine carefully the proof of Theorem 4.15, you will see that we have already proved (i) and (ii). They follow directly from Inequality (4.15) and the fact that the limit supremum equals the limit whenever the limit exists. We can show (iii) by using the ratio test. We leave the details as an exercise. $\square$

We now give an example illustrating each of these cases.

Example 4.21. The series $\sum\limits_{n=0}^{\infty} (\frac{n+2}{3n+1})^n (z-4)^n$ has radius of convergence 3 by Cauchy's root test because $\lim\limits_{n\to\infty} |c_n|^{\frac{1}{n}} = \lim\limits_{n\to\infty} \frac{n+2}{3n+1} = \frac{1}{3}$.

Example 4.22. The series $\sum\limits_{n=1}^{\infty} c_n z^n = 4z + 5^2 z^2 + 4^3 z^3 + 5^4 z^4 + 4^5 z^5 + \cdots$ has radius of convergence $\frac{1}{5}$ by the Cauchy-Hadamard formula because $\{|c_n|^{\frac{1}{n}}\} = \{4, 5, 4, 5, \ldots\}$, so $\lim\limits_{n\to\infty} \sup |c_n|^{\frac{1}{n}} = 5$.

Example 4.23. The series $\sum\limits_{n=0}^{\infty} \frac{1}{n!} z^n$ has radius of convergence ∞ by the ratio test because $\lim\limits_{n\to\infty} \left| \frac{n!}{(n+1)!} \right| = \lim\limits_{n\to\infty} \left| \frac{1}{n+1} \right| = 0$.

We come now to the main result of this section.

Theorem 4.17. Suppose that the function $f(z) = \sum\limits_{n=0}^{\infty} c_n (z-\alpha)^n$ has radius of convergence $\rho > 0$. Then

i. f is infinitely differentiable for all $z \in D_\rho(\alpha)$. In fact

ii. for all k, $f^{(k)}(z) = \sum\limits_{n=k}^{\infty} n(n-1)\cdots(n-k+1) c_n (z-\alpha)^{n-k}$; and

iii. $c_k = \dfrac{f^{(k)}(\alpha)}{k!}$, where $f^{(k)}$ denotes the kth derivative of f. (When $k = 0$, $f^{(k)}$ denotes the function f itself so that $f^{(0)}(z) = f(z)$ for all z.)

Proof. Remarkably, the entire proof hinges on verifying (ii) for the simple case when $k = 1$. The cases in (ii) for $k \geq 2$ follow by induction. For instance, we get the case when $k = 2$ by applying the result for $k = 1$ to the series $f'(z) = \sum\limits_{n=1}^{\infty} n c_n (z-\alpha)^{n-1}$. Also, (i) is an automatic consequence of (ii), because (ii) gives a formula for computing derivatives of all orders in addition to assuring us of their existence. Finally, (iii) follows by setting $z = \alpha$ in (ii), as all the terms drop out except when $n = k$, giving us $f^{(k)}(\alpha) = k(k-1)\cdots(k-k+1) c_k$. Solving for c_k gives the desired result.

Verifying (ii) when $k = 1$, however, is no simple task. We begin by defining the following:

$$g(z) = \sum_{n=1}^{\infty} nc_n(z - \alpha)^{n-1},$$

$$S_j(z) = \sum_{n=0}^{j} c_n(z - \alpha)^n,$$

$$R_j(z) = \sum_{n=j+1}^{\infty} c_n(z - \alpha)^n.$$

Here $S_j(z)$ is simply the $(j + 1)$st partial sum of the series $f(z)$, and $R_j(z)$ is the sum of the remaining terms of that series. We leave as an exercise to show that the radius of convergence for $g(z)$ is ρ, the same as that of $f(z)$. For a fixed $z_0 \in D_\rho(\alpha)$, we must prove that $f'(z_0) = g(z_0)$; that is, we must prove that $\lim_{z \to z_0} \frac{f(z) - f(z_0)}{z - z_0} = g(z_0)$. We do so by showing that for all $\varepsilon > 0$ there exists $\delta > 0$ such that, if $z \in D_\rho(\alpha)$ with $0 < |z - z_0| < \delta$, then $|\frac{f(z) - f(z_0)}{z - z_0} - g(z_0)| < \varepsilon$.

Let $z_0 \in D_\rho(\alpha)$ and $\varepsilon > 0$ be given. Choose $r < \rho$ so that $z_0 \in D_r(\alpha)$. We choose δ to be small enough so that $D_\delta(z_0) \subset D_r(\alpha) \subset D_\rho(\alpha)$ (see Figure 4.9) and also small enough to satisfy an additional restriction, which we shall specify in a moment.

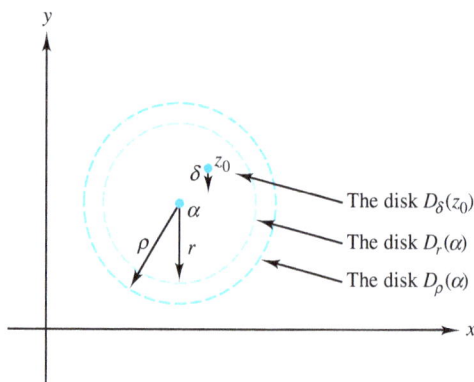

Figure 4.9: Choosing δ to prove that $f'(z_0) = g(z_0)$

Because $f(z) = S_j(z) + R_j(z)$, simplifying the right side of the following equation reveals that for all j,

$$\left[\frac{f(z) - f(z_0)}{z - z_0} - g(z_0)\right] = \left[\frac{S_j(z) - S_j(z_0)}{z - z_0} - S_j{}'(z_0)\right]$$
$$+ \left[S_j{}'(z_0) - g(z_0)\right] + \left[\frac{R_j(z) - R_j(z_0)}{z - z_0}\right], \tag{4.16}$$

where $S_j{}'(z_0)$ is the derivative of the function S_j evaluated at z_0. Equation (4.16) has the general form $A = B + C + D$. By the triangle inequality,

$$|A| = |B + C + D| \leq |B| + |C| + |D|,$$

so our proof will be complete if we can show that for a small enough value of δ, each of the expressions $|B|$, $|C|$, and $|D|$ is less than $\frac{\varepsilon}{3}$.

Calculation for $|D|$:

$$\left| \frac{R_j(z) - R_j(z_0)}{z - z_0} \right| = \left| \frac{1}{z - z_0} \left(\sum_{n=j+1}^{\infty} c_n[(z - \alpha)^n - (z_0 - \alpha)^n] \right) \right|$$

$$< \sum_{n=j+1}^{\infty} |c_n| \left| \frac{(z - \alpha)^n - (z_0 - \alpha)^n}{z - z_0} \right|,$$

where the last inequality follows from Exercise 12, Section 4.1.

As an exercise, we ask you to show that

$$\left| \frac{(z - \alpha)^n - (z_0 - \alpha)^n}{z - z_0} \right| < n r^{n-1}. \tag{4.17}$$

Assuming the validity of this inequality, we then get

$$\left| \frac{R_j(z) - R_j(z_0)}{z - z_0} \right| < \sum_{n=j+1}^{\infty} |c_n| n r^{n-1}. \tag{4.18}$$

Since $r < \rho$, the series $\sum_{n=1}^{\infty} |c_n| n r^{n-1}$ converges (can you explain why?). Thus the tail part of the series, which is the right side of Inequality (4.18), can certainly be made less than $\frac{\varepsilon}{3}$ if we choose j large enough—say, $j \geq N_1$.

Calculation for $|C|$:

Since $S_j{}'(z_0) = \sum_{n=1}^{j} n c_n (z_0 - \alpha)^{n-1}$, it is clear that $\lim_{j \to \infty} S_j(z_0) = g(z_0)$. Thus there is an integer N_2 such that if $j \geq N_2$, then $|S_j{}'(z_0) - g(z_0)| < \frac{\varepsilon}{3}$.

Calculation for $|B|$:

We define $N = \max\{N_1, N_2\}$. Because $S_N(z)$ is a polynomial, $S_N{}'(z_0)$ exists. This means we can find δ small enough that it complies with the restriction previously placed on it as well as ensuring that

$$\left| \frac{S_N(z) - S_N(z_0)}{z - z_0} - S_N{}'(z_0) \right| < \frac{\varepsilon}{3}$$

whenever $z \in D_\rho(\alpha)$, with $0 < |z - z_0| < \delta$. Using this value of N for j in Equation (4.16), together with our chosen δ, yields conclusion (ii) and hence the entire theorem. $\square$

Example 4.24. Show that $\sum_{n=0}^{\infty} (n + 1) z^n = \frac{1}{(1-z)^2}$ for all $z \in D_1(0)$.

Solution:

We know from Theorem 4.12 that $f(z) = \frac{1}{1-z} = \sum_{n=0}^{\infty} z^n$ for all $z \in D_1(0)$. If we set $k = 1$ in

Theorem 4.17, part (ii), then $f'(z) = \frac{1}{(1-z)^2} = \sum_{n=1}^{\infty} n z^{n-1} = \sum_{n=0}^{\infty} (n + 1) z^n$, for all $z \in D_1(0)$.

Example 4.25. The Bessel function of order zero is defined by

$$J_0(z) = \sum_{n=0}^{\infty} \frac{(-1)^n}{(n!)^2} \left(\frac{z}{2} \right)^{2n} = 1 - \frac{z^2}{2^2} + \frac{z^4}{2^4 4^4} - \frac{z^6}{2^2 4^2 6^2} + \cdots,$$

127

and termwise differentiation shows that its derivative is

$$J_0{}'(z) = \sum_{n=0}^{\infty} \frac{(-1)^{n+1}}{n!(n+1)!}\left(\frac{z}{2}\right)^{2n+1} = \frac{-z}{2} + \frac{1}{1!2!}\left(\frac{z}{2}\right)^3 - \frac{1}{2!3!}\left(\frac{z}{2}\right)^5 + \cdots.$$

We leave as an exercise to show that the radius of convergence of these series is infinity. The Bessel function $J_1(z)$ of order 1 is known to satisfy the differential equation $J_1(z) = -J_0{}'(z)$.

Exercises for Section 4.4 (Selected answers or hints are on page 439.)

1. Prove part (iii) of Theorem 4.16.

2. Consider the series $\sum\limits_{n=0}^{\infty} z^n$, $\sum\limits_{n=1}^{\infty} \frac{z^n}{n^2}$, and $\sum\limits_{n=1}^{\infty} \frac{z^n}{n}$.

    (a) Show that each series has radius of convergence 1.

    (b) Show that the first series converges nowhere on $C_1(0) = \{z : |z| = 1\}$.

    (c) Show that the second series converges everywhere on $C_1(0)$.

    (d) It turns out that the third series converges everywhere on $C_1(0)$, except at the point $z = 1$. This is not easy to prove. Give it a try.

3. Find the radius of convergence of the following.

    (a) $g(z) = \sum\limits_{n=0}^{\infty} (-1)^n \frac{z^n}{(2n)!}$.

    (b) $h(z) = \sum\limits_{n=0}^{\infty} n! z^n$.

    (c) $f(z) = \sum\limits_{n=0}^{\infty} \left(\frac{4n^2}{2n+1} - \frac{6n^2}{3n+4}\right)^n z^n$.

    (d) $g(z) = \sum\limits_{n=0}^{\infty} \frac{(n!)^2}{(2n)!} z^n$.

    (e) $h(z) = \sum\limits_{n=0}^{\infty} (2 - (-1)^n)^n z^n$.

    (f) $f(z) = \sum\limits_{n=0}^{\infty} \frac{n(n-1)z^n}{(3+4i)^n}$.

    (g) $g(z) = \sum\limits_{n=0}^{\infty} \left(\frac{3n+7}{4n+2}\right)^n z^n$.

    (h) $h(z) = \sum\limits_{n=0}^{\infty} \frac{2^n}{1+3^n} z^n$.

    (i) $g(z) = \sum\limits_{n=0}^{\infty} z^{2n}$.

    (j) $g(z) = \sum\limits_{n=0}^{\infty} \frac{n^n}{n!} z^n$. *Hint:* $\lim\limits_{n\to\infty} (1 + \frac{1}{n})^n = e$.

4. Show that $\sum\limits_{n=0}^{\infty} (n+1)^2 z^n = \frac{1+z}{(1-z)^3}$. For what values of z is this valid?

128

5. Suppose that $\sum\limits_{n=0}^{\infty} c_n z^n$ has radius of convergence R. Show that $\sum\limits_{n=0}^{\infty} c_n^2 z^n$ has radius of convergence R^2.

6. Does there exist a power series $\sum\limits_{n=0}^{\infty} c_n z^n$ that converges at $z_1 = 4 - i$ and diverges at $z_2 = 2 + 3i$? Why or why not?

7. Verify part (ii) of Theorem 4.17 for all k by using mathematical induction.

8. This exercise establishes that the radius of convergence for g given in Theorem 4.17 is ρ, the same as that of the function f.

   (a) Explain why the radius of convergence for g is $\dfrac{1}{\lim\limits_{n\to\infty} \sup |nc_n|^{\frac{1}{n-1}}}$.

   (b) Show that $\lim\limits_{n\to\infty} \sup |n|^{\frac{1}{n-1}} = 1$. *Hint:* The lim sup equals the limit. Show that $\lim\limits_{n\to\infty} \dfrac{\log n}{n-1} = 0$.

   (c) Assuming that $\lim\limits_{n\to\infty} \sup |c_n|^{\frac{1}{n-1}} = \lim\limits_{n\to\infty} \sup |c_n|^{\frac{1}{n}}$, show that the conclusion for this exercise follows.

   (d) Verify the truth of the assumption made in part (c).

9. Here we establish the validity of Inequality 4.17 in the proof of Theorem 4.17.

   (a) Show that

   $$\left| \frac{s^n - t^n}{s - t} \right| = |s^{n-1} + s^{n-2}t + s^{n-3}t^2 + \cdots + st^{n-2} + t^{n-1}|$$
   $$\leq |s^{n-1}| + |s^{n-2}t| + |s^{n-3}t^2| + \cdots + |st^{n-2}| + |t^{n-1}|,$$

   where s and t are arbitrary complex numbers, $s \neq t$.

   (b) Explain why, in Inequality 4.17, $|z - \alpha| < r$ and $|z_0 - \alpha| < r$.

   (c) Let $s = z - \alpha$ and $t = z_0 - \alpha$ in part (a) to establish Inequality 4.17.

10. Show that the radius of convergence of the series for $J_0(z)$ and $J_0'(z)$ in Example 4.25 is infinity.

11. Consider the series obtained by substituting for the complex number z the real number x in the Maclaurin series for $\sin x$. Where does this series converge?

12. Show that, for $|z - i| < \sqrt{2}$, $\dfrac{1}{1-z} = \sum\limits_{n=0}^{\infty} \dfrac{(z-i)^n}{(1-i)^{n+1}}$.

   Hint: $\dfrac{1}{1-z} = \dfrac{1}{(1-i)-(z-i)} = \dfrac{1}{1-i}\left[\dfrac{1}{1 - \frac{z-i}{1-i}} \right]$. Now use Theorem 4.12.

129

Chapter 5

Elementary Functions

Overview

How should complex-valued functions such as e^z, $\log z$, $\sin z$, and the like, be defined? Clearly, any responsible definition should satisfy the following criteria.

- The functions so defined must give the same values as the corresponding functions for real variables when the number z is a real number.

- As much as possible, the properties of these new functions must correspond with their real counterparts. For example, we would want $e^{z_1+z_2} = e^{z_1}e^{z_2}$ to be valid regardless of whether z were real or complex.

These requirements may seem like a tall order to fill. There is a procedure, however, that offers promising results. It is to put the expansion of the real functions e^x, $\sin x$, and so on, as power series in complex form. We use this strategy in this chapter.

5.1 The Complex Exponential Function

Recall that the real exponential function can be represented by the power series $e^x = \sum\limits_{n=0}^{\infty} \frac{1}{n!}x^n$. Thus it is only natural to define the complex exponential e^z, also written as $\exp(z)$, in the following way.

Definition 5.1 (The Complex Exponential).

$$e^z = \exp(z) = \sum_{n=0}^{\infty} \frac{1}{n!}z^n.$$

Clearly, this definition agrees with that of the real exponential function when z is a real number. We now show that this complex exponential has two of the key properties associated with its real counterpart and verify the identity $e^{i\theta} = \cos\theta + i\sin\theta$, which, back in Chapter 1 (see Identity (1.32) of Section 1.4) we promised to establish.

Theorem 5.1. *The function $\exp z$ is an entire function satisfying the following conditions.*

    i. $\exp{}'(z) = \exp(z) = e^z$ *(using Leibniz notation we write $\frac{d}{dz}e^z = e^z$).*

ii. $\exp(z_1 + z_2) = \exp(z_1)\exp(z_2)$ *(i.e.,* $e^{z_1+z_2} = e^{z_1}e^{z_2}$*).*

iii. If θ is a real number, then $e^{i\theta} = \cos\theta + i\sin\theta$.

Proof. By the ratio test (check Example 4.23), the series in Definition 5.1 has an infinite radius of convergence, so $\exp(z)$ is entire by Theorem 4.17, part (i).

Using Theorem 4.17, part (ii), we get

$$\exp{}'(z) = \sum_{n=1}^{\infty} \frac{n}{n!} z^{n-1} = \sum_{n=1}^{\infty} \frac{1}{(n-1)!} z^{n-1} = \sum_{n=0}^{\infty} \frac{1}{n!} z^n = \exp(z),$$

which gives us part (i) of Theorem 5.1

To prove part (ii), we let ζ be an arbitrary complex number and define $g(z)$ to be

$$g(z) = \exp(z)\exp(\zeta - z).$$

Using the product rule, chain rule, and part (i), we have

$$g'(z) = \exp(z)\exp(\zeta - z) + \exp(z)[-\exp(\zeta - z)] = 0 \text{ for all } z.$$

According to Theorem 3.7, this result implies that the function g must be constant. Thus, for all z, $g(z) = g(0)$. Since $\exp(0) = 1$ (verify!), we deduce

$$g(z) = g(0) = \exp(0)\exp(\zeta - 0) = \exp(\zeta).$$

Hence, for all z,

$$g(z) = \exp(z)\exp(\zeta - z) = \exp(\zeta).$$

Setting $z = z_1$ and letting $\zeta = z_1 + z_2$, we get

$$\exp(z_1)\exp(z_1 + z_2 - z_1) = \exp(z_1 + z_2),$$

which simplifies to our desired result.

To prove part (iii), we let θ be a real number. By Definition 5.1,

$$\begin{aligned}
e^{i\theta} &= \exp(i\theta) \\
&= \sum_{n=0}^{\infty} \frac{1}{n!}(i\theta)^n \\
&= \sum_{n=0}^{\infty} \left[\frac{1}{(2n)!}(i\theta)^{2n} + \frac{1}{(2n+1)!}(i\theta)^{2n+1} \right] \quad \text{(separating odd and even exponents)} \\
&= \sum_{n=0}^{\infty} \left[\frac{1}{(2n)!}(i^2)^n \theta^{2n} + \frac{1}{(2n+1)!} i(i^2)^n \theta^{2n+1} \right] \\
&= \sum_{n=0}^{\infty} (-1)^n \frac{\theta^{2n}}{(2n)!} + i \sum_{n=0}^{\infty} (-1)^n \frac{\theta^{2n+1}}{(2n+1)!} \\
&= \cos\theta + i\sin\theta \quad \text{(by the series representations for the real-valued sine and cosine).}
\end{aligned}$$

$\square$

Note that parts (ii) and (iii) of the Theorem 5.1 combine to verify DeMoivre's formula, which we introduced in Section 1.5—see Identity (1.40). Further, if $z = x + iy$, we have

$$\exp(z) = e^z = e^{x+iy} = e^x e^{iy} = e^x(\cos y + i \sin y). \tag{5.1}$$

Some texts start with Identity (5.1) as the definition for $\exp(z)$. In the exercises, we show that this is a natural approach from the standpoint of differential equations.

The notation $\exp(z)$ is preferred over e^z in some situations. For example, the number $\exp(\frac{1}{5}) = 1.22140275816017\ldots$ is the value of $\exp(z)$ when $z = \frac{1}{5}$ and equals the positive fifth root of $e = 2.71828182845904\ldots$. The notation $e^{\frac{1}{5}}$, however, is ambiguous and might be interpreted as any of the complex fifth roots of the number e that we discussed in Section 1.5:

$$e^{\frac{1}{5}} \approx 1.22140275816017 \left(\cos \frac{2\pi k}{5} + i \sin \frac{2\pi k}{5} \right), \quad \text{for} \quad k = 0,\, 1,\, \ldots,\, 4.$$

To prevent confusion, we often use $\exp(z)$ to denote the single-valued exponential function.

We now explore some additional properties of $\exp(z)$. From Identity 5.1 it follows that

$$\begin{aligned}
e^{z+i2n\pi} &= e^z, & &\text{for all } z, \text{ provided } n \text{ is an integer,} & (5.2)\\
e^z &= 1, & &\text{iff } z = i2n\pi, \text{ where } n \text{ is an integer, and} & (5.3)\\
e^{z_1} &= e^{z_2}, & &\text{iff } z_2 = z_1 + i2n\pi \text{ for some integer } n. & (5.4)
\end{aligned}$$

For example, because Identity (5.1) involves the periodic functions $\cos y$ and $\sin y$, any two points in the z plane that lie on the same vertical line with their imaginary parts differing by an integral multiple of 2π are mapped onto the same point in the w plane. Thus the complex exponential function is periodic with period $2\pi i$, which establishes Equation (5.2). We leave the verification of Equations (5.3) and (5.4) as exercises.

Example 5.1. For any integer n, the points

$$z_n = \frac{5}{4} + i\left(\frac{11\pi}{6} + 2n\pi \right)$$

in the z plane (*i.e.*, the xy plane) are mapped onto the single point

$$w_0 = \exp(z_n) = e^{\frac{5}{4}}\left(\cos \frac{11\pi}{6} + i \sin \frac{11\pi}{6} \right) = \frac{\sqrt{3}}{2} e^{\frac{5}{4}} - i\frac{1}{2}e^{\frac{5}{4}} \approx 3.02 - 1.75i$$

in the w plane (*i.e.*, the uv plane), as indicated in Figure 5.1.

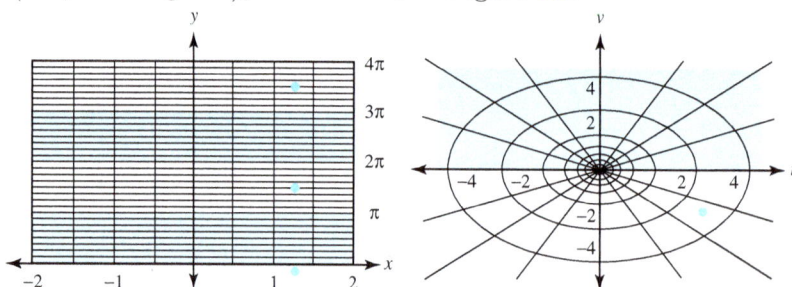

Figure 5.1: The points $\{z_n\}$ in the z plane and their image $w_0 = \exp(z_n)$ in the w plane

Let's look at the range of the exponential function. If $z = x + iy$, we see from Identity (5.1)—$e^z = e^x e^{iy} = e^x(\cos y + i \sin y)$—that e^z can never equal zero, as e^x is never zero, and the cosine and sine functions are never zero at the same point. Suppose, then, that $w = e^z \neq 0$. If we write w in its exponential form as $w = \rho e^{i\phi}$, Identity (5.1) gives

$$\rho e^{i\phi} = e^x e^{iy}.$$

Using Identity (5.1), and Property (1.41) of Section 1.5, we get

$$\rho = e^x \text{ and } \phi = y + 2n\pi, \text{ where } n \text{ is an integer. Therefore,} \tag{5.5}$$
$$\rho = |e^z| = e^x, \text{ and} \tag{5.6}$$
$$\phi \in \arg(e^z) = \{\text{Arg}(e^z) + 2n\pi : n \text{ is an integer}\}. \tag{5.7}$$

Solving Equations (5.5) for x and y yields

$$x = \ln \rho \quad \text{and} \quad y = \phi + 2n\pi, \tag{5.8}$$

where n is an integer. Thus for any complex number $w \neq 0$, there are infinitely many complex numbers $z = x + iy$ such that $w = e^z$. From Equations (5.8), the numbers z are

$$\begin{aligned} z = x + iy &= \ln \rho + i(\phi + 2n\pi) \\ &= \ln|w| + i(\text{Arg } w + 2n\pi). \end{aligned} \tag{5.9}$$

where n is an integer. Hence

$$\exp\left[|w| + i(\text{Arg } w + 2n\pi)\right] = w.$$

In summary, the transformation $w = e^z$ maps the complex plane (infinitely often) onto the set of nonzero complex numbers.

If we restrict the solutions to Equation (5.9) so that only the principal value of the argument, $-\pi < \text{Arg } w \leq \pi$, is used, the transformation $w = e^z = e^{x+iy}$ maps the horizontal strip $\{(x, y) : -\pi < y \leq \pi\}$ one-to-one and onto the range set $S = \{w : w \neq 0\}$. This strip is called the **fundamental period strip** and is shown in Figure 5.2.

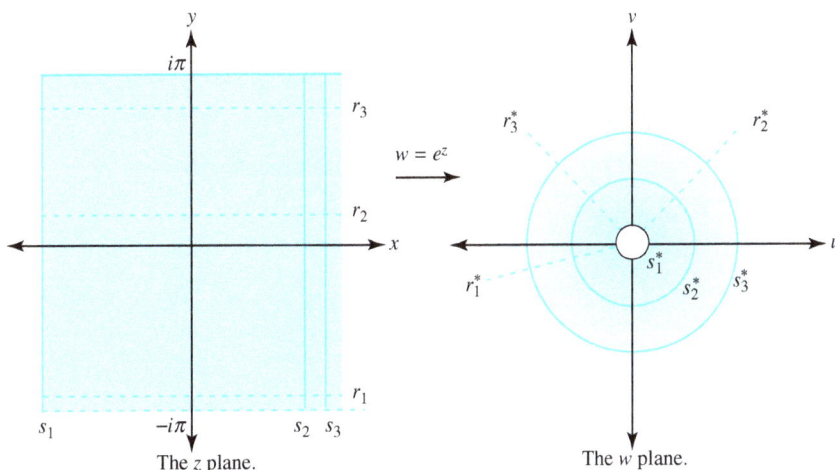

Figure 5.2: The fundamental period strip for the mapping $w = \exp(z)$

The horizontal line $z = t + ib$, for $-\infty < t < \infty$ in the z plane, is mapped onto the ray $w = e^t e^{ib} = e^t(\cos b + i \sin b)$ that is inclined at an angle $\phi = b$ in the w plane. The vertical segment $z = a + i\theta$, for $-\pi < \theta \leq \pi$ in the z plane, is mapped onto the circle centered at the origin with radius e^a in the w plane. That is, $w = e^a e^{i\theta} = e^a(\cos\theta + i\sin\theta)$. The lines r_1, r_2, and r_3, are mapped to the rays r_1^*, r_2^*, and r_3^*, respectively. Likewise, the segments s_1, s_2, and s_3 are mapped to the corresponding circles s_1^*, s_2^*, and s_3^*.

Example 5.2. Consider a rectangle $R = \{(x, y) : a \leq x \leq b \text{ and } c \leq y \leq d\}$, where $-\pi < c < d \leq \pi$. Show that the transformation $w = e^z = e^{x+iy}$ maps R onto a portion of an annular region bounded by two rays.

Solution:

The image points in the w plane satisfy the following relationships involving the modulus and argument of w:

$$e^a = |e^{a+iy}| \leq |e^{x+iy}| \leq |e^{b+iy}| = e^b, \quad \text{and}$$
$$c = \text{Arg}(e^{x+ic}) \leq \text{Arg}(e^{x+iy}) \leq \text{Arg}(e^{x+id}) \leq d.$$

which is a portion of the annulus $\{\rho e^{i\phi} : e^a \leq \rho \leq e^b\}$ in the w plane subtended by the rays $\phi = c$ and $\phi = d$. In Figure 5.3, we show the image of the rectangle

$$R = \left\{(x, y) : -1 \leq x \leq 1 \text{ and } -\frac{\pi}{4} \leq y \leq \frac{\pi}{3}\right\}.$$

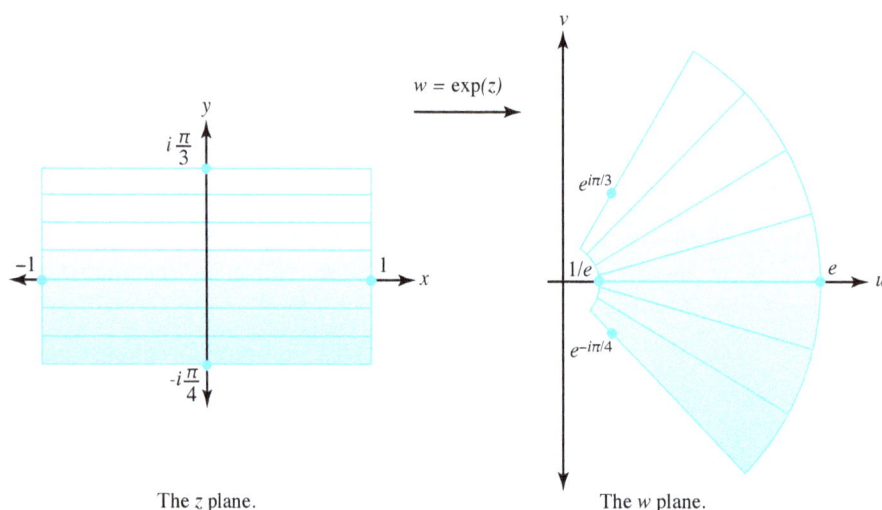

Figure 5.3: The image of R under the transformation $w = \exp(z)$

Exercises for Section 5.1 (Selected answers or hints are on page 439.)

1. Using Definition 5.1, explain why $\exp(0) = e^0 = 1$.

2. The questions for this problem relate to Figure 5.2. The shaded portion in the w plane indicates the image of the shaded portion in the z plane, with the lighter shading indicating expansion of the area of corresponding regions.

   (a) Why is there no shading inside the circle s_1^*?

   (b) Explain why the images of r_1, r_2, and r_3 appear to make, respectively, angles of $-\frac{7\pi}{8}$, $\frac{\pi}{4}$, and $\frac{3\pi}{4}$ radians with the positive u axis.

   (c) Precisely where should the images of the points $\pm i\pi$ be located?

3. Verify Equations (5.3) and (5.4).

4. Express e^z in the form $u + iv$ for the following values of z.

   (a) $-\frac{\pi}{3}$.

   (b) $\frac{1}{2} - i\frac{\pi}{4}$.

   (c) $-4 + 5i$.

   (d) $-1 + i\frac{3\pi}{2}$.

   (e) $1 + i\frac{5\pi}{4}$.

   (f) $\frac{\pi}{3} - 2i$.

5. Find all values of z for which the following equations hold.

   (a) $e^z = -4$.

   (b) $e^z = 2 + 2i$.

   (c) $e^z = \sqrt{3} - i$.

   (d) $e^z = -1 + i\sqrt{3}$.

6. Prove that $|\exp(z^2)| \le \exp(|z|^2)$ for all z. Where does equality hold?

7. Show that $\exp(z + i\pi) = \exp(z - i\pi)$ holds for all z.

8. Express $\exp(z^2)$ and $\exp\left(\frac{1}{z}\right)$ in the Cartesian form $u(x,y) + iv(x,y)$.

9. Explain why

   (a) $\exp(\overline{z}) = \overline{\exp z}$ holds for all z.

   (b) $\exp(\overline{z})$ is nowhere analytic.

10. Show that $|e^{-z}| < 1$ iff $\text{Re}(z) > 0$.

11. Verify that

   (a) $\lim\limits_{z \to 0} \frac{e^z - 1}{z} = 1$.

   (b) $\lim\limits_{z \to i\pi} \frac{e^z + 1}{z - i\pi} = -1$.

135

12. Show that $f(z) = ze^z$ is analytic for all z by showing that its real and imaginary parts satisfy the Cauchy-Riemann sufficient conditions for differentiability.

13. Find the derivatives of the following.

   (a) e^{iz}.

   (b) $z^4 \exp(z^3)$.

   (c) $e^{(a+ib)z}$.

   (d) $\exp(\frac{1}{z})$.

14. Let n be a positive integer. Show that

   (a) $(\exp z)^n = \exp(nz)$.

   (b) $\frac{1}{(\exp z)^n} = \exp(-nz)$.

15. Show that $\sum\limits_{n=0}^{\infty} e^{inz}$ converges for $\mathrm{Im}(z) > 0$.

16. Generalize Example 5.1, where the condition $-\pi < c < d \leq \pi$ is replaced by $d - c < 2\pi$. Illustrate what this means.

17. Use the fact that $\exp(z^2)$ is analytic to show that $e^{x^2 - y^2} \sin 2xy$ is a harmonic function.

18. Show the following concerning the exponential map.

   (a) The image of the line $\{(x, y) : x = t,\ y = 2\pi + t\}$, where $-\infty < t < \infty$ is a spiral.

   (b) The image of the first quadrant$\{(x, y) : x > 0,\ y > 0\}$ is the region $\{w : |w| > 1\}$.

   (c) If a is a real constant, the horizontal strip $\{(x, y) : a < y \leq a + 2\pi\}$ is mapped one-to-one and onto the nonzero complex numbers.

   (d) The image of the vertical line segment $\{(x, y) : x = 2,\ y = t\}$, where $\frac{\pi}{6} < t < \frac{7\pi}{6}$ is half a circle.

   (e) The image of the horizontal ray $\{(x, y) : x > 0,\ y = \frac{\pi}{3}\}$ is a ray.

19. Explain how the complex function e^z and the real function e^x are different. How are they similar?

20. Many texts give an alternative definition for $\exp(z)$, starting with Identity (5.1) as the definition for $f(z) = \exp(z)$. Recall that this identity states that $\exp(z) = \exp(x + iy) = e^x(\cos y + i \sin y)$. This exercise shows such a definition is a natural approach in terms of differential equations. We start by requiring $f(z)$ to be the solution to an initial-value problem satisfying three conditions: (1) f is entire, (2) $f'(z) = f(z)$ for all z, and (3) $f(0) = 1$. Suppose that $f(z) = f(x + iy) = u(x, y) + iv(x, y)$ satisfies conditions (1), (2), and (3).

   (a) Use the result $f'(z) = u_x(x, y) + iv_x(x, y)$ and the requirement $f'(z) = f(z)$ from condition (2) to show that $u_x(x, y) - u(x, y) = 0$, for all $z = (x, y)$.

   (b) Show that the result in part (a) implies that $\frac{\partial}{\partial x}[u(x, y)e^{-x}] = 0$. This means $u(x, y)e^{-x}$ is constant with respect to x, so $u(x, y)e^{-x} = p(y)$, where $p(y)$ is a function of y alone.

   (c) Using a similar procedure for $v(x, y)$, show we wind up getting a pair of solutions $u(x, y) = p(y)e^x$, and $v(x, y) = q(y)e^x$ where $p(y)$ and $q(y)$ are functions of y alone.

(d) Now use the Cauchy-Riemann equations to conclude from part (c) that $p(y) = q'(y)$ and $p'(y) = -q(y)$.

(e) Use part (d) to show that $p''(y) + p(y) = 0$ and $q''(y) + q(y) = 0$.

(f) Identify the general solutions to part (e). Then, given the initial conditions
$$f(0) = f(0 + 0i) = u(0, 0) + iv(0, 0) = 1 + 0i,$$
find the particular solutions and conclude that Identity (5.1) follows.

5.2   The Complex Logarithm

In Section 5.1, we showed that, if w is a nonzero complex number, then the equation $w = \exp z$ has infinitely many solutions. Because the function $\exp(z)$ is a many-to-one function, its inverse (the logarithm) is necessarily multivalued.

Definition 5.2 (Multivalued logarithm). *For $z \neq 0$, we define the multivalued function $\log$ as the inverse of the exponential function; that is,*

$$\log(z) = w \quad \textit{iff} \quad z = \exp(w). \tag{5.10}$$

If we go through the same steps as we did in Equations (5.8) and (5.9), we find that, for any complex number $z \neq 0$, the solutions w to Equation (5.10) take the form

$$w = \ln|z| + i\theta, \quad \text{for} \quad z \neq 0. \tag{5.11}$$

where $\theta \in \arg(z)$ and $\ln|z|$ denotes the natural logarithm of the positive number $|z|$. Because $\arg(z)$ is the set $\arg(z) = \{\text{Arg}(z) + 2n\pi : n \text{ is an integer}\}$, we can express the set of values comprising $\log(z)$ as

$$\log(z) = \{\ln|z| + i(\text{Arg}(z) + 2n\pi) : n \text{ is an integer}\} \tag{5.12}$$
$$= \ln|z| + i\arg(z), \tag{5.13}$$

where it is understood that Identity (5.13) refers to the same set of numbers per Identity (5.12).

Recall that Arg is defined so that for $z \neq 0$, $-\pi < \text{Arg}(z) \leq \pi$. We call any one of the values given in Identities (5.12) or (5.13) a logarithm of z. Note that the different values of $\log(z)$ all have the same real part and that their imaginary parts differ by the amount $2n\pi$, where n is an integer. When $n = 0$, we have a special situation.

Definition 5.3 (Principal value of the logarithm). *For $z \neq 0$, we define Log, the principal value of the logarithm, by*
$$\text{Log}(z) = \ln|z| + i\text{Arg}(z). \tag{5.14}$$

The domain for the function Log is the set of all nonzero complex numbers in the z plane, and its range is the horizontal strip $\{w : -\pi < \text{Im}(w) \leq \pi\}$ in the w plane. We stress again that Log is a single-valued function and corresponds to setting $n = 0$ in Equation (5.12). As we demonstrated in Chapter 2, the function Arg is discontinuous at each point along the negative x axis; hence so is the function Log. In fact, because any branch of the multivalued function arg is discontinuous along some ray, a corresponding branch of the logarithm will have a discontinuity along that same ray.

Example 5.3. Find the values of $\log(1+i)$ and $\log(i)$.

Solution:

By standard computations, we have

$$\log(1+i) = \left\{ \ln|1+i| + i(\mathrm{Arg}(1+i) + 2n\pi) : n \text{ is an integer} \right\}$$

$$= \left\{ \ln\sqrt{2} + i\left(\frac{\pi}{4} + 2n\pi\right) : n \text{ is an integer} \right\}, \quad \text{and}$$

$$\log(i) = \left\{ \ln|i| + i(\mathrm{Arg}(i) + 2n\pi) : n \text{ is an integer} \right\}$$

$$= \left\{ i\left(\frac{\pi}{2} + 2n\pi\right) : n \text{ is an integer} \right\}.$$

The principal values are

$$\mathrm{Log}(1+i) = \ln\sqrt{2} + i\frac{\pi}{4} = \frac{\ln 2}{2} + i\frac{\pi}{4}. \quad \text{and}$$

$$\mathrm{Log}(i) = i\frac{\pi}{2}.$$

We now investigate some of the properties of log and Log. From Equations (5.10), (5.12), and (5.14), it follows that

$$\exp(\mathrm{Log}\, z) = z \quad \text{for all} \quad z \neq 0, \quad \text{and} \tag{5.15}$$

$$\mathrm{Log}(\exp z) = z, \quad \text{provided} \quad -\pi < \mathrm{Im}(z) \leq \pi, \tag{5.16}$$

and that the mapping $w = \mathrm{Log}(z)$ is one-to-one from domain $D = \{z : |z| > 0\}$ in the z plane onto the horizontal strip $\{w : -\pi < \mathrm{Im}(w) \leq \pi\}$ in the w plane.

The following example illustrates that, even though Log is not continuous along the negative real axis, it is still defined there.

Example 5.4. Identity (5.14) reveals that

$$\mathrm{Log}(-e) = \ln|-e| + i\mathrm{Arg}(-e) = 1 + i\pi, \quad \text{and}$$

$$\mathrm{Log}(-1) = \ln|-1| + i\mathrm{Arg}(-1) = i\pi.$$

When $z = x + i0$, where x is a positive real number, the principal value of the complex logarithm of z is

$$\mathrm{Log}(x + i0) = \ln x + i\mathrm{Arg}(x) = \ln x + i0 = \ln x.$$

where $x > 0$. Hence Log is an extension of the real function $\ln$ to the complex case. Are there other similarities? Let's use complex function theory to find the derivative of Log. When we use polar coordinates for $z = re^{i\theta} \neq 0$, Equation (5.14) becomes

$$\mathrm{Log}(z) = \ln r + i\mathrm{Arg}(z)$$

$$= \ln r + i\theta, \quad \text{for} \quad r > 0 \quad \text{and} \quad -\pi < \theta \leq \pi$$

$$= U(r, \theta) + iV(r, \theta),$$

where $U(r, \theta) = \ln r$ and $V(r, \theta) = \theta$. Because $\mathrm{Arg}(z)$ is discontinuous only at points in its domain that lie on the negative real axis, U and V have continuous partials for any point (r, θ) in their domain, provided $re^{i\theta}$ is not on the negative real axis, that is, provided $-\pi < \theta < \pi$.

(Note the strict inequality for θ here.) In addition, the polar form of the Cauchy-Riemann equations holds in this region (see Equation (3.22) of Section 3.2), since

$$U_r(r, \theta) = \frac{1}{r}V_\theta(r, \theta) = \frac{1}{r} \quad \text{and} \quad V_r(r, \theta) = -\frac{1}{r}U_\theta(r, \theta) = 0.$$

Using Theorem 3.5 of Section 3.2, we see that

$$\frac{d}{dz}\text{Log}(z) = e^{-i\theta}(U_r + iV_r) = e^{-i\theta}\left(\frac{1}{r} + 0i\right) = \frac{1}{re^{i\theta}} = \frac{1}{z}.$$

provided $r > 0$ and $-\pi < \theta < \pi$. Thus the principal branch of the complex logarithm has the derivative we would expect. Other properties of the logarithm carry over, but only in specified regions of the complex plane.

Example 5.5. Show that the identity $\text{Log}(z_1 z_2) = \text{Log}(z_1) + \text{Log}(z_2)$ is not always valid.

Solution:

Let $z_1 = -\sqrt{3} + i$ and $z_2 = -1 + i\sqrt{3}$. Then

$$\text{Log}(z_1 z_2) = \text{Log}(-4i)$$
$$= \ln 4 + i\left(-\frac{\pi}{2}\right), \quad \text{but}$$
$$\text{Log}(z_1) + \text{Log}(z_2) = \ln 2 + i\frac{5\pi}{6} + \ln 2 + i\frac{2\pi}{3}$$
$$= \ln 4 + i\frac{3\pi}{2}.$$

Our next result explains why $\text{Log}(z_1 z_2) = \text{Log}(z_1) + \text{Log}(z_2)$ didn't hold for the particular numbers we chose.

Theorem 5.2. $\text{Log}(z_1 z_2) = \text{Log}(z_1) + \text{Log}(z_2) \quad$ *iff* $\quad -\pi < \text{Arg}(z_1) + \text{Arg}(z_2) \le \pi$.

Proof. Suppose first that $-\pi < \text{Arg}(z_1) + \text{Arg}(z_2) \le \pi$. By definition, $\text{Log}(z_1 z_2) = \ln|z_1 z_2| + i\text{Arg}(z_1 z_2) = \ln|z_1| + \ln|z_2| + i\text{Arg}(z_1 z_2)$. Because $-\pi < \text{Arg}(z_1) + \text{Arg}(z_2) \le \pi$, it follows that $\text{Arg}(z_1 z_2) = \text{Arg}(z_1) + \text{Arg}(z_2)$ (explain!), and so

$$\text{Log}(z_1 z_2) = \ln|z_1| + \ln|z_2| + i\text{Arg}(z_1) + i\text{Arg}(z_2) = \text{Log}(z_1) + \text{Log}(z_2).$$

The "only if" part is left as an exercise. $\qquad\square$

As Example 5.5 and Theorem 5.2 illustrate, properties of the complex logarithm don't carry over when arguments of products combine in such a way that they drop down to $-\pi$ or rise above π. This is because of the restrictions placed on the domain of the function Arg. From the set of numbers associated with the *multivalued* logarithm, however, we can formulate properties that look exactly the same as those corresponding with the real logarithm.

Theorem 5.3. *Let z_1 and z_2 be nonzero complex numbers. The multivalued function $\log$ obeys the familiar properties of logarithms:*

$$\log(z_1 z_2) = \log(z_1) + \log(z_2), \tag{5.17}$$

$$\log\left(\frac{z_1}{z_2}\right) = \log(z_1) - \log(z_2), \quad \text{and} \tag{5.18}$$

$$\log\left(\frac{1}{z}\right) = -\log(z). \tag{5.19}$$

Proof. Identity (5.17) is easy to establish: Using Identity (1.38) in Section 1.4 concerning the argument of a product (and keeping in mind we are dealing with sets of numbers), we write

$$\log(z_1 z_2) = \ln|z_1 z_2| + i \arg(z_1 z_2)$$
$$= \ln|z_1| + \ln|z_2| + i \arg(z_1) + i \arg(z_2)$$
$$= \big[\ln|z_1| + i \arg(z_1)\big] + \big[\ln|z_2| + i \arg(z_2)\big] = \log(z_1) + \log(z_2).$$

Identities (5.18) and (5.19) are left as exercises. $\qquad\square$

We can construct many different branches of the multivalued logarithm function that are continuous and differentiable except at points along any preassigned ray $\{re^{i\alpha} : r > 0\}$. If we let α denote a real fixed number and choose the value of $\theta \in \arg(z)$ that lies in the range $\alpha < \theta \le \alpha + 2\pi$, then the function $\log_\alpha$ defined by

$$\log_\alpha(z) = \ln r + i\theta, \tag{5.20}$$

where $z = re^{i\theta} \ne 0$, and $\alpha < \theta \le \alpha + 2\pi$, is a single-valued branch of the logarithm function. The branch cut for $\log_\alpha$ is the ray $\{re^{i\alpha} : r \ge 0\}$, and each point along this ray is a point of discontinuity of $\log_\alpha$. Because $\exp[\log_\alpha(z)] = z$, we conclude that the mapping $w = \log_\alpha(z)$ is a one-to-one mapping of the domain $|z| > 0$ onto the horizontal strip $\{w : \alpha < \operatorname{Im}(w) \le \alpha + 2\pi\}$. If $\alpha < c < d < \alpha + 2\pi$, then the function $w = \log_\alpha(z)$ maps the set $D = \{re^{i\theta} : a < r < b, \, c < \theta < d\}$ one-to-one and onto the rectangle $R = \{u + iv : \ln a < u < \ln b, \, c < v < d\}$. Figure 5.4 shows the mapping $w = \log_\alpha(z)$, its branch cut $\{re^{i\alpha} : r > 0\}$, the set D, and its image R.

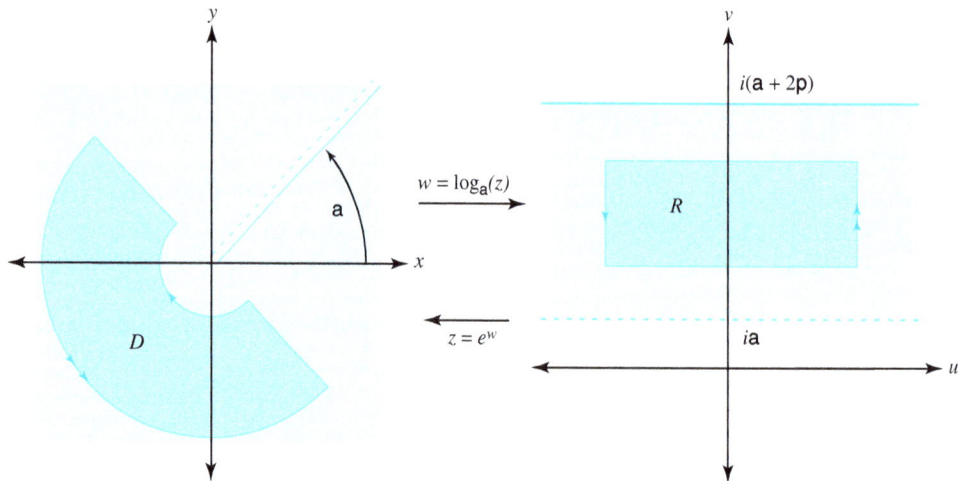

Figure 5.4: The branch $w = \log_\alpha(z)$ of the logarithm

We can easily compute the derivative of any branch of the multivalued logarithm. For a particular branch $w = \log_\alpha(z)$ for $z = re^{i\theta} \ne 0$, and $\alpha < \theta < \alpha + 2\pi$ (note the strict inequality for θ), we start with $z = \exp(w)$ in Equations (5.10) and differentiate both sides to get

$$1 = \frac{d}{dz} z = \frac{d}{dz} \exp\big(\log_\alpha(z)\big)$$
$$= \exp\big(\log_\alpha(z)\big) \frac{d}{dz} \log_\alpha(z)$$
$$= z \frac{d}{dz} \log_\alpha(z).$$

Solving for $\frac{d}{dz}\log_\alpha(z)$ gives

$$\frac{d}{dz}\log_\alpha(z) = \frac{1}{z}, \quad \text{for} \quad z = re^{i\theta} \neq 0, \quad \text{and} \quad \alpha < \theta < \alpha + 2\pi.$$

The Riemann surface for the multivalued function $w = \log(z)$ is similar to the one for the square root function. However, it requires infinitely many copies of the z plane cut along the negative x axis, which we label S_k for $k = \ldots, -n, \ldots, -2, -1, 0, 1, 2, \ldots, n, \ldots$. We stack these cut planes directly on each other so that the corresponding points have the same position, and join the sheet S_k to S_{k+1} as follows: For each integer k, the edge of the sheet S_k in the upper half-plane is joined to the edge of the sheet S_{k+1} in the lower half-plane. The Riemann surface for the domain of log looks like a spiral staircase that extends upward on the sheets $S_1, , S_2, \ldots$ and downward on the sheets $S_{-1}, S_{-2}, \ldots$, as shown in Figure 5.5. We use polar coordinates for z on each sheet. Thus, for S_k we have

$$z = r(\cos\theta + i\sin\theta), \quad \text{where}$$
$$r = |z|, \quad \text{and} \quad 2\pi k - \pi < \theta \leq \pi + 2\pi k.$$

Again, for S_k, the correct branch of $\log(z)$ on each sheet is

$$\log(z) = \ln r + i\theta, \quad \text{where}$$
$$r = |z|, \quad \text{and} \quad 2\pi k - \pi < \theta \leq \pi + 2\pi k.$$

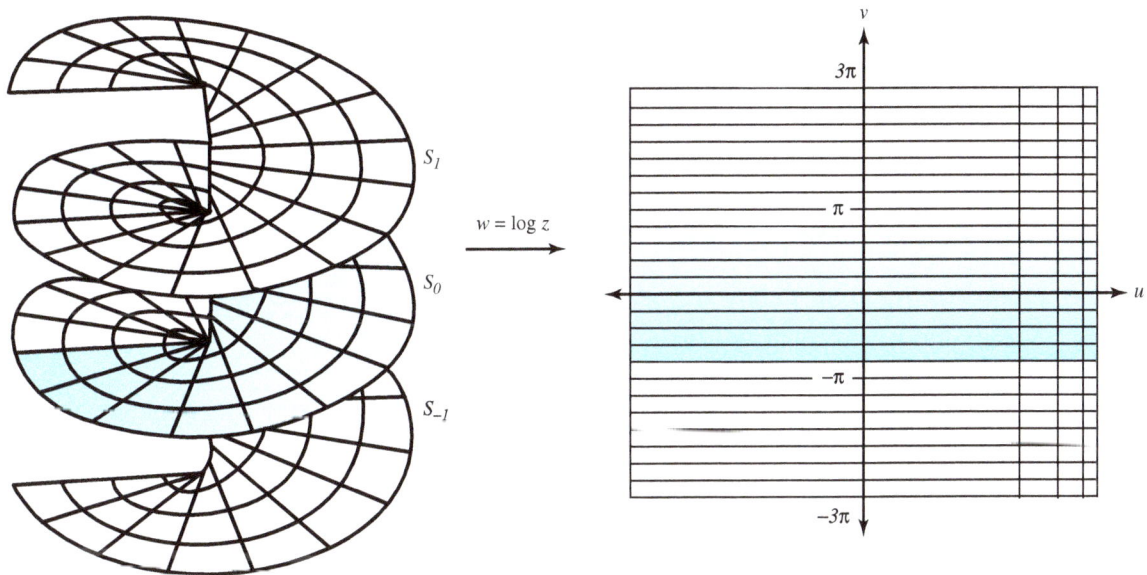

Figure 5.5: The Riemann surface for mapping $w = \log(z)$

Exercises for Section 5.2 (Selected answers or hints are on page 440.)

1. Find all values for

   (a) $\text{Log}(ie^2)$.

   (b) $\text{Log}(\sqrt{3} - i)$.

   (c) $\text{Log}(i\sqrt{2} - \sqrt{2})$.

   (d) $\text{Log}[(1 + i)^4]$.

   (e) $\log(-3)$.

   (f) $\log 8$.

   (g) $\log(4i)$.

   (h) $\log(-\sqrt{3} - i)$.

2. Use the properties of $\arg(z)$ in Section 1.4 to establish

   (a) Equation (5.18).

   (b) Equation (5.19).

3. Find all the values of z for which each equation holds.

   (a) $\text{Log}(z) = 1 - i\frac{\pi}{4}$.

   (b) $\text{Log}(z - 1) = i\frac{\pi}{2}$.

   (c) $\exp(z) = -ie$.

   (d) $\exp(z + 1) = i$.

4. Refer to Theorem 5.2.

   (a) Explain why $-\pi < \text{Arg}(z_1) + \text{Arg}(z_2) \le \pi$ implies that $\text{Arg}(z_1 z_2) = \text{Arg}(z_1) + \text{Arg}(z_2)$.

   (b) Prove the "only if" part.

5. Refer to Equation (5.20) and pick an appropriate value for α so that the branch of the logarithm $\log_\alpha(z)$ will *not* be analytic at $z = z_0$, where

   (a) $z_0 = 1$.

   (b) $z_0 = -1 + i\sqrt{3}$.

   (c) $z_0 = i$.

   (d) $z_0 = -i$.

   (e) $z_0 = -1 - i$.

   (f) $z_0 = \sqrt{3} - i$.

6. Show that $f(z) = \frac{\text{Log}(z+5)}{z^2+3z+2}$ is analytic everywhere except at the points -1, -2, and on the ray $\{(x, y) : x \le -5, \ y = 0\}$.

7. Show that the following are harmonic functions in the right half-plane $\{z : \text{Re} z > 0\}$.

   (a) $u(x, y) = \ln(x^2 + y^2)$.

   (b) $v(x, y) = \text{Arctan}(\frac{y}{x})$.

8. Show that $z^n = \exp[n \log_\alpha(z)]$, where n is an integer and $\log_\alpha$ is any branch of the logarithm.

9. Construct a branch of $f(z) = \log(z + 4)$ that is analytic at the point $z = -5$ and takes on the value $7\pi i$ there.

10. For what values of z is it true that

   (a) $\text{Log}(\frac{z_1}{z_2}) = \text{Log}(z_1) - \text{Log}(z_2)$? Why?
   (b) $\frac{d}{dz}\text{Log}(z) = \frac{1}{z}$? Why?
   (c) $\text{Log}(\frac{1}{z}) = -\text{Log}(z)$? Why?

11. Construct branches of $f(z) = \log(z+2)$ that are analytic at all points in the plane except at points on the following rays.

   (a) $\{(x,y) : x \geq -2, \, y = 0\}$.
   (b) $\{(x,y) : x = -2, \, y \geq 0\}$.
   (c) $\{(x,y) : x = -2, \, y \leq 0\}$.

12. Show that the mapping $w = \text{Log}(z)$ maps

   (a) the ray $\{z = re^{i\theta} : r > 0, \, \theta = \frac{\pi}{3}\}$ one-to-one and onto the horizontal line $\{(u,v) : v = \frac{\pi}{3}\}$.
   (b) the semicircle $\{z = 2e^{i\theta} : -\frac{\pi}{2} \leq \theta \leq \frac{\pi}{2}\}$ one-to-one and onto the vertical line segment $\{(\ln 2, \, v) : -\frac{\pi}{2} \leq v \leq \frac{\pi}{2}\}$.

13. Find specific values of z_1 and z_2 so that $\text{Log}(\frac{z_1}{z_2}) \neq \text{Log}(z_1) - \text{Log}(z_2)$.

14. Show why the solutions to Equation (5.10) are given by those in Equation (5.11). *Hint*: Mimic the process used in obtaining Identities (5.8) and (5.9).

15. Explain why no branch of the logarithm is defined when $z = 0$.

5.3   Complex Exponents

In Section 1.5 we indicated that it is possible to make sense out of expressions such as $\sqrt{1+i}$ or i^i without appealing to a number system beyond the framework of complex numbers. We now show how this is done by taking note of some rudimentary properties of the complex exponential and logarithm, and then using our imagination.

We begin by generalizing Identity (5.15). Equations (5.12) and 5.14 show that $\log(z)$ can be expressed as the set $\log(z) = \{\text{Log}(z) + i2n\pi : n \text{ is an integer}\}$. We can easily show (left as an exercise) that, for $z \neq 0$, $\exp[\log_\alpha(z)] = z$, where $\log_\alpha$ is *any* branch of the function log. But this means that, for any $\zeta \in \log(z)$, the identity $\exp \zeta = z$ holds true. Because $\exp[\log(z)]$ denotes the set $\{\exp \zeta : \zeta \in \log(z)\}$, we see that $\exp[\log(z)] = z$, for $z \neq 0$.

Next, note that Identity (5.17) gives $\log(z^n) = n \log(z)$, where n is any natural number, so that $\exp[\log(z^n)] = \exp[n \log(z)] = z^n$, for $z \neq 0$. With these preliminaries out of the way, we can now come up with a definition of a complex number raised to a complex power.

Definition 5.4 (Complex exponent). *Let c be a complex number. We define z^c as*

$$z^c = \exp\left[c \log(z)\right]. \tag{5.21}$$

143

The right side of Equation (5.21) is a set. This definition makes sense because, if both z and c are real numbers with $z > 0$, Equation (5.21) gives the familiar (real) definition for z^c, as the following example illustrates.

Example 5.6. Use Equation (5.21) to evaluate $4^{\frac{1}{2}}$.

Solution:

Calculating $4^{\frac{1}{2}} = \exp[\frac{1}{2}\log(4)]$ gives

$$\frac{1}{2}\log(4) = \{\ln 2 + in\pi : n \text{ is an integer}\}. \tag{5.22}$$

Thus $4^{\frac{1}{2}}$ is the set $\{\exp(\ln 2 + in\pi) : n \text{ is an integer}\}$. The distinct values occur when $n = 0, 1$. Plugging these values into Equation (5.22) gives $\exp(\ln 2) = 2$ and $\exp(\ln 2 + i\pi) = \exp(\ln 2)\exp(i\pi) = -2$. In other words, $4^{\frac{1}{2}} = \{-2, 2\}$.

The expression $4^{\frac{1}{2}}$ is different from $\sqrt{4}$, as the former represents the set $\{-2, 2\}$ and the latter gives only one value, $\sqrt{4} = 2$.

Because log is multivalued, the function z^c will, in general, be multivalued. If we want to focus on a single value for z^c, we can do so via the function defined for $z \neq 0$ by

$$f(z) = \exp[c\text{Log}(z)], \tag{5.23}$$

which is called the principal branch of the multivalued function z^c. Note that the principal branch of z^c is obtained from Equation (5.21) by replacing $\log(z)$ with the principal branch of the logarithm.

Example 5.7. Find the principal values of $\sqrt{1+i}$ and i^i.

Solution:

From Example 5.3

$$\text{Log}(1+i) = \frac{\ln 2}{2} + i\frac{\pi}{4} = \ln 2^{\frac{1}{2}} + i\frac{\pi}{4}, \quad \text{and}$$
$$\text{Log}(i) = i\frac{\pi}{2}.$$

Identity (5.23) yields the principal values of $\sqrt{1+i}$ and i^i:

$$\sqrt{1+i} = (1+i)^{\frac{1}{2}}$$
$$= \exp\left[\frac{1}{2}\text{Log}(1+i)\right]$$
$$= \exp\left[\frac{1}{2}(\ln 2^{\frac{1}{2}} + i\frac{\pi}{4})\right]$$
$$= \exp\left(\ln 2^{\frac{1}{4}} + i\frac{\pi}{8}\right)$$
$$= 2^{\frac{1}{4}}\left(\cos\frac{\pi}{8} + i\sin\frac{\pi}{8}\right)$$
$$\approx 1.09684 + 0.45509i, \quad \text{and}$$
$$i^i = \exp[i\text{Log}(i)]$$
$$= \exp\left[i\left(i\frac{\pi}{2}\right)\right]$$
$$= \exp\left(-\frac{\pi}{2}\right)$$
$$\approx 0.20788.$$

Note that the result of raising a complex number to a complex power may be a real number in a nontrivial way.

We now consider the possibilities that arise when we apply Equation (5.21).

Case(i): Suppose that $c = k$, where k is an integer. Then, if $z = re^{i\theta} \neq 0$,

$$k \log(z) = \{k \ln(r) + ik(\theta + 2n\pi) : n \text{ is an integer}\}.$$

Recalling that the complex exponential function has period $2\pi i$, we have

$$
\begin{aligned}
z^k &= \exp\left[k \log(z)\right] \\
&= \exp\left[k \ln(r) + ik(\theta + 2n\pi)\right] \\
&= \exp\left[\ln(r^k) + ik\theta + i2kn\pi\right] \\
&= \exp\left[\ln(r^k)\right] \exp(ik\theta) \exp(i2kn\pi) \\
&= r^k \exp(ik\theta) = r^k(\cos k\theta + i \sin k\theta),
\end{aligned}
$$

which is the single-valued kth power of z that we discussed in Section 1.5.

Case(ii): If $c = \frac{1}{k}$, where k is an integer, and $z = re^{i\theta} \neq 0$, then

$$\frac{1}{k} \log z = \left\{ \frac{1}{k} \ln r + \frac{i(\theta + 2n\pi)}{k} : n \text{ is an integer} \right\}.$$

Hence, Equation (5.21) becomes

$$z^{\frac{1}{k}} = \exp\left[\frac{1}{k} \log(z)\right] \tag{5.24}$$

$$= \exp\left[\frac{1}{k} \ln(r) + i\frac{\theta + 2n\pi}{k}\right]$$

$$= r^{\frac{1}{k}} \exp\left(i\frac{\theta + 2n\pi}{k}\right) \tag{5.25}$$

$$= r^{\frac{1}{k}} \left[\cos\left(\frac{\theta + 2n\pi}{k}\right) + i \sin\left(\frac{\theta + 2n\pi}{k}\right)\right].$$

When we again use the periodicity of the complex exponential function, Equation (5.24) gives k distance values corresponding to

$$n = 0, 1, \ldots, k - 1.$$

Therefore, as Example 5.6 illustrated, the fractional power $z^{\frac{1}{k}}$ is the multivalued kth root function.

Case(iii): If j and k are positive integers that have no common factors and $c = \frac{j}{k}$, then Equation (5.21) becomes

$$z^{\frac{j}{k}} = r^{\frac{j}{k}} \exp\left[i\frac{(\theta + 2n\pi)j}{k}\right] = r^{\frac{j}{k}} \left[\cos\left(\frac{(\theta + 2n\pi)j}{k}\right) + i \sin\left(\frac{(\theta + 2n\pi)}{k}\right)\right].$$

Again, there are k distance values that correspond with $n = 0, 1, \ldots, k - 1$.

Case(iv): If c is not a rational number, then there are infinitely many values for z^c, provided $z \neq 0$.

Example 5.8. The values of $2^{\frac{1}{9}+\frac{i}{50}}$ are

$$2^{\frac{1}{9}+\frac{i}{50}} = \exp\left[\left(\frac{1}{9}+\frac{i}{50}\right)(\ln 2 + i2n\pi)\right]$$

$$= \exp\left[\frac{\ln 2}{9} - \frac{n\pi}{25} + i\left(\frac{\ln 2}{50}+\frac{2n\pi}{9}\right)\right]$$

$$= 2^{\frac{1}{9}}e^{-\frac{n\pi}{25}}\left[\cos\left(\frac{\ln 2}{50}+\frac{2n\pi}{9}\right)+i\sin\left(\frac{\ln 2}{50}+\frac{2n\pi}{9}\right)\right],$$

where n is an integer. The principal value of $2^{\frac{1}{9}+\frac{i}{50}}$ is

$$2^{\frac{1}{9}+\frac{i}{50}} = 2^{\frac{1}{9}}\left[\cos\left(\frac{\ln 2}{50}\right)+i\sin\left(\frac{\ln 2}{50}\right)\right] \approx 1.079956 + 0.014972i.$$

Figure 5.6 shows the terms for this multivalued expression corresponding to

$$n = -9, -8, \ldots, -1, 0, 1, \ldots, 8, 9.$$

They exhibit a spiral pattern that is often present in complex powers.

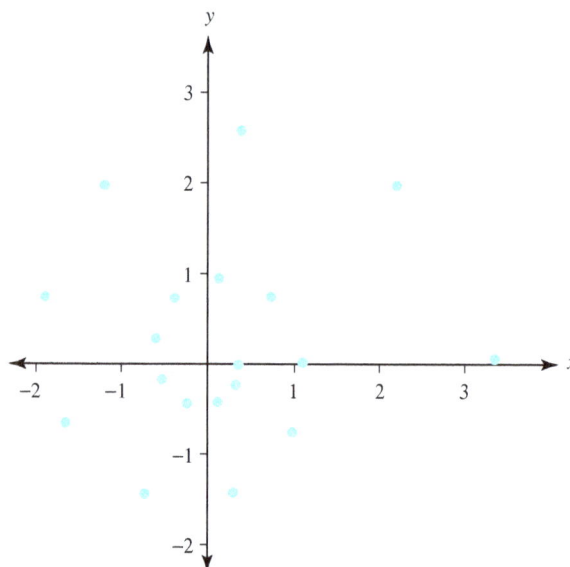

Figure 5.6: Some of the values of $2^{\frac{1}{9}+\frac{i}{50}}$

Some of the rules for exponents carry over from the real case. In the exercises we ask you to show that if c and d are complex numbers and $z \neq 0$, then

$$z^{-c} = \frac{1}{z^c}, \tag{5.26}$$

$$z^c z^d = z^{c+d}, \tag{5.27}$$

$$\frac{z^c}{z^d} = z^{c-d}, \tag{5.28}$$

$$(z^c)^n = z^{cn}, \tag{5.29}$$

where n is an integer.

The following example shows that Identity (5.29) does not hold if n is replaced with an arbitrary complex value.

Example 5.9.

$$(i^2)^i = \exp\left[i\log(-1)\right] = e^{-(1+2n)\pi}, \quad \text{where } n \text{ is an integer, and}$$

$$(i)^{2i} = \exp(2i\log i) = e^{-(1+4n)\pi}, \quad \text{where } n \text{ is an integer.}$$

Since these sets of solutions are not equal, Identity (5.29) does not always hold.

We can compute the derivative of the principal branch of z^c, which is the function $f(z) = \exp[c\mathrm{Log}(z)]$. By the chain rule,

$$f'(z) = \frac{c}{z}\exp\left[c\mathrm{Log}(z)\right]. \tag{5.30}$$

If we restrict z^c to the principal branch, $z^c = \exp\left[c\mathrm{Log}(z)\right]$, then Equation (5.30) can be written in the familiar form that you learned in calculus. That is, for $z \neq 0$ and z not a negative real number,

$$\frac{d}{dz}z^c = \frac{c}{z}z^c = cz^{c-1}.$$

We can use Identity (5.21) to define the exponential function with base b, where $b \neq 0$ is a complex number:

$$b^z = \exp\left[z\log(b)\right].$$

If we specify a branch of the logarithm, then b^z will be single-valued and we can use the rules of differentiation to show that the resulting branch of b^z is an analytic function. The derivative of b^z is then given by the familiar rule

$$\frac{d}{dz}b^z = b^z \log_\alpha(b), \tag{5.31}$$

where $\log_\alpha$ is any branch of the logarithm whose branch cut does not include the point b.

Exercises for Section 5.3 (Selected answers or hints are on page 441.)

1. Find the principal value of

   (a) 4^i.

   (b) $(1+i)^{\pi i}$.

   (c) $(-1)^{\frac{1}{\pi}}$.

   (d) $(1+i\sqrt{3})^{\frac{i}{2}}$.

2. Find *all* values of

   (a) i^i.

   (b) $(-1)^{\sqrt{2}}$.

   (c) $i^{\frac{2}{\pi}}$.

   (d) $(1+i)^{2-i}$.

   (e) $(-1)^{\frac{3}{4}}$.

   (f) $i^{\frac{2}{3}}$.

3. Show that, if $z \neq 0$, then z^0 has a unique value.

4. For $z = re^{i\theta} \neq 0$, show that the principal branch of

   (a) z^i is given by the equation

   $$z^i = e^{-\theta}\big[(\ln r) + i\sin(\ln r)\big],$$

   where $r > 0$ and $-\pi < \theta \leq \pi$.

   (b) z^α (α a real number) is given by the equation

   $$z^\alpha = r^\alpha \cos \alpha\theta + ir^\alpha \sin \alpha\theta,$$

   where $r > 0$ and $-\pi < \theta \leq \pi$.

5. Let $z_n = (1+i)^n$ for $n = 1, 2, \ldots$ Show that the sequence $\{z_n\}$ is a solution to the difference equation $z_n = 2z_{n-1} - 2z_{n-2}$ for $n \geq 3$.

6. Verify the following identities:

   (a) Identity (5.26).
   (b) Identity (5.27).
   (c) Identity (5.28).
   (d) Identity (5.29).
   (e) Identity (5.31).

7. Does 1 raised to any power always equal 1? Why or why not?

8. Construct an example that shows that the principal value of $(z_1 z_2)^{\frac{1}{3}}$ need not equal the product of the principal values of $z_1^{\frac{1}{3}}$ and $z_2^{\frac{1}{3}}$.

9. If c is a complex number, the expression i^c may be multivalued. Suppose all the values of $|i^c|$ are identical. What are these values, and what can be said about the number c? Justify your assertions.

5.4 Trigonometric and Hyperbolic Functions

Based on the success we had in using power series to define the complex exponential, we have reason to believe that this approach will also be fruitful for other elementary functions. The power series expansions for the real-valued sine and cosine functions are

$$\sin x = \sum_{n=0}^{\infty}(-1)^n \frac{x^{2n+1}}{(2n+1)!} \quad \text{and} \quad \cos x = \sum_{n=0}^{\infty}(-1)^n \frac{x^{2n}}{(2n)!}.$$

so it is natural to make the following definitions.

Definition 5.5 ($\sin z$ and $\cos z$).

$$\sin z = \sum_{n=0}^{\infty}(-1)^n \frac{z^{2n+1}}{(2n+1)!} \quad \text{and} \quad \cos z = \sum_{n=0}^{\infty}(-1)^n \frac{z^{2n}}{(2n)!}.$$

With these definitions in place, we can now easily create the other complex trigonometric functions, provided the denominators in the following expressions are not zero.

Definition 5.6 (Other Trigonometric functions).

$$\tan z = \frac{\sin z}{\cos z}, \quad \cot z = \frac{\cos z}{\sin z}, \quad \sec z = \frac{1}{\cos z}, \quad and \quad \csc z = \frac{1}{\sin z}.$$

The series for the complex sine and cosine agree with the real sine and cosine when z is real, so the remaining complex trigonometric functions likewise agree with their real counterparts. What additional properties are common? For starters, we have

Theorem 5.4. $\sin z$ *and* $\cos z$ *are entire functions, with* $\frac{d}{dz} \sin z = \cos z$ *and* $\frac{d}{dz} \cos z = -\sin z$.

Proof. The ratio test shows that the radius of convergence for both functions is infinity, so they are entire by Theorem 4.17, part (i). Part (iii) of that theorem gives

$$\begin{aligned}
\frac{d}{dz} \sin z &= \frac{d}{dz}\left[\sum_{n=0}^{\infty}(-1)^n \frac{z^{2n+1}}{(2n+1)!}\right] \\
&= \sum_{n=0}^{\infty}(-1)^n \frac{(2n+1)z^{2n}}{(2n+1)!} \quad \text{(Why does the index } n \text{ stay at 0 here?)} \\
&= \sum_{n=0}^{\infty}(-1)^n \frac{z^{2n}}{(2n)!} \\
&= \cos z.
\end{aligned}$$

We leave the proof that $\frac{d}{dz} \cos z = -\sin z$ as an exercise. $\square$

We now list several additional properties, providing proofs for some and leaving others as exercises.

- For all complex numbers z,

$$\begin{aligned}
\sin(-z) &= -\sin z, \\
\cos(-z) &= \cos z, \quad \text{and} \\
\sin^2 z + \cos^2 z &= 1.
\end{aligned}$$

The verification that $\sin(-z) = -\sin z$ and $\cos(-z) = \cos z$ comes from substituting $-z$ for z in Definition 5.5. We leave verification of the identity $\sin^2 z + \cos^2 z = 1$ as an exercise (with hints).

- For all complex numbers z for which the expressions are defined,

$$\frac{d}{dz} \tan z = \sec^2 z,$$

$$\frac{d}{dz} \cot z = -\csc^2 z,$$

$$\frac{d}{dz} \sec z = \sec z \tan z, \quad \text{and}$$

$$\frac{d}{dz} \csc z = -\csc z \cot z.$$

The proof that $\frac{d}{dz}\tan z = \sec^2 z$ uses the identity $\sin^2 z + \cos^2 z = 1$:

$$\frac{d}{dz}\tan z = \frac{d}{dz}\left(\frac{\sin z}{\cos z}\right) = \frac{\cos z \frac{d}{dz}\sin z - \sin z \frac{d}{dz}\cos z}{\cos^2 z}$$

$$= \frac{\cos^2 z + \sin^2 z}{\cos^2 z} = \frac{1}{\cos^2 z}$$

$$= \sec^2 z.$$

We leave the proofs of the other derivative formulas as exercises.

To establish additional properties, expressing $\cos z$ and $\sin z$ in the Cartesian form $u+iv$ will be useful. (Additionally, the applications in Chapters 10 and 11 will use these formulas.) We begin by observing that the argument given to prove part (iii) in Theorem 5.1 easily generalizes to the complex case with the aid of Definition 5.5. That is,

$$e^{iz} = \cos z + i\sin z, \tag{5.32}$$

for all z, whether z is real or complex. Hence

$$e^{-iz} = \cos(-z) + i\sin(-z) = \cos z - i\sin z. \tag{5.33}$$

Subtracting Equation (5.33) from Equation (5.32) and solving for $\sin z$ gives

$$\sin z = \frac{1}{2i}(e^{iz} - e^{-iz}) \tag{5.34}$$

$$= \frac{1}{2i}\left(e^{i(x+iy)} - e^{-i(x+iy)}\right)$$

$$= \frac{1}{2i}\left(e^{-y+ix} - e^{y-ix}\right)$$

$$= \frac{1}{2i}\left[e^{-y}(\cos x + i\sin x) - e^{y}(\cos x - i\sin x)\right]$$

$$= \sin x\left(\frac{e^y + e^{-y}}{2}\right) + i\cos x\left(\frac{e^y - e^{-y}}{2}\right)$$

$$= \sin x\cosh y + i\cos x\sinh y, \tag{5.35}$$

where $\cosh y = \frac{e^y + e^{-y}}{2}$ and $\sinh y = \frac{e^y - e^{-y}}{2}$, respectively, are the hyperbolic cosine and hyperbolic sine functions that you studied in calculus.

Similarly,

$$\cos z = \frac{1}{2}(e^{iz} + e^{-iz}) \tag{5.36}$$

$$= \frac{1}{2}\left(e^{i(x+iy)} + e^{-i(x+iy)}\right)$$

$$= \frac{1}{2}\left(e^{-y+ix} + e^{y-ix}\right)$$

$$= \frac{1}{2}\left[e^{-y}(\cos x + i\sin x) + e^{y}(\cos x - i\sin x)\right]$$

$$= \cos x\left(\frac{e^y + e^{-y}}{2}\right) - i\sin x\left(\frac{e^y - e^{-y}}{2}\right)$$

$$= \cos x\cosh y - i\sin x\sinh y. \tag{5.37}$$

Equipped with Identities (5.34)–(5.37), we can now establish many other properties of the trigonometric functions. We begin with some periodic results.

- For all complex numbers $z = x + iy$,

$$\sin(z + 2\pi) = \sin z,$$
$$\cos(z + 2\pi) = \cos z.$$
$$\sin(z + \pi) = -\sin z,$$
$$\cos(z + \pi) = -\cos z,$$
$$\tan(z + \pi) = \tan z, \quad \text{and}$$
$$\cot(z + \pi) = \cot z.$$

Clearly, $\sin(z + 2\pi) = \sin[(x + 2\pi) + iy]$. By Identity (5.35) this expression is

$$\sin(x + 2\pi)\cosh y + i\cos(x + 2\pi)\sinh y = \sin x \cosh y + i\cos x \sinh y$$
$$= \sin z.$$

Again, the proofs for the other periodic results are left as exercises.

- If z_1 and z_2 are any complex numbers, then

$$\sin(z_1 + z_2) = \sin z_1 \cos z_2 + \cos z_1 \sin z_2 \quad \text{and}$$
$$\cos(z_1 + z_2) = \cos z_1 \cos z_2 - \sin z_1 \sin z_2 \quad \text{so}$$
$$\sin 2z = 2\sin z \cos z.$$
$$\cos 2z = \cos^2 z - \sin^2 z, \quad \text{and}$$
$$\sin\left(\frac{\pi}{2} + z\right) = \sin\left(\frac{\pi}{2} - z\right) = \cos z.$$

We demonstrate that $\cos(z_1 + z_2) = \cos z_1 \cos z_2 - \sin z_1 \sin z_2$ by making use of Identities (5.34)–(5.37):

$$\cos z_1 \cos z_2 = \frac{1}{4}\left[e^{i(z_1+z_2)} + e^{i(z_1-z_2)} + e^{i(z_2-z_1)} + e^{-i(z_1+z_2)}\right], \quad \text{and}$$
$$-\sin z_1 \sin z_2 = \frac{1}{4}\left[e^{i(z_1+z_2)} - e^{i(z_1-z_2)} - e^{i(z_2-z_1)} + e^{-i(z_1+z_2)}\right].$$

Adding these expressions gives

$$\cos z_1 \cos z_2 - \sin z_1 \sin z_2 = \frac{1}{2}\left[e^{i(z_1+z_2)} + e^{-i(z_1+z_2)}\right] = \cos(z_1 + z_2),$$

which is what we wanted.

A solution to the equation $f(z) = 0$ is called a *zero* of the given function f. As we now show, the zeros of the sine and cosine function are exactly where you might expect them to be.

- We have $\sin z = 0$ iff $z = n\pi$, where n is any integer, and $\cos z = 0$ iff $z = (n + \frac{1}{2})\pi$, where n is any integer.

We show the result for $\cos z$ and leave the result for $\sin z$ as an exercise. When we use Identity (5.37), $\cos z = 0$ iff

$$0 = \cos x \cosh y - i \sin x \sinh y.$$

Equating the real and imaginary parts of this equation gives

$$0 = \cos x \cosh y, \quad \text{and} \quad 0 = \sin x \sinh y.$$

The real-valued function $\cosh y$ is never zero, so the equation $0 = \cos x \cosh y$ implies that $0 = \cos x$, from which we obtain $x = (n + \frac{1}{2})\pi$ for any integer n. Using the values for $z = x + iy = (n + \frac{1}{2})\pi + iy$ in the equation $0 = \sin x \sinh y$ yields

$$0 = \sin\left[(n + \frac{1}{2})\pi\right] \sinh y = (-1)^n \sinh y,$$

which implies that $y = 0$, so the only zeros for $\cos z$ are the values $z = (n + \frac{1}{2})\pi$ for any integer n.

What does the mapping $w = \sin z$ look like? We can get a graph of the mapping $w = \sin z = \sin(x+iy) = \sin x \cosh y + i \cos x \sinh y$ by using parametric methods. Let's consider the vertical line segments in the z plane obtained by successfully setting $x = -\frac{\pi}{2} + \frac{k\pi}{12}$ for $k = 0, 1, \ldots, 12$, and for each x value and letting y vary continuously, $-3 \le y \le 3$. In the exercises we ask you to show that the images of these vertical segments are hyperbolas in the uv plane, as Figure 5.7 illustrates. In Chapter 10, we give a more detailed analysis of the mapping $w = \sin z$.

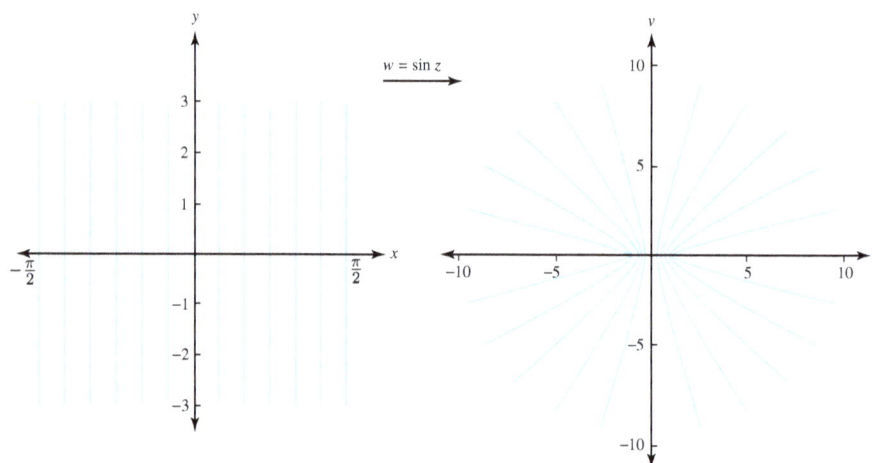

Figure 5.7: Vertical segments mapped onto hyperbolas by $w = \sin(z)$

Figure 5.7 suggests one big difference between the real and complex sine functions. The real sine has the property that $|\sin x| \le 1$ for all real x. In Figure 5.7, however, the modulus of the complex sine appears to be unbounded, which is indeed the case. Using Identity (5.35) gives

$$|\sin z|^2 = |\sin z \cosh y + i \cos x \sinh y|^2$$
$$= \sin^2 x \cosh^2 y + \cos^2 x \sinh^2 y$$
$$= \sin^2 x(\cosh^2 y - \sinh^2 y) + \sinh^2 y(\cos^2 x + \sin^2 x).$$

The identities $\cosh^2 y - \sinh^2 y = 1$ and $\cos^2 x + \sin^2 x = 1$ then yield

$$|\sin z|^2 = \sin^2 x + \sinh^2 y. \tag{5.38}$$

A similar derivation produces

$$|\cos z|^2 = \cos^2 x + \sinh^2 y. \tag{5.39}$$

If we set $z = x_0 + iy$ in Identity (5.38) and let $y \to \infty$, we get

$$\lim_{y \to \infty} |\sin(x_0 + iy)|^2 = \sin^2 x_0 + \lim_{y \to \infty} \sinh^2 y = \infty.$$

As advertised, we have shown that $\sin z$ is not a bounded function; it is also evident from Identity (5.39) that $\cos z$ is unbounded.

The periodic character of the trigonometric functions makes apparent that any point in their ranges is actually the image of infinitely many points.

Example 5.10. Find the values of z for which $\cos z = \cosh 2$.

Solution:

Starting with Identity (5.37), we write

$$\cos z = \cos x \cosh y - i \sin x \sinh y = \cosh 2.$$

If we equate real and imaginary parts, then we get

$$\cos x \cosh y = \cosh 2 \text{ and } \sin x \sinh y = 0.$$

The equation $\sin x \sinh y = 0$ implies either that $x = \pi n$, where n is an integer, or that $y = 0$. Using $y = 0$ in the equation $\cos x \cosh y = \cosh 2$ leads to the impossible situation $\cos x = \frac{\cosh 2}{\cosh 0} = \cosh 2 > 1$. Therefore $x = \pi n$, where n is an integer. Since $\cosh y \geq 1$ for all values of y, the term $\cos x$ in the equation $\cos x \cosh y = \cosh 2$ must also be positive. For this reason we eliminate the odd values of n and get $x = 2\pi k$, where k is an integer.

Finally, we solve the equation $\cos 2\pi k \cosh y = \cosh y = \cosh 2$ and use the fact that $\cosh y$ is an even function to conclude that $y = \pm 2$. Therefore the solutions to the equation $\cos z = \cosh 2$ are $z = 2\pi k \pm 2i$, where k is an integer.

The hyperbolic functions also have practical use in putting the tangent function into the Cartesian form $u + iv$. Using Definition 5.6, and Equations (5.35) and 5.37, we have

$$\tan z = \tan(x + iy) = \frac{\sin(x + iy)}{\cos(x + iy)} = \frac{\sin x \cosh y + i \cos x \sinh y}{\cos x \cosh y - i \sin x \sinh y}.$$

If we multiply each term on the right by the conjugate of the denominator, the simplified result is

$$\tan z = \frac{\cos x \sin x + i \cosh y \sinh y}{\cos^2 x \cosh^2 y + \sin^2 x \sinh^2 y}. \tag{5.40}$$

We leave it as an exercise to show that the identities $\cosh^2 y - \sinh^2 y = 1$ and $\sinh 2y = 2 \cosh y \sinh y$ can be used in simplifying Equation (5.40) to get

$$\tan z = \frac{\sin 2x}{\cos 2x + \cosh 2y} + i \frac{\sinh 2y}{\cos 2x + \cosh 2y}. \tag{5.41}$$

As with $\sin z$, we obtain a graph of the mapping $w = \tan z$ parametrically. Consider the vertical line segments in the z plane obtained by successively setting $x = -\frac{\pi}{4} + \frac{k\pi}{16}$ for $k = 0, 1, \ldots, 8$ and for each z value letting y vary continuously, $-3 \leq y \leq 3$. In the exercises we ask you to show that the images of these vertical segments are circular arcs in the uv plane, as Figure 5.8 shows. In Section 9.4 we give a more detailed investigation of the mapping $w = \tan z$.

How should we define the complex hyperbolic functions? We begin with

153

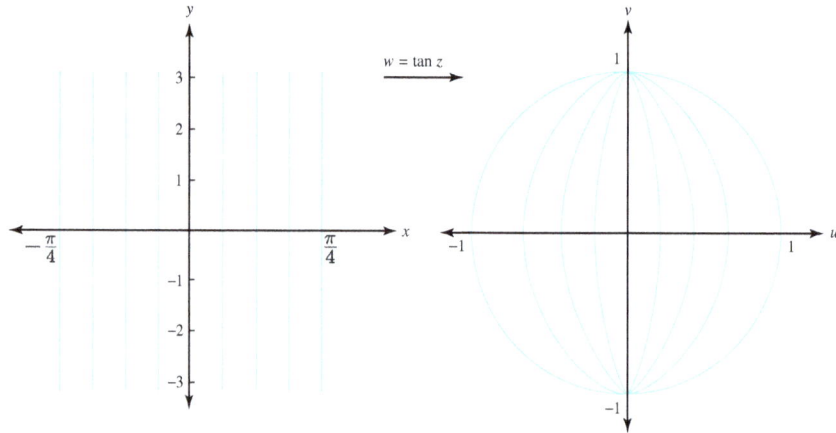

Figure 5.8: Vertical segments mapped onto circular arcs by $w = \tan z$

Definition 5.7 ($\cosh z$ and $\sinh z$).

$$\cosh z = \frac{1}{2}(e^z + e^{-z}) \quad and \quad \sinh z = \frac{1}{2}(e^z - e^{-z}).$$

With these definitions in place, we can now easily create the other complex hyperbolic trigonometric functions, provided the denominators in the following expressions are not zero.

Definition 5.8 (Complex hyperbolic functions).

$$\tanh z = \frac{\sinh z}{\cosh z}, \quad \coth z = \frac{\cosh z}{\sinh z}, \quad \operatorname{sech} z = \frac{1}{\cosh z} \quad and \quad \operatorname{csch} z = \frac{1}{\sinh z}.$$

As the series for the complex hyperbolic sine and cosine agree with the real hyperbolic sine and cosine when z is real, the remaining complex hyperbolic trigonometric functions likewise agree with their real counterparts. Many other properties are also shared. We state several results without proof, as they follow from the definitions we gave using standard operations, such as the quotient rule for derivatives. We ask you to establish some of these identities in the exercises.

The derivatives of the hyperbolic functions follow the same rules as in calculus:

$$\frac{d}{dz}\cosh z = \sinh z \quad and \quad \frac{d}{dz}\sinh z = \cosh z.$$
$$\frac{d}{dz}\tanh z = \operatorname{sech}^2 z \quad and \quad \frac{d}{dz}\coth z = -\operatorname{csch}^2 z.$$
$$\frac{d}{dz}\operatorname{sech} z = -\operatorname{sech} z \tanh z \quad and \quad \frac{d}{dz}\operatorname{csch} z = -\operatorname{csch} z \coth z.$$

The hyperbolic cosine and hyperbolic sine can be expressed as

$$\cosh z = \cosh x \cos y + i \sinh x \sin y, \quad and$$
$$\sinh z = \sinh x \cos y + i \cosh x \sin y.$$

The complex trigonometric and hyperbolic functions are all defined in terms of the exponential function, so we can easily show them to be related by

$$\cosh(iz) = \cos z \quad and \quad \sinh(iz) = i \sin z;$$
$$\cos(iz) = \cosh z \quad and \quad \sin(iz) = i \sinh z.$$

154

Some of the important identities involving the hyperbolic functions are

$$\cosh^2 z - \sinh^2 z = 1.$$
$$\sinh(z_1 + z_2) = \sinh z_1 \cosh z_2 + \cosh z_1 \sinh z_2,$$
$$\cosh(z_1 + z_2) = \cosh z_1 \cosh z_2 + \sinh z_1 \sinh z_2,$$
$$\cosh(z + 2\pi i) = \cosh z,$$
$$\sinh(z + 2\pi i) = \sinh z,$$
$$\cosh(-z) = \cosh z, \quad \text{and}$$
$$\sinh(-z) = -\sinh z.$$

We conclude this section with an example from electronics. In electric circuits, the voltage drop, E_R, across a resistance R obeys Ohm's law,

$$E_R = IR,$$

where I is the current flowing through the resistor. Additionally, the current and voltage drop across an inductor, L, obey the equation

$$E_L = L\frac{dI}{dt}.$$

The current and voltage across a capacitor, C, are related by

$$E_C = \frac{1}{C} \int_{t_0}^{t} I(\tau)d\tau.$$

The voltages E_L, E_R, and E_C and the impressed voltage $E(t)$ illustrated in Figure 5.9 satisfy the equation

$$E_L + E_R + E_C = E(t). \tag{5.42}$$

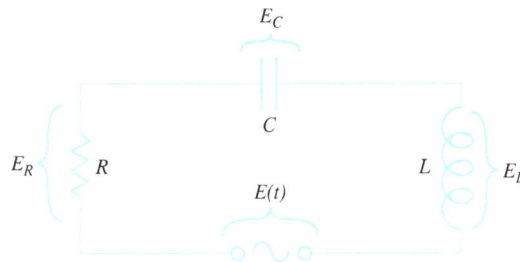

Figure 5.9: An LRC circuit

Suppose that the current $I(t)$ in the circuit is given by

$$I(t) = I_0 \sin \omega t.$$

Using this in the equations for E_R and E_L gives

$$E_R = RI_0 \sin \omega t, \quad \text{and} \tag{5.43}$$
$$E_L = \omega L I_0 \cos \omega t. \tag{5.44}$$

We then set $t_0 = \frac{\pi}{2}$ in the equation for E_C to obtain

$$E_C = \frac{1}{C}\int\limits_{\frac{\pi}{2}}^{t} I(\tau)d\tau = \frac{1}{C}\int\limits_{\frac{\pi}{2}}^{t} I_0 \sin\omega t d\tau = -\frac{1}{\omega C}I_0 \cos\omega t. \qquad (5.45)$$

We rewrite the equation $I(t) = I_0 \sin\omega t$ as a "complex current,"

$$I^* = I_0 e^{i\omega t}$$

with the understanding that the actual physical current I is the imaginary part of I^*. Similarly, we rewrite Equations (5.43)–(5.45) as

$$E_R^* = RI_0 e^{i\omega t} \quad = RI^*.$$
$$E_L^* = i\omega L I_0 e^{i\omega t} = i\omega L I^*, \quad \text{and}$$
$$E_C^* = \frac{1}{i\omega C} I_0 e^{i\omega t} = \frac{1}{i\omega C} I^*.$$

Substituting these terms leads to an extension of Equation (5.42),

$$E^* = E_R^* + E_L^* + E_C^* = \left[R + i\left(\omega L - \frac{1}{\omega C}\right)\right] I^*. \qquad (5.46)$$

The complex quantity Z defined by

$$Z = R + i\left(\omega L - \frac{1}{\omega C}\right)$$

is called the *complex impedance*. Substituting this expression into Equation (5.46) gives

$$E^* = ZI^*.$$

which is the complex extension of Ohm's law.

Exercises for Section 5.4 (Selected answers or hints are on page 441.)

1. Establish that $\frac{d}{dz}\cos z = -\sin z$ for all z.

2. Demonstrate that, for all z, $\sin^2 z + \cos^2 z = 1$, as follows.

   (a) Define the function $g(z) = \sin^2 z + \cos^2 z$. Explain why g is entire.
   (b) Show that g is constant. *Hint:* Look at $g'(z)$.
   (c) Use part (b) to establish that, for all z, $\sin^2 z + \cos^2 z = 1$.

3. Show that Equation (5.40) simplifies to Equation (5.41). *Hint:* Use the facts that $\cosh^2 y - \sinh^2 y = 1$ and $\sinh 2y = 2\cosh y \sinh y$.

4. Explain why the diagrams in Figures 5.8 and 5.9 came out they way they did.

5. Show that, for all z,

   (a) $\sin(\pi - z) = \sin z$.

(b) $\sin(\frac{\pi}{2} - z) = \cos z$.

(c) $\sinh(z + i\pi) = -\sinh z$.

(d) $\tanh(z + i\pi) = \tanh z$.

(e) $\sin(iz) = i \sinh z$.

(f) $\cosh(iz) = \cos z$.

6. Express the following quantities in $u + iv$ form.

(a) $\cos(1 + i)$.

(b) $\sin(\frac{\pi + 4i}{4})$.

(c) $\sin 2i$.

(d) $\cos(-2 + i)$.

(e) $\tan(\frac{\pi + 2i}{4})$.

(f) $\tan(\frac{\pi + i}{2})$.

(g) $\sinh(1 + i\pi)$.

(h) $\cosh \frac{i\pi}{2}$.

(i) $\cosh(\frac{4 - i\pi}{4})$.

7. Find the derivatives of the following and state where they are defined.

(a) $\sin(\frac{1}{z})$.

(b) $z \tan z$.

(c) $\sec z^2$.

(d) $z \csc^2 z$.

(e) $z \sinh z$.

(f) $\cosh z^2$.

(g) $z \tan z$.

8. Show that

(a) $\sin \overline{z} = \overline{\sin z}$ holds for all z.

(b) $\sin \overline{z}$ is nowhere analytic.

(c) $\cosh \overline{z} = \overline{\cosh z}$ holds for all z.

(d) $\cosh \overline{z}$ is nowhere analytic.

9. Show that

(a) $\lim_{z \to 0} \frac{\cos z - 1}{z} = 0$.

(b) $\lim_{y \to +\infty} \tan(x_0 + iy) = i$, where x_0 is any fixed real number.

10. Find all values of z for which each equation holds.

(a) $\sin z = \cosh 4$.

(b) $\cos z = 2$.

(c) $\sin z = i \sinh 1$.

157

(d) $\sinh z = \frac{i}{2}$.

(e) $\cosh z = 1$.

11. Show that the zeros of $\sin z$ are at $z = n\pi$ where n is an integer.

12. Use Equation (5.38) to show that, for $z = x + iy$,

$$|\sinh y| \le |\sin z| \le \cosh y.$$

13. Use Identities (5.38) and (5.39) to help establish the inequality

$$|\cos z|^2 + |\sin z|^2 \ge 1,$$

and show that equality holds iff z is a real number.

14. Show that the mapping $w = \sin z$

(a) maps the y axis one-to-one and onto the v axis;

(b) maps the ray given $\{(x, y) : x = \frac{\pi}{2}, y > 0\}$ one-to-one and onto the ray defined by $\{(u, v) : u > 1, v = 0\}$.

15. Given an elegant argument that explains why the following functions are harmonic.

(a) $h(x, y) = \sin x \cosh y$.

(b) $h(x, y) = \cos x \sinh y$.

(c) $h(x, y) = \sinh x \cos y$.

(d) $h(x, y) = \cosh x \sin y$.

16. Establish the following identities.

(a) $e^{iz} = \cos z + i \sin z$.

(b) $\cos z = \cos x \cosh y - i \sin x \sinh y$.

(c) $\sin(z_1 + z_2) = \sin z_1 \cos z_2 + \cos z_1 \sin z_2$.

(d) $|\cos z|^2 = \cos^2 x + \sinh^2 y$.

(e) $\cosh z = \cosh x \cos y + i \sinh x \sin y$.

(f) $\cosh^2 z - \sinh^2 z = 1$.

(g) $\cosh(z_1 + z_2) = \cosh z_1 \cosh z_2 + \sinh z_1 \sinh z_2$.

17. Find the complex impedance Z if

(a) $R = 10$, $L = 10$, $C = 0.05$, and $\omega = 2$.

(b) $R = 15$, $L = 10$, $C = 0.05$, and $\omega = 4$.

18. Explain how $\sin z$ and the function $\sin x$ that you studied in calculus are different. How are they similar?

5.5 Inverse Trigonometric and Hyperbolic Functions

We expressed trigonometric and hyperbolic functions in Section 5.4 in terms of the exponential function. In this section we look at their inverses. When we solve equations such as $w = \sin z$ for z, we obtain formulas that involve the logarithm. Because trigonometric and hyperbolic functions are all periodic, they are many-to-one; hence their inverses are necessarily multivalued. The formulas for the inverse trigonometric functions are

$$\arcsin z = -i \log \left[iz + (1 - z^2)^{\frac{1}{2}} \right], \tag{5.47}$$

$$\arccos z = -i \log \left[z + i(1 - z^2)^{\frac{1}{2}} \right], \quad \text{and}$$

$$\arctan z = \frac{i}{2} \log \left(\frac{i + z}{i - z} \right).$$

We can find the derivatives of any branch of these functions by using the chain rule:

$$\frac{d}{dz} \arcsin z = \frac{1}{(1 - z^2)^{\frac{1}{2}}}, \tag{5.48}$$

$$\frac{d}{dz} \arccos z = \frac{-1}{(1 - z^2)^{\frac{1}{2}}}, \quad \text{and}$$

$$\frac{d}{dz} \arctan z = \frac{1}{1 + z^2}.$$

We derive Equations (5.47) and (5.48) and leave the others as exercises. If we take a particular branch of the multivalued function, $w = \arcsin z$, we have

$$z = \sin w = \frac{1}{2i}(e^{iw} - e^{-iw}),$$

which we can also write as

$$e^{iw} - 2iz - e^{-iw} = 0.$$

Multiplying both sides of this equation by e^{iw} gives $(e^{iw})^2 - 2ize^{iw} - 1 = 0$, which is a quadratic equation in terms of e^{iw}. Using the quadratic equation to solve for e^{iw}, we obtain

$$e^{iw} = \frac{2iz + (4 - 4z^2)^{\frac{1}{2}}}{2} = iz + (1 - z^2)^{\frac{1}{2}},$$

where the square root is a multivalued function. Taking the logarithm of both sides of this last equation leads to the desired result:

$$w = \arcsin z = -i \log \left[iz + (1 - z^2)^{\frac{1}{2}} \right].$$

where the multivalued logarithm is used. To construct a specific branch of $\arcsin z$, we must first select a branch of the square root and then select a branch of the logarithm.

We get the derivative of $w = \arcsin z$ by starting with the equation $\sin w = z$ and differentiating both sides, using the chain rule:

$$\frac{d}{dz} \sin w = \frac{d}{dz} z,$$

$$\frac{d}{dw} \sin w \frac{dw}{dz} = 1,$$

$$\frac{dw}{dz} = \frac{1}{\cos w}.$$

When the principal value is used, $w = \arcsin z = -i\text{Log}\left[iz + (1 - z^2)^{\frac{1}{2}}\right]$ maps the upper half-plane $\{z : \text{Im}(z) > 0\}$ onto a portion of the upper half-plane $\{w : \text{Im}(w) > 0\}$ that lies in the vertical strip $\{w : \frac{-\pi}{2} < \text{Re}(w) < \frac{\pi}{2}\}$. The image of a rectangular grid in the z plane is a "spider web" in the w plane, as Figure 5.10 shows.

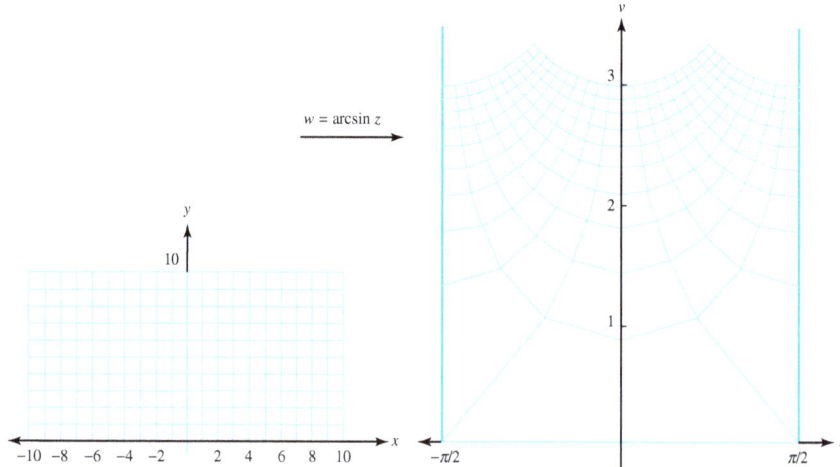

Figure 5.10: A rectangular grid is mapped onto a spider web by $w = \arcsin z$

Example 5.11. The values of $\arcsin \sqrt{2}$ are given by

$$\arcsin \sqrt{2} = -i \log\left[i\sqrt{2} + \left(1 - \left(\sqrt{2}\right)^2\right)^{\frac{1}{2}}\right] = -i\log(i\sqrt{2} \pm i). \tag{5.49}$$

Using straightforward techniques, we simplify this equation and obtain

$$\arcsin \sqrt{2} = -i\log\left[(\sqrt{2} \pm 1)i\right]$$
$$= i\left[\ln(\sqrt{2} \pm 1) + i\left(\frac{\pi}{2} + 2n\pi\right)\right]$$
$$= \frac{\pi}{2} + 2n\pi - i\ln(\sqrt{2} \pm 1), \quad \text{where } n \text{ is an integer.}$$

We observe that

$$\ln(\sqrt{2} - 1) = \ln\frac{(\sqrt{2} - 1)(\sqrt{2} + 1)}{\sqrt{2} + 1} = \ln\frac{1}{\sqrt{2} + 1} = -\ln(\sqrt{2} + 1)$$

and then write

$$\arcsin \sqrt{2} = \frac{\pi}{2} + 2n\pi \pm i\ln(\sqrt{2} + 1), \quad \text{where } n \text{ is an integer.}$$

Example 5.12. Suppose that we make specific choices in Equation (5.49) by selecting $+i$ as the value of the square root $[1 - (\sqrt{2})^2]^{\frac{1}{2}}$ and using the principal value of the logarithm. With $f(z) = \text{Arcsin } z$, The result is

$$f(\sqrt{2}) = \text{Arcsin}\sqrt{2} = -i\text{Log}(i\sqrt{2} + i) = \frac{\pi}{2} - i\ln(\sqrt{2} + 1),$$

and the corresponding value of the derivative is given by

$$f'(\sqrt{2}) = \frac{1}{\left[1 - \left(\sqrt{2}\right)^2\right]^{\frac{1}{2}}} = \frac{1}{i} = -i.$$

160

The inverse hyperbolic functions are

$$\text{arcsinh } z = \log \left[z + (z^2 + 1)^{\frac{1}{2}} \right],$$

$$\text{arccosh } z = \log \left[z + (z^2 - 1)^{\frac{1}{2}} \right], \quad \text{and}$$

$$\text{arctanh } z = \frac{1}{2} \log \left(\frac{1+z}{1-z} \right). \tag{5.50}$$

Their derivatives are

$$\frac{d}{dz} \text{arcsinh } z = \frac{1}{(z^2+1)^{\frac{1}{2}}},$$

$$\frac{d}{dz} \text{arccosh } z = \frac{1}{(z^2-1)^{\frac{1}{2}}}, \quad \text{and}$$

$$\frac{d}{dz} \text{arctanh } z = \frac{1}{1-z^2}.$$

To establish Identity (5.50), we start with $w = \text{arctanh} z$ and obtain

$$z = \tanh w = \frac{e^w - e^{-w}}{e^w + e^{-w}} = \frac{e^{2w} - 1}{e^{2w} + 1}.$$

which we solve for e^{2w}, getting $e^{2w} = \frac{1+z}{1-z}$. Taking the logarithms of both sides gives

$$w = \text{arctanh } z = \frac{1}{2} \log \left(\frac{1+z}{1-z} \right),$$

which is what we wanted to show.

Example 5.13. Calculation reveals that

$$\text{arctanh}(1 + 2i) = \frac{1}{2} \log \frac{1 + 1 + 2i}{1 - 1 - 2i} = \frac{1}{2} \log(-1 + i)$$

$$= \frac{1}{4} \ln 2 + i \left(\frac{3}{8} + n \right) \pi,$$

where n is an integer.

Exercises for Section 5.5 (Selected answers or hints are on page 442.)

1. Find *all* values of the following.

   (a) $\arcsin \frac{5}{4}$.

   (b) $\arccos \frac{5}{3}$.

   (c) $\arcsin 3$.

   (d) $\arccos 3i$.

   (e) $\arctan 2i$.

   (f) $\arctan i$.

   (g) $\text{arcsinh } i$.

(h) arcsinh $\frac{3}{4}$.

(i) arccosh i.

(j) arccosh $\frac{1}{2}$.

(k) arctanh i.

(l) arctanh $i\sqrt{3}$.

2. Establish the following identities.

(a) $\arccos z = -i \log \left[z + i(1 - z^2)^{\frac{1}{2}} \right]$.

(b) $\frac{d}{dz} \arccos z = \frac{-1}{(1-z^2)^{\frac{1}{2}}}$.

(c) $\arctan z = \frac{i}{2} \log(\frac{i+z}{i-z})$.

(d) $\frac{d}{dz} \arctan z = \frac{1}{1+z^2}$.

(e) $\arcsin z + \arccos z = \frac{\pi}{2} + 2n\pi$, where n is an integer.

(f) $\frac{d}{dz} \text{arctanh} z = \frac{1}{1-z^2}$.

(g) $\text{arcsinh} z = \log \left[z + (z^2 + 1)^{\frac{1}{2}} \right]$.

(h) $\frac{d}{dz} \text{arcsinh} z = \frac{1}{(z^2+1)^{1/2}}$.

(i) $\text{arccosh} z = \log \left[z + (z^2 - 1)^{\frac{1}{2}} \right]$.

(j) $\frac{d}{dz} \text{arccosh} z = \frac{1}{(z^2-1)^{\frac{1}{2}}}$.

Chapter 6

Complex Integration

Overview

Of the two main topics studied in calculus–differentiation and integration–we have so far only studied derivatives of complex functions. We now turn to the problem of integrating complex functions. The theory you will learn is elegant, powerful, and a useful tool for physicists and engineers. It also connects widely with other branches of mathematics. For example, even though the ideas presented here belong to the general area of mathematics known as analysis, you will see as an application of them one of the simplest proofs of the fundamental theorem of algebra.

6.1   Complex Integrals

We introduce the integral of a complex function by defining the integral of a complex-valued function of a *real* variable.

Definition 6.1 (Integral of $f(t)$). *Let $f(t) = u(t)+iv(t)$, where u and v are real-valued functions of the real variable t for $a \le t \le b$. Then*

$$\int_a^b f(t)\, dt = \int_a^b u(t)\, dt + i \int_a^b v(t)\, dt. \qquad (6.1)$$

We generally evaluate integrals of this type by finding the antiderivatives of u and v and evaluating the definite integrals on the right side of Equation (6.1). That is, if $U'(t) = u(t)$, and $V'(t) = v(t)$, for $a \le t \le b$, we have

$$\int_a^b f(t)\, dt = \big[U(t) + iV(t)\big]\Big|_{t=a}^{t=b} = U(b) - U(a) + i\big[V(b) - V(a)\big]. \qquad (6.2)$$

Example 6.1. Show that

$$\int_0^1 (t-i)^3 \, dt = -\frac{5}{4}.$$

Solution:

We write the integrand in terms of its real and imaginary parts, *i.e.*, $f(t) = (t-i)^3 = t^3 - 3t + i(-3t^2 + 1)$. Here, $u(t) = t^3 - 3t$ and $v(t) = -3t^2 + 1$. The integrals of u and v are

$$\int_0^1 (t^3 - 3t)\, dt = -\frac{5}{4}, \quad \text{and} \quad \int_0^1 (-3t^2 + 1)\, dt = 0.$$

163

Hence, by Definition 6.1,

$$\int_0^1 (t - i)^3 \, dt = \int_0^1 u(t) \, dt + i \int_0^1 v(t) \, dt = -\frac{5}{4}.$$

Example 6.2. Show that

$$\int_0^{\frac{\pi}{2}} \exp(t + it) \, dt = \frac{1}{2}(e^{\frac{\pi}{2}} - 1) + \frac{i}{2}(e^{\frac{\pi}{2}} + 1).$$

Solution:

We use the method suggested by Definitions 6.1 and 6.2.

$$\int_0^{\frac{\pi}{2}} \exp(t + it) \, dt = \int_0^{\frac{\pi}{2}} e^t e^{it} dt$$

$$= \int_0^{\frac{\pi}{2}} e^t (\cos t + i \sin t) \, dt$$

$$= \int_0^{\frac{\pi}{2}} e^t \cos t \, dt + i \int_0^{\frac{\pi}{2}} e^t \sin t \, dt.$$

We can evaluate each of the integrals via integration by parts. For example,

$$\int_0^{\frac{\pi}{2}} \underbrace{e^t}_{u} \underbrace{\cos t \, dt}_{dv} = (\underbrace{e^t}_{u} \underbrace{\sin t}_{v}) \Big|_{t=0}^{t=\frac{\pi}{2}} - \int_0^{\frac{\pi}{2}} \underbrace{\sin t}_{v} \underbrace{e^t dt}_{du}$$

$$= (e^{\frac{\pi}{2}} \sin \frac{\pi}{2} - e^0 \sin 0) - \int_0^{\frac{\pi}{2}} \underbrace{e^t}_{u} \underbrace{\sin t \, dt}_{dv}$$

$$= e^{\frac{\pi}{2}} - \int_0^{\frac{\pi}{2}} \underbrace{e^t}_{u} \underbrace{\sin t \, dt}_{dv}$$

$$= e^{\frac{\pi}{2}} - (\underbrace{e^t}_{u} \underbrace{[-\cos t])}_{v} \Big|_{t=0}^{t=\frac{\pi}{2}} + \int_0^{\frac{\pi}{2}} \underbrace{-\cos t}_{v} \underbrace{e^t \, dt}_{du}$$

$$= e^{\frac{\pi}{2}} - 1 - \int_0^{\frac{\pi}{2}} e^t \cos t \, dt.$$

Adding $\int_0^{\frac{\pi}{2}} e^t \cos t \, dt$ to both sides of this equation and then dividing by 2 gives $\int_0^{\frac{\pi}{2}} e^t \cos t \, dt = \frac{1}{2}(e^{\frac{\pi}{2}} - 1)$. A similar computation procedure yields $i \int_0^{\frac{\pi}{2}} e^t \sin t \, dt = \frac{i}{2}(e^{\frac{\pi}{2}} + 1)$. Therefore,

$$\int_0^{\frac{\pi}{2}} \exp(t + it) \, dt = \frac{1}{2}(e^{\frac{\pi}{2}} - 1) + \frac{i}{2}(e^{\frac{\pi}{2}} + 1).$$

Complex integrals have properties that are similar to those of real integrals. We now trace through several commonalities. Let $f(t) = u(t) + iv(t)$ and $g(t) = p(t) + iq(t)$ be continuous on $a \le t \le b$.

- Using Definition 6.1, we can easily show that the integral of their sum is the sum of their integrals, that is

$$\int_a^b [f(t) + g(t)] \, dt = \int_a^b f(t) \, dt + \int_a^b g(t) \, dt. \tag{6.3}$$

164

- If we divide the interval $a \leq t \leq b$ into $a \leq t \leq c$ and $c \leq t \leq b$ and integrate $f(t)$ over these subintervals by using Definition 6.1, then we get

$$\int_a^b f(t)\, dt = \int_a^c f(t)\, dt + \int_c^b f(t)\, dt. \tag{6.4}$$

- Similarly, if $c + id$ denotes a complex constant, then

$$\int_a^b (c + id) f(t)\, dt = (c + id) \int_a^b f(t)\, dt. \tag{6.5}$$

- If the limits of integration are reversed, then

$$\int_a^b f(t)\, dt = - \int_b^a f(t)\, dt. \tag{6.6}$$

- The integral of the product fg becomes

$$\int_a^b f(t) g(t)\, dt = \int_a^b \left[u(t) p(t) - v(t) q(t) \right] dt$$
$$+ i \int_a^b \left[u(t) q(t) + v(t) p(t) \right] dt. \tag{6.7}$$

Example 6.3. Let us verify Property (6.5). We start by writing

$$(c + id) f(t) = (c + id)\big(u(t) + iv(t)\big) = cu(t) - dv(t) + i\big[cv(t) + du(t)\big].$$

Using Definition 6.1, we write the left side of Equation 6.5 as

$$c \int_a^b u(t)\, dt - d \int_a^b v(t)\, dt + ic \int_a^b v(t)\, dt + id \int_a^b u(t)\, dt.$$

which is equivalent to

$$(c + id) \left[\int_a^b u(t)\, dt + i \int_u^b v(t)\, dt \right].$$

It is worthwhile to point out the similarity between Equation (6.2) and its counterpart in calculus. Suppose that U and V are differentiable on $a < t < b$ and $F(t) = U(t) + iV(t)$. Since $F'(t) = U'(t) + iV'(t) = u(t) + iv(t) = f(t)$, Equation (6.2) takes on the familiar form

$$\int_a^b f(t)\, dt = F(t) \Big|_{t=a}^{t=b} = F(b) - F(a). \tag{6.8}$$

where $F'(t) = f(t)$. We can view Equation (6.8) as an extension of the fundamental theorem of calculus. In Section 6.5 we show how to generalize this extension to analytic functions of a complex variable. For now, we simply note an important case of Equation (6.8):

$$\int_a^b f'(t)\, dt = f(b) - f(a). \tag{6.9}$$

Example 6.4. Use Equation (6.8) to show that

$$\int_0^{\frac{\pi}{2}} \exp(t + it)\, dt = \frac{1}{2}(e^{\frac{\pi}{2}} - 1) + \frac{i}{2}(e^{\frac{\pi}{2}} + 1).$$

***Solution*:**

We seek a function F with the property that $F'(t) = \exp(t+it)$. We note that $F(t) = \frac{1}{1+i}e^{t(1+i)}$ satisfies this requirement, so

$$\int_0^{\frac{\pi}{2}} \exp(t + it)\, dt = \frac{1}{1+i}e^{t(1+i)}\Big|_{t=0}^{t=\frac{\pi}{2}} = \frac{1}{1+i}(ie^{\frac{\pi}{2}} - 1)$$

$$= \frac{1}{2}(1 - i)(ie^{\frac{\pi}{2}} - 1)$$

$$\frac{1}{2}(e^{\frac{\pi}{2}} - 1) + \frac{i}{2}(e^{\frac{\pi}{2}} + 1),$$

which is the same result we obtained in Example 6.2, but with a lot less work!

Remark 6.1. *Example 6.4 illustrates the potential computational advantage we have when we lift our sights to the complex domain. Using ordinary calculus techniques to evaluate $\int_0^{\frac{\pi}{2}} e^t \cos t\, dt$, for example, would require a lengthy integration by parts procedure. When we recognize this expression as the real part of $\int_0^{\frac{\pi}{2}} \exp(t + it)\, dt$, however, the solution comes quickly. This is just one of the many reasons why good physicists and engineers, in addition to mathematicians, benefit from a thorough working knowledge of complex analysis.*

Exercises for Section 6.1 (Selected answers or hints are on page 442.)

1. Use Equations (6.1) and (6.2) to find

   (a) $\int_0^1 (3t - i)^2\, dt.$

   (b) $\int_0^1 (t + 2i)^3\, dt.$

   (c) $\int_0^{\frac{\pi}{2}} \cosh(it)\, dt.$

   (d) $\int_0^2 \frac{t}{t+i}\, dt.$

   (e) $\int_0^{\frac{\pi}{4}} t \exp(it)\, dt.$

2. Let m and n be integers. Show that

$$\int_0^{2\pi} e^{imt} e^{-int}\, dt = \begin{cases} 0, & \text{when } m \neq n; \\ 2\pi, & \text{when } m = n. \end{cases}$$

3. Show that $\int_0^\infty e^{-zt}\, dt = \frac{1}{z}$ provided $\text{Re}(z) > 0.$

4. Establish the following identities.

   (a) Identity (6.3).

   (b) Identity (6.4).

   (c) Identity (6.6).

166

(d) Identity (6.7).

5. Let $f(t) = u(t) + iv(t)$, where u and v are differentiable. Show that

$$\int_a^b f(t)f'(t)\,dt = \frac{1}{2}[f(b)]^2 - \frac{1}{2}[f(a)]^2\,.$$

6. Use integration by parts to verify that $i\int_0^{\frac{\pi}{2}} e^t \sin t\,dt = \frac{i}{2}(e^{\frac{\pi}{2}} + 1)$.

6.2 Contours and Contour Integrals

In Section 6.1 we showed how to evaluate integrals of the form $\int_a^b f(t)\,dt$, where f was complex-valued and $[a, b]$ was an interval on the real axis (so that t was real, with $t \in [a, b]$). In this section, we define and evaluate integrals of the form $\int_c f(z)\,dz$, where f is complex-valued and C is a contour in the plane (so that z is complex, with $z \in C$).Theorem 6.1 provides our main result, which shows how to transform the latter type of integral into the kind we investigated in Section 6.1.

We use concepts first introduced in Section 1.6. Recall that to represent a curve C in the plane we use the parametric notation

$$C : z(t) = x(t) + iy(t), \quad \text{for} \quad a \le t \le b, \tag{6.10}$$

where $x(t)$ and $y(t)$ are continuous functions. We now place a few more restrictions on the type of curve to be described. The following discussion leads to the concept of a contour, which is a type of curve that is adequate for the study of integration.

Recall that C is *simple* if it does not cross itself, which means that $z(t_1) \ne z(t_2)$ whenever $t_1 \ne t_2$, except possibly when $t_1 = a$ and $t_2 = b$. A curve C with the property $z(b) = z(a)$ is a **closed curve**. If $z(b) = z(a)$ is the only point of intersection, then we say that C is a **simple closed curve.** As the parameter t increases from the value a to the value b, the point $z(t)$ starts at the *initial point* $z(a)$, moves along the curve C, and ends up at the *terminal point* $z(b)$. If C is simple, then $z(t)$ moves continuously from $z(a)$ to $z(b)$ as t increases and the curve is given an *orientation*, which we indicate by drawing arrows along the curve. Figure 6.1 illustrates how the terms *simple* and *closed* describe a curve.

The complex-valued function $z(t) = x(t) + iy(t)$ is said to be *differentiable* on $[a, b]$ if both $x(t)$ and $y(t)$ are differentiable for $a \le t \le b$. Here we require the one-sided derivatives [1] of $x(t)$ and $y(t)$ to exist at the endpoints of the interval. As in Section 6.1, the derivative $z'(t)$ is

$$z'(t) = x'(t) + iy'(t), \quad \text{for} \quad a \le t \le b.$$

The curve C defined by Equation (6.10) is said to be a **smooth curve** if the function z' is continuous and nonzero on the interval. If C is a smooth curve, then C has a nonzero tangent vector at each point $z(t)$, which is given by the vector $z'(t)$. If $x'(t_0) = 0$, then the tangent vector $z'(t_0) = iy'(t_0)$ is vertical. If $x'(t_0) \ne 0$, then the slope $\frac{dy}{dx}$ of the tangent line to C at the point $z(t_0)$ is given by $\frac{y'(t_0)}{x'(t_0)}$. Hence for a smooth curve the angle of inclination $\theta(t)$ of its tangent vector $z'(t)$ is defined for all values of $t \in [a, b]$ and is continuous. Thus a smooth curve has no corners or cusps. Figure 6.2 illustrates this concept.

---

[1]The derivatives on the right, $x'(a^+)$, and on the left, $x'(b^-)$, are defined by the limits

$$x'(a^+) = \lim_{t \to a^+} \frac{x(t) - x(a)}{t - a} \quad \text{and} \quad x'(b^-) = \lim_{t \to b^-} \frac{x(t) - x(b)}{t - b}.$$

(a) A curve that is simple.

(b) A simple closed curve.

(c) A curve that is *not* simple
and *not* closed.

(d) A closed curve that is *not* simple.

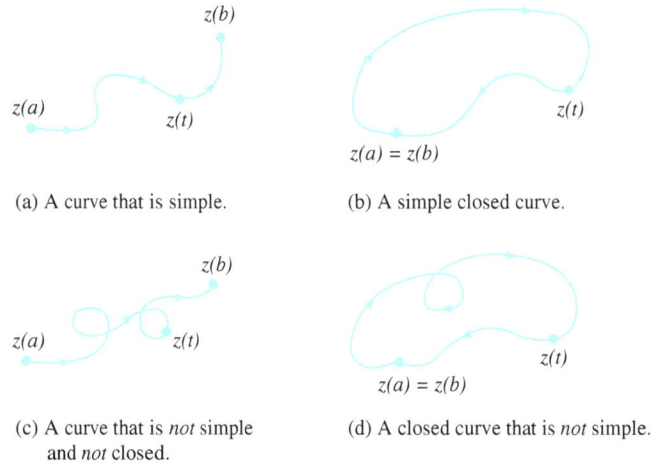

Figure 6.1: The terms *simple* and *closed* used to describe curves

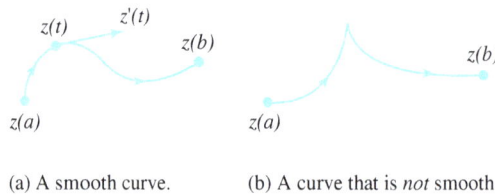

(a) A smooth curve.

(b) A curve that is *not* smooth.

Figure 6.2: The term *smooth* used to describe curves

If C is a smooth curve, then ds, the differential of arc length, is given by

$$ds = \sqrt{[x'(t)]^2 + [y'(t)]^2}\, dt = |z'(t)|\, dt.$$

The function $s(t) = \sqrt{x'(t)^2 + y'(t)^2}$ is continuous because x' and y' are continuous functions, so the length $L(C)$ of the curve C is

$$L(C) = \int_a^b \sqrt{[x'(t)]^2 + [y'(t)]^2}\, dt = \int_a^b |z'(t)|\, dt. \tag{6.11}$$

Now, consider C to be a curve with parametrization

$$C : z_1(t) = x(t) + iy(t) \quad \text{for} \quad a \le t \le b.$$

The **opposite curve** $-C$ traces out the same set of points in the plane, but in the reverse order, and has the parametrization

$$-C : z_2(t) = x(-t) + iy(-t) \quad \text{for} \quad -b \le t \le -a.$$

Since $z_2(t) = z_1(-t)$, $-C$ is merely C traversed in the opposite sense, as Figure 6.3 illustrates.

A curve C that is constructed by joining finitely many smooth curves end to end is called a **contour**. Let $C_1, C_2, \ldots, C_n$ denote n smooth curves such that the terminal point of the curve C_k coincides with the initial point of C_{k+1}, for $k = 1, 2, \ldots, n-1$. We express the contour C by the equation

$$C = C_1 + C_2 + \cdots + C_n.$$

A synonym for *contour* is *path*.

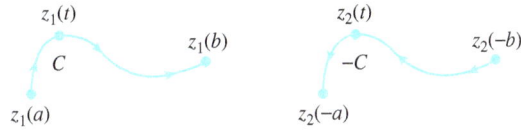

Figure 6.3: The curve C and its opposite curve $-C$

Example 6.5. Find a parametrization of the polygonal path C from $-1 + i$ to $3 - i$ shown in Figure 6.4.

Solution:

We express C as three smooth curves, or $C = C_1 + C_2 + C_3$. If we set $z_0 = -1 + i$ and $z_1 = -1$, we can use Equation 1.48 to get a formula for the straight-line segment joining two points:

$$C_1 : z_1(t) = z_0 + t(z_1 - z_0) = (-1 + i) + t[-1 - (-1 + i)], \quad \text{for} \quad 0 \le t \le 1.$$

When simplified, this formula becomes

$$C_1 : z_1(t) = -1 + i(1 - t), \quad \text{for} \quad 0 \le t \le 1.$$

Similarly, the segments C_2 and C_3 are given by

$$C_2 : z_2(t) = (-1 + 2t) + it, \quad \text{for} \quad 0 \le t \le 1, \quad \text{and}$$
$$C_3 : z_3(t) = (1 + 2t) + i(1 - 2t), \quad \text{for} \quad 0 \le t \le 1.$$

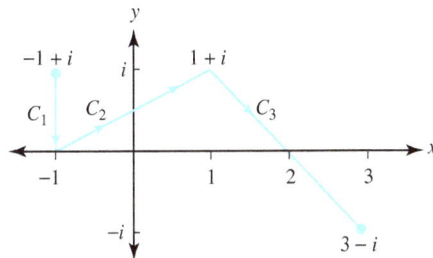

Figure 6.4: The polygonal path $C = C_1 + C_2 + C_3$ from $-1 + i$ to $3 - i$

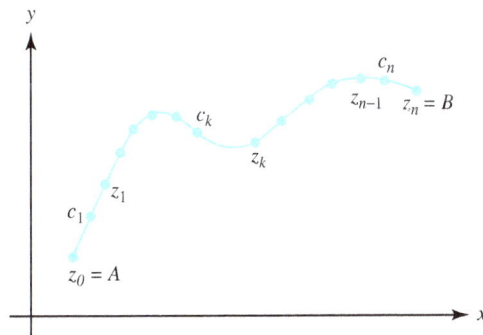

Figure 6.5: Partition points $\{z_k\}$ and function evaluation points $\{c_k\}$ for a Riemann sum along the contour C from $z = A$ to $z = B$

We are now ready to define the integral of a complex function along a contour C in the plane with initial point A and terminal point B. Our approach is to mimic what is done in

169

calculus. We create a partition $P_n = \{z_0 = A, z_1, z_2, \ldots, z_n = B\}$ of points that proceed along C from A to B and form the differences $\Delta z_k = z_k - z_{k-1}$, for $k = 1, 2, \ldots, n$. Between each pair of partition points z_{k-1} and z_k we select a point c_k on C, as shown in Figure 6.5, and evaluate the function f. We use these values to make a Riemann sum for the partition:

$$S(P_n) = \sum_{k=1}^{n} f(c_k)(z_k - z_{k-1}) = \sum_{k=1}^{n} f(c_k)\Delta z_k. \tag{6.12}$$

Assume now that there exists a unique complex number L that is the limit of every sequence $\{S(P_n)\}$ of Riemann sums given in Equation (6.12), where the maximum of $|\Delta z_k|$ tends toward 0 for the sequence of partitions. That number is the value of the integral of the function f along C.

Definition 6.2 (Complex integral). *Let C be a contour. Then*

$$\int_C f(z)\, dt = \lim_{n \to \infty} \sum_{k=1}^{n} f(c_k)\Delta z_k,$$

provided the limit exists in the sense previously discussed.

In Definition 6.2 the value of the integral depends on the contour. In Section 6.3 the Cauchy-Goursat theorem will establish the remarkable fact that, if f is analytic, then $\int_C f(z)dz$ is *independent* of the contour.

Example 6.6. Use a Riemann sum to approximate the integral $\int_C \exp z\, dt$, where C is the line segment joining the point $A = 0$ to $B = 2 + i\frac{\pi}{4}$.

Solution:

Set $n = 8$ in Equation (6.12) and form the partition $P_8 : z_k = \frac{k}{4} + i\frac{\pi k}{32}$ for $k = 0, 1, 2, \ldots, 8$. For this situation, we have a uniform increment $\Delta z_k = \frac{1}{4} + i\frac{\pi}{32}$. For convenience we select $c_k = \frac{z_{k-1}+z_k}{2} = \frac{2k-1}{8} + i\frac{\pi(2k-1)}{64}$, for $k = 1, 2, \ldots, 8$. Figure 6.6 shows the points $\{z_k\}$ and $\{c_k\}$.

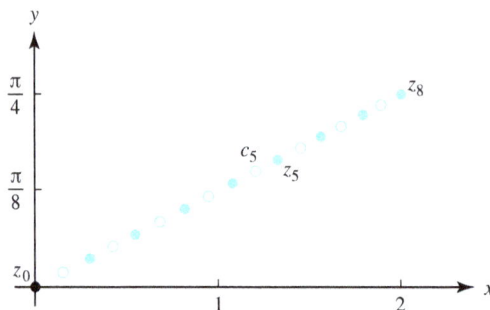

Figure 6.6: Partition and evaluation points for the Riemann sum $S(P_8)$

One possible Riemann sum, then, is

$$S(P_8) = \sum_{k=1}^{8} f(c_k)\Delta z_k = \sum_{k=1}^{8} \exp\left[\frac{2k-1}{8} + i\frac{\pi(2k-1)}{64}\right]\left(\frac{1}{4} + i\frac{\pi}{32}\right).$$

By rounding the terms in this Riemann sum to two decimal digits, we obtain an approximation for the integral:

$$S(P_8) \approx (0.28 + 0.13i) + (0.33 + 0.19i) + (0.41 + 0.29i) + (0.49 + 0.42i)$$
$$+ (0.57 + 0.6i) + (0.65 + 0.84i) + (0.72 + 1.16i) + (0.78 + 1.57i)$$
$$\approx 4.23 + 5.20i.$$

This result compares favorably with the precise value of the integral, which you will soon see equals

$$\exp\left(2 + i\frac{\pi}{4}\right) - 1 = -1 + e^2 \left(\frac{\sqrt{2}}{2}\right) + ie^2 \left(\frac{\sqrt{2}}{2}\right) \approx 4.22485 + 5.22485i.$$

In general, obtaining an exact value for an integral given by Definition 6.2 could be a daunting task. Fortunately, there is a beautiful theory that allows for an easy computation of many contour integrals. Suppose that we have a parametrization of the contour C given by the function $z(t)$ for $a \leq t \leq b$. That is, C is the range of the function $z(t)$ over the interval $[a, b]$, as Figure 6.7 shows.

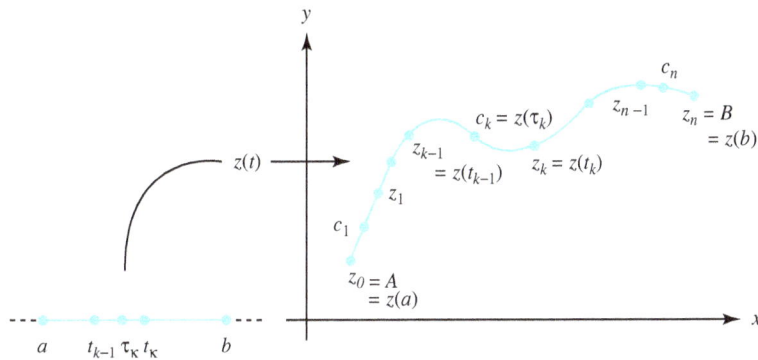

Figure 6.7: A parametrization of the contour C by $z(t)$, for $a \leq t \leq b$

It follows that

$$\lim_{n\to\infty} \sum_{k=1}^{n} f(c_k)\Delta z_k = \lim_{n\to\infty} \sum_{k=1}^{n} f(c_k)(z_k - z_{k-1})$$
$$= \lim_{n\to\infty} \sum_{k=1}^{n} f(z(\tau_k))\big[z(t_k) - z(t_{k-1})\big].$$

where τ_k and t_k are the points contained in the interval $[a, bt]$ with the property that $c_k = z(\tau_k)$ and $z_k = z(t_k)$, as is also shown in Figure 6.7. If for all k we multiply the kth term in the last sum by $\frac{t_k - t_{k-1}}{t_k - t_{k-1}}$, then we get

$$\lim_{n\to\infty} \sum_{k=1}^{n} f(z(\tau_k)) \left[\frac{z(t_k) - z(t_{k-1})}{t_k - t_{k-1}}\right] (t_k - t_{k-1}) =$$
$$\lim_{n\to\infty} \sum_{k=1}^{n} f(z(\tau_k)) \left[\frac{z(t_k) - z(t_{k-1})}{t_k - t_{k-1}}\right] \Delta t_k.$$

The quotient inside the last summation looks suspiciously like a derivative, and the entire quantity looks like a Riemann sum. Assuming no difficulties, this last expression should equal

171

$\int_a^b f(z(t))z'(t)\,dt$, as defined in Section 6.1. Of course, if we're to have any hope of this happening, we would have to get the same limit *regardless of how we parametrize the contour C*. As Theorem 6.1 states, this is indeed the case.

Theorem 6.1. *Suppose that $f(z)$ is a continuous complex-valued function defined on a set containing the contour C. Let $z(t)$ be any parametrization of C for $a \leq t \leq b$. Then*

$$\int_C f(z)\,dz = \int_a^b f\big(z(t)\big)z'(t)\,dt.$$

We omit the proof of Theorem 6.1 because it involves ideas (*e.g.*, the theory of the Riemann-Stieltjes integral) that are beyond the scope of this book. A more rigorous development of the contour integral based on Riemann sums is presented in advanced texts such as L.V. Ahlfors, *Complex Analysis*, 3rd ed. (New York: McGraw-Hill, 1979).

Two important facets of Theorem 6.1 are worth mentioning. First, Theorem 6.1 makes the problem of evaluating complex-valued functions along contours easy, as it reduces the task to the evaluation of complex-valued functions over real intervals—a procedure that you studied in Section 6.1. Second, according to Theorem 6.1, this transformation yields the same answer regardless of the parametrization we choose for C.

Example 6.7. Give an exact calculation of the integral in Example 6.6.

Solution:

We must compute $\int_C \exp z\,dz$, where C is the line segment joining $A = 0$ to $B = 2 + i\frac{\pi}{4}$. According to Equation 1.48, we can parametrize C by $z(t) = (2 + i\frac{\pi}{4})t$, for $0 \leq t \leq 1$. As $z'(t) = (2 + i\frac{\pi}{4})$, Theorem 6.1 guarantees that

$$
\begin{aligned}
\int_C \exp z\,dz &= \int_0^1 \exp[z(t)]z'(t)\,dt \\
&= \int_0^1 \exp\left[\left(2 + i\frac{\pi}{4}\right)t\right]\left(2 + i\frac{\pi}{4}\right)dt \\
&= \left(2 + i\frac{\pi}{4}\right)\int_0^1 e^{2t}e^{i\frac{\pi}{4}t}\,dt \\
&= \left(2 + i\frac{\pi}{4}\right)\int_0^1 e^{2t}\left(\cos\frac{\pi t}{4} + i\sin\frac{\pi t}{4}\right)dt \\
&= \left(2 + i\frac{\pi}{4}\right)\left(\int_0^1 e^{2t}\cos\frac{\pi t}{4}\,dt + i\int_0^1 e^{2t}\sin\frac{\pi t}{4}\,dt\right).
\end{aligned}
$$

Each integral in the last expression can be done using integration by parts. (There is a simpler way—see Remark 6.1.) We leave as an exercise to show that the final answer simplifies to $\exp(2 + i\frac{\pi}{4}) - 1$, as we claimed in Example 6.6.

Example 6.8. Evaluate $\int_C \frac{1}{z-2}\,dz$, where C is the upper semicircle with radius 1 centered at $x = 2$ and oriented in a positive direction (*i.e.*, counterclockwise).

Solution:

The function $z(t) = 2 + e^{it}$, for $0 \leq t \leq \pi$ is a parametrization for C. We apply Theorem 6.1 with $f(z) = \frac{1}{z-2}$. (Note: $f\big(z(t)\big) = \frac{1}{z(t)-2}$, and $z'(t) = ie^{it}$.) Hence

$$\int_C \frac{1}{z-2}\,dz = \int_0^\pi \frac{1}{(2 + e^{it}) - 2}ie^{it}\,dt = \int_0^\pi i\,dt = i\pi.$$

To help convince yourself that the value of the integral is independent of the parametrization chosen for the given contour, try working through Example 6.8 with $z(t) = 2 + e^{i\pi t}$, for $0 \leq t \leq 1$.

A convenient bookkeeping device can help you remember how to apply Theorem 6.1. Because $\int_C f(z) \, dz = \int_a^b f(z(t)) z'(t) \, dt$, you can symbolically equate z with $z(t)$ and dz with $z'(t) \, dt$. These identities should be easy to remember because z is supposed to be a point on the contour C parametrized by $z(t)$, and $\frac{dz}{dt} = z'(t)$, according to the Leibniz notation for the derivative.

If $z(t) = x(t) + iy(t)$, then by the preceding paragraph we have

$$dz = z'(t) \, dt = \left[x'(t) + iy'(t) \right] dt = dx + i \, dy. \tag{6.13}$$

where dx and dy are the differentials for $x(t)$ and $y(t)$, respectively (*i.e.*, dx is equated with $x'(t) \, dt$, etc.). The expression dz is often called the *complex differential* of z. Just as dx and dy are intuitively considered to be small segments along the x and y axes in real variables, we can think of dz as representing a tiny piece of the contour C. Moreover, if we write

$$|dz| = \left| [x'(t) + iy'(t)] \right| = \left| x'(t) + iy'(t) \right| dt = \sqrt{\left[x'(t) \right]^2 + \left[y'(t) \right]^2} \, dt,$$

we can put Equation (6.11) into the form

$$L(C) = \int_a^b \sqrt{\left[x'(t) \right]^2 + \left[y'(t) \right]^2} \, dt = \int_C |dz|. \tag{6.14}$$

so we can think of $|dz|$ as representing the length of dz.

Suppose that $f(z) = u(z) + iv(z)$ and that $z(t) = x(t) + iy(t)$ is a parametrization for the contour C. Then

$$
\begin{aligned}
\int_C f(z) \, dz &= \int_a^b f(z(t)) z'(t) \, dt \\
&= \int_a^b \left[u(z(t)) + iv(z(t)) \right] \left[x'(t) + iy'(t) \right] dt \\
&= \int_a^b \left[u(z(t)) x'(t) - v(z(t)) y'(t) \right] dt \\
&\qquad + i \int_a^b \left[v(z(t)) x'(t) + u(z(t)) y'(t) \right] dt \\
&= \int_a^b (ux' - vy') \, dt + i \int_a^b (vx' + uy') \, dt, \tag{6.15}
\end{aligned}
$$

where Equation (6.15) replaced $u(z(t))$ with u, $x'(t)$ with x', and so on.

If we use the differentials given in Equation (6.13), then we can write Equation (6.15) in terms of line integrals of the real-valued functions u and v, giving

$$\int_C f(z) \, dz = \int_C u \, dx - v \, dy + i \int_C v \, dx + u \, dy. \tag{6.16}$$

which is easy to remember if we recall that, symbolically,

$$f(z) \, dz = (u + iv)(dx + i \, dy).$$

We emphasize that Equation (6.16) is merely a notational device for applying Theorem 6.1. You should carefully apply Theorem 6.1, as illustrated in Examples 6.7 and 6.8, before using any shortcuts suggested by Equation (6.16).

Example 6.9. Show that

$$\int_{C_1} z\, dz = \int_{C_2} z\, dz = 4 + 2i,$$

where C_1 is the line segment from $-1-i$ to $3+i$ and C_2 is the portion of the parabola $x = y^2 + 2y$ joining $-1 - i$ to $3 + i$, as indicated in Figure 6.8.

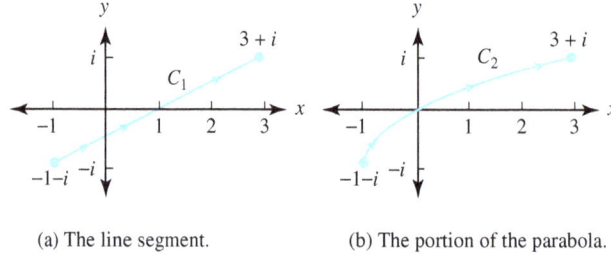

(a) The line segment.   (b) The portion of the parabola.

Figure 6.8: Two contours C_1 and C_2 joining $-1 - i$ to $3 + i$

The line segment joining $(-1, -1)$ to $(3, 1)$ is given by the slope intercept formula $y = \frac{1}{2}x - \frac{1}{2}$, which can be written as $x = 2y + 1$. If we choose the parametrization $y = t$ and $x = 2t + 1$, we can write segment C_1 as

$$C_1 : z(t) = 2t + 1 + it \quad \text{and} \quad dz = (2 + i)\, dt, \quad \text{for} \quad -1 \le t \le 1.$$

Along C_1 we have $f(z(t)) = 2t + 1 + it$. Applying Theorem 6.1 gives

$$\int_{C_1} z\, dz = \int_{-1}^{1} (2t + 1 + it)(2 + i)\, dt.$$

We now multiply out the integrand and put it into its real and imaginary parts:

$$\int_{C_1} z\, dz = \int_{-1}^{1} (3t + 2)\, dt + i \int_{-1}^{1} (4t + 1)\, dt = 4 + 2i.$$

Similarly, we can parametrize the portion of the parabola $x = y^2 + 2y$ joining $(-1, -1)$ to $(3, 1)$ by $y = t$ and $x = t^2 + 2t$ so that

$$C_2 : z(t) = t^2 + 2t + it \quad \text{and} \quad dz = (2t + 2 + i)\, dt, \quad \text{for} \quad -1 \le t \le 1.$$

Along C_2 we have $f(z(t)) = t^2 + 2t + it$. Theorem 6.1 now gives

$$\int_{C_2} z\, dz = \int_{-1}^{1} (t^2 + 2t + it)(2t + 2 + i)\, dt$$

$$= \int_{-1}^{1} (2t^3 + 6t^2 + 3t)\, dt + i \int_{-1}^{1} (3t^2 + 4t)\, dt = 4 + 2i.$$

In Example 6.9, the value of the two integrals is the same. This outcome doesn't hold in general, as Example 6.10 shows.

Example 6.10. Show that

$$\int_{C_1} \bar{z}\, dz = -\pi i, \quad \text{but that} \quad \int_{C_2} \bar{z}\, dz = -4i.$$

174

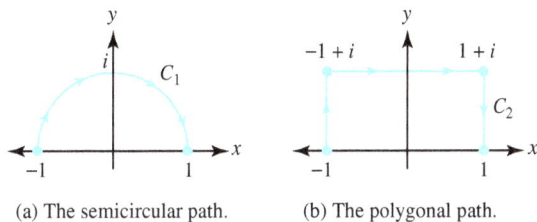

(a) The semicircular path.　　(b) The polygonal path.

Figure 6.9: Two contours C_1 and C_2 joining -1 to 1

where C_1 is the semicircular path from -1 to 1 and C_2 is the polygonal path from -1 to 1, respectively, shown in Figure 6.9.

Solution:

We parametrize the semicircle C_1 as

$$C_1 : z(t) = -\cos t + i \sin t, \quad \text{and} \quad dz = (\sin t + i \cos t)\, dt, \quad \text{for} \quad 0 \le t \le \pi.$$

Applying Theorem 6.1, we have $f(z) = \overline{z}$, so

$$f(z(t)) = \overline{z(t)} = \overline{(-\cos t + i \sin t)} = -\cos t - i \sin t, \quad \text{and}$$

$$\int_{C_1} \overline{z}\, dz = \int_0^\pi (-\cos t - i \sin t)(\sin t + i \cos t)\, dt$$

$$= -i \int_0^\pi (\cos^2 t + \sin^2 t)\, dt = -\pi i.$$

We parametrize C_2 in three parts, one for each line segment:

$$z_1(t) = -1 + it, \qquad dz_1 = i\, dt, \qquad \text{and} \quad f(z_1(t)) = -1 - it.$$
$$z_2(t) = -1 + 2t + i, \quad dz_2 = 2\, dt, \quad \text{and} \quad f(z_2(t)) = -1 + 2t - i.$$
$$z_3(t) = 1 + i(1 - t), \quad dz_3 = -i\, dt, \quad \text{and} \quad f(z_3(t)) = 1 - i(1 - t).$$

where $0 \le t \le 1$ in each case. We get our answer by adding the three integrals along the three segments:

$$\int_{C_2} \overline{z}\, dz = \int_0^1 (-1 - it)i\, dt + \int_0^1 (-1 + 2t - i)2\, dt + \int_0^1 [1 - i(1 - t)](-i)\, dt.$$

Separating the right side of this equation into its real and imaginary parts gives

$$\int_{C_2} \overline{z}\, dz = \int_0^1 (6t - 3)\, dt + i \int_0^1 (-4)\, dt = -4i.$$

Note that the value of the contour integral along C_1 isn't the same as the value of the contour integral along C_2, although both integrals have the same initial and terminal points.

Contour integrals have properties that are similar to those of integrals of a complex function of a real variable, which you studied in Section 6.1. If C is given by Equation (6.10), then the integral for the opposite contour $-C$ is

$$\int_{-C} f(z)\, dz = \int_{-b}^{-a} f(z(-\tau))\big[-z'(-\tau)\big]\, d\tau.$$

175

Using the change of variable $t = -\tau$ in this last equation and the property that $\int_a^b f(t)\,dt = -\int_b^a f(t)\,dt$, we obtain

$$\int_{-C} f(z)\,dz = -\int_C f(z)\,dz. \tag{6.17}$$

If two functions f and g can be integrated over the same path of integration C, then their sum can be integrated over C, and we have the familiar result

$$\int_C [f(z) + g(z)]\,dz = \int_C f(z)\,dz + \int_C g(z)\,dz.$$

Constant multiples also behave as we would expect:

$$\int_C (a + ib)f(z)\,dz = (a + ib)\int_C f(z)\,dz.$$

If two contours C_1 and C_2 are placed end to end so that the terminal point of C_1 coincides with the initial point of C_2, then the contour $C = C_1 + C_2$ is a *continuation* of C_1, and

$$\int_{C_1 + C_2} f(z)\,dz = \int_{C_1} f(z)\,dz + \int_{C_2} f(z)\,dz. \tag{6.18}$$

If the contour C has two parametrizations

$$C : z_1(t) = x_1(t) + iy_1(t), \quad \text{for} \quad a \le t \le b, \quad \text{and}$$
$$C : z_2(\tau) = x_2(\tau) + iy_2(\tau), \quad \text{for} \quad c \le \tau \le d.$$

and there exists a differentiable function $\tau = \phi(t)$ such that

$$c = \phi(a), \quad d = \phi(b), \quad \text{and} \quad \phi'(t) > 0, \quad \text{for} \quad a < t < b. \tag{6.19}$$

then we say that $z_2(\tau)$ is a *reparametrization* of the contour C. If f is continuous on C, then we have

$$\int_a^b f(z_1(t))z_1'(t)\,dt = \int_c^d f(z_2(\tau))z_2'(\tau)\,d\tau. \tag{6.20}$$

Equation (6.20) shows that the value of a contour integral is invariant under a change in the parametric representation of its contour if the reparametrization satisfies Equations (6.19).

We now give two important inequalities relating to complex integrals.

Theorem 6.2 (Absolute value inequality). *If $f(t) = u(t) + iv(t)$ is a continuous function of the real parameter t, then*

$$\left| \int_a^b f(t)\,dt \right| \le \int_a^b |f(t)|\,dt. \tag{6.21}$$

Proof. If $\int_a^b f(t)\,dt = 0$, then Equation (6.21) is obviously true. If the integral is not zero, we write its value in polar form, say $\int_a^b f(t)\,dt = r_0 e^{i\theta_0}$, so that $\left| \int_a^b f(t)\,dt \right| = |r_0 e^{i\theta_0}| = r_0$, and $r_0 = \int_a^b e^{-i\theta_0} f(t)\,dt$. Taking the real part of both sides of this last equation gives

$$r_0 = \mathrm{Re}(r_0) = \mathrm{Re}\left[\int_a^b e^{-i\theta_0} f(t)\,dt \right] = \int_a^b \mathrm{Re}[e^{-i\theta_0} f(t)]\,dt.$$

where the last equality is justified because an integral is a limit of a sum, and its real part is the same as the limit of the sum of its real parts.

Now, using Equation (1.21), we obtain

$$\operatorname{Re}\left[e^{-i\theta_0} f(t)\right] \leq \left|e^{-i\theta_0} f(t)\right| = |f(t)|. \tag{6.22}$$

Recall that, if g and h are real functions, with $g(t) \leq h(t)$, for all $t \in [a, b]$, then $\int_a^b g(t\,dt \leq \int_a^b h(t)\,dt$. Applying this fact to the left and right sides of Equation (6.22) (with $g(t) = \operatorname{Re}\left[e^{-i\theta_0} f(t)\right]$ and $h(t) = |f(t)|$) yields

$$r_0 = \int_a^b \operatorname{Re}\left[e^{-i\theta_0} f(t)\right] dt \leq \int_a^b |f(t)|\,dt.$$

Since $r_0 = \left|\int_a^b f(t)\,dt\right|$, this last inequality establishes our result. $\qquad \square$

Theorem 6.3 (ML inequality). *If $f(z) = u(x, y) + iv(x, y)$ is continuous on the contour C, then*

$$\left|\int_C f(z)\,dz\right| \leq ML, \tag{6.23}$$

where L is the length of the contour C and M is an upper bound for the modulus $|f(z)|$ on C; that is, $|f(z)| \leq M$ for all $z \in C$.

Proof. Using Inequality (6.21) with Theorem 6.1 gives

$$\left|\int_C f(z)\,dz\right| = \left|\int_a^b f(z(t))z'(t)\,dt\right| \leq \int_a^b |f(z(t))z'(t)|\,dt. \tag{6.24}$$

Equations 6.13, 6.14, and Inequality (6.24) imply that

$$\left|\int_C f(z)\,dz\right| \leq \int_a^b M|z'(t)|\,dt = ML.$$

$\qquad \square$

Example 6.11. Use Inequality (6.23) to show that

$$\left|\int_C \frac{1}{z^2 + 1}\,dz\right| \leq \frac{1}{2\sqrt{5}}.$$

where C is the straight-line segment from 2 to $2 + i$.

Solution:
Here $|z^2 + 1| = |z - i|\,|z + i|$, and the terms $|z - i|$ and $|z + i|$ represent the distance from the point z to the points i and $-i$, respectively. Referring to Figure 6.10 and using a geometric argument, we get

$$|z - i| \geq 2, \quad \text{and} \quad |z + i| \geq \sqrt{5}, \quad \text{for} \quad z \text{ on } C.$$

Thus we have

$$|f(z)| = \frac{1}{|z - i|\,|z + i|} \leq \frac{1}{2\sqrt{5}} = M.$$

Because L, the length of C, equals 1, Inequality (6.23) implies that

$$\left|\int_C \frac{1}{z^2 + 1}\,dz\right| \leq ML = \frac{1}{2\sqrt{5}}.$$

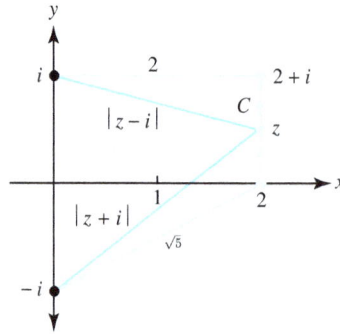

Figure 6.10: The distances $|z - i|$ and $|z + i|$ for z on C

Exercises for Section 6.2 (Selected answers or hints are on page 443.)

1. Give a parametrization of each contour.

    (a) $C = C_1 + C_2$, as indicated in Figure 6.11.
    (b) $C = C_1 + C_2 + C_3$, as indicated in Figure 6.12.

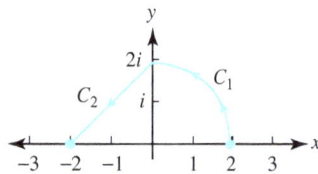

Figure 6.11: For exercise 1(a)

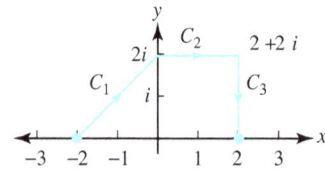

Figure 6.12: For exercise 1(b)

2. Sketch the following curves.

    (a) $z(t) = t^2 - 1 + i(t + 4)$, for $1 \leq t \leq 3$.
    (b) $z(t) = \sin t + i \cos 2t$, for $-\frac{\pi}{2} \leq t \leq \frac{\pi}{2}$.
    (c) $z(t) = 5 \cos t - i3 \sin t$, for $\frac{\pi}{2} \leq t \leq 2\pi$.

3. Consider the integral $\int_C z^2 \, dz$, where C is the positively oriented upper semicircle of radius 1, centered at 0.

    (a) Give a Riemann sum approximation for the integral by selecting $n = 4$ and the points $z_k = e^{i \frac{k\pi}{4}}$ $(k = 0, \ldots 4)$ and $c_k = e^{i \frac{(2k-1)\pi}{8}}$ $(k = 1, \ldots 4)$.
    (b) Compute the integral exactly by selecting a parametrization for C and applying Theorem 6.1.

4. Show that the integral of Example 6.7 simplifies to $\exp(2 + i\frac{\pi}{4}) - 1$.

5. Evaluate $\int_C x \, dz$ from -4 to 4 along the following contours, as shown in Figures 6.13(a) and 6.13(b).

    (a) The polygonal path C with vertices -4, $-4 + 4i$, $4 + 4i$, and 4.
    (b) The contour C that is the upper half of the circle $|z| = 4$, oriented clockwise.

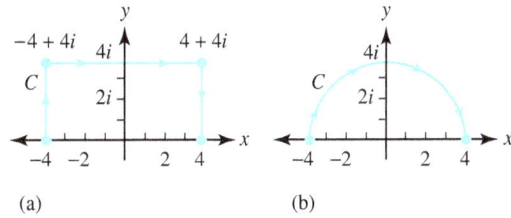

Figure 6.13: For Exercise 5 (a) and (b)

6. Evaluate $\int_C y\,dz$ for $-i$ to i along the following contours, as shown in Figures 6.14(a) and 6.14(b).

   (a) The polygonal path C with vertices $-i$, $-1-i$, -1, and i.

   (b) The contour C that is oriented clockwise, as shown.

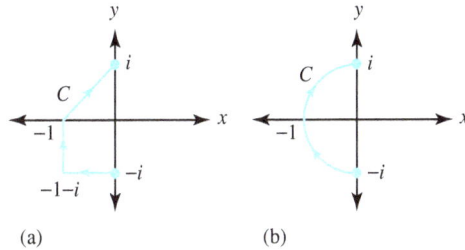

Figure 6.14: For Exercise 6 (a) and (b)

7. Recall $C_r^+(a)$ is the circle of radius r that is centered at a and oriented counter-clockwise.

   (a) Evaluate $\int_{C_4^+(0)} z\,dz$.

   (b) Evaluate $\int_{C_4^+(0)} \overline{z}\,dz$.

   (c) Evaluate $\int_{C_2^-(0)} \frac{1}{z}\,dz$. (The minus sign means clockwise orientation.)

   (d) Evaluate $\int_{C_2^-(0)} \frac{1}{\overline{z}}\,dz$.

   (e) Evaluate $\int_C (z+1)\,dz$, where C is $C_1^+(0)$ in the first quadrant.

   (f) Evaluate $\int_C (x^2 - iy^2)\,dz$, where C is the upper half of $C_1^+(0)$.

   (g) Evaluate $\int_C |z-1|^2\,dz$, where C is the upper half of $C_1^+(0)$.

8. Let f be a continuous function on the circle $\{z : |z - z_0| = R\}$.
   Show that $\int_{C_R^+(z_0)} f(z)\,dz = iR \int_0^{2\pi} f(z_0 + Re^{i\theta})e^{i\theta}\,d\theta$.

9. Use the results of Exercise 8 to evaluate

   (a) $\int_{C_R^+(z_0)} \frac{1}{z - z_0}\,dz$.

   (b) $\int_{C_R^+(z_0)} \frac{1}{(z - z_0)^n}\,dz$, where $n \neq 1$ is an integer.

10. Use the techniques of Example 6.11 to show that

    (a) $\left| \int_C \frac{1}{z^2 - 1}\,dz \right| \leq \frac{\pi}{3}$, where C is the first quadrant portion of $C_2^+(0)$.

179

(b) $\left| \int_{C_R^+(0)} \frac{\mathrm{Log}(z)}{z^2} \, dz \right| \le 2\pi \left(\frac{\pi + \ln R}{R} \right).$

11. Evaluate $\int_C z^2 \, dz$, where C is the line segment from 1 to $1 + i$.

12. Evaluate $\int_C |z^2| \, dz$, where C given by $C : z(t) = t + it^2$, for $0 \le t \le 1$.

13. Evaluate $\int_C \exp z \, dz$, where C is the straight-line segment joining 1 to $1 + i\pi$.

14. Evaluate $\int_C \bar{z} \exp z \, dz$, where C is the square with vertices 0, 1, $1 + i$, and i taken with the counterclockwise orientation.

15. Evaluate $\int_C \exp z \, dz$, where C is the straight-line segment joining 0 to $1 + i$.

16. Let $z(t) = x(t) + iy(t)$, for $a \le t \le b$, be a smooth curve. Give a meaning for each of the following expressions.

   (a) $z'(t)$.
   (b) $|z'(t)| \, dt$.
   (c) $\int_a^b z'(t) \, dt$.
   (d) $\int_a^b |z'(t)| \, dt$.

17. Evaluate $\int_C \cos z \, dz$, where C is the polygonal path from 0 to $1 + i$ that consists of the line segments from 0 to 1 and 1 to $1 + i$.

18. Let $f(t) = e^{it}$ be defined on $a \le t \le b$, where $a = 0$, and $b = 2\pi$. Show that there is no number $c \in (a, b)$ such that $f(c)(b - a) = \int_a^b f(t) \, dt$. In other words, the mean value theorem for definite integrals that you learned in calculus does not hold for complex functions.

19. Use the ML inequality to show that $|P_n(x)| \le 1$, where P_n is the nth Legendre polynomial defined on $-1 \le x \le 1$ by

$$P_n(x) = \frac{1}{\pi} \int_0^\pi \left(x + i\sqrt{1 - x^2} \cos \theta \right)^n d\theta.$$

20. Explain how contour integrals in complex analysis and line integrals in calculus are different. How are they similar?

6.3   The Cauchy-Goursat Theorem

The Cauchy-Goursat theorem states that within certain domains the integral of an analytic function over a simple closed contour is zero. An extension of this theorem allows us to replace integrals over certain complicated contours with integrals over contours that are easy to evaluate. We demonstrate how to use the technique of partial fractions with the Cauchy-Goursat theorem to evaluate certain integrals. In Section 6.4 we show that the Cauchy-Goursat theorem implies that an analytic function has an antiderivative. To begin, we need to introduce some new concepts.

   Recall from Section 1.6 that each simple closed contour C divides the plane into two domains. One domain is bounded and is called the **interior** of C; the other domain is unbounded and

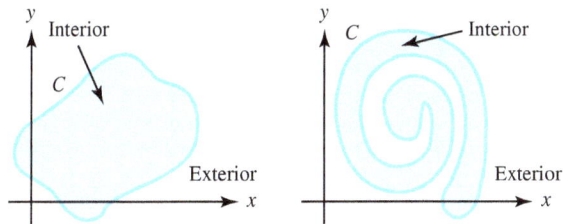

Figure 6.15: The interior and exterior of simple closed contours

is called the **exterior** of C. Figure 6.15 illustrates this concept, which is known as the Jordan curve theorem.

Recall also that a domain D is a connected open set. In particular, if z_1 and z_2 are any pair of points in D, then they can be joined by a curve that lies entirely in D. A domain D is said to be a **simply connected domain** if the interior of any simple closed contour C contained in D is contained in D. In other words, there are no "holes" in a simply connected domain. A domain that is not simply connected is called a **multiply connected domain**. Figure 6.16 illustrates uses of the terms *simply connected* and *multiply connected*.

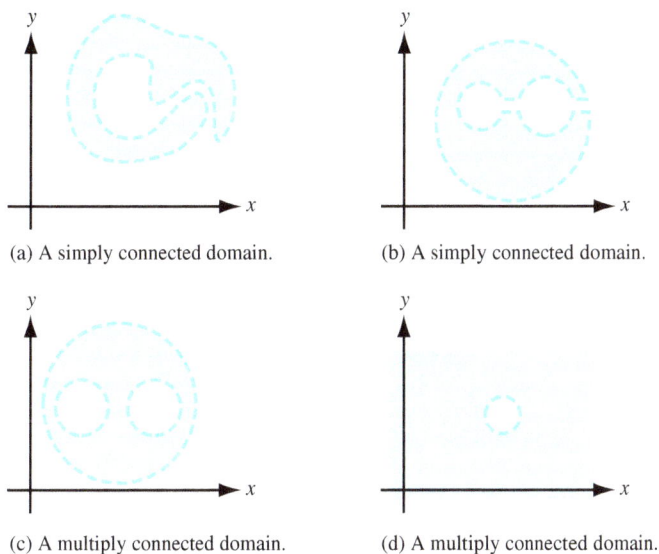

(a) A simply connected domain.

(b) A simply connected domain.

(c) A multiply connected domain.

(d) A multiply connected domain.

Figure 6.16: Simply connected and multiply connected domains

Let the simple closed contour C have the parametrization $C : z(t) = x(t) + iy(t)$ for $a \leq t \leq b$. Recall that if C is parametrized so that the interior of C is kept on the left as $z(t)$ moves around C, then we say that C is oriented *positively* (counterclockwise); otherwise, C is oriented *negatively* (clockwise). If C is positively oriented, then $-C$ is negatively oriented. Figure 6.17 illustrates the concept of positive and negative orientation.

Green's theorem is an important result from the calculus of real variables. It tells you how to evaluate the line integral of real-valued functions.

Theorem 6.4 (Green's theorem). *Let C be a simple closed contour with positive orientation and let R be the domain that forms the interior of C. If P and Q are continuous and have*

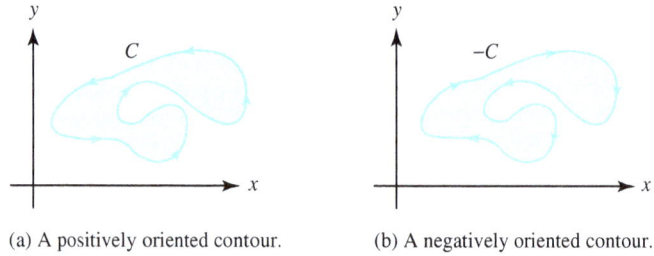

(a) A positively oriented contour. (b) A negatively oriented contour.

Figure 6.17: Positively and negatively oriented simple closed contours

continuous partial derivatives P_x, P_y, Q_x, and Q_y at all points on C and R then

$$\int_C P(x,y)\,dx + Q(x,y)\,dy = \iint_R \left[Q_x(x,y) - P_y(x,y) \right] dx\,dy. \qquad (6.25)$$

Proof. (We give a proof for a *standard region*, which is a region that is bounded by a contour C tha can be expressed in the two forms $C = C_1 + C_2$ and $C = C_3 + C_4$.) If R is a standard region, then there exist functions $y = g_1(x)$, and $y = g_2(x)$, for $a \le x \le b$, whose graphs form the lower and upper portions of C, respectively, as indicated in Figure 6.18.

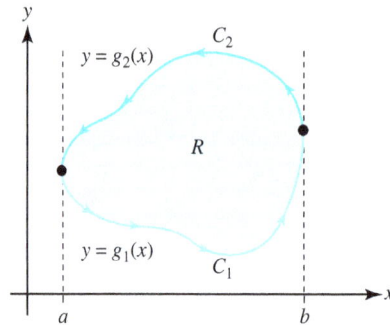

Figure 6.18: Integration over a standard region, where $C = C_1 + C_2$

As C is positively oriented, these functions can be used to express C as the sum of two contours C_1 and C_2, where

$$C_1 : z_1(t) = t + ig_1(t), \qquad \text{for} \quad a \le t \le b, \quad \text{and}$$
$$C_2 : z_2(t) = -t + ig_2(-t), \quad \text{for} \quad -b \le t \le -a.$$

We now use the functions $g_1(x)$ and $g_2(x)$ to express the double integral of $-P_y(x,y)$ over R as an iterated integral, first with respect to y and second with respect to x:

$$-\iint_R P_y(x,y)\,dx\,dy = -\int_a^b \left[\int_{g_1(x)}^{g_2(x)} P_y(x,y)\,dy \right] dx.$$

Computing the first iterated integral on the right side gives

$$-\iint_R P_y(x,y)\,dx\,dy = \int_a^b P\big(x,\,g_1(x)\big)\,dx - \int_a^b P\big(x,\,g_2(x)\big)\,dx.$$

182

In the second integral on the right side of this equation we can use the change of variable $x = -t$ to obtain

$$-\iint_R P_y(x, y)\, dx\, dy = \int_a^b P\big(x,\, g_1(x)\big)\, dx + \int_{-b}^{-a} P\big(-t,\, g_2(-t)\big)(-1)\, dt.$$

Interpreting the two integrals on the right side of this equation as contour integrals along C_1 and C_2, respectively, gives

$$-\iint_R P_y(x, y)\, dx\, dy = \int_{C_1} P(x, y)\, dx + \int_{C_2} P(x, y)\, dx = \int_C P(x, y)\, dx. \qquad (6.26)$$

To complete the proof, we rely on the fact that for a standard region there exist functions $x = h_1(y)$ and $x = h_2(y)$ for $c \le y \le d$ whose graphs form the left and right portions of C, respectively, as indicated in Figure 6.19.

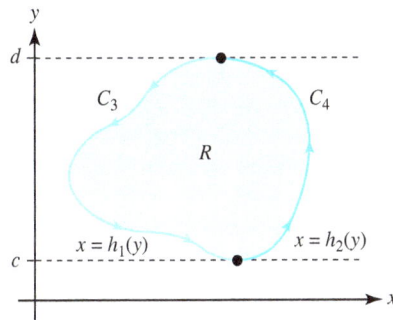

Figure 6.19: Integration over a standard region, where $C = C_3 + C_4$

Because C has positive orientation, it can be expressed as the sum of two contours C_3 and C_4, where

$$C_3 : z_3(t) = h_1(-t) - it, \quad \text{for} \quad -d \le t \le -c, \quad \text{and}$$
$$C_4 : z_4(t) = h_2(t) + it, \quad \text{for} \quad c \le t \le d.$$

Using the functions $h_1(y)$ and $h_2(y)$, we express the double integral of $Q_x(x.y)$ over R as an iterated integral:

$$\iint_R Q_x(x, y)\, dx\, dy = \int_c^d \left[\int_{h_1(y)}^{h_2(y)} Q_x(x, y)\, dx \right] dy.$$

A derivation similar to that which led to Equation (6.26) shows that

$$\iint_R Q_x(x.y)\, dx\, dy = \int_C Q(x, y)\, dy. \qquad (6.27)$$

Adding Equations (6.26) and (6.27) gives Equation (6.25). $\qquad\qquad \square$

We are now ready to state the main result of this section.

Theorem 6.5 (Cauchy-Goursat theorem). *Let f be analytic in a simply connected domain D. If C is a simple closed contour that lies in D, then*

$$\int_C f(z)\, dz = 0.$$

183

We give two proofs. The first, by Augustin Cauchy, is more intuitive but requires the additional hypothesis that f' is continuous.

Proof. (Cauchy's proof of Theorem 6.5.) If we suppose that f' is continuous, then with C oriented positively we use Equation (6.16) to write

$$\int_C f(z)\,dz = \int_C u\,dx - v\,dy + i\int_C v\,dx + u\,dy. \tag{6.28}$$

If we use Green's theorem on the real part of the right side of Equation (6.28) (with $P = u$ and $Q = -v$), we obtain

$$\int_C u\,dx - v\,dy = \iint_R (-v_x - u_y)\,dx\,dy, \tag{6.29}$$

where R is the region that is the interior of C. If we use Green's theorem on the imaginary part, we get

$$\int_C v\,dx + u\,dy = \iint_R (u_x - v_y)\,dx\,dy. \tag{6.30}$$

If we use the Cauchy-Riemann equations $u_x = v_y$ and $u_y = -v_x$ in Equations (6.29) and (6.30), Equation (6.28) becomes

$$\int_C f(z)\,dz = \iint_R 0\,dx\,dy + i\iint_R 0\,dx\,dy = 0,$$

which completes the proof. $\qquad\square$

In 1883, Edward Goursat (1858-1936) produced a proof that does not require the continuity of f'.

Goursat's proof of Theorem 6.5. We first establish the result for a triangular contour C with positive orientation. We then construct four positively oriented contours C^1, C^2, C^3, and C^4 that are the triangles obtained by joining the midpoints of the sides of C, as Figure 6.20 shows.

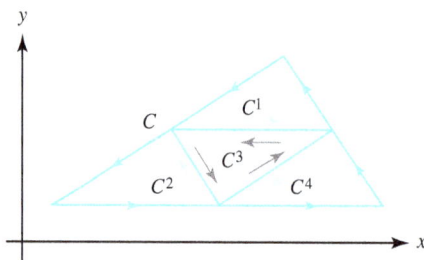

Figure 6.20: The triangular contours C and C^1, C^2, C^3, and C^4

Each contour is positively oriented, so if we sum the integrals along the four triangular contours, the integrals along the segments interior to C cancel out in pairs, giving

$$\int_C f(z)\,dz = \sum_{k=1}^{4} \int_{C^k} f(z)\,dz. \tag{6.31}$$

Let C_1 be selected from C^1, C^2, C^3, and C^4 so that the following holds:

$$\left| \int_C f(z)\, dz \right| \le \sum_{k=1}^{4} \left| \int_{C^k} f(z)\, dz \right| \le 4 \left| \int_{C_1} f(z)\, dz \right|.$$

Proceeding inductively, we carry out a similar subdivision process to obtain a sequence of triangular contours $\{C_n\}$, where the interior of C_{n+1} lies in the interior of C_n and the following inequality holds:

$$\left| \int_{C_n} f(z)\, dz \right| \le 4 \left| \int_{C_{n+1}} f(z)\, dz \right|, \quad \text{for} \quad n = 1, 2, \ldots \tag{6.32}$$

We let T_n denote the closed region that consists of C_n and its interior. The length of the sides of C_n go to zero as $n \to \infty$, so there exists a unique point z_0 that belongs to all the closed triangular regions $\{T_n\}$. Since D is simply connected, $z_0 \in D$, so f is analytic at the point z_0. Thus, there exists a function $\eta(z)$ such that

$$f(z) = f(z_0) + f'(z_0)(z - z_0) + \eta(z)(z - z_0), \quad \text{where} \quad \lim_{z \to z_0} \eta(z) = 0. \tag{6.33}$$

Using Equation (6.33) and integrating f along C_n, we get

$$\int_{C_n} f(z)\, dz = \int_{C_n} f(z_0)\, dz + \int_{C_n} f'(z_0)(z - z_0)\, dz + \int_{C_n} \eta(z)(z - z_0)\, dz$$

$$= [f(z_0) - f'(z_0)z_0] \int_{C_n} 1\, dz + f'(z_0) \int_{C_n} z\, dz$$

$$+ \int_{C_n} \eta(z)(z - z_0)\, dz$$

$$= \int_{C_n} \eta(z)(z - z_0)\, dz.$$

Since $\lim_{z \to z_0} \eta(z) = 0$, we know that given $\varepsilon > 0$, we can find $\delta > 0$ such that

$$|z - z_0| < \delta \quad \text{implies that} \quad |\eta(z)| < \frac{2}{L^2}\varepsilon, \tag{6.34}$$

where L is the length of the original contour C. We can now choose an integer n so that C_n lies in the neighborhood $z - z_0| < \delta$, as shown in Figure 6.21.

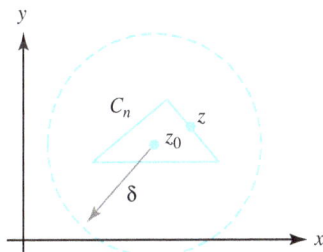

Figure 6.21: The contour C_n that lies in the neighborhood $|z - z_0| < \delta$

Since the distance between any point z on a triangle and a point z_0 interior to the triangle is less than half the perimeter of the triangle, it follows that

$$|z - z_0| < \frac{1}{2}L_n, \quad \text{for all } z \text{ on } C_n,$$

185

where L_n is the length of the triangle C_n. From the preceding construction process, it follows that

$$L_n = \left(\frac{1}{2}\right)^n L, \quad \text{and} \quad |z - z_0| < \left(\frac{1}{2}\right)^{n+1} L, \quad \text{for } z \text{ on } C_n. \tag{6.35}$$

We can use Equations (6.32), (6.34), and 6.35 and Theorem 6.3 to conclude

$$\left|\int_C f(z)\,dz\right| \leq 4^n \int_{C_n} |\eta(z)(z - z_0)|\,|dz|$$

$$\leq 4^n \int_{C_n} \frac{2}{L^2}\varepsilon \left(\frac{1}{2}\right)^{n+1} L\,|dz|$$

$$= \frac{2^n \varepsilon}{L} \int_{C_n} |dz|$$

$$= \frac{2^n \varepsilon}{L} \left(\frac{1}{2}\right)^n L$$

$$= \varepsilon.$$

Because ε was arbitrary, it follows that our theorem holds for the triangular contour C. If C is a polygonal contour, then we can add interior edges until the interior is subdivided into a finite number of triangles. The integral around each triangle is zero, and the sum of all these integrals equals the integral around the polygonal contour C. Therefore our theorem also holds for polygonal contours. The proof for an arbitrary simple closed contour is established by approximating the contour "sufficiently close" with a polygonal contour. We omit the details of this last step. $\square$

Example 6.12. Recall that $\exp z$, $\cos z$, and z^n (where n is a positive integer) are all entire functions. The Cauchy-Goursat theorem implies that, for any simple closed contour,

$$\int_C \exp z\,dz = 0, \quad \int_C \cos z\,dz = 0, \quad \text{and} \quad \int_C z^n\,dz = 0.$$

Example 6.13. Let n be an integer. If C is a simple closed contour such that the origin does not lie interior to C, then there is a simply connected domain D that contains C in which $f(z) = \frac{1}{z^n}$ is analytic, as is indicated in Figure 6.22. The Cauchy-Goursat theorem implies that $\int_C \frac{1}{z^n}\,dz = 0$.

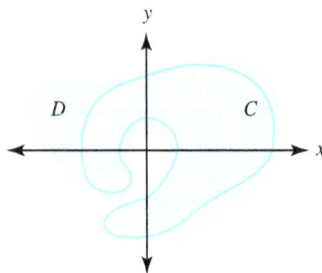

Figure 6.22: A simple connected domain D containing the simple closed contour C that does not contain the origin

We want to be able to replace integrals over certain complicated contours with integrals that are easy to evaluate. If C_1 is a simple closed contour that can be "continuously deformed" into another simple closed contour C_2 without passing through a point where f is not analytic,

186

then the value of the contour integral of f over C_1 is the same as the value of the integral of f over C_2. To be precise, we state the following result.

Theorem 6.6 (Deformation of contour). *Let C_1 and C_2 be two simple closed positively oriented contours such that C_1 lies interior to C_2. If f is analytic in a domain D that contains both C_1 and C_2 and the region between them, as shown in Figure 6.23, then*

$$\int_{C_1} f(z)\,dz = \int_{C_2} f(z)\,dz.$$

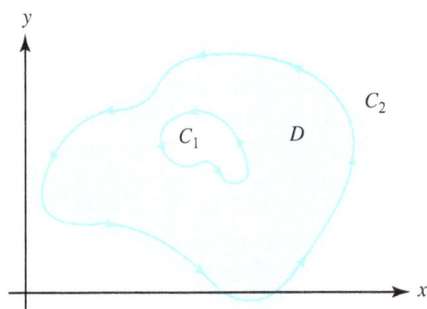

Figure 6.23: The domain D that contains the simple closed contours C_1 and C_2 and the region between them

Proof. Assume that both C_1 and C_2 have positive (counterclockwise) orientation. We construct two disjoint contours or *cuts*, L_1 and L_2, that join C_1 to C_2. The contour C_1 is cut into two contours C_1^* and C_1^{**}, and the contour C_2 is cut into C_2^* and C_2^{**}. We now form two new contours:

$$K_1 = -C_1^* + L_1 + C_2^* - L_2 \quad \text{and} \quad K_2 = -C_1^{**} + L_2 + C_2^{**} - L_1,$$

which are shown in Figure 6.24. The function f will be analytic on a simply connected domain D_1 that contains K_1, and f will be analytic on the simply connected domain D_2 that contains K_2, as illustrated in Figure 6.24.

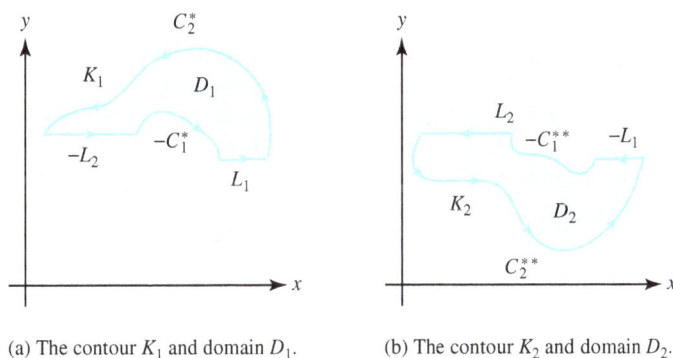

(a) The contour K_1 and domain D_1. (b) The contour K_2 and domain D_2.

Figure 6.24: The cuts L_1 and L_2 and the contours K_1 and K_2 used to prove the deformation of contour theorem

We apply the Cauchy-Goursat theorem to the contours K_1 and K_2, giving

$$\int_{K_1} f(z)\,dz = 0 \quad \text{and} \quad \int_{K_2} f(z)\,dz = 0. \tag{6.36}$$

Adding contours gives

$$
\begin{aligned}
K_1 + K_2 &= -C_1^* + L_1 + C_2^* - L_2 - C_1^{**} + L_2 + C_2^{**} - L_1 \\
&= C_2^* + C_2^{**} - C_1^* - C_1^{**} \\
&= C_2 - C_1.
\end{aligned}
\tag{6.37}
$$

We use Identities (6.17) and (6.18) of Section 6.2 and Equations (6.36) and (6.37) in this proof to conclude that

$$
\int_{C_2} f(z)\,dz - \int_{C_1} f(z)\,dz = \int_{K_1} f(z)\,dz + \int_{K_2} f(z)\,dz = 0,
$$

which establishes the theorem. $\qquad\square$

We now state as a corollary an important result that is implied by the deformation of contour theorem. This result occurs several times in the theory to be developed and is an important tool for computations. You may want to compare the proof of Corollary 6.1 with your solution to Exercise 9 from Section 6.2.

Corollary 6.1. *Let z_0 denote a fixed complex value. If C is a simple closed contour with positive orientation such that z_0 lies interior to C, then*

$$
\int_C \frac{1}{z - z_0}\,dz = 2\pi i, \quad and \quad \int_C \frac{1}{(z - z_0)^n}\,dz = 0,
$$

where n is any integer except $n = 1$.

Proof. Since z_0 lies interior to C, we can choose R so that the circle C_R with center z_0 and radius R lies interior to C. Hence $f(z) = \frac{1}{(z-z_0)^n}$ is analytic in a domain D that contains both C and C_R and the region between them, as shown in Figure 6.25.

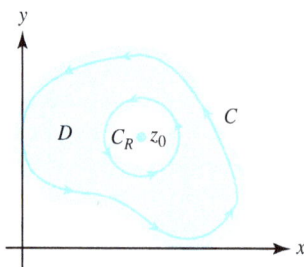

Figure 6.25: The domain D that contains both C and C_R

We let C_R have the parametrization

$$
C_R : z(\theta) = z_0 + Re^{i\theta} \quad and \quad dz = iRe^{i\theta}d\theta, \quad for \quad 0 \le \theta \le 2\pi.
$$

The deformation of contour theorem implies that the integral of f over C_R has the same value as the integral of f over C, so

$$
\int_C \frac{1}{z - z_0}\,dz = \int_{C_R} \frac{1}{z - z_0}\,dz = \int_0^{2\pi} \frac{iRe^{i\theta}}{Re^{i\theta}}\,d\theta = i \int_0^{2\pi} d\theta = 2\pi i
$$

and

$$\int_C \frac{1}{(z-z_0)^n}\,dz = \int_{C_R} \frac{1}{(z-z_0)^n}\,dz = \int_0^{2\pi} \frac{iRe^{i\theta}}{R^n e^{in\theta}}\,d\theta = iR^{1-n}\int_0^{2\pi} e^{i(1-n)\theta}\,d\theta$$

$$= \frac{R^{1-n}}{1-n}e^{i(1-n)\theta}\Big|_{\theta=0}^{\theta=2\pi} = \frac{R^{1-n}}{1-n} - \frac{R^{1-n}}{1-n} = 0.$$

$\square$

The deformation of contour theorem is an extension of the Cauchy-Goursat theorem to a doubly connected domain in the following sense. We let D be a domain that contains C_1 and C_2 and the region between them, as shown in Figure 6.23. Then the contour $C = C_2 - C_1$ is a parametrization of the boundary of the region R that lies between C_1 and C_2 so that the points of R lie to the left of C as a point $z(t)$ moves around C. Hence C is a positive orientation of the boundary of R, and Theorem 6.6 implies that $\int_C f(z)\,dz = 0$.

We can extend Theorem 6.6 to multiply connected domains with more than one "hole." The proof, which we leave for you, involves the introduction of several cuts and is similar to the proof of Theorem 6.6.

Theorem 6.7 (Extended Cauchy-Goursat theorem). *Let $C, C_1, \ldots, C_n$ be simple closed positively oriented contours with the properties that C_k lies interior to C for $k = 1, 2, \ldots, n$, and the interior of C_k has no points in common with the interior of C_j if $k \neq j$. Let f be analytic on a domain D that contains all the contours and the region between C and $C_1 + C_2 + \cdots + C_n$, as shown in Figure 6.26. Then*

$$\int_C f(z)\,dz = \sum_{k=1}^{n}\int_{C_k} f(z)\,dz.$$

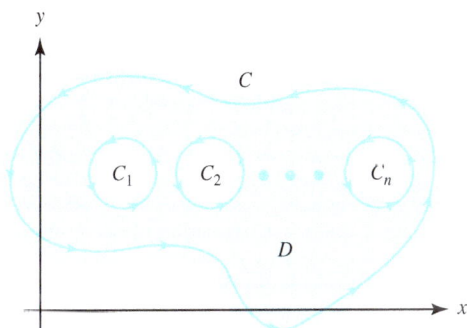

Figure 6.26: The multiply connected domain D and the contours C and $C_1, C_2, \ldots, C_n$ in the statement of the extended Cauchy-Goursat theorem

Example 6.14. Show that $\int_{C_2^+(0)} \frac{2z}{z^2+2}\,dz = 4\pi i$.

Solution:

Recall that $C_2^+(0)$ is the circle $\{z : |z = 2|\}$ with positive orientation. Using partial fraction decomposition gives

$$\frac{2z}{z^2+2} = \frac{2z}{(z+i\sqrt{2})(z-i\sqrt{2})} = \frac{1}{z+i\sqrt{2}} + \frac{1}{z-i\sqrt{2}}, \quad \text{so}$$

189

$$\int_{C_2^+(0)} \frac{2z}{z^2+2}\, dz = \int_{C_2^+(0)} \frac{1}{z+i\sqrt{2}}\, dz + \int_{C_2^+(0)} \frac{1}{z-i\sqrt{2}}\, dz. \tag{6.38}$$

The points $z = \pm i\sqrt{2}$ lie interior to $C_2^+(0)$, so Corollary 6.1 implies that

$$\int_{C_2^+(0)} \frac{1}{z+i\sqrt{2}}\, dz = 2\pi i \quad \text{and} \quad \int_{C_2^+(0)} \frac{1}{z-i\sqrt{2}}\, dz = 2\pi i.$$

Substituting these values into Equation (6.38) yields

$$\int_{C_2^+(0)} \frac{2z}{z^2+2}\, dz = 2\pi i + 2\pi i = 4\pi i.$$

Example 6.15. Show that $\int_{C_1^+(i)} \frac{2z}{z^2+2}\, dz = 2\pi i$.

Solution:

Recall that $C_1^+(i)$ is the circle $\{z : |z-i| = 1\}$ having positive orientation. Using partial fractions again, we have

$$\int_{C_1^+(i)} \frac{2z}{z^2+2}\, dz = \int_{C_1^+(i)} \frac{1}{z+i\sqrt{2}}\, dz + \int_{C_1^+(i)} \frac{1}{z-i\sqrt{2}}\, dz.$$

In this case, $z = i\sqrt{2}$ lies interior to $C_1^+(i)$ but $z = -i\sqrt{2}$ does not, as shown in Figure 6.27.

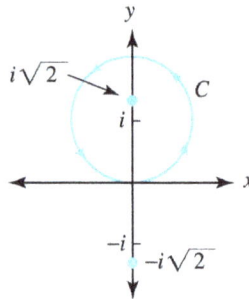

Figure 6.27: The circle $C_1^+(i)$ and the points $z = \pm i\sqrt{2}$

By Corollary 6.1, the second integral on the right side of this equation has the value $2\pi i$. The first integral equals zero by the Cauchy-Goursat theorem because the function $f(z) = \frac{1}{z+i\sqrt{2}}$ is analytic on a simply connected domain that contains $C_1^+(i)$. Thus

$$\int_{C_1^+(i)} \frac{2z}{z^2+2}\, dz = 0 + 2\pi i = 2\pi i.$$

Example 6.16. Show that $\int_C \frac{z-2}{z^2-z}\, dz = -6\pi i$, where C is the "figure eight" contour shown in Figure 6.28(a).

Solution:

Again, we use partial fractions to express the integral:

$$\int_C \frac{z-2}{z^2-z}\, dz = 2\int_C \frac{1}{z}\, dz - \int_C \frac{1}{z-1}\, dz. \tag{6.39}$$

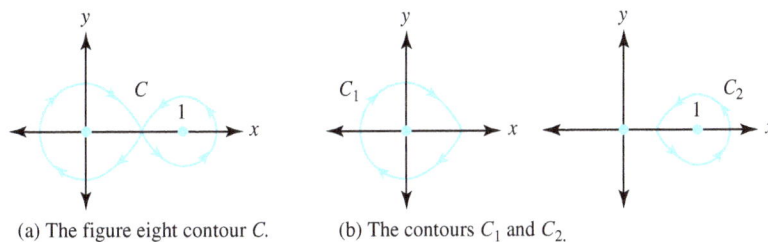

(a) The figure eight contour C.   (b) The contours C_1 and C_2.

Figure 6.28: The contour $C = C_1 + C_2$

Using the Cauchy-Goursat theorem, Property (6.17), and Corollary 6.1 (with $z_0 = 0$), we compute the value of the first integral on the right side of Equation (6.39):

$$2 \int_C \frac{1}{z} \, dz = 2 \int_{C_1} \frac{1}{z} \, dz + 2 \int_{C_2} \frac{1}{z} \, dz$$
$$= -2 \int_{-C_1} \frac{1}{z} \, dz + 0$$
$$= -2(2\pi i)$$
$$= -4\pi i.$$

Similarly, we find that

$$-\int_C \frac{1}{z-1} \, dz = -\int_{C_1} \frac{1}{z-1} \, dz - \int_{C_2} \frac{1}{z-1} \, dz$$
$$= 0 - 2\pi i$$
$$= -2\pi i.$$

If we substitute the results of the last two equations into Equation (6.39) we get

$$\int_C \frac{z-2}{z^2-z} \, dz = -4\pi i - 2\pi i = -6\pi i.$$

Exercises for Section 6.3 (Selected answers or hints are on page 443.)

1. Determine the domain of analyticity for the following functions and evaluate $\int_{C_1^+(0)} f(z) \, dz$.

    (a) $f(z) = \frac{z}{2z^2+1}$.

    (b) $f(z) = \frac{1}{z^2+2z+2}$.

    (c) $f(z) = \tan z$.

    (d) $f(z) = \text{Log}(z+5)$.

2. Show that $\int_C z^{-1} dz = 2\pi i$, where C is the square with vertices $1 \pm i$, and $-1 \pm i$ and having positive orientation.

3. Show that $\int_{C_1^+(0)} (4z^2 - 4z + 5)^{-1} \, dz = 0$.

4. Find $\int_C (z^2 - z)^{-1} \, dz$ for

    (a) circle $C = C_2^+(1) = \{z : |z - 1| = 2\}$ having positive orientation.

191

(b) circle $C = C^+_{\frac{1}{2}}(1) = \{z : |z - 1| = \frac{1}{2}\}$ having positive orientation.

5. Find $\int_C (2z - 1)(z^2 - z)^{-1} \, dz$ for the following:

   (a) The circle $C = C^+_2(0) = \{z : |z| = 2\}$ having positive orientation.
   (b) The circle $C = C^+_{\frac{1}{2}}(0) = \{z : |z| = \frac{1}{2}\}$ having positive orientation.

6. Let C be the triangle with vertices 0, 1, and i and having positive orientation. Parametrize C and show that

   (a) $\int_C 1 \, dz = 0$.
   (b) $\int_C z \, dz = 0$.

7. Evaluate $\int_C (4z^2 + 4z - 3)^{-1} \, dz = \int_C (2z - 1)^{-1}(2z + 3)^{-1} \, dz$ for the following:

   (a) The circle $C = C^+_1(0)$.
   (b) The circle $C = C^+_1(-\frac{2}{3}) = \{z : |z + \frac{2}{3}| = 1\}$.
   (c) the circle $C = C^+_3(0)$.

8. Use Green's theorem to show that the area enclosed by a simple closed contour C is $\frac{1}{2} \int_C x \, dy - y \, dx$.

9. Parametrize $C^+_1(0)$ with $z(t) = \cos t + i \sin t$, for $-\pi \le t \le \pi$. Use the principal branch of the square root function: $z^{\frac{1}{2}} = r^{\frac{1}{2}} \cos \frac{\theta}{2} + i r^{\frac{1}{2}} \sin \frac{\theta}{2}$, for $-\pi < \theta \le \pi$, to find $\int_{C^+_1(0)} z^{\frac{1}{2}} \, dz$. *Hint*: Take limits as $t \to -\pi$.

10. Evaluate $\int_C (z^2 - 1)^{-1} dz$ for the contours shown in Figure 6.29.

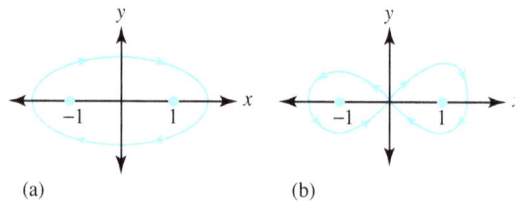

Figure 6.29: For Exercise 10

11. Evaluate $\int_{C^+_1(0)} |z|^2 \exp z \, dz$.

12. Suppose that $f(z) = u(r, \theta) + iv(r, \theta)$ is analytic for all values of $z = re^{i\theta}$. Show that

$$\int_0^{2\pi} [u(r, \theta) \cos \theta - v(r, \theta) \sin \theta] \, d\theta = 0.$$

   Hint: Integrate f around the circle $C^+_1(0)$.

13. If C is the figure eight contour shown in Figure 6.28(a),

   (a) evaluate $\int_C (z^2 - z)^{-1} \, dz$.
   (b) evaluate $\int_C (2z - 1)(z^2 - z)^{-1} \, dz$.

14. Compare the various methods for evaluating contour integrals. What are the limitations of each method?

192

6.4 The Fundamental Theorems of Integration

Let f be analytic in the simply connected domain D. The theorems in this section show that an antiderivative F can be constructed by contour integration. A consequence will be the fact that in a simply connected domain, the integral of an analytic function f along any contour joining z_1 to z_2 is the same, and its value is given by $F(z_2) - F(z_1)$. As a result, we can use the antiderivative formulas from calculus to compute the value of definite integrals.

Theorem 6.8 (Indefinite integrals, or antiderivatives). *Let f be analytic in the simply connected domain D. If z_0 is a fixed value in D, and if C is any contour in D with initial point z_0 and terminal point z, then the function*

$$F(z) = \int_C f(\xi)\, d\xi = \int_{z_0}^{z} f(\xi)\, d\xi \tag{6.40}$$

is well-defined and analytic in D, with its derivative given by $F'(z) = f(z)$.

Proof. We first establish that the integral is independent of the path of integration. This will show that the function F is well-defined, which in turn will justify the notation $F(z) = \int_{z_0}^{z} f(\xi)\, d\xi$.

We let C_1 and C_2 be two contours in D, both with initial point z_0 and terminal point z, as shown in Figure 6.30.

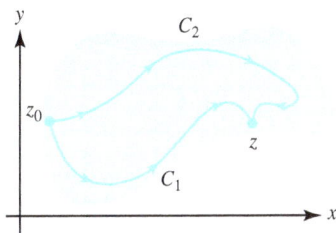

Figure 6.30: The contours C_1 and C_2 joining z_0 to z

Then $C_1 - C_2$ is a simple closed contour, and the Cauchy-Goursat theorem implies that

$$\int_{C_1} f(\xi)\, d\xi - \int_{C_2} f(\xi)\, d\xi = \int_{C_1 - C_2} f(\xi)\, d\xi = 0.$$

Therefore, the contour integral in Equation (6.40) is independent of path. Here we have taken the liberty of drawing contours that intersect only at the endpoints. A slight modification of the proof shows that a finite number of other points of intersection are permitted.

We now show that $F'(z) = f(z)$. Let z be held fixed, and let $|\Delta z|$ be chosen small enough so that the point $z + \Delta z$ also lies in the domain D. Since z is held fixed, $f(z) = K$, where K is a constant, and Equation (6.9) implies that

$$\int_{z}^{z+\Delta z} f(z)\, d\xi = \int_{z}^{z+\Delta z} K\, d\xi = K \Delta z = f(z) \Delta z. \tag{6.41}$$

Using the additive property of contours and the definition of F given in Equation (6.40), we have

$$F(z + \Delta z) - F(z) = \int_{z_0}^{z+\Delta z} f(\xi)\, d\xi - \int_{z_0}^{z} f(\xi)\, d\xi$$

$$= \int_{\Gamma_2} f(\xi)\, d\xi - \int_{\Gamma_1} f(\xi)\, d\xi = \int_{\Gamma} f(\xi)\, d\xi. \qquad (6.42)$$

where the contour Γ is the straight-line segment joining z to $z + \Delta z$, and Γ_1 and Γ_2 join z_0 to z, and z_0 to $z + \Delta z$, respectively, as shown in Figure 6.31.

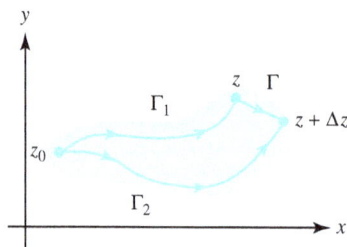

Figure 6.31: The contours Γ_1 and Γ_2 and the line segment Γ

Since f is continuous at z, for any $\varepsilon > 0$ there is a $\delta > 0$ so that

$$|f(\xi) - f(z)| < \varepsilon \quad \text{when} \quad |\xi - z| < \delta.$$

If we require that $|\Delta z| < \delta$ and combine this last inequality with Equations (6.41) and (6.42), and Theorem 6.23, we get

$$\left| \frac{F(z + \Delta z) - F(z)}{\Delta z} - f(z) \right| = \frac{1}{|\Delta z|} \left| \int_{\Gamma} f(\xi)\, d\xi - \int_{\Gamma} f(z)\, d\xi \right|$$

$$\leq \frac{1}{|\Delta z|} \int_{\Gamma} |f(\xi) - f(z)||d\xi|$$

$$< \frac{1}{|\Delta z|} \varepsilon |\Delta z|$$

$$= \varepsilon.$$

Thus $\left| \frac{F(z+\Delta z)-F(z)}{\Delta z} - f(z) \right|$ tends to 0 as $\Delta z \to 0$, so $F'(z) = f(z)$. $\qquad \square$

Remark 6.2. *It is important to stress that the line integral of an analytic function is independent of path. In Example 6.9 we showed that $\int_{C_1} z\, dz = \int_{C_2} z\, dz = 4 + 2i$, where C_1 and C_2 were different contours joining $-1 - i$ to $3 + i$. Because the integrand $f(z) = z$ is an analytic function, Theorem 6.8 lets us know ahead of time that the value of the two integrals is the same; hence one calculation would have sufficed. If you ever have to compute a line integral of an analytic function over a difficult contour, change the contour to something easier. You are guaranteed to get the same answer. Of course, you must be sure that the function you're dealing with is analytic in a simply connected domain containing your original and new contours.*

If we set $z = z_1$ in Theorem 6.8, then we obtain the following familiar result for evaluating a definite integral of an analytic function.

Theorem 6.9 (Definite integrals). *Let f be analytic in a simply connected domain D. If z_0 and z_1 are any two points in D joined by a contour C, then*

$$\int_C f(z)\, dz = \int_{z_0}^{z_1} f(z)\, dz = F(z_1) - F(z_0), \qquad (6.43)$$

where F is any antiderivative of f in D.

Proof. If we choose F to be the function defined by Formula (6.40), then Equation (6.43) holds. If G is any other antiderivative of f in D, then $G'(z) = F'(z)$ for all $z \in D$. Thus the function $H(z) = G(z) - F(z)$ is analytic in D, and $H'(z) = G'(z) - F'(z) = 0$, for all $z \in D$. Thus, by Theorem 3.7, this means $H(z) = K$, for all $z \in D$, where K is some complex constant. Therefore $G(z) = F(z) + K$, so $G(z_1) - G(z_0) = F(z_1) - F(z_0)$, which establishes our theorem. $\qquad\square$

Theorem 6.9 gives an important method for evaluating definite integrals when the integrand is an analytic function in a simply connected domain. In essence, it permits you to use all the rules of integration that you learned in calculus. When the conditions of Theorem 6.9 are met, applying it is generally much easier than parametrizing a contour.

Example 6.17. Show that $\int_C \dfrac{1}{2z^{\frac{1}{2}}}\, dz = 1 + i$, where $z^{\frac{1}{2}}$ is the principal branch of the square root function and C is the line segment joining 4 to $8 + 6i$.

Solution:

We showed in Chapter 3 that if $F(z) = z^{\frac{1}{2}}$, then $F'(z) = \dfrac{1}{2z^{\frac{1}{2}}}$, where the principal branch of the square root function is used in both the formulas for F and F'. We note that C is contained in the simply connected domain $D_4(6 + 3i)$, which is the open disk of radius 4 centered at the midpoint of the segment C. Since $f(z) = \dfrac{1}{2z^{\frac{1}{2}}}$ is analytic in $D_4(6 + 3i)$, Theorem 6.9 implies

$$\int_4^{8+6i} \frac{1}{2z^{\frac{1}{2}}}\, dz = (8 + 6i)^{\frac{1}{2}} - 4^{\frac{1}{2}} = 3 + i - 2 = 1 + i.$$

Example 6.18. Show that $\int_C \cos z\, dz = -\sin 1 + i \sinh 1$, where C is the line segment between 1 and i.

Solution:

An antiderivative of $f(z) = \cos z$ is $F(z) = \sin z$. Since F is entire, Theorem 6.9 yields

$$\int_C \cos z\, dz = \int_1^i \cos z\, dz = \sin i - \sin 1 = -\sin 1 + i \sinh i.$$

Example 6.19. We let $D = \{z = re^{i\theta} : r > 0 \text{ and } -\pi < \theta < \pi\}$ be the simply connected domain shown in Figure 6.32. We know that $f(z) = \frac{1}{z}$ is analytic in D and has an antiderivative

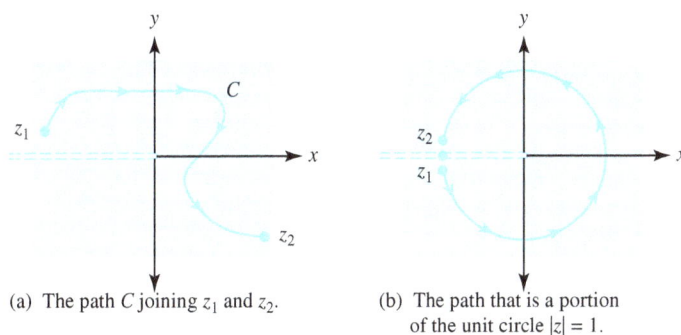

(a) The path C joining z_1 and z_2.

(b) The path that is a portion of the unit circle $|z| = 1$.

Figure 6.32: The simply connected domain D from Examples 6.19 and 6.20

$F(z) = \text{Log}(z)$, for all $z \in D$. If C is a contour in D that joins the point z_1 to the point z_2, then Theorem 6.9 implies that

$$\int_C \frac{1}{z}\, dz = \int_{z_1}^{z_2} \frac{1}{z}\, dz = \text{Log}(z_2) - \text{Log}(z_1).$$

Example 6.20. Show that $\int_{C_1^+(0)} \frac{dz}{z} = 2\pi i$.

Solution:

Recall that $C_1^+(0)$ is the unit circle with positive orientation. We let C be that circle with the point -1 omitted, as shown in Figure 6.32(b). The contour C is contained in the simply connected domain D of Example 6.19. We know that $f(z) = \frac{1}{z}$ is analytic in D, and has an antiderivative $F(z) = \text{Log}(z)$, for all $z \in D$. Therefore, if we let z_2 approach -1 on C through the upper half-plane and z_1 approach -1 on C through the lower half-plane,

$$\int_{C_1^+(0)} \frac{1}{z}\, dz = \lim_{\substack{z_2 \to -1\,(z_2 \in C,\, \text{Im} z_2 > 0)\\ z_1 \to -1\,(z_1 \in C,\, \text{Im} z_1 < 0)}} \int_{z_1}^{z_2} \frac{1}{z}\, dz$$

$$= \lim_{\substack{z_2 \to -1\,(z_2 \in C,\, \text{Im} z_2 > 0)}} \text{Log}(z_2) - \lim_{\substack{z_1 \to -1\,(z_1 \in C,\, \text{Im} z_1 < 0)}} \text{Log}(z_1)$$

$$= i\pi - (-i\pi)$$

$$= 2\pi i.$$

Exercises for Section 6.4 (Selected answers or hints are on page 444.)

For Exercises 1–14, find the value of the definite integral using Theorem 6.9 and explain why you are justified in using it.

1. $\int_C z^2\, dz$, where C is the line segment from $1 + i$ to $2 + i$.

2. $\int_C \cos z\, dz$, where C is the line segment from $-i$ to $1 + i$.

3. $\int_C \exp z\, dz$, where C is the line segment from 2 to $i\frac{\pi}{2}$.

4. $\int_C z \exp z\, dz$, where C is the line segment from $-1 - i\frac{\pi}{2}$ to $2 + i\pi$.

5. $\int_C \frac{1+z}{z}\, dz$, where C is the line segment from 1 to i.

6. $\int_C \sin \frac{z}{2}\, dz$, where C is the line segment from 0 to $\pi - 2i$.

7. $\int_C (z^2 + z^{-2})dz$, where C is the line segment from i to $1 + i$.

8. $\int_C z \exp(z^2)dz$, where C is the line segment from $1 - 2i$ to $1 + 2i$.

9. $\int_C z \cos z\, dz$, where C is the line segment from 0 to i.

10. $\int_C \sin^2 z\, dz$, where C is the line segment from 0 to i.

11. $\int_C \text{Log}\, z\, dz$, where C is the line segment from 1 to $1 + i$.

12. $\int_C \frac{dz}{z^2 - z}$, where C is the line segment from 2 to $2 + i$.

13. $\int_C \frac{2z-1}{z^2 - z}\, dz$, where C is the line segment from 2 to $2 + i$.

14. $\int_C \frac{z-2}{z^2 - z}\, dz$, where C is the line segment from 2 to $2 + i$.

15. Show that $\int_C 1\, dz = z_2 - z_1$, where C is the line segment from z_1 to z_2, by parametrizing C.

16. Let z_1 and z_2 be points in the right half-plane and let C be the line segment joining them. Show that $\int_C \frac{1}{z^2}\, dz = \frac{1}{z_1} - \frac{1}{z_2}$.

17. Let $z^{\frac{1}{2}}$ be the principal branch of the square root function.

   (a) Evaluate $\int_C \frac{1}{2z^{\frac{1}{2}}}\, dz$, where C is the line segment joining 9 to $3 + 4i$.

   (b) Evaluate $\int_C z^{\frac{1}{2}}\, dz$, where C is the right half of the circle $C_2^+(0)$ joining $-2i$ to $2i$.

18. Using partial fraction decomposition, show that if z lies in the right half-plane and C is the line segment joining 0 to z, then

$$\int_C \frac{1}{\xi^2 + 1}\, d\xi = \text{Arctan}(z) = \frac{i}{2}\text{Log}(z + i) - \frac{i}{2}\text{Log}(z - i) + \frac{\pi}{2}.$$

19. Let f' and g' be analytic for all z, and let C be any contour joining the points z_1 and z_2. Show that

$$\int_C f(z)g'(z)\, dz = f(z_2)g(z_2) - f(z_1)g(z_1) - \int_C f'(z)g(z)\, dz.$$

20. Compare the various methods for evaluating contour integrals. What are the limitations of each method?

21. Explain how the fundamental theorem of calculus studied in complex analysis and the fundamental theorem of calculus studied in calculus are different. How are they similar?

22. Show that $\int_C z^i dz = (i - 1)\frac{1 + e^{-\pi}}{2}$, where C is the upper half of $C_1^+(0)$.

6.5   Integral Representations

We now present some major results in the theory of functions of a complex variable. The first one is known as Cauchy's integral formula. It shows that the value of an analytic function f can be represented by a certain contour integral. The nth derivative, $f^{(n)}(z)$, has a similar representation. In Chapter 7, we use these results to prove Taylor's theorem and also establish the power series representation for analytic functions. The Cauchy integral formulas are a convenient tool for evaluating certain contour integrals.

Theorem 6.10 (Cauchy's integral formula). *Let f be analytic in the simply connected domain D and let C be a simple closed positively oriented contour that lies in D. If z_0 is a point that lies interior to C, then*

$$f(z_0) = \frac{1}{2\pi i} \int_C \frac{f(z)}{z - z_0}\, dz. \tag{6.44}$$

Proof. Because f is continuous at z_0, if $\varepsilon > 0$ is given there is a $\delta > 0$ such that the positively oriented circle $C_0 = \{z : |z - z_0| = \frac{1}{2}\delta\}$ lies interior to C (as Figure 6.33 shows) and such that

$$|f(z) - f(z_0)| < \varepsilon \quad \text{whenever} \quad |z - z_0| < \delta. \tag{6.45}$$

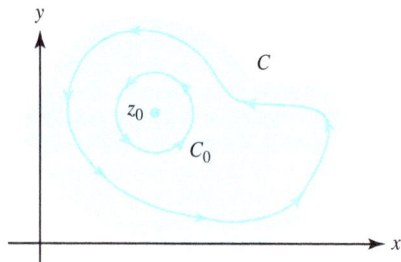

Figure 6.33: The contours C and C_0 in the proof of Cauchy's integral formula

Since $f(z_0)$ is a fixed value, we can use the result of Corollary 6.1 to conclude that

$$f(z_0) = \frac{f(z_0)}{2\pi i} \int_{C_0} \frac{dz}{z - z_0} = \frac{1}{2\pi i} \int_{C_0} \frac{f(z_0)}{z - z_0} \, dz. \tag{6.46}$$

By the deformation of contour theorem (Theorem 6.6),

$$\frac{1}{2\pi i} \int_C \frac{f(z)}{z - z_0} \, dz = \frac{1}{2\pi i} \int_{C_0} \frac{f(z)}{z - z_0} \, dz. \tag{6.47}$$

Using Inequality (6.45) and Equations (6.46) and (6.47) above, together with the *ML* inequality (Theorem 6.3), we obtain the estimate:

$$\left| \frac{1}{2\pi i} \int_C \frac{f(z)}{z - z_0} \, dz - f(z_0) \right| = \left| \frac{1}{2\pi i} \int_{C_0} \frac{f(z)}{z - z_0} \, dz - \frac{1}{2\pi i} \int_{C_0} \frac{f(z_0)}{z - z_0} \, dz \right|$$

$$\leq \frac{1}{2\pi} \int_{C_0} \frac{|f(z) - f(z_0)|}{|z - z_0|} \, |dz|$$

$$\leq \frac{1}{2\pi} \frac{\varepsilon}{(\frac{1}{2})\delta} \pi\delta$$

$$= \varepsilon.$$

This proves the theorem because ε can be made arbitrarily small. $\qquad\square$

Example 6.21. Show that $\int_{C_2^+(0)} \frac{\exp z}{z-1} \, dz = i2\pi e$.

Solution:

Recall that $C_2^+(0)$ is the circle centered at 0 with radius 2 and having positive orientation. We have $f(z) = \exp z$ and $f(1) = \exp(1) = e$. The point $z_0 = 1$ lies interior to the circle, so Cauchy's integral formula implies that

$$e = f(1) = \frac{1}{2\pi i} \int_{C_2^+(0)} \frac{\exp z}{z - 1} \, dz.$$

Multiplication by $2\pi i$ establishes the desired result.

Example 6.22. Show that $\int_{C_1^+(0)} \frac{\sin z}{4z+\pi} \, dz = i\left(-\frac{\sqrt{2}\pi}{4}\right)$.

Solution:

Here we have $f(z) = \sin z$. We manipulate the integral and use Cauchy's integral formula to

198

obtain

$$\int_{C_1^+(0)} \frac{\sin z}{4z + \pi} \, dz = \frac{1}{4} \int_{C_1^+(0)} \frac{\sin z}{z + (\frac{\pi}{4})} \, dz$$

$$= \frac{1}{4} \int_{C_1^+(0)} \frac{f(z)}{z - (-\frac{\pi}{4})} \, dz$$

$$= \frac{1}{4} (2\pi i) f\left(-\frac{\pi}{4}\right)$$

$$= \frac{\pi i}{2} \sin\left(-\frac{\pi}{4}\right) = -\frac{\sqrt{2}\pi i}{4}.$$

Example 6.23. Show that $\int_{C_1^+(0)} \frac{\exp(i\pi z)}{2z^2 - 5z + 2} \, dz = \frac{2\pi}{3}$.

Solution:

We see that $2z^2 - 5z + 2 = (2z - 1)(z - 2) = \frac{1}{2}(z - \frac{1}{2})(z - 2)$. The only zero of this expression that lies in the interior of $C_1(0)$ is $z_0 = \frac{1}{2}$. We set $f(z) = \frac{\exp(i\pi z)}{z - 2}$ and use Theorem 6.10 to conclude that

$$\int_{C_1^+(0)} \frac{\exp(i\pi z)}{2z^2 - 5z + 2} \, dz = \frac{1}{2} \int_{C_1^+(0)} \frac{f(z)}{z - \frac{1}{2}} \, dz$$

$$= \frac{1}{2} (2\pi i) f\left(\frac{1}{2}\right)$$

$$= \pi i \frac{\exp(\frac{i\pi}{2})}{\frac{1}{2} - 2}$$

$$= \frac{2\pi}{3}.$$

We now state a general result that shows how to accomplish differentiation under the integral sign. The proof is in some advanced texts. See, for instance, Rolf Nevanlinna and V. Paatero, *Introduction to Complex Analysis* (Reading, Mass.: Addison-Wesley, 1969), Section 9.7.

Theorem 6.11 (Leibniz's rule). *Let G be an open set and let $I : a \le t \le b$ be an interval of real numbers. Let $g(z, t)$ and its partial derivative $g_z(z, t)$ with respect to z be continuous functions for all z in G and all t in I. Then $F(z) = \int_a^b g(z, t) \, dt$ is analytic for z in G, and $F'(z) = \int_a^b g_z(z, t) \, dt$.*

We now generalize Theorem 6.10 to give an integral representation for the nth derivative, $f^{(n)}(z)$. We use Leibniz's rule in the proof and note that this method of proof serves as a mnemonic device for remembering Theorem 6.12.

Theorem 6.12 (Cauchy's integral formulas for derivatives). *Let f be analytic in the simply connected domain D and let C be a simple closed positively oriented contour that lies in D. If z_0 is a point that lies interior to C, then for any integer $n \ge 0$,*

$$f^{(n)}(z_0) = \frac{n!}{2\pi i} \int_C \frac{f(z)}{(z - z_0)^{n+1}} \, dz. \tag{6.48}$$

Proof. Because $f^{(0)}(z_0) = f(z_0)$, the case for $n = 0$ reduces to Theorem 6.10. We now establish the theorem for the case $n = 1$. We start by using the parametrization

$$C : z = z(t) \quad \text{and} \quad dz = z'(t) \, dt, \quad \text{for} \quad a \le t \le b.$$

199

We use Theorem 6.10 and write

$$f(z_0) = \frac{1}{2\pi i} \int_C \frac{f(z)}{z - z_0}\, dz = \frac{1}{2\pi i} \int_a^b \frac{f(z(t))z'(t)}{z(t) - z_0}\, dt. \tag{6.49}$$

The integrand on the right side of Equation (6.49) is a function $g(z_0, t)$ of the two variables z_0 and t, where

$$g(z_0, t) = \frac{f(z(t))z'(t)}{z(t) - z_0} \quad \text{and} \quad \frac{\partial g}{\partial z_0}(z_0, t) = g_{z_0}(z_0,\ t) = \frac{f(z(t))z'(t)}{(z(t) - z_0)^2}.$$

Moreover, $g(z_0, t)$ and $g_{z_0}(z_0, t)$ are continuous on the interior of C, which is an open set. Applying Leibniz's rule to Equations (6.49) gives

$$f'(z_0) = \frac{1}{2\pi i} \int_a^b \frac{f(z(t))z'(t)}{(z(t) - z_0)^2}\, dt = \frac{1}{2\pi i} \int_C \frac{f(z)}{(z - z_0)^2}\, dz,$$

which completes the proof in the case when $n = 1$. We then apply the same argument to the analytic function f' and show that its derivative f'' is also represented by Equation (6.48) for $n = 2$. The principle of mathematical induction then establishes the theorem for all integers. $\square$

Example 6.24. Let z_0 denote a fixed complex value. Show that, if C is a simple closed positively oriented contour such that z_0 lies interior to C, then for any integer $n \geq 1$,

$$\int_C \frac{1}{z - z_0}\, dz = 2\pi i \quad \text{and} \quad \int_C \frac{1}{(z - z_0)^{n+1}}\, dz = 0. \tag{6.50}$$

Solution:

We let $f(z) = 1$. Then $f^{(n)}(z) = 0$ for $n \geq 1$. Theorem 6.10 implies that the value of the first integral in Equations (6.50) is

$$\int_C \frac{1}{z - z_0}\, dz = 2\pi i f(z_0) = 2\pi i.$$

and Theorem 6.12 further implies that

$$\int_C \frac{1}{(z - z_0)^{n+1}}\, dz = \frac{2\pi i}{n!} f^{(n)}(z_0) = 0.$$

This result is the same as that proven earlier in Corollary 6.1. Obviously, though, the technique of using Theorems 6.10 and 6.12 is easier.

Example 6.25. Show that $\displaystyle\int_{C_2^+(0)} \frac{\exp z^2}{(z - i)^4}\, dz = \frac{-4\pi}{3e}$.

Solution:

If we let $f(z) = \exp z^2$, then a straightforward calculation shows that $f^{(3)}(z) = (12z + 8z^3)\exp z^2$. Using Cauchy's integral formulas with $n = 3$, we conclude that

$$\int_C \frac{\exp z^2}{(z - i)^4}\, dz = \frac{2\pi i}{3!} f^{(3)}(i) = \frac{2\pi i}{6} \frac{4i}{e} = -\frac{4\pi}{3e}.$$

We now state two important corollaries of Theorem 6.12.

Corollary 6.2. *If f is analytic in the domain D, then, for integers $n \geq 0$, all derivatives $f^{(n)}(z)$ exist for $z \in D$ (and therefore are analytic in D).*

Proof. For each point z_0 in D, there exists a closed disk $|z - z_0| \leq R$ that is contained in D. We use the circle $C = C_R(z_0) = \{z : |z - z_0| = R\}$ in Theorem 6.12 to show that $f^{(n)}(z_0)$ exists for all integers $n \geq 0$. $\qquad\square$

Remark 6.3. *This result is interesting, as it illustrates a big difference between real and complex functions. A real function f can have the property that f' exists everywhere in a domain D, but f'' exists nowhere. Corollary 6.2 states that if a complex function f has the property that f' exists everywhere in a domain D, then, remarkably, all derivatives of f exist in D.*

Corollary 6.3. *If u is a harmonic function at each point (x, y) in the domain D, then all partial derivatives u_x, u_y, u_{xx}, u_{xy}, and u_{yy} exist and are harmonic functions.*

Proof. For each point $z_0 = (x_0, y_0)$ in D there exists a disk $D_R(z_0)$ that is contained in D. In this disk, a conjugate harmonic function v exists, so the function $f(z) = u + iv$ is analytic. We use the Cauchy-Riemann equations to get $f'(z) = u_x + iv_x = v_y - iu_y$, for $z \in D_R(z_0)$. Since f' is analytic in $D_R(z_0)$, the functions u_x and u_y are harmonic there. Again, we can use the Cauchy-Riemann equations to obtain, for $z \in D_R(z_0)$,

$$f''(z) = u_{xx} + iv_{xx} = v_{yx} - iu_{yx} = -u_{yy} - iv_{yy}.$$

Because f'' is analytic in $D_R(z_0)$, the functions u_{xx}, u_{xy}, and u_{yy} are harmonic there. $\qquad\square$

Exercises for Section 6.5 (Selected answers or hints are on page 444.)

Recall that $C_\rho^+(z_0)$ denotes the positively oriented circle $\{z : |z - z_0| = \rho\}$.

1. Evaluate $\int_{C_1^+(0)} (\exp z + \cos z) z^{-1}\, dz$.

2. Evaluate $\int_{C_1^+(1)} (z + 1)^{-1} (z - 1)^{-1}\, dz$.

3. Evaluate $\int_{C_1^+(1)} (z + 1)^{-1} (z - 1)^{-2}\, dz$.

4. Evaluate $\int_{C_1^+(1)} (z^3 - 1)^{-1}\, dz$.

5. Evaluate $\int_{C_1^+(0)} z^{-4} \sin z\, dz$.

6. Evaluate $\int_{C_1^+(0)} (z \cos z)^{-1}\, dz$.

7. Evaluate $\int_{C_1^+(0)} z^{-3} \sinh(z^2)\, dz$.

8. Evaluate $\int_C z^{-2} \sin z\, dz$ along the following contours:

   (a) The circle $C_1^+(\frac{\pi}{2})$.

   (b) The circle $C_1^+(\frac{\pi}{4})$.

9. Evaluate $\int_{C_1^+(0)} z^{-n} \exp z\, dz$, where n is a positive integer.

10. Evaluate $\int_C z^{-2}(z^2 - 16)^{-1} \exp z\, dz$ along the following contours:

(a) The circle $C_1^+(0)$.

(b) The circle $C_1^+(4)$.

11. Evaluate $\int_{C_1^+(1+i)} (z^4 + 4)^{-1} \, dz$.

12. Evaluate $\int_C z^{-1}(z-1)^{-1} \exp z \, dz$ along the following contours:

(a) The circle $C_{\frac{1}{2}}^+(0)$.

(b) The circle $C_2^+(0)$.

13. Evaluate $\int_C (z^2 + 1)^{-1} \sin z \, dz$ along the following contours:

(a) The circle $C_1^+(i)$.

(b) The circle $C_1^+(-i)$.

14. Evaluate $\int_{C_1^+(i)} (z^2 + 1)^{-2} \, dz$.

15. Evaluate $\int_C (z^2 + 1)^{-1} \, dz$ along the following contours:

(a) The circle $C_1^+(i)$.

(b) The circle $C_1^+(-i)$.

16. Let $P(z) = a_0 + a_1 z + a_2 z^2 + a_3 z^3$. Evaluate $\int_{C_1^+(0)} P(z) z^{-n} \, dz$, where n is a positive integer.

17. Let z_1 and z_2 be two complex numbers that lie interior to the simple closed contour C with positive orientation. Evaluate

$$\int_C (z - z_1)^{-1} (z - z_2)^{-1} \, dz.$$

18. Let f be analytic in the simply connected domain D and let z_1 and z_2 be two complex numbers that lie interior to the simple closed contour C having positive orientation that lies in D. Show that

$$\frac{f(z_2) - f(z_1)}{z_2 - z_1} = \frac{1}{2\pi i} \int_C \frac{f(z)}{(z - z_1)(z - z_2)} \, dz.$$

What happens when $z_2 \to z_1$? Why?

19. The *Legendre polynomial* $P_n(z)$ is defined by

$$P_n(z) = \frac{1}{2^n n!} \frac{d^n}{dz^n} [(z^2 - 1)^n].$$

Use Cauchy's integral formula to show that

$$P_n(z) = \frac{1}{2\pi i} \int_C \frac{(\xi^2 - 1)^n}{2^n (\xi - z)^{n+1}} \, d\xi,$$

where C is a simple closed contour having positive orientation and z lies inside C.

20. Discuss the importance of being able to define an analytic function $f(z)$ with the contour integral in Formula (6.44). How does this definition differ from other definitions of a function that you have learned?

6.6 The Theorems of Morera and Liouville

In this section we investigate some of the qualitative properties of analytic and harmonic functions. Our first result shows that the existence of an antiderivative for a continuous function is equivalent to the statement that the integral of f is independent of the path of integration. This result is stated in a form that will serve as a converse of the Cauchy-Goursat theorem.

Theorem 6.13 (Morera's theorem). *Let f be a continuous function in a simply connected domain D. If $\int_C f(z)\,dz = 0$ for every closed contour C in D, then f is analytic in D.*

Proof. We select a point z_0 in D and define $F(z)$ by

$$F(z) = \int_{z_0}^z f(\xi)\,d\xi,$$

where the notation indicates the integral is taken on *any* contour that begins at z_0 and ends at z. The function $F(z)$ is well defined because, if C_1 and C_2 are two contours in D—both with initial point z_0 and terminal point z, then $C = C_1 - C_2$ is a closed contour in D, and by hypothesis,

$$0 = \int_C f(\xi)\,d\xi = \int_{C_1} f(\xi)\,d\xi - \int_{C_2} f(\xi)\,d\xi.$$

Since f is continuous, we know that for any $\varepsilon > 0$ there exists $\delta > 0$ such that $|f(\xi) - f(z)| < \varepsilon$ whenever $|\xi - z| < \delta$. Now we can use the same steps as those in the proof of Theorem 6.8 to show that $F'(z) = f(z)$. Hence $F(z)$ is analytic on D, and Corollary 6.2 implies that $F'(z)$ and $F''(z)$ are also analytic. Therefore $f'(z) = F''(z)$ exists for all z in D, proving that $f(z)$ is analytic in D. $\qquad\square$

Cauchy's integral formula shows how the value $f(z_0)$ can be represented by a contour integral. If we choose the contour of integration C to be a circle with center z_0, then we can show that the value $f(z_0)$ is the integral average of the values of $f(z)$ at points z on the circle C, a fact that the following theorem elucidates.

Theorem 6.14 (Gauss's mean value theorem). *If f is analytic in a simply connected domain D that contains the circle $C_R(z_0) = \{z : |z - z_0| = R\}$, then*

$$f(z_0) = \frac{1}{2\pi} \int_0^{2\pi} f(z_0 + Re^{i\theta})\,d\theta.$$

Proof. We parametrize the circle $C_R(z_0)$ by

$$C_R(z_0) : z(\theta) = z_0 + Re^{i\theta}, \quad \text{and} \quad dz = iRe^{i\theta}\,d\theta, \quad \text{for} \quad 0 \le \theta \le 2\pi,$$

and use this parametrization along with Cauchy's integral formula to obtain

$$f(z_0) = \frac{1}{2\pi i} \int_0^{2\pi} \frac{f(z_0 + Re^{i\theta})iRe^{i\theta}}{Re^{i\theta}}\,d\theta = \frac{1}{2\pi} \int_0^{2\pi} f(z_0 + Re^{i\theta})\,d\theta.$$

$\qquad\square$

We now prove an important result concerning the modulus of an analytic function.

Theorem 6.15 (Maximum modulus principle). *Let f be analytic and nonconstant in the domain D. Then $|f(z)|$ does not attain a maximum value at any point z_0 in D.*

Proof. (By contraposition): Assume the contrary and suppose that there exists a point z_0 in D such that

$$|f(z)| \le |f(z_0)| \tag{6.51}$$

holds for all z in D. If $C_R(z_0)$ is any circle contained in D, Theorems 6.14 and 6.3 imply that

$$|f(z_0)| = \left| \frac{1}{2\pi} \int_0^{2\pi} f(z_0 + re^{i\theta})\, d\theta \right| \le \frac{1}{2\pi} \int_0^{2\pi} |f(z_0 + re^{i\theta})|\, d\theta, \tag{6.52}$$

for $0 \le r \le R$. We now treat $|f(z)| = |f(z_0 + re^{i\theta})|$ as a real-valued function of the real variable θ and use Inequality (6.51) to get

$$\frac{1}{2\pi} \int_0^{2\pi} |f(z_0 + re^{i\theta})|\, d\theta \le \frac{1}{2\pi} \int_0^{2\pi} |f(z_0)|\, d\theta = |f(z_0)|, \tag{6.53}$$

for $0 \le r \le R$. Combining Inequalities (6.52) and (6.53) gives

$$|f(z_0)| = \frac{1}{2\pi} \int_0^{2\pi} |f(z_0 + re^{i\theta})|\, d\theta,$$

which we rewrite as

$$\frac{1}{2\pi i} \int_0^{2\pi} \left(|f(z_0)| - |f(z_0 + re^{i\theta})| \right) d\theta = 0, \quad \text{for} \quad 0 \le r \le R. \tag{6.54}$$

A theorem from calculus states that if the integral of a nonnegative continuous function taken over an interval is zero, then that function must be identically zero. Since Inequality (6.51) implies that the integrand in Equation (6.54) is a nonnegative real-valued function, we conclude that it is identically zero; that is,

$$\left| f(z_0) \right| = \left| f(z_0 + re^{i\theta}) \right|, \quad \text{for} \quad 0 \le r \le R \quad \text{and} \quad 0 \le \theta \le 2\pi. \tag{6.55}$$

If the modulus of an analytic function is constant in a closed disk, then the function is constant in that closed disk by Theorem 3.6. Therefore we conclude from Identity (6.55) that

$$f(z) = f(z_0), \quad \text{for all } z \text{ in the closed disk } \overline{D}_R(z_0), \tag{6.56}$$

where $\overline{D}_R(z_0) = \{z : |z - z_0| \le R\}$. Now we let ζ denote an arbitrary point in D, C be a contour in the original domain D that joins z_0 to ζ, and $2d$ denote the minimum distance from C to the boundary of D. We can find consecutive points $z_0, z_1, z_2, \ldots, z_n = \zeta$ along C, with $|z_{k+1} - z_k| \le d$, such that the disks $D_k = \{z : |z - z_k| \le d\}$, for $k = 0, 1, \ldots, n$, are contained in D and cover C as illustrated in Figure 6.34.

Each disk D_k contains the center z_{k+1} of the next disk D_{k+1}, so it follows that z_1 lies in D_0 and, from Equation (6.56) $|f(z)|$ also reaches its maximum value at z_1. An identical argument to the one given above will show that

$$f(z) = f(z_1) = f(z_0), \quad \text{for all} \quad z \in D_1. \tag{6.57}$$

We proceed inductively to get

$$f(z) = f(z_{k+1}) = f(z_k), \quad \text{for all} \quad z \in D_{k+1},\ 0 \le k < n - 1,$$

from which it follows that $f(\zeta) = f(z_0)$. Therefore f is constant in D, which contradicts the assumption of our theorem. With this contraposition, the proof is complete. $\qquad \square$

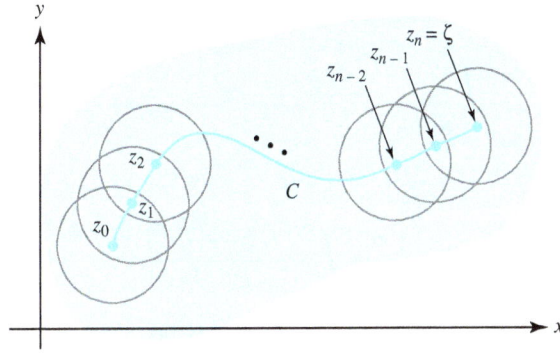

Figure 6.34: The "chain of disks" $D_0, D_1, \ldots, D_n$ that cover C

We sometimes state the maximum modulus principle in the following form.

Theorem 6.16 (Maximum modulus principle). *Let f be analytic and nonconstant in the bounded domain D. If f is continuous on the closed region R that consists of D and all its boundary points B, then $|f(z)|$ assumes its maximum value, and does so only at point(s) z_0 on the boundary B.*

Example 6.26. Let $f(z) = az + b$. If we set our domain D to be $D_1(0)$, then f is continuous on the closed region $\overline{D}_1(0) = \{z : |z| \leq 1\}$. Prove that

$$\max_{|z| \leq 1} |f(z)| = |a| + |b|,$$

and that this value is assumed by f at a point $z_0 = e^{i\theta_0}$ on the boundary of $D_1(0)$.

***Solution*:**
From the triangle inequality and the fact that $|z| \leq 1$ in $\overline{D}_1(0)$, it follows that

$$|f(z)| = |az + b| \leq |az| + |b| \leq |a| + |b|. \tag{6.58}$$

If we choose $z_0 = e^{i\theta_0}$, where $\theta_0 \in \arg b - \arg a$, then

$$
\begin{aligned}
\arg(az_0) - \arg a + \arg z_0 \\
= \arg a + (\arg b - \arg a) \\
= \arg b,
\end{aligned}
$$

so the vectors az_0 and b lie on the same ray through the origin. This is the requirement for the Inequality (6.58) to be an equality (see Exercise 21, Section 1.3). Hence $|az_0 + b| = |az_0| + |b| = |a| + |b|$, and the result is established.

Theorem 6.17 (Cauchy's inequalities). *Let f be analytic in the simply connected domain D that contains the circle $C_R(z_0) = \{z : |z - z_0| = R\}$. If $|f(z)| \leq M$ holds for all points $z \in C_R(z_0)$, then*

$$\left| f^{(n)}(z_0) \right| \leq \frac{n!M}{R^n}, \quad \text{for} \quad n = 1, 2, \ldots.$$

Proof. Let $C_R(z_0)$ have the parametrization

$$C_R(z_0) : z(\theta) = z_0 + Re^{i\theta} \quad \text{and} \quad dz = iRe^{i\theta}\,d\theta, \quad \text{for} \quad 0 \leq \theta \leq 2\pi.$$

We use Cauchy's integral formula and write

$$f^{(n)}(z_0) = \frac{n!}{2\pi i} \int_{C_R(z_0)} \frac{f(z)}{(z - z_0)^{n+1}} \, dz = \frac{n!}{2\pi i} \int_0^{2\pi} \frac{f(z_0 + Re^{i\theta})iRe^{i\theta}}{R^{n+1}e^{i(n+1)\theta}} \, d\theta.$$

Combining this result with the *ML* inequality (Theorem 6.3), we obtain

$$\left| f^{(n)}(z_0) \right| = \left| \frac{n!}{2\pi i} \int_0^{2\pi} \frac{f(z_0 + Re^{i\theta})iRe^{i\theta}}{R^{n+1}e^{i(n+1)\theta}} \, d\theta \right|$$

$$\leq \frac{n!}{2\pi} \int_0^{2\pi} \left| f(z_0 + Re^{i\theta}) \right| \left| \frac{iRe^{i\theta}}{R^{n+1}e^{i(n+1)\theta}} \right| \, d\theta$$

$$\leq \frac{n!}{2\pi} \int_0^{2\pi} M \frac{1}{R^n} \, d\theta$$

$$= \frac{n!}{2\pi R^n} M 2\pi$$

$$= \frac{n! M}{R^n}.$$

$\square$

Theorem 6.18 shows that a nonconstant entire function cannot be a bounded function.

Theorem 6.18 (Liouville's theorem). *If f is an entire function and is bounded for all values of z in the complex plane, then f is constant.*

Proof. Suppose that $|f(z)| \leq M$ holds for all values of z, and let z_0 denote an arbitrary point. Using the circle $C_R(z_0) = \{z : |z - z_0| = R\}$ and Cauchy's Inequality with $n = 1$ yields

$$\left| f'(z_0) \right| \leq \frac{M}{R}.$$

Because R can be arbitrarily large we must have $f'(z_0) = 0$. But z_0 was arbitrary, so $f'(z) = 0$ for all z. If the derivative of an analytic function is zero for all z, then by Theorem 3.7 the function must be constant. Therefore, f is constant. $\square$

Example 6.27. Show that the function $f(z) = \sin z$ is *not* a bounded function.

Solution:

We established this characteristic with a somewhat tedious argument in Section 5.4. All we need do now is observe that f is not constant, and hence it is not bounded.

We can use Liouville's theorem to establish an important theorem of elementary algebra.

Theorem 6.19 (The fundamental theorem of algebra). *If P is a polynomial of degree $n \geq 1$, then P has at least one zero.*

Proof. (By contraposition): We will show that if $P(z) \neq 0$ for all z, then the degree of P must be zero. Suppose that $P(z) \neq 0$ for all z. This supposition implies that the function $f(z) = \frac{1}{P(z)}$ is an entire function. Our strategy for the rest of the proof is as follows: We will show that f is bounded. Then Liouville's theorem will imply that f is constant, and since $f = \frac{1}{P}$, this conclusion will imply that the polynomial P is constant, which will mean that its degree must be zero.

First we write $P(z) = a_n z^n + a_{n-1} z^{n-1} + \cdots + a_1 z + a_0$ and consider the equation

$$|f(z)| = \frac{1}{|P(z)|} = \frac{1}{|z|^n} \frac{1}{\left| a_n + \frac{a_{n-1}}{z} + \frac{a_{n-2}}{z^2} + \cdots + \frac{a_1}{z^{n-1}} + \frac{a_0}{z^n} \right|}. \qquad (6.59)$$

For $k = 1, \ldots n$, $\frac{|a_{n-k}|}{|z^k|} \to 0$ as $|z| \to \infty$, so

$$a_n + \frac{a_{n-1}}{z} + \frac{a_{n-2}}{z^2} + \cdots + \frac{a_0}{z^n} \to a_n, \quad \text{as} \quad |z| \to \infty.$$

Combining this result with Equation (6.59) gives

$$|f(z)| \to 0, \quad \text{as} \quad |z| \to \infty.$$

In particular, we can find a value of R such that

$$|f(z)| \leq 1 \quad \text{for all} \quad |z| \geq R. \qquad (6.60)$$

If $f(z) = u(x, y) + iv(x, y)$, we have

$$|f(z)| = \left([u(x, y)]^2 + [v(x, y)]^2 \right)^{\frac{1}{2}}.$$

which is a continuous function of the two real variables x and y. A result from calculus regarding real functions says that a continuous function on a closed and bounded set is bounded. Hence $|f(z)|$ is a bounded function on the closed disk $\overline{D}_R(0)$. Thus there exists a positive real number K such that

$$|f(z)| \leq K, \quad \text{for all} \quad |z| \leq R.$$

Combining this with Inequality (6.60) gives

$$|f(z)| \leq M \quad \text{for all } z,$$

where $M = \max\{K, 1\}$. By Liouville's theorem, f is constant, so that the degree of f is zero. This observation completes the argument. $\qquad \square$

Corollary 6.4. *Let P be a polynomial of degree $n > 1$. Then P can be expressed as the product of linear factors. That is,*

$$P(z) = A(z - z_1)(z - z_2) \cdots (z - z_n)$$

where $z_1, z_2, \ldots, z_n$ are the zeros of P, counted according to multiplicity, and A is a constant.

Exercises for Section 6.6 (Selected answers or hints are on page 445.)

1. Factor each polynomial as a product of linear factors.

   (a) $P(z) = z^4 + 4$.
   (b) $P(z) = z^2 + (1 + i)z + 5i$.
   (c) $P(z) = z^4 - 4z^3 + 6z^2 - 4z + 5$.
   (d) $P(z) = z^3 - (3 + 3i)z^2 + (-1 + 6i)z + 3 - i$. *Hint*: Show that $P(i) = 0$.

2. Let $f(z) = az^n + b$, where the region is the disk $R = \{z : |z| \leq 1\}$. Show that $\max_{|z| \leq 1} |f(z)| = |a| + |b|$.

3. Show that $\cos z$ is *not* a bounded function.

4. Let $f(z) = z^2$. Evaluate the following, where R represents the rectangular region defined by the set
$$R = \{z = x + iy : 2 \leq x \leq 3 \quad \text{and} \quad 1 \leq y \leq 3\}.$$

   (a) $\max_{z \in R} |f(z)|$.

   (b) $\min_{z \in R} |f(z)|$.

   (c) $\max_{z \in R} \text{Re}[f(z)]$.

   (d) $\min_{z \in R} \text{Im}[f(z)]$.

5. Let f be analytic in the disk $D_5(0)$ and suppose that $|f(z)| \leq 10$ for $z \in C_3(1)$.

   (a) Find a bound for $|f^{(4)}(1)|$.

   (b) Find a bound for $|f^{(4)}(0)|$.
   Hint: $\overline{D}_2(0) \subseteq \overline{D}_3(1)$. Use Theorems 6.16 and 6.17.

6. Let f be an entire function such that $|f(z)| \leq M|z|$ for all z.

   (a) Show that, for $n \geq 2$, $f^{(n)}(z) = 0$ for all z.

   (b) Use part (a) to show that $f(z) = az + b$.

7. Establish the following *minimum modulus principle*.

   (a) Let f be analytic and nonconstant in the domain D, and continuous on the closed region R that consists of D and all its boundary points B. Show that, if $f(z) \neq 0$ throughout R, then $|f(z)|$ assumes its *minimum* value, but does so only at point(s) z_0 on the boundary B.

   (b) Show that the requirement $f(z) \neq 0$ in part (a) is necessary by finding a function for which the requirement fails, and whose minimum modulus is attained at some place other than the boundary.

8. Let $u(x, y)$ be harmonic for all (x, y). Show that
$$u(x_0, y_0) = \frac{1}{2\pi} \int_0^{2\pi} u(x_0 + R\cos\theta, \ y_0 + R\sin\theta) \, d\theta,$$
where $R > 0$. *Hint:* Let $f(z) = u(x, y) + iv(x, y)$, where v is a harmonic conjugate of u.

9. Establish the following maximum principle for harmonic functions: Let $u(x, y)$ be harmonic and nonconstant in the simply connected domain D. Then u does not have a maximum value at any point (x_0, y_0) in D.

10. Let f be an entire function with the property that $|f(z)| \geq 1$ for all z. Show that f is constant.

11. Let f be nonconstant and analytic in the closed disk $\overline{D}_1(0)$. Suppose that $|f(z)|$ is constant for $z \in C_1(0)$, *i.e.*, that there is some number K such that $|f(z)| = K$ for all $z \in C_1(0)$. Show that f has a zero in $\overline{D}_1(0)$, *i.e.*, that there exists some $z_0 \in \overline{D}_1(0)$ such that $f(z_0) = 0$. *Hint*: use both the minimum modulus principle (see Exercise 7) and maximum modulus principle .

12. Why is it important to study the fundamental theorem of algebra in a complex analysis course?

Chapter 7

Taylor and Laurent Series

Overview

Throughout this book we have compared and contrasted properties of complex functions with functions whose domain and range lie entirely within the real numbers. There are many similarities, such as the standard differentiation formulas. However, there are also some surprises, and in this chapter you will encounter one of the hallmarks that distinguishes complex functions from their real counterparts: It is possible for a function defined on the real numbers to be differentiable everywhere and yet not be expressible as a power series (see Exercise 20 of Section 7.2). For a complex function, however, things are much simpler! You will soon learn that if a complex function is analytic in the disk $D_r(\alpha)$, its Taylor series about α converges to the function at every point in this disk. Thus, analytic functions are locally nothing more than glorified polynomials.

7.1   Uniform Convergence

Complex functions are the key to unlocking many of the mysteries encountered when power series are first introduced in a calculus course. We begin by discussing an important property associated with power series–uniform convergence.

Recall that, for a function f defined on a set T, the sequence o f functions $\{S_n\}$ converges to f at the point $z_0 \in T$, provided $\lim_{n\to\infty} S_n(z_0) = f(z_0)$. Thus, for the particular point z_0, we know that for each $\varepsilon > 0$, there exists a positive integer N_{ε,z_0} (depending on both ε and z_0) such that

$$\text{if} \quad n \geq N_{\varepsilon,z_0}, \quad \text{then} \quad |S_n(z_0) - f(z_0)| < \varepsilon. \tag{7.1}$$

If $S_n(z)$ is the nth partial sum of the series $\sum_{k=0}^{\infty} c_k(z-\alpha)^k$, Statement (7.1) becomes

$$If \quad n \geq N_{\varepsilon,z_0}, \quad \text{then} \quad \left| \sum_{k=0}^{n-1} c_k(z_0 - \alpha)^k - f(z_0) \right| < \varepsilon.$$

For a given value of ε, the integer N_{ε,z_0} needed to satisfy Statement (7.1) often depends on our choice of z_0. This is not the case if the sequence $\{S_n\}$ converges uniformly. For a uniformly convergent sequence, it is possible to find an integer N_ε (depending *only* on ε) that guarantees Statement (7.1) no matter what value for $z_0 \in T$ we pick. In other words, if n is large enough, the function S_n is *uniformly close* to the function f for all $z \in T$. Formally, we have the following definition.

Definition 7.1 (Uniform convergence). *The sequence $\{S_n(z)\}$ **converges uniformly** to $f(z)$ on the set T if for every $\varepsilon > 0$, there exists a positive integer N_ε (depending only on ε) such that*

$$\text{if} \quad n \geq N_\varepsilon, \quad \text{then} \quad |S_n(z) - f(z)| < \varepsilon \quad \text{for all} \quad z \in T. \tag{7.2}$$

*If $S_n(z)$ is the nth partial sum of the series $\sum\limits_{k=0}^{\infty} c_k(z-\alpha)^k$, we say that the series $\sum\limits_{k=0}^{\infty} c_k(z-\alpha)^k$ **converges uniformly** to $f(z)$ on the set T.*

Example 7.1. The sequence $\{S_n(z)\} = \{e^z + \frac{1}{n}\}$ converges uniformly to the function $f(z) = e^z$ on the entire complex plane because for any $\varepsilon > 0$, Statement (7.2) is satisfied for all z for $n \geq N_\varepsilon$, where N_ε is any integer greater than $\frac{1}{\varepsilon}$. We leave the details of showing this result as an exercise.

A good example of a sequence of functions that does not converge uniformly is the sequence of partial sums forming the geometric series. Recall that the geometric series has $S_n(z) = \sum\limits_{k=0}^{n-1} z^k$ converging to $f(z) = \frac{1}{1-z}$ for $z \in D_1(0)$. Because the real numbers are a subset of the complex numbers, we can show that Statement (7.2) is not satisfied by demonstrating that it does not hold when we restrict our attention to the real numbers. In that context, $D_1(0)$ becomes the open interval $(-1, 1)$, and the inequality $|S_n(z) - f(z)| < \varepsilon$ becomes $|S_n(x) - f(x)| < \varepsilon$, which for real variables is equivalent to the inequality $f(x) - \varepsilon < S_n(x) < f(x) + \varepsilon$. If Statement (7.2) were to be satisfied, then given $\varepsilon > 0$, $S_n(x)$ would be within an ε-bandwidth of $f(x)$ *for all x* in the interval $(-1, 1)$ provided n were large enough. Figure 7.1 illustrates that there is an ε such that, no matter how large n is, we can find $x_0 \in (-1, 1)$ with the property that $S_n(x_0)$ lies outside this bandwidth. In other words, Figure 7.1 illustrates the negation of Statement (7.2), which in technical terms we state as:

> There exists $\varepsilon > 0$ such that, for all positive integers N,
> there is some $n \geq N$ and some $z_0 \in T$
> such that $|S_n(z_0) - f(z_0)| \geq \varepsilon$. $\hspace{2cm}$ (7.3)

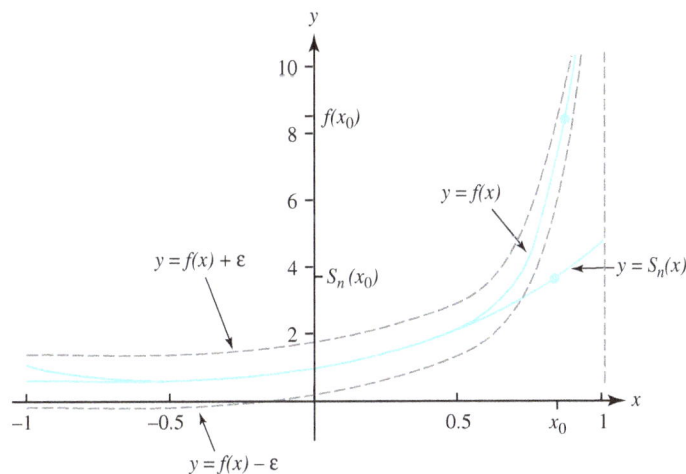

Figure 7.1: The geometric series does not converge uniformly on $(-1, 1)$

In the exercises, we ask you to use Statement (7.3) to show that the partial sums of the geometric series do not converge uniformly to $f(z) = \frac{1}{1-z}$ for $z \in D_1(0)$.

A useful procedure known as the Weierstrass M-test can help determine whether an infinite series is uniformly convergent.

Theorem 7.1 (Weierstrass M-test). *Suppose that the infinite series $\sum_{k=0}^{\infty} u_k(z)$ has the property that for each k, $|u_k(z)| \leq M_k$ for all $z \in T$. If $\sum_{k=0}^{\infty} M_k$ converges, then $\sum_{k=0}^{\infty} u_k(z)$ converges uniformly on T.*

Proof. Let $S_n(z) = \sum_{k=0}^{n-1} u_k(z)$ be the nth partial sum of the series. Note that, If $n > M$, then

$$\left| S_n(z) - S_m(z) \right| = \left| u_m(z) + u_{m+1}(z) + \cdots + u_{n-1}(z) \right| \leq \sum_{k=m}^{n-1} M_k.$$

Because the series $\sum_{k=0}^{\infty} M_k$ converges, we can make the last expression as small as we want to by choosing a large enough m. Thus, for $\varepsilon > 0$, there is a positive integer N_ε such that if n, $m > N_\varepsilon$, then $|S_n(z) - S_m(z)| < \varepsilon$. But this means that for all $z \in T$, $\{S_n(z)$ is a Cauchy sequence. According to Theorem 4.2, this sequence must converge to a number, which we might as well designate by $f(z)$. That is, $f(z) = \lim_{n \to \infty} S_n(z) = \sum_{k=0}^{\infty} u_k(z)$. This observation gives us a function to which the series $\sum_{k=0}^{\infty} u_k(z)$ converges. However, we still must show that the convergence is uniform. Let $\varepsilon > 0$ be given. Again, since $\sum_{k=0}^{\infty} M_k$ converges, there exists N_ε such that if $n \geq N_\varepsilon$, then $\sum_{k=n}^{\infty} M_k < \varepsilon$. Thus, if $n \geq N_\varepsilon$ and $z \in T$, then

$$\left| f(z) - S_n(z) \right| = \left| \sum_{k=0}^{\infty} u_k(z) - \sum_{k=0}^{n-1} u_k(z) \right|$$

$$= \left| \sum_{k=n}^{\infty} u_k(z) \right|$$

$$\leq \sum_{k=n}^{\infty} M_k$$

$$< \varepsilon,$$

which completes the argument. $\square$

Theorem 7.2 gives an interesting application of the Weierstrass M-test.

Theorem 7.2. *Suppose that the power series $\sum_{k=0}^{\infty} c_k(z - \alpha)^k$ has radius of convergence $\rho > 0$. Then for each r, $0 < r < \rho$, the series converges uniformly on the closed disk $\overline{D}_r(\alpha)$, where (by way of reminder) $\overline{D}_r(\alpha) = \{z : |z - \alpha| \leq r\}$.*

Proof. Given r, with $0 < r < \rho$, choose $z_0 \in D_\rho(\alpha)$ such that $|z_0 - \alpha| = r$. The proof of Theorem 4.15 part (ii) reveals that $\sum_{k=0}^{\infty} c_k(z - \alpha)^k$ converges absolutely for $z \in D_\rho(\alpha)$, from which it follows that $\sum\limits_{k=0}^{\infty} |c_k(z_0 - \alpha)^k| = \sum\limits_{k=0}^{\infty} |c_k| r^k$ converges. Moreover, for all $z \in \overline{D}_r(\alpha)$,

$$|c_k(z - \alpha)^k| = |c_k||z - \alpha|^k \le |c_k| r^k.$$

The conclusion follows from the Weierstrass M-test with $M_k = |c_k| r^k$. $\qquad\square$

An immediate consequence of Theorem 7.2 is Corollary 7.1.

Corollary 7.1. *For each r, $0 < r < 1$, the geometric series converges uniformly on the closed disk $\overline{D}_r(0)$.*

The following theorem gives important properties of uniformly convergent sequences.

Theorem 7.3. *Suppose that $\{S_k\}$ is a sequence of continuous functions defined on a set T containing the contour C. If $\{S_k\}$ converges uniformly to f on the set T, then*

    i. f is continuous on T, and

    ii. $\lim\limits_{k \to \infty} \int_C S_k(z)\,dz = \int_C \lim\limits_{k \to \infty} S_k(z)\,dz = \int_C f(z)\,dz.$

Proof. Given $z_0 \in T$, we must prove $\lim\limits_{z \to z_0} f(z) = f(z_0)$. Let $\varepsilon > 0$ be given. Since $\{S_k\}$ converges uniformly to f on T, there exists a positive integer N_ε such that for all $z \in T$, $|f(z) - S_k(z)| < \frac{\varepsilon}{3}$ whenever $k \ge N_\varepsilon$. And, as S_{N_ε} is continuous at z_0, there exists $\delta > 0$ such that if $|z - z_0| < \delta$, then $|S_{N_\varepsilon}(z) - S_{N_\varepsilon}(z_0)| < \frac{\varepsilon}{3}$. Hence, if $|z - z_0| < \delta$, we have

$$
\begin{aligned}
|f(z) - f(z_0)| &= \left| f(z) - S_{N_\varepsilon}(z) + S_{N_\varepsilon}(z) - S_{N_\varepsilon}(z_0) + S_{N_\varepsilon}(z_0) - f(z_0) \right| \\
&\le \left| f(z) - S_{N_\varepsilon}(z) \right| + \left| S_{N_\varepsilon}(z) - S_{N_\varepsilon}(z_0) \right| + \left| S_{N_\varepsilon}(z_0) - f(z_0) \right| \\
&< \frac{\varepsilon}{3} + \frac{\varepsilon}{3} + \frac{\varepsilon}{3} \\
&= \varepsilon,
\end{aligned}
$$

which completes part (i).

To prove part (ii), let $\varepsilon > 0$ be given and let L be the length of the contour C. Because $\{S_k\}$ converges uniformly to f on T, there exists a positive integer N_ε such that, if $k \ge N_\varepsilon$, then $|S_k(z) - f(z)| < \frac{\varepsilon}{L}$ for all $z \subset T$. Because C is contained in T, $\max\limits_{z \in C} |S_k(z) - f(z)| < \frac{\varepsilon}{L}$ if $k \ge N_\varepsilon$, and we can use the ML inequality (Theorem 6.3) to get

$$
\begin{aligned}
\left| \int_C S_k(z)\,dz - \int_C f(z)\,dz \right| &= \left| \int_C [S_k(z) - f(z)]\,dz \right| \\
&\le \max_{z \in C} |S_k(z) - f(z)| L \\
&< \left(\frac{\varepsilon}{L} \right) L \\
&= \varepsilon.
\end{aligned}
$$

$\qquad\square$

Corollary 7.2. *If the series $\sum\limits_{n=0}^{\infty} c_n(z - \alpha)^n$ converges uniformly to $f(z)$ on the set T and C is a contour contained in T, then*

$$\sum_{n=0}^{\infty} \int_C c_n(z - \alpha)^n\,dz = \int_C \sum_{n=0}^{\infty} c_n(z - \alpha)^n\,dz = \int_C f(z)\,dz.$$

213

Example 7.2. Show that $\text{Log}(1-z) = \sum_{n=1}^{\infty} \frac{1}{n} z^n$, for all $z \in D_1(0)$.

Solution:

For $z_0 \in D_1(0)$, we choose r and R so that $0 \le |z_0| < r < R < 1$, thus ensuring that $z_0 \in \overline{D}_r(0)$ and that $\overline{D}_r(0) \subset D_R(0)$. By Corollary 7.1, the geometric series $\sum_{n=0}^{\infty} z^n$ converges uniformly to $\frac{1}{1-z}$ on $\overline{D}_r(0)$. If C is any contour contained in $\overline{D}_r(0)$, Corollary 7.2 gives

$$\int_C \frac{1}{1-z}\, dz = \sum_{n=0}^{\infty} \int_C z^n\, dz. \tag{7.4}$$

Clearly, the function $f(z) = \frac{1}{1-z}$ is analytic in the simply connected domain $D_R(0)$, and $F(z) = -\text{Log}(1-z)$ is an antiderivative of $f(z)$ for all $z \in D_R(0)$, where Log is the principal branch of the logarithm. Likewise, $g(z) = z^n$ is analytic in the simply connected domain $D_R(0)$, and $G(z) = \frac{1}{n+1} z^{n+1}$ is an antiderivative of $g(z)$ for all $z \in D_R(0)$. Hence, if C is the straight-line segment joining 0 to z_0, we can apply Theorem 6.9 to Equation (7.4) to get

$$-\text{Log}(1-z)\Big|_{z=0}^{z=z_0} = \sum_{n=0}^{\infty} \left(\frac{1}{n+1} z^{n+1} \right)\Big|_{z=0}^{z=z_0},$$

which becomes

$$-\text{Log}(1-z_0) = \sum_{n=0}^{\infty} \frac{1}{n+1} z_0^{n+1} = \sum_{n=1}^{\infty} \frac{1}{n} z_0^n.$$

The point $z_0 \in D_1(0)$ was arbitrary, so the solution is complete.

Exercises for Section 7.1 (Selected answers or hints are on page 445.)

1. This exercise relates to Figure 7.1.

   (a) For x near -1, is the graph of $S_n(x)$ above or below $f(x)$? Explain.

   (b) Is the index n in $S_n(x)$ odd or even? Explain.

   (c) Assuming that the graph is accurate to scale, what is the value of n in $S_n(x)$? Explain.

2. Complete the details to verify the claim of Example. 7.1.

3. Prove that the following series converge uniformly on the sets indicated.

   (a) $\sum_{k=1}^{\infty} \frac{1}{k^2} z^k$ on $\overline{D}_1(0) = \{z : |z| \le 1\}$

   (b) $\sum_{k=0}^{\infty} \frac{1}{(z^2-1)^k}$ on $\{z : |z| \ge 2\}$

   (c) $\sum_{k=0}^{\infty} \frac{z^k}{z^{2k}+1}$ on $\overline{D}_r(0)$, where $0 < r < 1$.

4. Show that $S_n(z) = \sum_{k=0}^{n-1} z^k = \frac{1-z^n}{1-z}$ does not converge uniformly to $f(z) = \frac{1}{1-z}$ on the set $T = D_1(0)$ by appealing to Statement (7.3). *Hint*: Given $\varepsilon > 0$ and a positive integer n, let $z_n = \varepsilon^{\frac{1}{n}}$.

5. Why can't we use the arguments of Theorem 7.2 to prove that the geometric series converges uniformly on *all* of $D_1(0)$?

6. By starting with the series for the complex cosine given in Section 5.4, choose an appropriate contour and use the method in Example 7.2 to obtain the series for the complex sine.

7. Suppose that the sequences of functions $\{f_n\}$ and $\{g_n\}$ converge uniformly on the set T.

   (a) Show that the sequence $\{f_n + g_n\}$ converges uniformly on T.

   (b) Show by example that it is not necessarily the case that $\{f_n\, g_n\}$ converges uniformly on T.

8. On what portion of $D_1(0)$ does the sequence $\{nz^n\}_{n=1}^{\infty}$ converge, and on what portion does it converge uniformly?

9. Consider the function $\zeta(z) = \sum\limits_{n=1}^{\infty} n^{-z}$, where $n^{-z} = \exp(-z \ln n)$.

   (a) Show that $\zeta(z)$ converges uniformly on $A = \{z : \text{Re}(z) \geq 2\}$.

   (b) Let D be a closed disk contained in $\{z : \text{Re}(z) > 1\}$. Show that $\zeta(z)$ converges uniformly on D.

7.2 Taylor Series Representations

In Section 4.4 we showed that functions defined by power series have derivatives of all orders (Theorem 4.17). In Section 6.5 we demonstrated that analytic functions also have derivatives of all orders (Corollary 6.2). It seems natural, therefore, that there would be some connection between analytic functions and power series. As you might guess, the connection exists via the Taylor and Maclaurin series of analytic functions.

Definition 7.2 (Taylor series). *If $f(z)$ is analytic at $z = \alpha$, then the series*

$$\sum_{k=0}^{\infty} \frac{f^{(k)}(\alpha)}{k!}(z - \alpha)^k = f(\alpha) + f'(\alpha)(z - \alpha) + \frac{f^{(2)}(\alpha)}{2!}(z - \alpha)^2$$

$$+ \frac{f^{(3)}(\alpha)}{3!}(z - \alpha)^3 + \cdots$$

*is called the **Taylor series for f centered at α**. When the center is $\alpha = 0$, the series is called the **Maclaurin series for f**.*

To investigate when these series converge we need the following lemma.

Lemma 7.1. *If z, z_0, and α are complex numbers with $z \neq z_0$ and $z \neq \alpha$, then*

$$\frac{1}{z - z_0} = \frac{1}{z - \alpha} + \frac{z_0 - \alpha}{(z - \alpha)^2} + \frac{(z_0 - \alpha)^2}{(z - \alpha)^3} + \cdots$$

$$\cdots + \frac{(z_0 - \alpha)^n}{(z - \alpha)^{n+1}} + \frac{1}{z - z_0} \frac{(z_0 - \alpha)^{n+1}}{(z - \alpha)^{n+1}},$$

where n is a positive integer.

Proof. $\dfrac{1}{z-z_0} = \dfrac{1}{(z-\alpha)-(z_0-\alpha)} = \dfrac{1}{z-\alpha}\left(\dfrac{1}{1-\frac{z_0-\alpha}{z-\alpha}}\right)$. The result now follows from Corollary 4.3 if in it we replace z with $\left(\frac{z_0-\alpha}{z-\alpha}\right)$. We leave verification of the details as an exercise. $\square$

We are now ready for the main result of this section.

Theorem 7.4 (Taylor's theorem). *Suppose that f is analytic in a domain G and that $D_R(\alpha)$ is any disk contained in G. Then the Taylor series for f converges to $f(z)$ for all z in $D_R(\alpha)$; that is,*

$$f(z) = \sum_{k=0}^{\infty} \frac{f^{(k)}(\alpha)}{k!}(z-\alpha)^k \quad \text{for all} \quad z \in D_R(\alpha). \tag{7.5}$$

Furthermore, for any r, $0 < r < R$, the convergence is uniform on the closed subdisk $\overline{D}_r(\alpha) = \{z : |z-\alpha| \le r\}$.

Proof. If we can establish Equation (7.5), the uniform convergence on $\overline{D}_r(\alpha)$ for $0 < r < R$ will follow immediately from Theorem 7.2 by equating the c_k of that theorem with $\frac{f^{(k)}(\alpha)}{k!}$.

Let $z_0 \in D_R(\alpha)$ and let r designate the distance between z_0 and α so that $|z_0 - \alpha| = r$. Note that $0 \le r < R$ because z_0 belongs to the *open* disk $D_R(\alpha)$. We choose ρ such that $0 \le r < \rho < R$, and let $C = C_\rho^+(\alpha)$ be the positively oriented circle centered at α with radius ρ as shown in Figure 7.2.

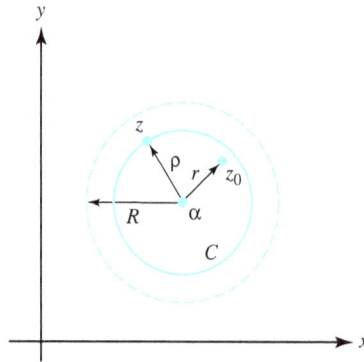

Figure 7.2: The constructions for Taylor's theorem

With C contained in G, we can use the Cauchy integral formula to get

$$f(z_0) = \frac{1}{2\pi i} \int_C \frac{1}{z-z_0} f(z)\, dz.$$

Replacing $\frac{1}{z-z_0}$ in the integrand by its equivalent expression in Lemma 7.1 gives

216

$$f(z_0) = \frac{1}{2\pi i} \int_C \left[\frac{1}{z-\alpha} + \frac{z_0-\alpha}{(z-\alpha)^2} + \cdots + \frac{(z_0-\alpha)^n}{(z-\alpha)^{n+1}} \right.$$
$$\left. + \frac{1}{z-z_0} \frac{(z_0-\alpha)^{n+1}}{(z-\alpha)^{n+1}} \right] f(z)\, dz$$
$$= \frac{1}{2\pi i} \int_C \frac{f(z)}{z-\alpha}\, dz + \frac{z_0-\alpha}{2\pi i} \int_C \frac{f(z)}{(z-\alpha)^2}\, dz + \cdots$$
$$+ \frac{(z_0-\alpha)^n}{2\pi i} \int_C \frac{f(z)}{(z-\alpha)^{n+1}}\, dz$$
$$+ \frac{(z_0-\alpha)^{n+1}}{2\pi i} \int_C \frac{f(z)}{(z-z_0)(z-\alpha)^{n+1}}\, dz, \tag{7.6}$$

where n is a positive integer. We can put the last term in Equation (7.6) in the form

$$E_n(z_0) = \frac{1}{2\pi i} \int_C \frac{(z_0-\alpha)^{n+1} f(z)}{(z-z_0)(z-\alpha)^{n+1}}\, dz. \tag{7.7}$$

Recall also by the Cauchy integral formulas that

$$\frac{2\pi i}{k!} f^{(k)}(\alpha) = \int_C \frac{f(z)}{(z-\alpha)^{k+1}}\, dz, \quad \text{for} \quad k = 0,\, 1,\, 2,\, \ldots.$$

Using these last two identities reduces Equation (7.6) to

$$f(z_0) = \sum_{k=0}^{n} \frac{f^{(k)}(\alpha)}{k!} (z_0-\alpha)^k + E_n(z_0).$$

The summation on the right-hand side of this last expression is the first $n+1$ terms of the Taylor series. Verification of Equation (7.5) relies on our ability to show that we can make the remainder term, $E_n(z_0)$, as small as we please by making n sufficiently large. We will use the ML inequality (Theorem 6.3) to get a bound for $|E_n(z_0)|$. According to the constructions shown in Figure 7.2, we have

$$|z_0 - a| = r \quad \text{and} \quad |z-\alpha| = \rho. \tag{7.8}$$

By Property (1.24) of Section 1.3, we also have

$$|z-z_0| = |(z-\alpha) - (z_0-\alpha)|$$
$$\geq |z-\alpha| - |z_0-\alpha|$$
$$= \rho - r. \tag{7.9}$$

If we set $M = \max\limits_{z \in C} |f(z)|$, Equations (7.8) and (7.9) allow us to conclude that, for all $z \in C$,

$$\left| \frac{(z_0-\alpha)^{n+1} f(z)}{(z-z_0)(z-\alpha)^{n+1}} \right| = \left| \frac{(z_0-\alpha)^{n+1}}{(z-\alpha)^{n+1}} \right| \left| \frac{f(z)}{(z-z_0)} \right|$$
$$\leq \left(\frac{r}{\rho} \right)^{n+1} \left(\frac{1}{\rho-r} \right) M. \tag{7.10}$$

217

The length of the circle C is $2\pi\rho$, so the ML inequality in conjunction with Equations (7.7) and (7.10) gives

$$|E_n(z_0)| \le \frac{1}{2\pi} \left(\frac{r}{\rho}\right)^{n+1} \left(\frac{1}{\rho - r}\right) M(2\pi\rho). \tag{7.11}$$

Because $0 \le r < \rho < R$, the fraction $\frac{r}{\rho}$ is less than 1, so $(\frac{r}{\rho})^{n+1}$ (and hence the right side of Equation (7.11)) goes to zero as n goes to infinity. Thus, for any $\varepsilon > 0$, we can find an integer N_ε such that $|E_n(z_0)| < \varepsilon$ for $n \ge N_\varepsilon$, and this fact completes the proof. $\square$

A singular point of a function is a point at which the function fails to be analytic. You will see in Section 7.4 that singular points of a function can be classified according to how badly the function behaves at those points. Loosely speaking, a *nonremovable* singular point of a function has the property that it is impossible to redefine the value of the function at that point so as to make it analytic there. For example, the function $f(z) = \frac{1}{1-z}$ has a nonremovable singularity at $z = 1$. We give a formal definition of this concept in Section 7.4, but with this language we can nuance Taylor's theorem a bit.

Corollary 7.3. *Suppose that f is analytic in the domain G that contains the point α. Let z_0 be a nonremovable singular point of minimum distance to the point α. If $|z_0 - \alpha| = R$, then*

i. the Taylor series $\sum\limits_{k=0}^{\infty} \frac{f^{(k)}(\alpha)}{k!}(z - \alpha)^k$ converges to $f(z)$ on all of $D_R(\alpha)$, and

ii. if $|z_1 - \alpha| = S > R$, the Taylor series $\sum\limits_{k=0}^{\infty} \frac{f^{(k)}(\alpha)}{k!}(z_1 - \alpha)^k$ does not converge to $f(z_1)$.

Proof. Taylor's theorem gives us part (i) immediately. To establish part (ii), we note that if $|z_0 - \alpha| = R$, then $z_0 \in D_S(\alpha)$ whenever $S > R$. If for some z_1, with $|z_1 - \alpha| = S > R$, the Taylor series converged to $f(z_1)$, then according to Theorem 4.17, the radius of convergence of the series $\sum\limits_{k=0}^{\infty} \frac{f^{(k)}(\alpha)}{k!}(z - \alpha)^k$ would be at least equal to S. We could then make f differentiable at z_0 by redefining $f(z_0)$ to equal the value of the series at z_0, thus contradicting the fact that z_0 is a nonremovable singular point. $\square$

Example 7.3. Show that $\frac{1}{(1-z)^2} = \sum\limits_{n=0}^{\infty} (n+1)z^n$ is valid for $z \in D_1(0)$.

Solution:

In Example 4.24 we established this identity with the use of Theorem 4.17. We now do so via Theorem 7.4. If $f(z) = \frac{1}{(1-z)^2}$, then a standard induction argument (which we leave as an exercise) will show that $f^{(n)}(z) = \frac{(n+1)!}{(1-z)^{n+2}}$ for $z \in D_1(0)$. Thus $f^{(n)}(0) = (n+1)!$, and Taylor's theorem gives

$$f(z) = \frac{1}{(1-z)^2} = \sum_{n=0}^{\infty} \frac{f^{(n)}(0)}{n!} z^n = \sum_{n=0}^{\infty} \frac{(n+1)!}{n!} z^n = \sum_{n=0}^{\infty} (n+1)z^n,$$

and since f is analytic in $D_1(0)$, this series expansion is valid for all $z \in D_1(0)$.

Example 7.4. Show that, for $z \in D_1(0)$,

$$\frac{1}{1-z^2} = \sum_{n=0}^{\infty} z^{2n} \quad \text{and} \quad \frac{1}{1+z^2} = \sum_{n=0}^{\infty} (-1)^n z^{2n}. \tag{7.12}$$

Solution:

For $z \in D_1(0)$,

$$\frac{1}{1-z} = \sum_{n=0}^{\infty} z^n. \tag{7.13}$$

If we let z^2 take the role of z in Equation (7.13), we get that $\frac{1}{1-z^2} = \sum_{n=0}^{\infty} (z^2)^n = \sum_{n=0}^{\infty} z^{2n}$ for $z^2 \in D_1(0)$. But $z^2 \in D_1(0)$ iff $z \in D_1(0)$. Letting $-z^2$ take the role of z in Equation (7.13) gives the second part of Equations (7.12).

Remark 7.1. *Corollary 7.3 clears up what often seems to be a mystery when series are first introduced in calculus. The calculus analog of Equations (7.12) is*

$$\frac{1}{1-x^2} = \sum_{n=0}^{\infty} x^{2n} \quad and \quad \frac{1}{1+x^2} = \sum_{n=0}^{\infty} (-1)^n x^{2n} \quad for \quad x \in (-1, 1). \tag{7.14}$$

For many students, it makes sense that the first series in Equations (7.14) converges only on the interval $(-1, 1)$ because $\frac{1}{1-x^2}$ is undefined at the points $x = \pm 1$. It seems unclear as to why this should also be the case for the series representing $\frac{1}{1+x^2}$, since the real-valued function $f(x) = \frac{1}{1+x^2}$ is defined everywhere. The explanation, of course, comes from the complex domain. The complex function $f(z) = \frac{1}{1+z^2}$ is not defined everywhere. In fact, the singularities of f are at the points $\pm i$, and the distance between them and the point $\alpha = 0$ equals 1. According to Corollary 7.3, therefore, Equations (7.12) are valid only for $z \in D_1(0)$, so Equations (7.14) are valid only for $x \in (-1, 1)$.

Alas, there is a potential fly in this ointment: Corollary 7.3 applies to Taylor series. To form the Taylor series of a function, we must compute its derivatives. We didn't get the series in Equations (7.12) by computing derivatives, so how do we know that they are indeed the Taylor series centered at $\alpha = 0$? Perhaps the Taylor series would give completely different expressions from those given by Equations (7.12). Fortunately, Theorem 7.5 removes this possibility.

Theorem 7.5 (Uniqueness of power series). *Suppose that in some disk $D_r(\alpha)$ we have*

$$f(z) = \sum_{n=0}^{\infty} a_n (z - \alpha)^n = \sum_{n=0}^{\infty} b_n (z - \alpha)^n.$$

Then $a_n = b_n$, for $n = 0, 1, 2, \ldots$.

Proof. By Theorem 4.17 part (*ii*), $a_n = \frac{f^{(n)}(\alpha)}{n!} = b_n$, for $n = 0, 1, 2, \ldots$. $\square$

Thus, any power series representation of $f(z)$ is automatically the Taylor series.

Example 7.5. Find the Maclaurin series of $f(z) = \sin^3 z$.

Solution:

Computing derivatives for $f(z)$ would be an onerous task. Fortunately, we can make use of the trigonometric identity

$$\sin^3 z = \frac{3}{4} \sin z - \frac{1}{4} \sin 3z.$$

219

Recall that the series for $\sin z$ (valid for all z) is $\sin z = \sum\limits_{n=0}^{\infty} (-1)^n \frac{z^{2n+1}}{(2n+1)!}$. Using the identity for $\sin^3 z$, we obtain

$$\sin^3 z = \frac{3}{4} \sum_{n=0}^{\infty} (-1)^n \frac{z^{2n+1}}{(2n+1)!} - \frac{1}{4} \sum_{n=0}^{\infty} (-1)^n \frac{(3z)^{2n+1}}{(2n+1)!}$$

$$= \frac{3}{4} \sum_{n=0}^{\infty} (-1)^n \frac{z^{2n+1}}{(2n+1)!} - \frac{3}{4} \sum_{n=0}^{\infty} (-1)^n \frac{9^n z^{2n+1}}{(2n+1)!}$$

$$= \sum_{n=0}^{\infty} (-1)^n \frac{3(1 - 9^n)}{4(2n+1)!} z^{2n+1}.$$

By the uniqueness of power series, this last expression is the Maclaurin series for $\sin^3 z$.

In the preceding argument we used some obvious results of power series representations that we haven't yet formally stated. The requisite results are part of Theorem 7.6.

Theorem 7.6. *Let f and g have the power series representations*

$$f(z) = \sum_{n=0}^{\infty} a_n (z - \alpha)^n, \quad \text{for} \quad z \in D_{r_1}(\alpha);$$

$$g(z) = \sum_{n=0}^{\infty} b_n (z - \alpha)^n, \quad \text{for} \quad z \in D_{r_2}(\alpha).$$

If $r = \min\{r_1, r_2\}$ and β is any complex constant, then

$$\beta f(z) = \sum_{n=0}^{\infty} \beta a_n (z - \alpha)^n, \quad \text{for} \quad z \in D_{r_1}(\alpha), \tag{7.15}$$

$$f(z) + g(z) = \sum_{n=0}^{\infty} (a_n + b_n)(z - \alpha)^n, \quad \text{for} \quad z \in D_r(\alpha), \quad \text{and} \tag{7.16}$$

$$f(z)g(z) = \sum_{n=0}^{\infty} c_n (z - \alpha)^n, \quad \text{for} \quad z \in D_r(\alpha), \tag{7.17}$$

where

$$c_n = \sum_{k=0}^{n} a_k b_{n-k}. \tag{7.18}$$

*Identity (7.17) is known as the **Cauchy product** of the series for f and g.*

Proof. We leave the details of establishing Equations (7.15) and (7.16) for you to do as an exercise. To establish Equation (7.17), we note that the function $h(z) = f(z)g(z)$ is analytic in $D_r(\alpha)$. Thus, for $z \in D_r(\alpha)$,

$$h'(z) = f(z)g'(z) + f'(z)g(z);$$
$$h''(z) = f''(z)g(z) + 2f'(z)g'(z) + f(z)g''(z).$$

By mathematical induction, we can generalize the preceding pattern to the n th derivative, giving Leibniz's formula for the derivative of a product of functions:

$$h^{(n)}(z) = \sum_{k=0}^{n} \frac{n!}{k!(n-k)!} f^{(k)}(z) g^{(n-k)}(z). \tag{7.19}$$

(We will ask you to show this result in an exercise.) By Theorem 4.17 we know that

$$\frac{f^{(k)}(\alpha)}{k!} = a_k, \quad \text{and} \quad \frac{g^{(n-k)}(\alpha)}{(n-k)!} = b_{n-k},$$

so Equation (7.19) becomes

$$\frac{h^{(n)}(\alpha)}{n!} = \sum_{k=0}^{n} \frac{f^{(k)}(\alpha)}{k!} \frac{g^{(n-k)}(\alpha)}{(n-k)!} = \sum_{k=0}^{n} a_k b_{n-k}. \tag{7.20}$$

Now, according to Taylor's theorem,

$$h(z) = \sum_{k=0}^{\infty} \frac{h^{(n)}(\alpha)}{n!} (z-\alpha)^n.$$

Substituting Equation (7.20) into this equation gives Equation (7.17) because of the uniqueness of power series. $\quad\square$

Example 7.6. Use the Cauchy product of series to show that

$$\frac{1}{(1-z)^2} = \sum_{n=0}^{\infty} (n+1)z^n, \quad \text{for} \quad z \in D_1(0).$$

Solution:

We let $f(z) = g(z) = \frac{1}{1-z} = \sum_{n=0}^{\infty} z^n$, for $z \in D_1(0)$. In terms of Theorem 7.6, we have $a_n = b_n = 1$, for all n, and thus Equation (7.17) gives

$$\frac{1}{(1-z)^2} = h(z) = f(z)g(z) = \sum_{n=0}^{\infty} \left(\sum_{k=0}^{n} a_k b_{n-k} \right) z^n = \sum_{n=0}^{\infty} (n+1)z^n.$$

Exercises for Section 7.2 (Selected answers or hints are on page 446.)

1. By computing derivatives, find the Maclaurin series for each function and state where it is valid.

   (a) $\sinh z$.
   (b) $\cosh z$.
   (c) $\text{Log}(1+z)$.

2. Using methods other than computing derivatives, find the Maclaurin series for

   (a) $\cos^3 z$. *Hint*: Use the trigonometric identity

   $$4\cos^3 z = \cos 3z + 3\cos z.$$

   (b) $\text{Arctan } z$. *Hint*: Choose an appropriate contour and integrate second series in Equations (7.12).
   (c) $f(z) = (z^2 + 1)\sin z$.

221

(d) $f(z) = e^z \cos z$. *Hint*: $\cos z = \frac{1}{2}(e^{iz} + e^{-iz})$, so

$$f(z) = \frac{1}{2} e^{(1+i)z} + \frac{1}{2} e^{(1-i)z}.$$

Now use the Maclaurin series for e^z.

3. Find the Taylor series centered at $a = 1$ and state where it converges for

(a) $f(z) = \frac{1-z}{z-2}$.

(b) $f(z) = \frac{1-z}{z-3}$. *Hint*: $\frac{1-z}{z-3} = \left(\frac{1}{2}\right)\left(\frac{z-1}{1-\frac{z-1}{2}}\right) = \left(\frac{1}{2}\right)(z-1)\left(\frac{1}{1-\frac{z-1}{2}}\right)$.

4. Let $f(z) = \frac{\sin z}{z}$ and set $f(0) = 1$.

(a) Explain why f is analytic at $z = 0$.

(b) Find the Maclaurin series for $f(z)$.

(c) Find the Maclaurin series for $g(z) = \int_C f(\zeta)\, d\zeta$, where C is the straight-line segment from 0 to z.

5. Show that $f(z) = \frac{1}{1-z}$ has its Taylor series representation about the point $\alpha = i$ given by

$$f(z) = \sum_{n=0}^{\infty} \frac{(z-i)^n}{(1-i)^{n+1}}, \quad \text{for all} \quad z \in D_{\sqrt{2}}(i).$$

6. Let $f(z) = (1+z)^\beta = \exp[\beta \mathrm{Log}(1+z)]$ be the principal branch of $(1+z)^\beta$, where β is a fixed complex number. Establish the validity for $z \in D_1(0)$ of the binomial expansion

$$(1+z)^\beta = 1 + \beta z + \frac{\beta(\beta-1)}{2!} z^2 + \frac{\beta(\beta-1)(\beta-2)}{3!} z^3 + \cdots$$

$$= 1 + \sum_{n=1}^{\infty} \frac{\beta(\beta-1)(\beta-2)\cdots(\beta-n+1)}{n!} z^n.$$

7. Find $f^{(3)}(0)$ for

(a) $f(z) = \sum_{n=0}^{\infty} (3+(-1)^n)^n z^n$.

(b) $g(z) = \sum_{n=1}^{\infty} \frac{(1+i)^n}{n} z^n$.

(c) $h(z) = \sum_{n=0}^{\infty} \frac{z^n}{(\sqrt{3}+i)^n}$.

8. Suppose that $f(z) = \sum_{n=0}^{\infty} c_n z^n$ is an entire function.

(a) Find a series representation for $\overline{f(\bar{z})}$, using powers of $\bar{z}$.

(b) Show that $\overline{f(\bar{z})}$ is an entire function.

(c) Does $\overline{f(\bar{z})} = f(z)$? Why or why not?

9. Let $f(z) = \sum\limits_{n=0}^{\infty} c_n z^n = 1 + z + 2z^2 + 3z^3 + 5z^4 + 8z^5 + 13z^6 + \cdots$, where the coefficients c_n are the Fibonacci numbers defined by $c_0 = 1$, $c_1 = 1$, and $c_n = c_{n-1} + c_{n-2}$, for $n \geq 2$.

   (a) Show that $f(z) = \frac{1}{1-z-z^2}$, for all $z \in D_R(0)$ for some number R.

   (b) Find the value of R in part (a) for which the series representation is valid.
      Hint: Find the singularities of $f(z)$ and use Corollary 7.3.

10. Complete the details in the verification of Lemma 7.1.

11. We used Lemma 7.1 in establishing Identity (7.6). However, Lemma 7.1 is valid provided $z \neq z_0$ and $z \neq \alpha$. Explain why these conditions are indeed the case in Identity (7.6).

12. Prove by mathematical induction that $f^{(n)}(z) = \frac{(n+1)!}{(1-z)^{n+2}}$ in Example 7.3.

13. Establish the validity of Identities (7.15) and (7.16).

14. Use the Maclaurin series and the Cauchy product in Identity (7.17) to verify that $\sin 2z = 2\cos z \sin z$ up to terms involving z^5.

15. Compute the Taylor series for the principal logarithm $f(z) = \operatorname{Log} z$ expanded about the point $z_0 = -1 + i$.

16. The Fresnel integrals $C(z)$ and $S(z)$ are defined by

$$C(z) = \int_0^z \cos(\xi^2)\, d\xi \quad \text{and} \quad S(z) = \int_0^z \sin(\xi^2)\, d\xi.$$

   Define $F(z)$ by $F(z) = C(z) + iS(z)$ and complete the following:

   (a) Verify the identity $F(z) = \int_0^z \exp(i\xi^2)\, d\xi$.

   (b) Integrate the power series for $\exp(i\xi^2)$ and obtain the power series for $F(z)$.

   (c) Use the partial sum involving terms up to z^9 to find approximations to $C(1.0)$ and $S(1.0)$.

17. Let f be defined in a domain that contains the origin. The function f is said to be even if $f(-z) = f(z)$, and it is called odd if $f(-z) = -f(z)$.

   (a) Show that the derivative of an odd function is an even function.

   (b) Show that the derivative of an even function is an odd function. *Hint*: Use limits.

   (c) If $f(z)$ is even, show that all the coefficients of the odd powers of z in the Maclaurin series are zero.

   (d) If $f(z)$ is odd, show that all the coefficients of the even powers of z in the Maclaurin series are zero.

18. Verify Identity (7.18) by using mathematical induction.

19. Consider the function

$$f(z) = \begin{cases} \frac{1}{1-z}, & \text{when } z \neq \frac{1}{2}; \\ 0, & \text{when } z = \frac{1}{2}. \end{cases}$$

   (a) Use Theorem 7.4, Taylor's theorem, to show that the Maclaurin series for $f(z)$ equals
   $$\sum_{n=0}^{\infty} z^n.$$

(b) Obviously, the radius of convergence of this series equals 1 (ratio test). However, the distance between 0 and the nearest singularity of f equals $\frac{1}{2}$. Explain why this condition does not contradict Corollary 7.3.

20. Consider the real-valued function f defined on the real numbers as

$$f(x) = \begin{cases} e^{-\frac{1}{x^2}}, & \text{when } x \neq 0; \\ 0, & \text{when } x = 0. \end{cases}$$

(a) Show that, for all $n > 0$, $f^{(n)}(0) = 0$, where $f^{(n)}$ is the nth derivative of f. *Hint:* Use the limit definition for the derivative to establish the case for $n = 1$ and then use mathematical induction to complete your argument.

(b) Explain why the function f is an example of a function that, although differentiable everywhere on the real line, is not expressible as a Taylor series about 0 that is valid for any interval $(-\varepsilon, \varepsilon)$, no matter how small ε is. *Hint:* Evaluate the Taylor series representation for $f(x)$ when $x \neq 0$, and show that the series does not equal $f(x)$.

(c) Explain why a similar argument could not be made for the complex-valued function g defined on the complex numbers as

$$g(z) = \begin{cases} e^{-\frac{1}{z^2}}, & \text{when } z \neq 0; \\ 0, & \text{when } z = 0. \end{cases}$$

Hint: Show that $g(z)$ is not even continuous at $z = 0$ by taking limits along the real and imaginary axes.

7.3 Laurent Series Representations

Suppose that $f(z)$ is not analytic in $D_R(\alpha)$ but *is* analytic in the punctured disk $D_R^*(\alpha) = \{z : 0 < |z - \alpha| < R\}$. For example, the function $f(z) = \frac{1}{z^3}e^z$ is not analytic when $z = 0$ but is analytic for $|z| > 0$. Clearly, this function does not have a Maclaurin series representation. If we use the Maclaurin series for $g(z) = e^z$, however, and formally divide each term in that series by z^3, we obtain the representation

$$f(z) = \frac{1}{z^3}e^z = \frac{1}{z^3} + \frac{1}{z^2} + \frac{1}{2!z} + \frac{1}{3!} + \frac{z}{4!} + \frac{z^2}{5!} + \frac{z^3}{6!} + \cdots,$$

which is valid for all z such that $|z| > 0$.

This example raises the question as to whether it might be possible to generalize the Taylor series method to functions analytic in an annulus

$$A(\alpha, r, R) = \{z : r < |z - \alpha| < R\}.$$

Perhaps we can represent these functions with a series that involves negative powers of z in some way as we did with $f(z) = \frac{1}{z^3}e^z$. As you will see shortly, we can indeed. We begin by defining a series that allows for negative powers of z.

Definition 7.3 (Laurent Series). *Let c_n be a complex number for $n = 0, \pm1, \pm2, \pm3, \ldots$ The doubly infinite series $\sum_{n=-\infty}^{\infty} c_n(z - \alpha)^n$, called a **Laurent series**, is defined by*

$$\sum_{n=-\infty}^{\infty} c_n(z - \alpha)^n = \sum_{n=1}^{\infty} c_{-n}(z - \alpha)^{-n} + c_0 + \sum_{n=1}^{\infty} c_n(z - \alpha)^n, \tag{7.21}$$

provided the series on the right-hand side of this equation converge.

224

Remark 7.2. *Recall that* $\sum\limits_{n=0}^{\infty} c_n(z-\alpha)^n$ *is a simplified expression for the sum* $c_0 + \sum\limits_{n=1}^{\infty} c_n(z-\alpha)^n$.

At times it will be convenient to write $\sum\limits_{n=-\infty}^{\infty} c_n(z-\alpha)^n$ *as* $\sum\limits_{n=-\infty}^{\infty} c_n(z-\alpha)^n = \sum\limits_{n=-\infty}^{-1} c_n(z-\alpha)^n +$

$\sum\limits_{n=0}^{\infty} c_n(z-\alpha)^n$ *rather than using the expression given in Equation* (7.21).

Definition 7.4 (Annulus). *Given* $0 \le r < R$, *we define the* **annulus** *centered at* α *with radii* r *and* R *by*

$$A(\alpha, r, R) = \{z : r < |z-\alpha| < R\}.$$

The **closed annulus** *centered at* α *with radii* r *and* R *is denoted by*

$$\overline{A}(\alpha, r, R) = \{z : r \le |z-\alpha| \le R\}.$$

Figure 7.3 illustrates an open annulus.

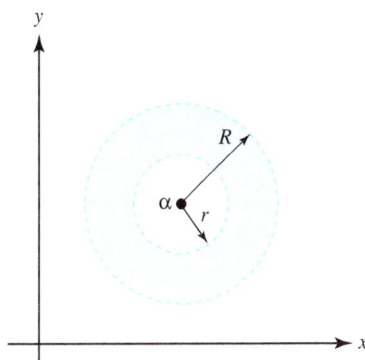

Figure 7.3: The open annulus $A(\alpha, r, R)$ (shaded)

Theorem 7.7. *Suppose that the Laurent series* $\sum\limits_{n=-\infty}^{\infty} c_n(z - \alpha)^n$ *converges on the annulus* $A(\alpha, r, R)$. *Then the series converges uniformly on any closed subannulus* $\overline{A}(\alpha, s, t)$, *where* $r < s < t < R$.

Proof. According to Equation (7.21),

$$\sum_{n=-\infty}^{\infty} c_n(z - \alpha)^n = \sum_{n-1}^{\infty} c_{-n}(z - \alpha)^{-n} + \sum_{n=0}^{\infty} c_n(z - \alpha)^n.$$

By Theorem 7.2, the series $\sum\limits_{n=0}^{\infty} c_n(z - \alpha)^n$ must converge uniformly on the closed disk $\overline{D}_t(\alpha)$.

By the Weierstrass M-test, we can show that the series $\sum\limits_{n=1}^{\infty} c_{-n}(z - \alpha)^{-n}$ converges uniformly on $\{z : |z - \alpha| \ge s\}$ (we leave the details as an exercise). Combining these two facts yields the required result. $\qquad\square$

The main result of this section specifies how functions analytic in an annulus can be expanded in a Laurent series. In it, we use symbols of the form $C_\rho^+(\alpha)$, which—we remind you—designate the positively oriented circle with radius ρ and center α. That is, $C_\rho^+(\alpha) = \{z : |z - \alpha| = \rho\}$, oriented counterclockwise.

Theorem 7.8 (Laurent's theorem). *Suppose that $0 \leq r < R$, and that f is analytic in the annulus $A = A(\alpha, r, R)$. If ρ is any number such that $r < \rho < R$, then for all $z_0 \in A$, f has the Laurent series representation*

$$f(z_0) = \sum_{n=-\infty}^{\infty} c_n(z_0 - \alpha)^n = \sum_{n=1}^{\infty} c_{-n}(z_0 - \alpha)^{-n} + \sum_{n=0}^{\infty} c_n(z_0 - \alpha)^n, \tag{7.22}$$

where for $n = 0, 1, 2, \ldots$, the coefficients c_{-n} and c_n are given by

$$c_{-n} = \frac{1}{2\pi i} \int_{C_\rho^+(\alpha)} \frac{f(z)}{(z-\alpha)^{-n+1}} \, dz \quad and \quad c_n = \frac{1}{2\pi i} \int_{C_\rho^+(\alpha)} \frac{f(z)}{(z-\alpha)^{n+1}} \, dz. \tag{7.23}$$

Moreover, the convergence in Equation (7.22) is uniform on any closed subannulus $\overline{A}(\alpha, s, t)$, where $r < s < t < R$.

Proof. If we can establish Equation (7.22), the uniform convergence on $\overline{A}(\alpha, s, t)$ will follow from Theorem 7.7. Let z_0 be an arbitrary point of A. Choose r_0 small enough so that the circle $C_0 = C_{r_0}^+(z_0)$ is contained in A. Since f is analytic in $D_{r_0}(z_0)$, the Cauchy integral formula gives

$$f(z_0) = \frac{1}{2\pi i} \int_{C_0} \frac{f(z)}{(z-z_0)} \, dz. \tag{7.24}$$

Let $C_1 = C_{r_1}^+(\alpha)$ and $C_2 = C_{r_2}^+(\alpha)$, where we choose r_1 and r_2 so that C_0 lies in the region between C_1 and C_2, and $r < r_1 < r_2 < R$ as shown in Figure 7.4, where the annulus A is the shaded region.

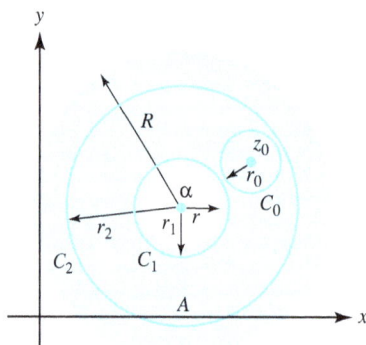

Figure 7.4: The annulus A and, in its interior, the circles C_0, C_1, and C_2

Now let D be the domain consisting of the annulus A except for the point z_0. The domain D includes the contours C_0, C_1, and C_2, as well as the region between C_2 and $C_0 + C_1$. In addition, since z_0 does not belong to D, the function $\frac{f(z)}{z-z_0}$ is analytic on D, so by the extended Cauchy-Goursat theorem we obtain

$$\frac{1}{2\pi i} \int_{C_2} \frac{f(z)}{(z-z_0)} \, dz = \frac{1}{2\pi i} \int_{C_0} \frac{f(z)}{(z-z_0)} \, dz + \frac{1}{2\pi i} \int_{C_1} \frac{f(z)}{(z-z_0)} \, dz. \tag{7.25}$$

Subtracting the last integral from both sides of Equation (7.25) and using the identity for $f(z_0)$ in Equation (7.24) give

$$f(z_0) = -\frac{1}{2\pi i} \int_{C_1} \frac{f(z)}{z-z_0} \, dz + \frac{1}{2\pi i} \int_{C_2} \frac{f(z)}{z-z_0} \, dz. \tag{7.26}$$

226

Now, if $z \in C_1$, then $|z - \alpha| < |z_0 - \alpha|$, so $\left| \frac{z-\alpha}{z_0-\alpha} \right| < 1$ and we can use the geometric series (Theorem 4.12) to get

$$
\begin{aligned}
\frac{1}{z - z_0} &= -\frac{1}{(z_0 - \alpha) - (z - \alpha)} \\
&= -\frac{1}{(z_0 - \alpha)} \frac{1}{\left(1 - \frac{z-\alpha}{z_0-\alpha}\right)} \\
&= -\frac{1}{(z_0 - \alpha)} \sum_{n=0}^{\infty} \left(\frac{z - \alpha}{z_0 - \alpha}\right)^n \\
&= -\sum_{n=0}^{\infty} \frac{(z - \alpha)^n}{(z_0 - \alpha)^{n+1}}.
\end{aligned}
\tag{7.27}
$$

Moreover, one can show by using the Weierstrass M-test that the preceding series converges uniformly for $z \in C_1$. We leave the details as an exercise. Likewise, using techniques similar to the ones just discussed, one can show that, for $z \in C_2$,

$$
\frac{1}{z - z_0} = \sum_{n=0}^{\infty} \frac{(z_0 - \alpha)^n}{(z - \alpha)^{n+1}},
\tag{7.28}
$$

and that the convergence is uniform for $z \in C_2$. Again, we leave the details as an exercise.

Taking the series for $\frac{1}{z-z_0}$ as given by Equations (7.27) and (7.28) and substituting into the two integrals, respectively, of Equation (7.26) yields

$$
f(z_0) = \frac{1}{2\pi i} \int_{C_1} \sum_{n=0}^{\infty} \frac{(z - \alpha)^n}{(z_0 - \alpha)^{n+1}} f(z)\,dz + \frac{1}{2\pi i} \int_{C_2} \sum_{n=0}^{\infty} \frac{(z_0 - \alpha)^n}{(z - \alpha)^{n+1}} f(z)\,dz.
$$

Because the series in this equation converge uniformly on C_2 and C_1, respectively, we can interchange the summations and the integrals, in accordance with Corollary 7.2 to obtain

$$
\begin{aligned}
f(z_0) = &\sum_{n=0}^{\infty} \left[\frac{1}{2\pi i} \int_{C_2} \frac{f(z)}{(z - \alpha)^{n+1}}\,dz \right] (z_0 - \alpha)^n \\
&+ \sum_{n=0}^{\infty} \left[\frac{1}{2\pi i} \int_{C_1} f(z)(z - \alpha)^n\,dz \right] \frac{1}{(z_0 - \alpha)^{n+1}}.
\end{aligned}
$$

If we move some terms around in the second series of this equation and reindex, we get

$$
\begin{aligned}
f(z_0) = &\sum_{n=0}^{\infty} \left[\frac{1}{2\pi i} \int_{C_2} \frac{f(z)}{(z - \alpha)^{n+1}}\,dz \right] (z_0 - \alpha)^n \\
&+ \sum_{n=1}^{\infty} \left[\frac{1}{2\pi i} \int_{C_1} \frac{f(z)}{(z - \alpha)^{-n+1}}\,dz \right] (z_0 - \alpha)^{-n}.
\end{aligned}
\tag{7.29}
$$

We apply the extended Cauchy-Goursat theorem once more to conclude that the integrals taken over C_2 and C_1 in Equation (7.29) give the same result if they are taken over the contour

227

$C_\rho^+(\alpha)$, where ρ is any number such that $r < \rho < R$. This yields

$$f(z_0) = \sum_{n=0}^{\infty} \left[\frac{1}{2\pi i} \int_{C_\rho^+(\alpha)} \frac{f(z)}{(z-\alpha)^{n+1}} \, dz \right] (z_0 - \alpha)^n$$
$$+ \sum_{n=1}^{\infty} \left[\frac{1}{2\pi i} \int_{C_\rho^+(\alpha)} \frac{f(z)}{(z-\alpha)^{-n+1}} \, dz \right] (z_0 - \alpha)^{-n}.$$

Finally, writing the second series first in our last equation gives

$$f(z_0) = \sum_{n=1}^{\infty} \left[\frac{1}{2\pi i} \int_{C_\rho^+(\alpha)} \frac{f(z)}{(z-\alpha)^{-n+1}} \, dz \right] (z_0 - \alpha)^{-n}$$
$$+ \sum_{n=0}^{\infty} \left[\frac{1}{2\pi i} \int_{C_\rho^+(\alpha)} \frac{f(z)}{(z-\alpha)^{n+1}} \, dz \right] (z_0 - \alpha)^n.$$

Because $z_0 \in A$ was arbitrary, this result establishes Equations (7.22) and (7.23), completing the proof. □

What happens to the Laurent series if f is analytic in the disk $D_R(\alpha)$? Looking at Equation (7.29), we see that the coefficient for the positive power $(z_0 - \alpha)^n$ equals $\frac{f^{(n)}(z_0)}{n!}$ by using Cauchy's integral formula for derivatives. Hence the series in Equation 7.22 involving the positive powers of $(z_0 - \alpha)$ is actually the Taylor series for f. The Cauchy-Goursat theorem shows that the coefficients for the negative powers of $(z_0 - \alpha)$ equal zero. In this case, therefore, there are no negative powers involved, and the Laurent series reduces to the Taylor series.

Theorem 7.9 delineates two important properties of the Laurent series.

Theorem 7.9 (Properties of Laurent Series). *Suppose that f is analytic in the annulus $A(\alpha, r, R)$ and has the Laurent series $f(z) = \sum_{n=-\infty}^{\infty} c_n(z-\alpha)^n$, for all $z \in A(\alpha, r, R)$.*

i. If $f(z) = \sum_{n=-\infty}^{\infty} b_n(z-\alpha)^n$ for all $z \in A(\alpha, r, R)$, then $b_n = c_n$ for all n. (In other words, the Laurent series for f in a given annulus is unique.)

ii. For all $z \in A(\alpha, r, R)$, the derivatives of $f(z)$ may be obtained by termwise differentiation of its Laurent series.

Proof. We prove part (i) only because the proof for part (ii) involves no new ideas beyond those in the proof of Theorem 4.17. The series $\sum_{n=-\infty}^{\infty} b_n(z-\alpha)^n$ converges pointwise on $A(\alpha, r, R)$, so Theorem 7.7 guarantees that this series converges uniformly on $C_\rho^+(\alpha)$, for $0 \le r < \rho < R$. By Laurent's theorem, then,

$$c_n = \frac{1}{2\pi i} \int\limits_{C_\rho^+(\alpha)} \frac{f(z)}{(z-\alpha)^{n+1}} \, dz$$

$$= \frac{1}{2\pi i} \int\limits_{C_\rho^+(\alpha)} \left[(z-\alpha)^{-n-1} \sum_{m=-\infty}^{\infty} b_m (z-\alpha)^m \right] dz$$

$$= \sum_{m=-\infty}^{\infty} \frac{b_m}{2\pi i} \int\limits_{C_\rho^+(\alpha)} (z-\alpha)^{m-n-1} \, dz.$$

Since $(z-\alpha)^{m-n-1}$ has an antiderivative for all z except when $m = n$, all the terms in the preceding expression drop out except when $m = n$, giving

$$c_n = \frac{b_n}{2\pi i} \int_{C_\rho^+(\alpha)} (z-\alpha)^{-1} dz = b_n.$$

$\square$

The uniqueness of the Laurent series is an important property because the coefficients in the Laurent expansion of a function are seldom found by using Equation (7.23). The following examples illustrate some methods for finding Laurent series coefficients.

Example 7.7. Find three different Laurent series representations for $f(z) = \frac{3}{2+z-z^2}$ involving powers of z.

Solution:

The function f has singularities at $z = -1, 2$ and is analytic in the disk $D : |z| < 1$, in the annulus $A : 1 < |z| < 2$, and in the region $R : |z| > 2$. We want to find a different Laurent series for f in each of the three domains D, A, and R. We start by writing f in its partial fraction form:

$$f(z) = \frac{3}{(1+z)(2-z)} = \frac{1}{1+z} + \frac{1}{2}\left(\frac{1}{1-\frac{z}{2}} \right). \tag{7.30}$$

We use Theorem 4.12 and Corollary 4.2 to obtain the following representations for the terms on the right side of Equation (7.30):

$$\frac{1}{1+z} = \sum_{n=0}^{\infty} (-1)^n z^n, \qquad \text{valid for } |z| < 1, \tag{7.31}$$

$$\frac{1}{1+z} = \sum_{n=1}^{\infty} \frac{(-1)^{n+1}}{z^n}, \qquad \text{valid for } |z| > 1, \tag{7.32}$$

$$\frac{1}{2}\left(\frac{1}{1-\frac{z}{2}} \right) = \sum_{n=0}^{\infty} \frac{z^n}{2^{n+1}}, \qquad \text{valid for } |z| < 2, \quad \text{and} \tag{7.33}$$

$$\frac{1}{2}\left(\frac{1}{1-\frac{z}{2}} \right) = \sum_{n=1}^{\infty} \frac{-2^{n-1}}{z^n}, \qquad \text{valid for } |z| > 2. \tag{7.34}$$

Representations (7.31) and (7.33) are both valid in the disk D, and thus we have

$$f(z) = \sum_{n=0}^{\infty} \left[(-1)^n + \frac{1}{2^{n+1}} \right] z^n, \quad \text{valid for } |z| < 1,$$

which is a Laurent series that reduces to a Maclaurin series. In the annulus A, Representations 7.32 and 7.33 are valid; hence we get

$$f(z) = \sum_{n=1}^{\infty} \frac{(-1)^{n+1}}{z^n} + \sum_{n=0}^{\infty} \frac{z^n}{2^{n+1}}, \quad \text{valid for } 1 < |z| < 2.$$

Finally, in the region R we use Representations 7.32 and 7.34 to obtain

$$f(z) = \sum_{n=1}^{\infty} \frac{(-1)^{n+1} - 2^{n-1}}{z^n}, \quad \text{valid for } |z| > 2.$$

Example 7.8. Find the Laurent series representation for $f(z) = \frac{\cos z - 1}{z^4}$ that involves powers of z.

Solution:

We use the Maclaurin series for $\cos z - 1$ to write

$$f(z) = \frac{-\frac{1}{2!}z^2 + \frac{1}{4!}z^4 - \frac{1}{6!}z^6 + \cdots}{z^4}.$$

We formally divide each term by z^4 to obtain the Laurent series

$$f(z) = \frac{-1}{2z^2} + \frac{1}{24} - \frac{z^2}{720} + \cdots, \quad \text{valid for } z \neq 0.$$

Example 7.9. Find the Laurent series for $\exp\left(-\frac{1}{z^2}\right)$ centered at $z_0 = 0$.

Solution:

The Maclaurin series for $\exp z$ is $\exp z = \sum_{n=0}^{\infty} \frac{z^n}{n!}$, which is valid for all z. We let $-z^{-2}$ take the role of z in this equation to get

$$\exp\left(-\frac{1}{z^2}\right) = \sum_{n=0}^{\infty} \frac{(-1)^n}{n! z^{2n}},$$

which is valid for $|z| > 0$.

Exercises for Section 7.3 (Selected answers or hints are on page 447.)

1. Find two Laurent series expansions for $f(z) = \frac{1}{z^3 - z^4}$ that involve powers of z.

2. Show that $f(z) = \frac{1}{1-z} = \frac{1}{1-i}\left(\frac{1}{1-\frac{z-i}{1-i}}\right)$ has a Laurent series representation about the point $z_0 = i$ given by

$$f(z) = \frac{1}{1-z} = -\sum_{n=1}^{\infty} \frac{(1-i)^{n-1}}{(z-i)^n}, \quad \text{valid for } |z - i| > \sqrt{2}.$$

3. Find the Laurent series for $f(z) = \frac{\sin 2z}{z^4}$ that involves powers of z.

4. Show that $\frac{1-z}{z-2} = -\sum_{n=0}^{\infty} \frac{1}{(z-1)^n}$ is valid for $|z - 1| > 1$.

   Hint: Refer to the solution for Exercise 3a Section 7.2.

5. Find the Laurent series for $\sin(\frac{1}{z})$ centered at $\alpha = 0$. Where is the series valid?

6. Show that $\frac{1-z}{z-3} = -\sum_{n=0}^{\infty} \frac{2^n}{(z-1)^n}$ is valid for $|z - 1| > 2$.
   Hint: Use the hint for Exercise 3b, Section 7.2.

7. Find the Laurent series for $f(z) = \frac{\cosh z - \cos z}{z^5}$ that involves powers of z.

8. Find the Laurent series for $f(z) = \frac{1}{z^4(1-z)^2}$ that involves powers of z and is valid for
   $|z| > 1$. *Hint*: $\frac{1}{(1-\frac{1}{z})^2} = \frac{z^2}{(1-z)^2}$.

9. Find two Laurent series for $z^{-1}(4 - z)^{-2}$ involving powers of z, and state where they are valid.

10. Find three Laurent series for $(z^2 - 5z + 6)^{-1}$ centered at $\alpha = 0$.

11. Find the Laurent series for $\mathrm{Log}(\frac{z-a}{z-b})$, where a and b are positive real numbers with
    $b > a > 1$, and state where the series is valid. *Hint*: For these conditions, show that
    $\mathrm{Log}(\frac{z-a}{z-b}) = \mathrm{Log}(1 - \frac{a}{z}) - \mathrm{Log}(1 - \frac{b}{z})$.

12. Can $\mathrm{Log}\, z$ be represented by a Maclaurin series or a Laurent series about the point $\alpha = 0$?
    Explain your answer.

13. Use the Maclaurin series for $\sin z$ and then long division to get the Laurent series for $\csc z$
    with $\alpha = 0$.

14. Show that $\cosh(z + \frac{1}{z}) = \sum_{n=-\infty}^{\infty} a_n z^n$, where the coefficients can be expressed in the form
    $a_n = \frac{1}{2\pi} \int_0^{2\pi} \cos n\theta \cosh(2\cos\theta)\, d\theta$. *Hint*: Let the path of integration be the circle $C_1^+(0)$.

15. Consider the real-valued function $u(\theta) = \frac{1}{5-4\cos\theta}$.

    (a) Use the substitution $\cos\theta = \frac{1}{2}(z + \frac{1}{z})$ and obtain

    $$u(\theta) = f(z) = \frac{-z}{(z-2)(2z-1)} = \frac{1}{3}\left(\frac{1}{1-\frac{z}{2}}\right) - \frac{1}{3}\left(\frac{1}{1-2z}\right).$$

    (b) Expand the function $f(z)$ in part (a) in a Laurent series that is valid in the annulus
    $A(0, \frac{1}{2}, 2)$.

    (c) Use the substitutions $\cos(n\theta) = \frac{1}{2}(z^n + z^{-n})$ in part (b) and obtain the Fourier series
    for $u(\theta)$, where $u(\theta) = \frac{1}{3} + \frac{1}{3}\sum_{n=1}^{\infty} 2^{-n+1}\cos(n\theta)$.

16. The *Bessel function* $J_n(z)$ is sometimes defined by the generating function

    $$\exp\left[\frac{z}{2}\left(t - \frac{1}{t}\right)\right] = \sum_{n=-\infty}^{\infty} J_n(z)t^n.$$

    Use the circle $C_1^+(0)$ as the contour of integration and show that

    $$J_n(z) = \frac{1}{\pi}\int_0^{\pi} \cos(n\theta - z\sin\theta)\, d\theta.$$

17. Suppose that the Laurent expansion $f(z) = \sum\limits_{n=-\infty}^{\infty} a_n z^n$ converges in the annulus $A(0, r_1, r_2)$, where $r_1 < 1 < r_2$. Consider the real-valued function $u(\theta) = f(e^{i\theta})$ and show that $u(\theta)$ has the Fourier series expansion

$$u(\theta) = f(e^{i\theta}) = \sum_{n=-\infty}^{\infty} a_n e^{in\theta},$$

where

$$a_n = \frac{1}{2\pi} \int_0^{2\pi} e^{-in\phi} f(e^{i\phi}) \, d\phi.$$

18. (*The Z-transform*): Let $\{a_n\}$ be a sequence of complex numbers satisfying the growth condition $|a_n| \le MR^n$ for $n = 0, 1, \ldots$ and for some fixed positive values M and R. Then the Z-transform of the sequence $\{a_n\}$ is the function $F(z)$ defined by $Z(\{a_n\}) = F(z) = \sum\limits_{n=0}^{\infty} a_n z^{-n}$.

   (a) Prove that $F(z)$ converges for $|z| > R$.

   (b) Find $Z(\{a_n\})$ for

      i. $a_n = 2$.
      ii. $a_n = \frac{1}{n!}$.
      iii. $a_n = \frac{1}{n+1}$.
      iv. $a_n = 1$, when n is even, and $a_n = 0$ when n is odd.

   (c) Prove that $Z(\{a_{n+1}\}) = z\big[Z(\{a_n\}) - a_0\big]$. This relation is known as the *shifting property* for the Z-transform.

19. Use the Weierstrass M-test to show that the series $\sum\limits_{n=1}^{\infty} c_{-n}(z - \alpha)^{-n}$ of Theorem 7.7 converges uniformly on the set $\{z : |z - \alpha| \ge s\}$ as claimed.

20. Verify the following claims made in this section.

   (a) The series in Equation (7.27) converges uniformly for $z \in C_2$.

   (b) The validity of Equation (7.28), according to Corollary 4.2.

   (c) The series in Equation (7.28) converges uniformly for $z \in C_1$.

7.4 Singularities, Zeros, and Poles

Recall that the point α is called a **singular point,** or **singularity**, of the complex function f if f is not analytic at the point α, but every neighborhood $D_R(\alpha)$ of α contains at least one point at which f is analytic. For example, the function $f(z) = \frac{1}{1-z}$ is not analytic at $\alpha = 1$ but is analytic for all other values of z. Thus the point $\alpha = 1$ is a singular point of f. As another example, consider the function $g(z) = \text{Log } z$. We showed in Section 5.2 that g is analytic for all z except at the origin and at the points on the negative real axis. Thus the origin and each point on the negative real axis are singularities of g.

The point α is called an **isolated singularity** of a complex function f if f is not analytic at α but there exists a real number $R > 0$ such that f is analytic everywhere in the punctured disk $D_R^*(\alpha)$. The function $f(z) = \frac{1}{1-z}$ has an isolated singularity at $\alpha = 1$. The function $g(z) =$

232

Log z, however, has a singularity at $\alpha = 0$ (or at any point of the negative real axis) that is not isolated, because any neighborhood of α contains points on the negative real axis, and g is not analytic at those points. Functions with isolated singularities have a Laurent series because the punctured disk $D_R^*(\alpha)$ is the same as the annulus $A(\alpha, 0, R)$. We now look at this special case of Laurent's theorem in order to classify three types of isolated singularities.

Definition 7.5 (Classification of singularities). *Let f have an isolated singularity at α with Laurent series*

$$f(z) = \sum_{n=-\infty}^{\infty} c_n(z-\alpha)^n, \quad valid \text{ for all } z \in A(\alpha, 0, R).$$

Then we distinguish the following types of singularities at α.

   i. *If $c_n = 0$, for $n = -1, -2, -3, \ldots$, then f has a **removable singularity** at α.*

   ii. *If k is a positive integer such that $c_{-k} \neq 0$, but $c_n = 0$ for $n < -k$, then f has a **pole of order** k at α.*

   iii. *If $c_n \neq 0$ for infinitely many negative integers n, then f has an **essential singularity** at α.*

Let's investigate some examples of these three cases.

   i. If f has a removable singularity at α, then it has a Laurent series

$$f(z) = \sum_{n=0}^{\infty} c_n(z-\alpha)^n, \quad \text{valid for all } z \in A(\alpha, 0, R).$$

Theorem 4.17 implies that the power series for f defines an analytic function in the disk $D_R(\alpha)$. If we use this series to define $f(\alpha) = c_0$, then the function f becomes analytic at $z = \alpha$, removing the singularity. For example, consider the function $f(z) = \frac{\sin z}{z}$. It is undefined at $z = 0$ and has an isolated singularity at $z = 0$, as the Laurent series for f is

$$f(z) = \frac{\sin z}{z} = \frac{1}{z}\left(z - \frac{z^3}{3!} + \frac{z^5}{5!} - \frac{z^7}{7!} + \cdots\right)$$

$$= 1 - \frac{z^2}{3!} + \frac{z^4}{5!} - \frac{z^6}{7!} + \cdots, \quad \text{valid for} \quad |z| > 0.$$

We can remove this singularity if we define $f(0) = 1$, for then f will be analytic at 0 in accordance with Theorem 4.17.

Another example is $g(z) = \frac{\cos z - 1}{z^2}$, which has an isolated singularity at the point 0, as the Laurent series for g is

$$g(z) = \frac{1}{z^2}\left(-\frac{z^2}{2!} + \frac{z^4}{4!} - \frac{z^6}{6!} + \cdots\right)$$

$$= -\frac{1}{2} + \frac{z^2}{4!} - \frac{z^4}{6!} + \cdots, \quad \text{valid for} \quad |z| > 0.$$

If we define $g(0) = -\frac{1}{2}$, then g will be analytic for all z.

ii. If f has a pole of order k at α, the Laurent series for f is

$$f(z) = \sum_{n=-k}^{\infty} c_n(z-\alpha)^n, \quad \text{valid for all } z \in A(\alpha, 0, R),$$

where $c_{-k} \neq 0$. For example,

$$f(z) = \frac{\sin z}{z^3} = \frac{1}{z^2} - \frac{1}{3!} + \frac{z^2}{5!} - \frac{z^4}{7!} + \cdots$$

has a pole of order 2 at 0.

If f has a pole of order 1 at α, we say that f has a **simple pole** at α. For example,

$$f(z) = \frac{1}{z}e^z = \frac{1}{z} + 1 + \frac{z}{2!} + \frac{z^2}{3!} + \cdots,$$

which has a simple pole at 0.

iii. If infinitely many negative powers of $(z-\alpha)$ occur in the Laurent series, then f has an essential singularity at α. For example,

$$f(z) = z^2 \sin\frac{1}{z} = z - \frac{1}{3!}z^{-1} + \frac{1}{5!}z^{-3} - \frac{1}{7!}z^{-5} + \cdots$$

has an essential singularity at the origin.

Definition 7.6 (Zero of order k). *A function f analytic in $D_R(\alpha)$ has a **zero of order** k at the point α iff*

$$f^{(n)}(\alpha) = 0, \quad for \quad n = 0, 1, \ldots, \quad k-1, \quad but \quad f^{(k)}(\alpha) \neq 0.$$

*A zero of order 1 is called a **simple zero.***

Theorem 7.10. *A function f analytic in $D_R(\alpha)$ has a zero of order k at the point α iff its Taylor series given by $f(z) = \sum_{n=0}^{\infty} c_n(z-\alpha)^n$ has*

$$c_0 = c_1 = \cdots = c_{k-1} = 0, \quad but \quad c_k \neq 0.$$

Proof. The conclusion follows immediately from Definition 7.6, because we have $c_n = \frac{f^{(n)}(\alpha)}{n!}$ according to Taylor's theorem. $\qquad \square$

Example 7.10. From Theorem 7.10 we see that the function

$$f(z) = z \sin z^2 = z^3 - \frac{z^7}{3!} + \frac{z^{11}}{5!} - \frac{z^{15}}{7!} + \cdots$$

has a zero of order 3 at $z = 0$. Definition 7.6 confirms this fact because

$$f'(z) = 2z^2 \cos z^2 + \sin z^2;$$
$$f''(z) = 6z \cos z^2 - 4z^3 \sin z^2$$
$$f'''(z) = 6 \cos z^2 - 8z^4 \cos z^2 - 24z^2 \sin z^2$$

Then, $f(0) = f'(0) = f''(0) = 0$, but $f'''(0) = 6 \neq 0$.

Theorem 7.11. *Suppose that the function f is analytic in $D_R(\alpha)$. Then f has a zero of order k at the point α iff f can be expressed in the form*

$$f(z) = (z - \alpha)^k g(z), \tag{7.35}$$

where g is analytic at the point α and $g(\alpha) \neq 0$.

Proof. Suppose that f has a zero of order k at the point α, and that $f(z) = \sum\limits_{n=0}^{\infty} c_n(z - \alpha)^n$ for $z \in D_R(\alpha)$. Theorem 7.10 assures us that $c_n = 0$ for $0 \leq n \leq k - 1$, and that $c_k \neq 0$, so that we can write f as

$$
\begin{aligned}
f(z) &= \sum_{n=k}^{\infty} c_n(z - \alpha)^n \\
&= \sum_{n=0}^{\infty} c_{n+k}(z - \alpha)^{n+k} \\
&= (z - \alpha)^k \sum_{n=0}^{\infty} c_{n+k}(z - \alpha)^n, \tag{7.36}
\end{aligned}
$$

where $c_k \neq 0$. The series on the right side of Equation (7.36) defines a function, which we denote by g. That is,

$$g(z) = \sum_{n=0}^{\infty} c_{n+k}(z - \alpha)^n = c_k + \sum_{n=1}^{\infty} c_{n+k}(z - \alpha)^n, \quad \text{valid for all } z \in D_R(\alpha).$$

By Theorem 4.17, g is analytic in $D_R(\alpha)$, and $g(\alpha) = c_k \neq 0$.

Conversely, suppose that f has the form given by Equation (7.35). Since g is analytic at α, it has the power series representation $g(z) = \sum\limits_{n=0}^{\infty} b_n(z - \alpha)^n$, where $g(\alpha) = b_0 \neq 0$ by assumption. If we multiply both sides of the expression defining $g(z)$ by $(z - \alpha)^k$ we get

$$f(z) = g(z)(z - \alpha)^k = \sum_{n=0}^{\infty} b_n(z - \alpha)^{n+k} = \sum_{n=k}^{\infty} b_{n-k}(z - \alpha)^n.$$

By Theorem 7.10, f has a zero of order k at the point α. $\qquad\square$

An immediate consequence of Theorem 7.11 is Corollary 7.4. The proof is left as an exercise.

Corollary 7.4. *If $f(z)$ and $g(z)$ are analytic at $z = \alpha$ and have zeros of orders m and n, respectively, at $z = \alpha$, then their product $h(z) = f(z)g(z)$ has a zero of order $m + n$ at $z = \alpha$.*

Example 7.11. Let $f(z) = z^3 \sin z$. Then $f(z)$ can be factored as the product of z^3 and $\sin z$, which have zeros of orders $m = 3$ and $n = 1$, respectively, at $z = 0$. Hence $z = 0$ is a zero of order 4 of $f(z)$.

Theorem 7.12 gives a useful way to characterize a pole.

Theorem 7.12. *A function f analytic in the punctured disk $D_R^*(\alpha)$ has a pole of order k at the point α iff f can be expressed in the form*

$$f(z) = \frac{h(z)}{(z - \alpha)^k}, \tag{7.37}$$

where the function h is analytic at the point α, and $h(\alpha) \neq 0$.

Proof. Suppose that f has a pole of order k at the point α. We can then write the Laurent series for f as

$$f(z) = \frac{1}{(z-\alpha)^k} \sum_{n=0}^{\infty} c_{n-k}(z-\alpha)^n,$$

where $c_{-k} \neq 0$. The series on the right side of this equation defines a function, which we denote by $h(z)$. That is,

$$h(z) = \sum_{n=0}^{\infty} c_{n-k}(z-\alpha)^n, \quad \text{for all} \quad z \in D_R^*(\alpha) = \{z : 0 < |z-\alpha| < R\}.$$

If we specify that $h(\alpha) = c_{-k}$, then h is analytic in all of $D_R(\alpha)$, with $h(\alpha) \neq 0$.

Conversely, suppose that Equation (7.37) is satisfied. Because h is analytic at the point α with $h(\alpha) \neq 0$, it has a power series representation

$$h(z) = \sum_{n=0}^{\infty} b_n(z-\alpha)^n,$$

where $b_0 \neq 0$. If we divide both sides of this equation by $(z-\alpha)^k$, we obtain the following Laurent series representation for f:

$$f(z) = \sum_{n=0}^{\infty} b_n(z-\alpha)^{n-k}$$

$$= \sum_{n=-k}^{\infty} b_{n+k}(z-\alpha)^n$$

$$= \sum_{n=-k}^{\infty} c_n(z-\alpha)^n,$$

where $c_n = b_{n+k}$. Since $c_{-k} = b_0 \neq 0$, f has a pole of order k at α. $\square$

The following corollaries (7.5–7.7) are useful in determining the order of a zero or a pole. The proofs follow easily from Theorems 7.10 and 7.12, and are left as exercises.

Corollary 7.5. *If f is analytic and has a zero of order k at the point α, then $g(z) = \frac{1}{f(z)}$ has a pole of order k at α.*

Corollary 7.6. *If f has a pole of order k at the point α, then $g(z) = \frac{1}{f(z)}$ has a removable singularity at α. If we define $g(\alpha) = 0$, then $g(z)$ has a zero of order k at α.*

Corollary 7.7. *If f and g have poles of orders m and n, respectively, at the point α, then their product $h(z) = f(z)g(z)$ has a pole of order $m+n$ at α.*

Corollary 7.8. *Let f and g be analytic with zeros of orders m and n, respectively, at α. Then their quotient $h(z) = \frac{f(z)}{g(z)}$ has the following behavior:*

  i. If $m > n$, then h has a removable singularity at α. If we define $h(\alpha) = 0$, then h has a zero of order $m - n$ at α.

  ii. If $m < n$, then h has a pole of order $n - m$ at α.

iii. If $m = n$, then h has a removable singularity at α and can be defined so that h is analytic at α by $h(\alpha) = \lim\limits_{z \to \alpha} h(z)$.

Example 7.12. Locate the zeros and poles of $h(z) = \frac{\tan z}{z}$ and determine their order.

Solution:

In Section 5.4 we saw that the zeros of $f(z) = \sin z$ occur at the points $n\pi$, where n is an integer. Because $f'(n\pi) = \cos n\pi \neq 0$, the zeros of f are simple. Similarly, the function $g(z) = z \cos z$ has simple zeros at the points 0 and $(n + \frac{1}{2})\pi$, where n is an integer. From the information given, we find that $h(z) = \frac{f(z)}{g(z)}$ behaves as follows:

    i. h has simple zeros at $n\pi$, where $n = \pm 1, \pm 2, \ldots$;

    ii. h has simple poles at $(n + \frac{1}{2})\pi$, where n is an integer; and

    iii. h is analytic at 0 if we define $h(0) = \lim\limits_{z \to 0} h(z) = 1$.

Example 7.13. Locate the poles of $g(z) = \frac{1}{5z^4 + 26z^2 + 5}$ and specify their order.

Solution:

The roots of the quadratic equation $5z^2 + 26z + 5 = 0$ occur at the points -5 and $-\frac{1}{5}$. If we replace z with z^2 in this equation, the function $f(z) = 5z^4 + 26z^2 + 5$ has simple zeros at the points $\pm i\sqrt{5}$ and $\pm\frac{i}{\sqrt{5}}$. Corollary 7.5 implies that g has simple poles at $\pm i\sqrt{5}$ and $\pm\frac{i}{\sqrt{5}}$.

Example 7.14. Locate the poles of $g(z) = \frac{\pi \cot(\pi z)}{z^2}$ and specify their order.

Solution:

The function $f(z) = z^2 \sin \pi z$ has a zero of order 3 at $z = 0$ and simple zeros at the points $z = \pm 1, \pm 2, \ldots$. Corollary 7.5 implies that g has a pole of order 3 at the point 0 and simple poles at the points $\pm 1, \pm 2, \ldots$.

Exercises for Section 7.4 (Selected answers or hints are on page 448.)

1. Locate the zeros of the following functions and determine their order.

    (a) $(1 + z^2)^4$.

    (b) $\sin^2 z$.

    (c) $z^2 + 2z + 2$.

    (d) $\sin z^2$.

    (e) $z^4 + 10z^2 + 9$.

    (f) $1 + \exp z$.

    (g) $z^6 + 1$.

    (h) $z^3 \exp(z - 1)$.

    (i) $z^6 + 2z^3 + 1$.

    (j) $z^3 \cos^2 z$.

    (k) $z^8 + z^4$.

    (l) $z^2 \cosh z$.

2. Locate the poles of the following functions and determine their order.

   (a) $(z^2 + 1)^{-3}(z - 1)^{-4}$.

   (b) $z^{-1}(z^2 - 2z + 2)^{-2}$.

   (c) $(z^6 + 1)^{-1}$.

   (d) $(z^4 + z^3 - 2z^2)^{-1}$.

   (e) $(3z^4 + 10z^2 + 3)^{-1}$.

   (f) $(i + \frac{2}{z})^{-1}(3 + \frac{4}{z})^{-1}$.

   (g) $z \cot z$.

   (h) $z^{-5} \sin z$.

   (i) $(z^2 \sin z)^{-1}$.

   (j) $z^{-1} \csc z$.

   (k) $(1 - \exp z)^{-1}$.

   (l) $z^{-5} \sinh z$.

3. Locate the singularities of the following functions and determine their type.

   (a) $\frac{z^2}{z - \sin z}$.

   (b) $\sin(\frac{1}{z})$.

   (c) $z \exp(\frac{1}{z})$.

   (d) $\tan z$.

   (e) $(z^2 + z)^{-1} \sin z$.

   (f) $\frac{z}{\sin z}$.

   (g) $\frac{(\exp z) - 1}{z}$.

   (h) $\frac{\cos z - \cos(2z)}{z^4}$.

4. Suppose that f has a removable singularity at z_0. Show that the function $\frac{1}{f}$ has either a removable singularity or a pole at z_0.

5. Let f be analytic and have a zero of order k at z_0. Show that f' has a zero of order $k - 1$ at z_0.

6. Let f and g be analytic at z_0 and have zeros of order m and n, respectively, at z_0. What can you say about the zero of $f + g$ at z_0?

7. Let f and g have poles of order m and n, respectively, at z_0. Show that $f + g$ has either a pole or a removable singularity at z_0

8. Let f be analytic and have a zero of order k at z_0. Show that the function $\frac{f'}{f}$ has a simple pole at z_0.

9. Let f have a pole of order k at z_0. Show that f' has a pole of order $k + 1$ at z_0.

10. Prove the following corollaries.

    (a) Corollary 7.4.

    (b) Corollary 7.5.

(c) Corollary 7.6.

(d) Corollary 7.7.

(e) Corollary 7.8.

11. Find the singularities of the following functions.

   (a) $\frac{1}{\sin(\frac{1}{z})}$.

   (b) $\text{Log}(z^2)$.

   (c) $\cot z - \frac{1}{z}$.

12. How are the definitions of singularity in complex analysis and asymptote in calculus different? How are they similar?

7.5 Applications of Taylor and Laurent Series

In this section we show how you can use Taylor and Laurent series to derive important properties of analytic functions. We begin by showing that the zeros of an analytic function must be isolated unless the function is identically zero. A point α of a set T is called *isolated* if there exists a disk $D_R(\alpha)$ about α that does not contain any other points of T.

Theorem 7.13. *Suppose that f is analytic in a domain D containing α and that $f(\alpha) = 0$. If f is not identically zero in D, then there exists a punctured disk $D_R^*(\alpha)$ in which f has no zeros.*

Proof. By Taylor's theorem, there exists some disk $D_R(\alpha)$ about α such that

$$f(z) = \sum_{n=0}^{\infty} \frac{f^{(n)}(\alpha)}{n!}(z-\alpha)^n \quad \text{for all} \quad z \in D_R(\alpha).$$

If all the Taylor coefficients $\frac{f^{(n)}(\alpha)}{n!}$ of f were zero, then f would be identically zero on $D_R(\alpha)$. A proof similar to the proof of the maximum modulus principle (Theorem 6.15) would then show that f is identically zero in D, contradicting our assumption about f.

Thus, not all the Taylor coefficients of f are zero, and we may select the smallest integer k such that $\frac{f^{(k)}(\alpha)}{k!} \neq 0$. According to the results in Section 7.4, f has a zero of order k at α and can be written in the form
$$f(z) = (z-\alpha)^k g(z),$$

where g is analytic at α and $g(\alpha) \neq 0$. Since g is a continuous function, there exists a disk $D_r(\alpha)$ throughout which g is nonzero. Therefore $f(z) \neq 0$ in the punctured disk $D_r^*(\alpha)$. $\square$

The proofs of the following two corollaries are given as exercises.

Corollary 7.9. *Suppose that f is analytic in the domain D and that $\alpha \in D$. If there exists a sequence of points $\{z_n\} \in D$ such that $z_n \to \alpha$, and $f(z_n) = 0$, then $f(z) = 0$ for all $z \in D$.*

Corollary 7.10. *Suppose that f and g are analytic in the domain D, where $\alpha \in D$. If there exists a sequence $\{z_n\} \in D$ such that $z_n \to \alpha$, and $f(z_n) = g(z_n)$ for all n, then $f(z) = g(z)$ for all $z \in D$.*

Theorem 7.13 also allows us to give a simple argument for one version of L'Hôpital's rule.

239

Corollary 7.11. *(L'Hôpital's rule) Suppose that f and g are analytic at α. If $f(\alpha) = 0$ and $g(\alpha) = 0$, but $g'(\alpha) \neq 0$, then*

$$\lim_{z \to \alpha} \frac{f(z)}{g(z)} = \frac{f'(\alpha)}{g'(\alpha)}.$$

Proof. Because $g'(\alpha) \neq 0$, g is not identically zero and, by Theorem 7.13, there is a punctured disk $D_r^*(\alpha)$ in which $g(z) \neq 0$. Thus the quotient $\frac{f(z)}{g(z)} = \frac{f(z)-f(\alpha)}{g(z)-g(\alpha)}$ is defined for all $z \in D_r^*(\alpha)$, and we can write

$$\lim_{z \to \alpha} \frac{f(z)}{g(z)} = \lim_{z \to \alpha} \frac{f(z) - f(\alpha)}{g(z) - g(\alpha)} = \lim_{z \to \alpha} \frac{[f(z) - f(\alpha)]/(z - \alpha)}{[g(z) - g(\alpha)]/(z - \alpha)} = \frac{f'(\alpha)}{g'(\alpha)}.$$

$\square$

We can use the following Theorem to get Taylor series for quotients of analytic functions. Its proof involves ideas from Section 7.2, and we leave it as an exercise.

Theorem 7.14 (Division of power series). *Suppose that f and g are analytic at α with the power series representations*

$$f(z) = \sum_{n=0}^{\infty} a_n (z - \alpha)^n \quad and \quad g(z) = \sum_{n=0}^{\infty} b_n (z - \alpha)^n, \quad for\ all \quad z \in D_R(\alpha).$$

If $g(\alpha) \neq 0$, then the quotient $\frac{f}{g}$ has the power series representation

$$\frac{f(z)}{g(z)} = \sum_{n=0}^{\infty} c_n (z - \alpha)^n,$$

where the coefficients satisfy the equation

$$a_n = b_0 c_n + \cdots + b_{n-1} c_1 + b_n c_0.$$

In other words, we can obtain the series for the quotient $\frac{f(z)}{g(z)}$ by the familiar process of dividing the series for $f(z)$ by the series for $g(z)$, using the standard long division algorithm.

Example 7.15. Find the first few terms of the Maclaurin series for the function $f(z) = \sec z$, if $|z| < \frac{\pi}{2}$, and compute $f^{(4)}(0)$.

Solution:

Using long division, we see that

$$\sec z = \frac{1}{\cos z} = \frac{1}{1 - \frac{z^2}{2!} + \frac{z^4}{4!} - \frac{z^6}{6!} + \cdots} = 1 + \frac{1}{2} z^2 + \frac{5}{24} z^4 + \cdots.$$

Using Taylor's theorem, we see that if $f(z) = \sec z$, then $\frac{f^{(4)}(0)}{4!} = \frac{5}{24}$, so that $f^{(4)}(0) = 5$.

We close this section with some results concerning the behavior of complex functions at points near the different types of isolated singularities. The following is due to the German mathematician G. F. Bernhard Riemann (1826–1866).

Theorem 7.15 (Riemann). *Suppose that f is analytic in $D_r^*(\alpha)$. If f is bounded in $D_r^*(\alpha)$, then either f is analytic at α or f has a removable singularity at α.*

Proof. Consider the function g, defined as

$$g(z) = \begin{cases} (z-\alpha)^2 f(z), & \text{when } z \neq \alpha; \\ 0, & \text{when } z = \alpha. \end{cases} \tag{7.38}$$

Clearly, g is analytic in at least $D_r^*(\alpha)$. Straightforward calculation yields

$$g'(\alpha) = \lim_{z \to \alpha} \frac{g(z) - g(\alpha)}{z - \alpha} = \lim_{z \to \alpha} (z - \alpha) f(z) = 0.$$

The last equation follows because f is bounded. Thus g is also analytic at α, with $g(\alpha) = g'(\alpha) = 0$.

By Taylor's theorem, g has the representation

$$g(z) = \sum_{n=2}^{\infty} \frac{g^{(n)}(\alpha)}{n!} (z - \alpha)^n, \quad \text{for all} \quad z \in D_r(\alpha). \tag{7.39}$$

We divide both sides of Equation (7.39) by $(z - \alpha)^2$ and use Equation (7.38) to obtain the following power series representation for f:

$$f(z) = \sum_{n=2}^{\infty} \frac{g^{(n)}(\alpha)}{n!} (z - \alpha)^{n-2} = \sum_{n=0}^{\infty} \frac{g^{(n+2)}(\alpha)}{(n+2)!} (z - \alpha)^n.$$

By Theorem 4.17, f is analytic at α if we define $f(\alpha) = \frac{g^{(2)}(\alpha)}{2!}$. This observation completes the proof. $\square$

The proof of the following Corollary is given as an exercise.

Corollary 7.12. *If f is analytic in $D_r^*(\alpha)$, then f can be defined to be analytic at α iff $\lim_{z \to \alpha} f(z)$ exists and is finite.*

Theorem 7.16. *Suppose that f is analytic in $D_r^*(\alpha)$. The function f has a pole of order k at α, iff $\lim_{z \to \alpha} |f(z)| = \infty$.*

Proof. Suppose, first, that f has a pole of order k at α. Using Theorem 7.12, we can say that $f(z) = \frac{h(z)}{(z-\alpha)^k}$, where h is analytic at α, and $h(\alpha) \neq 0$. Because $\lim_{z \to a} |h(z)| = |h(\alpha)| \neq 0$ and $\lim_{z \to a} |(z - \alpha)| = 0$, we conclude that $\lim_{z \to a} |f(z)| - \lim_{z \to a} |h(z)| \lim_{z \to a} \frac{1}{|(z-\alpha)^k|} = \infty$.

Conversely, suppose that $\lim_{z \to \alpha} |f(z)| = \infty$. By the definition of a limit, there must be some $\delta > 0$ such that $|f(z)| > 1$ if $z \in D_\delta^*(\alpha)$. Thus the function $g(z) = \frac{1}{f(z)}$ is analytic and bounded (because $|g(z)| = |\frac{1}{f(z)}| \leq 1$) in $D_\delta^*(\alpha)$. By Theorem 7.15, we may define g at α so that g is analytic in all of $D_\delta(\alpha)$. In fact, $|g(\alpha)| = \lim_{z \to a} \frac{1}{|f(z)|} = 0$, so α is a zero of g. We claim that α must be of finite order; otherwise, we would have $g^{(n)}(\alpha) = 0$, for all n, and hence $g(z) = \sum_{n=0}^{\infty} \frac{g^{(n)}(\alpha}{n!} (z - \alpha)^n = 0$, for all $z \in D_\delta(\alpha)$. As $g(z) = \frac{1}{f(z)}$ is analytic in $D_\delta^*(\alpha)$, this result is impossible, so we can let k be the order of the zero of g at α. By Corollary 7.5, f has a pole of order k. $\square$

Theorem 7.17. *The function f has an essential singularity at α iff $\lim\limits_{z \to \alpha} |f(z)|$ does not exist.*

Proof. From Corollary 7.12 and Theorem 7.16, the conclusion of Theorem 7.17 is the only option possible. □

Example 7.16. Show that the function g defined by

$$g(z) = \begin{cases} e^{-\left(\frac{1}{z^2}\right)}, & \text{when } z \neq 0, \quad \text{and} \\ 0, & \text{when} \quad z = 0, \end{cases}$$

is not continuous at $z = 0$.

Solution:

In Exercise 20, Section 7.2, we asked you to show this relation by computing limits along the real and imaginary axes. Note, however, that the Laurent series for $g(z)$ in the annulus $D_r^*(0)$ is

$$g(z) = 1 + \sum_{n=1}^{\infty} (-1)^n \frac{1}{z^{2n}},$$

so that 0 is an essential singularity for g. According to Theorem 7.17, $\lim\limits_{z \to 0} |g(z)|$ doesn't exist, so g is not continuous at 0.

Exercises for Section 7.5 (Selected answers or hints are on page 449.)

1. Determine whether there exists a function f that is analytic at 0 with the property that, for $n = 1, 2, 3, \ldots$,

   (a) $f(\frac{1}{2n}) = 0$ and $f(\frac{1}{2n-1}) = 1$.
   (b) $f(\frac{1}{n}) = f(-\frac{1}{n}) = \frac{1}{n^2}$.
   (c) $f(\frac{1}{n}) = f(-\frac{1}{n}) = \frac{1}{n^3}$.

2. Prove the following results.

   (a) Corollary 7.9.
   (b) Corollary 7.10.
   (c) Theorem 7.14.
   (d) Corollary 7.12.

3. Consider the function $f(z) = z \sin(\frac{1}{z})$.

   (a) Show that there is a sequence $\{z_n\}$ of points converging to 0 such that $f(z_n) = 0$ for $n = 1, 2, 3, \ldots$.
   (b) Does this result contradict Corollary 7.9? Why or why not?

4. Let $f(z) = \tan z$.

   (a) Use Theorem 7.14 to find the first few terms of the Maclaurin series for $f(z)$, if $|z| < \frac{\pi}{2}$.
   (b) What are the values of $f^{(6)}(0)$ and $f^{(7)}(0)$?

5. Show that the real function f defined by

$$f(x) = \begin{cases} x\sin(\frac{1}{x}), & \text{when } x \neq 0, \quad \text{and} \\ 0, & \text{when } \quad x = 0, \end{cases}$$

is continuous at $x = 0$ but that the corresponding function $g(z)$ defined by

$$g(z) = \begin{cases} z\sin(\frac{1}{z}), & \text{when } \quad z \neq 0, \quad \text{and} \\ 0, & \text{when } \quad z = 0, \end{cases}$$

is *not* continuous at $z = 0$.

6. Use L'Hôpital's rule to find the following limits.

(a) $\displaystyle\lim_{z \to 1+i} \frac{z-1-i}{z^4+4}$.

(b) $\displaystyle\lim_{z \to i} \frac{z^2-2iz-1}{z^4+2z^2+1}$.

(c) $\displaystyle\lim_{z \to i} \frac{1+z^6}{1+z^2}$.

(d) $\displaystyle\lim_{z \to 0} \frac{\sin z + \sinh z - 2z}{z^5}$.

Chapter 8

Residue Theory

Overview

We now have the necessary machinery to see some amazing applications of the tools we developed in the last few chapters. You will how Laurent expansions can give useful information concerning seemingly unrelated properties of complex functions. You will also learn how the ideas of complex analysis make the solution of very complicated integrals of real-valued functions as easy–literally–as the computation of residues. We begin with a theorem relating residues to the evaluation of complex integrals

8.1 The Residue Theorem

The Cauchy integral formulas given in Section 6.5 are useful in evaluating contour integrals over a simple closed contour C where the integrand has the form $\frac{f(z)}{(z-z_0)^k}$ and f is an analytic function. In this case, the singularity of the integrand is at worst a pole of order k at z_0. We begin this section by extending this result to integrals that have a finite number of isolated singularities inside the contour C. This new method can be used in cases where the integrand has an essential singularity at z_0 and is an important extension of the previous method.

Definition 8.1 (Residue). *Let f have a nonremovable isolated singularity at the point z_0. Then f has the Laurent series representation for all z in some disk $D_R^*(z_0)$ given by $f(z) = \sum_{n=-\infty}^{\infty} a_n(z - z_0)^n$. The coefficient a_{-1} is called the **residue of f at z_0**. We use the notation*

$$\text{Res}[f, z_0] = a_{-1}.$$

Example 8.1. If $f(z) = \exp(\frac{2}{z})$, then the Laurent series of f about the point 0 has the form

$$f(z) = \exp\left(\frac{2}{z}\right) = 1 + \frac{2}{z} + \frac{2^2}{2!z^2} + \frac{2^3}{3!z^3} + \cdots.$$

and $\text{Res}[f, 0] = a_{-1} = 2$.

Example 8.2. Find $\text{Res}[g, 0]$ if $g(z) = \frac{3}{2z+z^2-z^3}$.

Solution:

Using the techniques of Example 7.7, we find that g has three Laurent series representations involving powers of z. In the punctured disk $D_1^*(0)$, $g(z) = \sum_{n=0}^{\infty} [(-1)^n + \frac{1}{2^{n+1}}] z^{n-1}$. Computing the first few coefficients, we obtain

$$g(z) = \frac{3}{2} \frac{1}{z} - \frac{3}{4} + \frac{9}{8} z - \frac{15}{16} z^2 + \cdots .$$

Therefore, $\text{Res}[g, 0] = a_{-1} = \frac{3}{2}$.

Recall that, for a function f analytic in $D_R^*(z_0)$ and for any r with $0 < r < R$, the Laurent series coefficients of f are given by

$$a_n = \frac{1}{2\pi i} \int_{C_r^+(z_0)} \frac{f(z)}{(z - z_0)^{n+1}} dz \quad \text{for} \quad n = 0, \pm 1, \pm 2, \ldots, \tag{8.1}$$

where $C_r^+(z_0)$ denotes the circle $\{z : |z - z_0| = r\}$ with positive orientation. This result gives us an important fact concerning $\text{Res}[f, z_0]$. If we set $n = -1$ in Equation (8.1) and replace $C_r^+(z_0)$ with any positively oriented simple closed contour C containing z_0, provided z_0 is the still only singularity of f that lies inside C, then we obtain

$$\int_C f(z) \, dz = 2\pi i a_{-1} = 2\pi i \text{Res}[f, z_0]. \tag{8.2}$$

If we are able to find the Laurent series expansion for f, then Equation (8.2) gives us an important tool for evaluating contour integrals.

Example 8.3. Evaluate $\int_{C_1^+(0)} \exp(\frac{2}{z}) \, dz$.

Solution:

Example 8.1 showed that the residue of $f(z) = \exp\left(\frac{2}{z}\right)$ at $z_0 = 0$ is $\text{Res}[f, 0] = 2$. Using Equation (8.2), we get

$$\int_{C_1^+(0)} \exp\left(\frac{2}{z}\right) dz = 2\pi i \text{Res}[f, 0] = 4\pi i.$$

Theorem 8.1 (Cauchy's residue theorem). *Let D be a simply connected domain and let C be a simple closed positively oriented contour that lies in D. If f is analytic inside C and on C, except at the points $z_1, z_2, \ldots, z_n$ that lie inside C, then*

$$\int_C f(z) \, dz = 2\pi i \sum_{k=1}^{n} \text{Res}[f, z_k].$$

The situation is illustrated in Figure 8.1.

Proof. Since there are a finite number of singular points inside C, there exists an $r > 0$ such that the positively oriented circles $C_k = C_r^+(z_k)$, for $k = 1, 2, \ldots, n$, are mutually disjoint and all lie inside C. From the extended Cauchy-Goursat theorem (Theorem 6.7 on page 189), it follows that

$$\int_C f(z) \, dz = \sum_{k=1}^{n} \int_{C_k} f(z) \, dz.$$

245

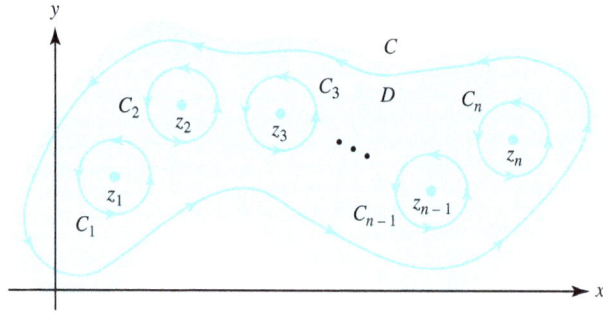

Figure 8.1: The domain D and contour C and the singular points z_1, z_2, ..., z_n in the statement of Cauchy's residue theorem

The function f is analytic in a punctured disk with center z_k that contains the circle C_k, so we can use Equation (8.2) to obtain

$$\int_{C_k} f(z)\, dz = 2\pi i \operatorname{Res}[f, z_k], \quad \text{for} \quad k = 1, 2, \ldots, n.$$

Combining the last two equations gives the desired result. $\qquad\square$

The domain D and contour C and the singular points z_1, z_2, ..., z_n in the statement of Cauchy's residue theorem.

The calculation of a Laurent series expansion is tedious in most circumstances. As the residue at z_0 involves only the coefficient a_{-1} in the Laurent expansion, we seek a method to calculate the residue from special information about the nature of the singularity at z_0.

If f has a removable singularity at z_0, then $a_{-n} = 0$, for $n = 1, 2, \ldots$. Therefore, $\operatorname{Res}[f, z_0] = 0$. The following theorem gives methods for evaluating residues at poles.

Theorem 8.2 (Residues at Poles).

i. If f has a simple pole at z_0, then

$$\operatorname{Res}[f, z_0] = \lim_{z \to z_0} (z - z_0) f(z).$$

ii. If f has a pole of order 2 at z_0, then

$$\operatorname{Res}[f, z_0] = \lim_{z \to z_0} \frac{d}{dz} (z - z_0)^2 f(z).$$

iii. If f has a pole of order k at z_0, then

$$\operatorname{Res}[f, z_0] = \frac{1}{(k-1)!} \lim_{z \to z_0} \frac{d^{k-1}}{dz^{k-1}} (z - z_0)^k f(z).$$

Proof. If f has a simple pole at z_0, then the Laurent series is

$$f(z) = \frac{a_{-1}}{z - z_0} + a_0 + a_1(z - z_0) + a_2(z - z_0)^2 + \cdots.$$

246

If we multiply both sides of this equation by $(z - z_0)$ and take the limit as $z \to z_0$, we obtain

$$\lim_{z \to z_0} (z - z_0) f(z) = \lim_{z \to z_0} [a_{-1} + a_0(z - z_0) + a_1(z - z_0)^2 + \cdots]$$
$$= a_{-1} = \text{Res}[f, z_0].$$

which establishes part (i). We proceed to part (iii), as part (ii) is a special case of it. Suppose that f has a pole of order k at z_0. Then f can be written as

$$f(z) = \frac{a_{-k}}{(z - z_0)^k} + \frac{a_{-k+1}}{(z - z_0)^{k-1}} + \cdots + \frac{a_{-1}}{z - z_0} + a_0 + a_1(z - z_0) + \cdots .$$

Multiplying both sides of this equation by $(z - z_0)^k$ gives

$$(z - z_0)^k f(z) = a_{-k} + \cdots + a_{-1}(z - z_0)^{k-1} + a_0(z - z_0)^k + \cdots .$$

If we differentiate both sides $k - 1$ times we get

$$\frac{d^{k-1}}{dz^{k-1}}[(z - z_0)^k f(z)] = (k - 1)!a_{-1} + k!a_0(z - z_0)$$
$$+ (k + 1)!\, a_1(z - z_0)^2 + \cdots .$$

and when we let $z \to z_0$ the result is

$$\lim_{z \to z_0} \frac{d^{k-1}}{dz^{k-1}}[(z - z_0)^k f(z)] = (k - 1)!a_{-1} = (k - 1)!\text{Res}[f, z_0],$$

which establishes part (iii). $\qquad\qquad\square$

Example 8.4. Find the residue of $f(z) = \frac{\pi \cot(\pi z)}{z^2}$ at $z_0 = 0$.

Solution:

We write $f(z) = \frac{\pi \cos(\pi z)}{z^2 \sin(\pi z)}$. Because $z^2 \sin \pi z$ has a zero of order 3 at $z_0 = 0$ and $\pi \cos(\pi z_0) \neq 0$, f has a pole of order 3 at z_0. By part (iii) of Theorem 8.2, we have

$$\text{Res}[f, 0] = \frac{1}{2!} \lim_{z \to 0} \frac{d^2}{dz^2} \pi z \cot(\pi z)$$
$$= \frac{1}{2} \lim_{z \to 0} \frac{d}{dz} [\pi \cot(\pi z) - \pi^2 z \csc^2(\pi z)]$$
$$= \frac{1}{2} \lim_{z \to 0} [-\pi^2 \csc^2(\pi z) - \pi^2 \{\csc^2(\pi z) - 2\pi z \csc^2(\pi z) \cot(\pi z)\}]$$
$$= \pi^2 \lim_{z \to 0} (\pi z \cot(\pi z) - 1) \csc^2(\pi z)$$
$$= \pi^2 \lim_{z \to 0} \frac{\pi z \cos(\pi z) - \sin(\pi z)}{\sin^3(\pi z)}.$$

This last limit involves an indeterminate form, which we evaluate by using L'Hôpital's rule:

$$\text{Res}[f, 0] = \pi^2 \lim_{z \to 0} \frac{-\pi^2 z \sin(\pi z) + \pi \cos(\pi z) - \pi \cos(\pi z)}{3\pi \sin^2(\pi z) \cos(\pi z)}$$
$$= \pi^2 \lim_{z \to 0} \frac{-\pi z}{3 \sin(\pi z) \cos(\pi z)}$$
$$= -\frac{\pi^2}{3} \lim_{z \to 0} \frac{\pi z}{\sin(\pi z)} \lim_{z \to 0} \frac{1}{\cos(\pi z)}$$
$$= -\frac{\pi^2}{3}.$$

Example 8.5. Find $\int\limits_{C_3^+(0)} \frac{1}{z^4+z^3-2z^2}\, dz$.

Solution:

We write the integrand as $f(z) = \frac{1}{z^2(z+2)(z-1)}$. The singularities of f that lie inside $C_3(0)$ are simple poles at the points 1 and -2, and a pole of order 2 at the origin. We compute the residues as follows:

$$\operatorname{Res}[f, 0] = \lim_{z \to 0} \frac{d}{dz}[z^2 f(z)] = \lim_{z \to 0} \frac{-2z-1}{(z^2+z-2)^2} = -\frac{1}{4}.$$

$$\operatorname{Res}[f, 1] = \lim_{z \to 1}(z-1)f(z) = \lim_{z \to 1} \frac{1}{z^2(z+2)} = \frac{1}{3}, \quad \text{and}$$

$$\operatorname{Res}[f, -2] = \lim_{z \to -2}(z+2)f(z) = \lim_{z \to -2} \frac{1}{z^2(z-1)} = -\frac{1}{12}.$$

Finally, the residue theorem yields

$$\int\limits_{C_3^+(0)} \frac{1}{z^4+z^3-2z^2}\, dz = 2\pi i \left[-\frac{1}{4} + \frac{1}{3} - \frac{1}{12} \right] = 0.$$

The answer $\int\limits_{C_3^+(0)} \frac{1}{z^4+z^3-2z^2}\, dz = 0$ is not at all obvious, and all the preceding calculations are required to get it.

Example 8.6. Find $\int\limits_{C_2^+(1)} (z^4+4)^{-1}\, dz$.

Solution:

The singularities of the integrand $f(z) = \frac{1}{z^4+4}$ that lie inside $C_2(1)$ are simple poles occurring at the points $1 \pm i$, as the points $-1 \pm i$ lie outside $C_2(1)$. Factoring the denominator is tedious, so we use a different approach. If z_0 is any one of the singularities of f, then we can use L'Hôpital's rule to compute $\operatorname{Res}[f, z_0]$:

$$\operatorname{Res}[f, z_0] = \lim_{z \to z_0} \frac{z-z_0}{z^4+4} = \lim_{z \to z_0} \frac{1}{4z^3} = \frac{1}{4z_0^3}.$$

Since $z_0^4 = -4$, we can simplify this expression further to yield $\operatorname{Res}[f, z_0] = -\frac{1}{16}z_0$. Hence $\operatorname{Res}[f, 1+i] = \frac{-1-i}{16}$, and $\operatorname{Res}[f, 1-i] = \frac{-1+i}{16}$. We now use the residue theorem to get

$$\int\limits_{C_2^+(1)} \frac{1}{z^4+4}\, dz = 2\pi i \left(\frac{-1-i}{16} + \frac{-1+i}{16} \right) = -\frac{\pi i}{4}.$$

The theory of residues can be used to expand the quotient of two polynomials into its *partial fraction* representation.

Example 8.7. Let $P(z)$ be a polynomial of degree at most 2. Show that if a, b, and c are distinct complex numbers, then

248

$$f(z) = \frac{P(z)}{(z-a)(z-b)(z-c)}$$

$$= \frac{A}{z-a} + \frac{B}{z-b} + \frac{C}{z-c}, \quad \text{where}$$

$$A = \text{Res}[f, a] = \frac{P(a)}{(a-b)(a-c)}.$$

$$B = \text{Res}[f, b] = \frac{P(b)}{(b-a)(b-c)} \quad \text{and}$$

$$C = \text{Res}[f, c] = \frac{P(c)}{(c-a)(c-b)}.$$

Solution:

It will suffice to prove that $A = \text{Res}[f, a]$. We expand f in its Laurent series about the point a by writing the three terms $\frac{A}{z-a}$, $\frac{B}{z-b}$, and $\frac{C}{z-c}$ in their Laurent series about the point a and adding them. The term $\frac{A}{z-a}$ is itself a one-term Laurent series about the point a. The term $\frac{B}{z-b}$ is analytic at the point a, and its Laurent series is actually a Taylor series given by

$$\frac{B}{z-b} = \frac{-B}{b-a} \frac{1}{1 - \frac{z-a}{b-a}} = -\sum_{n=0}^{\infty} \frac{B}{(b-a)^{n+1}} (z-a)^n,$$

which is valid for $|z-a| < |b-a|$. Likewise, the expansion of the term $\frac{C}{z-c}$ is

$$\frac{C}{z-c} = -\sum_{n=0}^{\infty} \frac{C}{(c-a)^{n+1}} (z-a)^n,$$

which is valid for $|z-a| < |c-a|$. Thus the Laurent series of f about the point a is

$$f(z) = \frac{A}{z-a} - \sum_{n=0}^{\infty} \left[\frac{B}{(b-a)^{n+1}} + \frac{C}{(c-a)^{n+1}} \right] (z-a)^n,$$

which is valid for $|z-a| < R$, where $R = \min\{|b-a|, |c-a|\}$. Therefore $A = \text{Res}[f, a]$, and calculation reveals that

$$\text{Res}[f, a] = A = \lim_{z \to a} \frac{P(z)}{(z-b)(z-c)} = \frac{P(a)}{(a-b)(a-c)}.$$

Example 8.8. Express $f(z) = \frac{3z+2}{z(z-1)(z-2)}$ in partial fractions.

Solution:

Computing the residues, we obtain

$$\text{Res}[f, 0] = 1, \quad \text{Res}[f, 1] = -5, \quad \text{and} \quad \text{Res}[f, 2] = 4.$$

Example 8.7 gives us

$$\frac{3z+2}{z(z-1)(z-2)} = \frac{1}{z} - \frac{5}{z-1} + \frac{4}{z-2}.$$

Remark 8.1. *If a repeated root occurs, then the process is similar, and we can easily show that if $P(z)$ has degree of at most 2, then*

$$f(z) = \frac{P(z)}{(z-a)^2(z-b)} = \frac{A}{(z-a)^2} + \frac{B}{z-a} + \frac{C}{z-b},$$

where $A = \text{Res}[(z-a)f(z), a]$, $B = \text{Res}[f, a]$, and $C = \text{Res}[f, b]$.

Example 8.9. Express $f(z) = \frac{z^2+3z+2}{z^2(z-1)}$ in partial fractions.

Solution:

Using the previous remark, we have

$$f(z) = \frac{A}{(z-a)^2} + \frac{B}{z-a} + \frac{C}{z-b},$$

where

$$A = \text{Res}[z f(z), 0] = \lim_{z \to 0} \frac{z^2+3z+2}{z-1} = -2.$$

$$B = \text{Res}[f, 0] = \lim_{z \to 0} \frac{d}{dz} \frac{z^2+3z+2}{z-1}$$

$$= \lim_{z \to 0} \frac{(2z+3)(z-1) - (z^2+3z+2)}{(z-1)^2} = -5, \quad \text{and}$$

$$C = \text{Res}[f, 1] = \lim_{z \to 1} \frac{z^2+3z+2}{z^2} = 6.$$

Thus,

$$\frac{z^2+3z+2}{z^2(z-1)} = \frac{-2}{z^2} - \frac{5}{z} + \frac{6}{z-1}.$$

Exercises for Section 8.1 (Selected answers or hints are on page 449.)

1. Find $\text{Res}[f, 0]$ for each of the following :

   (a) $f(z) = z^{-1} \exp z$.

   (b) $f(z) = z^{-3} \cosh 4z$.

   (c) $f(z) = \csc z$.

   (d) $f(z) = \frac{z^2+4z+5}{z^2+z}$.

   (e) $f(z) = \cot z$.

   (f) $f(z) = z^{-3} \cos z$.

   (g) $f(z) = z^{-1} \sin z$.

   (h) $f(z) = \frac{z^2+4z+5}{z^3}$.

   (i) $f(z) = \exp(1 + \frac{1}{z})$.

   (j) $f(z) = z^4 \sin(\frac{1}{z})$.

   (k) $f(z) = z^{-1} \csc z$.

   (l) $f(z) = z^{-2} \csc z$.

   (m) $f(z) = \frac{\exp(4z)-1}{\sin^2 z}$.

   (n) $f(z) = z^{-1} \csc^2 z$.

2. Let f and g have an isolated singularity at z_0. Show that

$$\text{Res}[f + g, z_0] = \text{Res}[f, z_0] + \text{Res}[g, z_0].$$

250

3. Evaluate the following:

(a) $\int\limits_{C_1^+(-1+i)} \frac{1}{z^4+4}\, dz.$

(b) $\int\limits_{C_2^+(i)} \frac{1}{z(z^2-2z+2)}\, dz.$

(c) $\int\limits_{C_2^+(0)} \frac{\exp z}{z^3+z}\, dz.$

(d) $\int\limits_{C_2^+(0)} \frac{\sin z}{4z^2-\pi^2}\, dz.$

(e) $\int\limits_{C_2^+(0)} \frac{\sin z}{z^2+1}\, dz.$

(f) $\int\limits_{C_1^+(0)} \frac{1}{z^2 \sin z}\, dz.$

(g) $\int\limits_{C_1^+(0)} \frac{1}{z \sin^2 z}\, dz.$

4. Let f and g be analytic at z_0. If $f(z_0) \neq 0$ and g has a simple zero at z_0, then show that
$\text{Res}\left[\frac{f}{g}, z_0\right] = \frac{f(z_0)}{g'(z_0)}.$

5. Find $\int\limits_C (z-1)^{-2}(z^2+4)^{-1}\, dz$ when

(a) $C = C_1^+(1).$
(b) $C = C_4^+(0).$

6. Find $\int\limits_C (z^6+1)^{-1} dz$ when

(a) $C = C_{\frac{1}{2}}^+(i).$
(b) $C = C_1^+(\frac{1+i}{2}).$
   Hint: If z_0 is a singularity of $f(z) = \frac{1}{z^6+1}$, show that $\text{Res}[f, z_0] = -\frac{1}{6}z_0.$

7. Find $\int\limits_C (3z^4+10z^2+3)^{-1}\, dz$ when

(a) $C = C_1^+(i\sqrt{3}).$
(b) $C = C_1^+(\frac{i}{\sqrt{3}}).$

8. Find $\int\limits_C (z^4-z^3-2z^2)^{-1} dz$ when

(a) $C = C_{\frac{1}{2}}^+(0).$
(b) $C = C_{\frac{3}{2}}^+(0).$

9. Use residues to find the partial fraction representations of:

(a) $\frac{1}{z^2+3z+2}.$
(b) $\frac{3z-3}{z^2-z-2}.$

(c) $\frac{z^2-7z+4}{z^2(z+4)}$.

(d) $\frac{10z}{(z^2+4)(z^2+9)}$.

(e) $\frac{2z^2-3z-1}{(z-1)^3}$.

(f) $\frac{z^3+3z^2-z+1}{z(z+1)^2(z^2+1)}$.

10. Let f be analytic in a simply connected domain D, and let C be a simple closed positively oriented contour in D. If z_0 is the only zero of f in D and z_0 lies interior to C, then show that $\frac{1}{2\pi i}\int_C \frac{f'(z)}{f(z)}\,dz = k$, where k is the order of the zero at z_0.

11. Let f be analytic at the points $0, \pm 1, \pm 2, \dots$. If $g(z) = \pi f(z)\cot \pi z$, then show that $\mathrm{Res}[g, n] = f(n)$ for $n = 0, \pm 1, \pm 2, \dots$.

8.2 Trigonometric Integrals

As indicated at the beginning of this chapter, we can evaluate certain definite *real* integrals with the aid of the residue theorem. One way to do this is by interpreting the definite integral as the parametric form of an integral of an analytic function along a simple closed contour. Suppose that we want to evaluate an integral of the form

$$\int_0^{2\pi} F(\cos\theta, \sin\theta)\,d\theta. \tag{8.3}$$

where $F(u, v)$ is a function of the two real variables u and v. Consider the unit circle $C_1(0)$ with parametrization

$$C_1^+(0) : z = \cos\theta + i\sin\theta = e^{i\theta}, \quad \text{for} \quad 0 \le \theta \le 2\pi,$$

which gives the symbolic differentials

$$dz = (-\sin\theta + i\cos\theta)d\theta = ie^{i\theta}\,d\theta, \quad \text{and}$$

$$d\theta = \frac{dz}{ie^{i\theta}} = \frac{dz}{iz}. \tag{8.4}$$

Combining $z = \cos\theta + i\sin\theta$ with $\frac{1}{z} = \cos\theta - i\sin\theta$, we obtain

$$\cos\theta = \frac{1}{2}\left(z + \frac{1}{z}\right), \quad \text{and} \quad \sin\theta = \frac{1}{2i}\left(z - \frac{1}{z}\right). \tag{8.5}$$

Using the substitutions for $\cos\theta$, $\sin\theta$, and $d\theta$ in Expression 8.3 transforms the definite integral into the contour integral

$$\int_0^{2\pi} F(\cos\theta, \sin\theta)\,d\theta = \int_{C_1^+(0)} f(z)\,dz.$$

where the new integrand is $f(z) = \dfrac{F\left(\frac{1}{2}(z + \frac{1}{z}), \frac{1}{2i}(z - \frac{1}{z})\right)}{iz}$.

Suppose that f is analytic inside and on the unit circle $C_1(0)$, except at the points $z_1, z_2, \dots, z_n$ that lie interior to $C_1(0)$. Then the residue theorem gives

$$\int_0^{2\pi} F(\cos\theta, \sin\theta)\,d\theta = 2\pi i\sum_{k=1}^{n}\mathrm{Res}[f, z_k]. \tag{8.6}$$

The situation is illustrated in Figure 8.2.

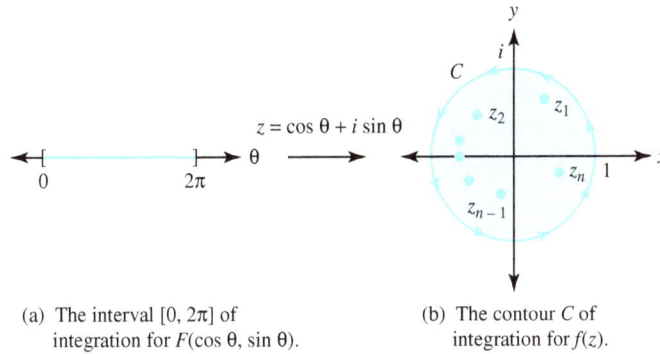

(a) The interval $[0, 2\pi]$ of integration for $F(\cos\theta, \sin\theta)$.

(b) The contour C of integration for $f(z)$.

Figure 8.2: Changing variables: a definite integral on $[0, 2\pi]$ to a contour integral around C

Example 8.10. Evaluate $\int_0^{2\pi} \frac{1}{1+3\cos^2\theta}\, d\theta$ by using complex analysis.

Solution:

Using Substitutions 8.4 and 8.5, we transform the integral to

$$\int_{C_1^+(0)} \frac{1}{1 + 3(\frac{z+z^{-1}}{2})^2} \left(\frac{1}{iz}\right) dz = \int_{C_1^+(0)} \frac{-i4z}{3z^4 + 10z^2 + 3}\, dz = \int_{C_1^+(0)} f(z)\, dz.$$

where $f(z) = \frac{-i4z}{3z^4+10z^2+3}$. The singularities of f are poles located at the points where $3(z^2)^2 + 10(z^2) + 3 = 0$. Using the quadratic formula, we see that the singular points satisfy the relation $z^2 = \frac{-10 \pm \sqrt{100-36}}{6} = \frac{-5 \pm 4}{3}$. Hence the only singularities that lie inside the unit circle are simple poles corresponding to the solutions of $z^2 = -\frac{1}{3}$, which are the two points $z_1 = \frac{i}{\sqrt{3}}$ and $z_2 = -\frac{i}{\sqrt{3}}$. We use Theorem 8.2 and L'Hôpital's rule to get the residues at z_k, for $k = 1, 2$:

$$\begin{aligned}
\text{Res}[f, z_k] &= \lim_{z \to z_k} \frac{-i4z(z - z_k)}{3z^4 + 10z^2 + 3} \\
&= \lim_{z \to z_k} \frac{-i4(2z - z_k)}{12z^3 + 20z} \\
&= \frac{-i4z_k}{12z_k^3 + 20z_k} \\
&= \frac{-i}{3z_k^2 + 5}.
\end{aligned}$$

Since $z_k = \frac{\pm i}{\sqrt{3}}$ and $z_k^2 = -\frac{1}{3}$, the residues for each z_k are given by $\text{Res}[f, z_k] = -\frac{i}{3(-\frac{1}{3})+5} = -\frac{i}{4}$. Now use Equation (8.6) to compute the value of the integral:

$$\int_0^{2\pi} \frac{1}{1 + 3\cos^2\theta}\, d\theta = 2\pi i \left(\frac{-i}{4} + \frac{-i}{4}\right) = \pi.$$

Example 8.11. Evaluate $\int_0^{2\pi} \frac{1}{1+3\cos^2 t}\, dt$ by using a computer algebra system.

Solution:

We can obtain the antiderivative of $\frac{1}{1+3\cos^2 t}$ by using software such as Mathematica or MAPLE. It is $\int \frac{1}{1+3\cos^2 t}\, dt = \frac{-\text{Arctan}(2\cot t)}{2} = g(t)$. Since $\cot 0$ and $\cot 2\pi$ are not defined, the computations for both $g(0)$ and $g(2\pi)$ are indeterminate. The graph $s = g(t)$ shown in Figure 8.3 reveals another problem: The integrand $\frac{1}{1+3\cos^2 t}$ is a continuous function for all t, but the function g

253

has a discontinuity at π. This condition appears to be a violation of the fundamental theorem of calculus, which asserts that the integral of a continuous function must be differentiable and hence continuous. The problem is that $g(t)$ is not an antiderivative of $\frac{1}{1+3\cos^2 t}$ for *all* t in the interval $[0, 2\pi]$. Oddly, it is the antiderivative at all points *except* 0, π, and 2π, which you can verify by computing $g'(t)$ and showing that it equals $\frac{1}{1+3\cos^2 t}$ whenever $g(t)$ is defined.

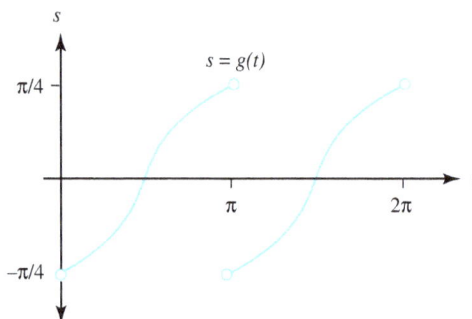

Figure 8.3: Graph of $g(t) = \int \frac{1}{1+3\cos^2 t}\, dt = \frac{-\mathrm{Arctan}(2\cot t)}{2}$

The integration algorithm used by computer algebra systems here (the Risch-Norman algorithm) gives the antiderivative $g(t) = \frac{-\mathrm{Arctan}(2\cot t)}{2}$, and we must take great care in using this information.

We get the proper value of the integral by using $g(t)$ on the open subintervals $(0, \pi)$ and $(\pi, 2\pi)$ where it is continuous, and taking appropriate limits:

$$\int_0^{2\pi} \frac{1}{1+3\cos^2 t}\, dt = \left[\int_0^{\pi} \frac{1}{1+3\cos^2 t}\, dt\right] + \left[\int_{\pi}^{2\pi} \frac{1}{1+3\cos^2 t}\, dt\right]$$

$$= \left[\lim_{t\to\pi^-,\, s\to 0^+} \int_s^t \frac{1}{1+3\cos^2 t}\, dt\right] + \left[\lim_{t\to 2\pi^-,\, s\to\pi^+} \int_s^t \frac{1}{1+3\cos^2 t}\, dt\right]$$

$$= \left[\lim_{t\to\pi^-} g(t) - \lim_{s\to 0^+} g(s)\right] + \left[\lim_{t\to 2\pi^-} g(t) - \lim_{s\to\pi^+} g(s)\right]$$

$$= \left[\frac{\pi}{4} - \frac{-\pi}{4}\right] + \left[\frac{\pi}{4} - \frac{-\pi}{4}\right]$$

$$= \pi.$$

Example 8.12. Evaluate $\int_0^{2\pi} \frac{\cos 2\theta}{5 - 4\cos\theta}\, d\theta$.

Solution:

For values of z that lie on the unit circle $C_1(0)$, we have

$$z^2 = \cos 2\theta + i\sin 2\theta, \quad \text{and} \quad z^{-2} = \cos 2\theta - i\sin 2\theta.$$

We solve for $\cos 2\theta$ and $\sin 2\theta$ to obtain the substitutions

$$\cos 2\theta = \frac{1}{2}(z^2 + z^{-2}), \quad \text{and} \quad \sin 2\theta = \frac{1}{2i}(z^2 - z^{-2}).$$

Using the identity for $\cos 2\theta$ along with Substitutions 8.4 and 8.5, we rewrite the integral as

$$\int_{C_1^+(0)} \frac{\frac{1}{2}(z^2 + z^{-2})}{5 - 4\left(\frac{z+z^{-1}}{2}\right)} \left(\frac{1}{iz}\right) dz = \int_{C_1^+(0)} \frac{i(z^4 + 1)}{2z^2(z - 2)(2z - 1)}\, dz = \int_{C_1^+(0)} f(z)\, dz,$$

254

where $f(z) = \frac{i(z^4+1)}{2z^2(z-2)(2z-1)}$. The singularities of f lying inside C are poles located at the points 0 and $\frac{1}{2}$. We use Theorem 8.2 to get the residues:

$$\text{Res}[f,0] = \lim_{z \to 0} \frac{d}{dz}\left[z^2 f(z)\right] = \lim_{z \to 0} \frac{d}{dz}\left[i\frac{(z^4+1)}{2(2z^2-5z+2)}\right]$$

$$= \lim_{z \to 0}\left[i\frac{4z^3(2z^2-5z+2)-(4z-5)(z^4+1)}{2(2z^2-5z+2)^2}\right]$$

$$= \frac{5i}{8},$$

and

$$\text{Res}\left[f,\frac{1}{2}\right] = \lim_{z \to \frac{1}{2}}\left(z-\frac{1}{2}\right)f(z) = \lim_{z \to \frac{1}{2}}\frac{i(z^4+1)}{4z^2(z-2)} = -\frac{17i}{24}.$$

Therefore we conclude that

$$\int_0^{2\pi} \frac{\cos 2\theta}{5-4\cos\theta}\,d\theta = 2\pi i\left(\frac{5i}{8}-\frac{17i}{24}\right) = \frac{\pi}{6}.$$

Exercises for Section 8.2 (Selected answers or hints are on page 450.)

Use residues to evaluate the following integrals:

1. $\int_0^{2\pi} \frac{1}{3\cos\theta+5}\,d\theta.$

2. $\int_0^{2\pi} \frac{1}{4\sin\theta+5}\,d\theta.$

3. $\int_0^{2\pi} \frac{1}{15\sin^2\theta+1}\,d\theta.$

4. $\int_0^{2\pi} \frac{1}{5\cos^2\theta+4}\,d\theta.$

5. $\int_0^{2\pi} \frac{\sin^2\theta}{5+4\cos\theta}\,d\theta.$

6. $\int_0^{2\pi} \frac{\sin^2\theta}{5-3\cos\theta}\,d\theta.$

7. $\int_0^{2\pi} \frac{1}{(5+3\cos\theta)^2}\,d\theta.$

8. $\int_0^{2\pi} \frac{1}{(5+4\cos\theta)^2}\,d\theta.$

9. $\int_0^{2\pi} \frac{\cos 2\theta}{5+3\cos\theta}\,d\theta.$

10. $\int_0^{2\pi} \frac{\cos 2\theta}{13-12\cos\theta}\,d\theta.$

11. $\int_0^{2\pi} \frac{1}{(1+3\cos^2\theta)^2}\,d\theta.$

12. $\int_0^{2\pi} \frac{1}{(1+8\cos^2\theta)^2}\,d\theta.$

13. $\int_0^{2\pi} \frac{\cos^2 3\theta}{5-4\cos 2\theta}\,d\theta.$

14. $\int_0^{2\pi} \frac{\cos^2 3\theta}{5-3\cos 2\theta}\,d\theta.$

15. $\int_0^{2\pi} \frac{1}{a\cos\theta+b\sin\theta+d}\,d\theta$, where a, b, and d are real, and $a^2+b^2 < d^2$.

16. $\int_0^{2\pi} \frac{1}{a\cos^2\theta+b\sin^2\theta+d}\,d\theta$, where a, b, and d are real, $a > d$, and $b > d$.

255

8.3 Improper Integrals of Rational Functions

An important application of the theory of residues is the evaluation of certain types of improper integrals. We let f be a continuous function of the real variable x on the interval $0 \le x < \infty$. Recall from calculus that the improper integral f over $[0, \infty)$ is defined by

$$\int_0^\infty f(x)\,dx = \lim_{b\to\infty} \int_0^b f(x)\,dx,$$

provided the limit exists. If f is defined for all real x, then the integral of f over $(-\infty, \infty)$ is defined by

$$\int_{-\infty}^\infty f(x)\,dx = \lim_{a\to-\infty} \int_a^0 f(x)\,dx + \lim_{b\to\infty} \int_0^b f(x)\,dx, \tag{8.7}$$

provided both limits exist. If the integral in Equation (8.7) exists, we can obtain its value by taking a single limit:

$$\int_{-\infty}^\infty f(x)\,dx = \lim_{R\to\infty} \int_{-R}^R f(x)\,dx. \tag{8.8}$$

The following example shows that, for some functions, the limit on the right side of Equation (8.8) exists, but the limit on the right side of Equation (8.7) doesn't exist.

Example 8.13. $\lim\limits_{R\to\infty} \int_{-R}^R x\,dx = \lim\limits_{R\to\infty} \left[\frac{R^2}{2} - \frac{(-R)^2}{2}\right] = 0$, but Equation (8.7) tells us that the improper integral of $f(x) = x$ over $(-\infty, \infty)$ doesn't exist. Therefore we can use Equation (8.8) to extend the notion of the value of an improper integral, as Definition 8.2 indicates.

Definition 8.2 (Cauchy principal value). *Let $f(x)$ be a continuous real-valued function for all x. The **Cauchy principal value** (P.V.) of the integral $\int_{-\infty}^\infty f(x)\,dx$ is defined by*

$$P.V. \int_{-\infty}^\infty f(x)\,dx = \lim_{R\to\infty} \int_{-R}^R f(x)\,dx,$$

provided the limit exists.

Example 8.13 showed that P.V. $\int_{-\infty}^\infty x\,dx = 0$.

Example 8.14. The Cauchy principal value of $\int_{-\infty}^\infty \frac{1}{x^2+1}\,dx$ is

$$P.V. \int_{-\infty}^\infty \frac{1}{x^2+1}\,dx = \lim_{R\to\infty} \int_{-R}^R \frac{1}{x^2+1}\,dx$$
$$= \lim_{R\to\infty} \left[\mathrm{Arctan}(R) - \mathrm{Arctan}(-R)\right]$$
$$= \frac{\pi}{2} - \left(-\frac{\pi}{2}\right)$$
$$= \pi.$$

If $f(x) = \frac{P(x)}{Q(x)}$, where P and Q are polynomials, then f is called a **rational function**. In calculus you probably learned techniques for integrating certain types of rational functions. We now show how to use the residue theorem to obtain the Cauchy principal value of the integral of f over $(-\infty, \infty)$.

Theorem 8.3. *Let* $f(z) = \frac{P(z)}{Q(z)}$, *where P and Q are polynomials of degree m and n, respectively. If $Q(x) \neq 0$ for all real x and $n \geq m + 2$, then*

$$P.V. \int_{-\infty}^{\infty} \frac{P(x)}{Q(x)}\, dx = 2\pi i \sum_{j=1}^{k} \operatorname{Res}\left[\frac{P}{Q}, z_j\right],$$

where $z_1, z_2, \ldots, z_{k-1}, z_k$ are the poles of $\frac{P}{Q}$ that lie in the upper half-plane. The situation is illustrated in Figure 8.4.

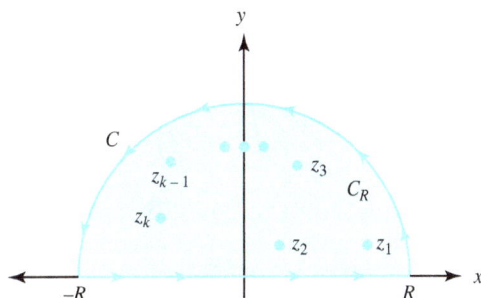

Figure 8.4: The poles $z_1, z_2, \ldots, z_{k-1}, z_k$ of $\frac{P}{Q}$ that lie in the upper half-plane

Proof. There are a finite number of poles of $\frac{P}{Q}$ that lie in the upper half-plane, so we can find a real number R such that the poles all lie inside the contour C, which consists of the segment $-R \leq x \leq R$ of the x-axis and the upper semicircle C_R of radius R shown in Figure 8.4. By properties of integrals,

$$\int_{-R}^{R} \frac{P(x)}{Q(x)}\, dx = \int_{C} \frac{P(z)}{Q(z)}\, dz - \int_{C_R} \frac{P(z)}{Q(z)}\, dz.$$

Using the residue theorem, we rewrite this equation as

$$\int_{-R}^{R} \frac{P(x)}{Q(x)}\, dx = 2\pi i \sum_{j-1}^{k} \operatorname{Res}\left[\frac{P}{Q}, z_j\right] - \int_{C_R} \frac{P(z)}{Q(z)}\, dz. \qquad (8.9)$$

Our proof will be complete if we can show that $\int_{C_R} \frac{P(z)}{Q(z)}\, dz$ tends to zero as $R \to \infty$. Suppose that

$$P(z) = a_m z^m + a_{m-1} z^{m-1} + \cdots + a_1 z + a_0, \quad \text{and}$$
$$Q(z) = b_n z^n + b_{n-1} z^{n-1} + \cdots + b_1 z + b_0.$$

Then

$$\frac{zP(z)}{Q(z)} = \frac{z^{m+1}\left(a_m + a_{m-1} z^{-1} + \cdots + a_1 z^{-m+1} + a_0 z^{-m}\right)}{z^n\left(b_n + b_{n-1} z^{-1} + \cdots + b_1 z^{-n+1} + b_0 z^{-n}\right)},$$

so

$$\lim_{|z| \to \infty} \frac{zP(z)}{Q(z)} = \lim_{|z| \to \infty} \frac{z^{m+1}\left(a_m + a_{m-1} z^{-1} + \cdots + a_1 z^{-m+1} + a_0 z^{-m}\right)}{z^n\left(b_n + b_{n-1} z^{-1} + \cdots + b_1 z^{-n+1} + b_0 z^{-n}\right)}$$
$$= \left(\lim_{|z| \to \infty} \frac{z^{m+1}}{z^n}\right)\left(\lim_{|z| \to \infty} \frac{a_m + a_{m-1} z^{-1} + \cdots + a_1 z^{-m+1} + a_0 z^{-m}}{b_n + b_{n-1} z^{-1} + \cdots + b_1 z^{-n+1} + b_0 z^{-n}}\right).$$

Since $n \geq m + 2$, this limit reduces to $0(\frac{a_m}{b_n}) = 0$. Therefore, for any $\varepsilon > 0$, we may choose R large enough so that $\left|\frac{zP(z)}{Q(z)}\right| < \frac{\varepsilon}{\pi}$ whenever z lies on C_R. But this means that

$$\left|\frac{P(z)}{Q(z)}\right| < \frac{\varepsilon}{\pi|z|} = \frac{\varepsilon}{\pi R} \qquad (8.10)$$

whenever z lies on C_R. Using the ML inequality (Theorem 6.3) and Inequality (8.10), we get

$$\left|\int_{C_R} \frac{P(z)}{Q(z)} \, dz\right| \leq \int_{C_R} \frac{\varepsilon}{\pi R} |dz| = \frac{\varepsilon}{\pi R} \pi R = \varepsilon.$$

Since $\varepsilon > 0$ was arbitrary, we conclude that

$$\lim_{R \to \infty} \int_{C_R} \frac{P(z)}{Q(z)} \, dz = 0. \qquad (8.11)$$

If we let $R \to \infty$ and combine Equations (8.9) and (8.11), we arrive at the desired conclusion. $\square$

Example 8.15. Evaluate $\int_{-\infty}^{\infty} \frac{1}{(x^2+1)(x^2+4)} \, dx$.

Solution:

We write the integrand as $f(z) = \frac{1}{(z+i)(z-i)(z+2i)(z-2i)}$. We see that f has simple poles at the points i and $2i$ in the upper half-plane. Computing the residues, we obtain

$$\text{Res}[f, i] = -\frac{i}{6}, \quad \text{and} \quad \text{Res}[f, 2i] = \frac{i}{12}.$$

Using Theorem 8.3, we conclude that

$$\int_{-\infty}^{\infty} \frac{1}{(x^2+1)(x^2+4)} \, dx = 2\pi i \left(-\frac{i}{6} + \frac{i}{12}\right) = \frac{\pi}{6}.$$

Example 8.16. Evaluate $\int_{-\infty}^{\infty} \frac{dx}{(x^2+4)^3}$.

Solution:

The integrand $f(z) = \frac{1}{(z^2+4)^3}$ has a pole of order 3 at the point $2i$, which is the only singularity of f in the upper half-plane. Computing the residue, we get

$$\begin{aligned}
\text{Res}[f, 2i] &= \frac{1}{2} \lim_{z \to 2i} \frac{d^2}{dz^2} \left[\frac{1}{(z+2i)^3}\right] \\
&= \frac{1}{2} \lim_{z \to 2i} \frac{d}{dz} \left[\frac{-3}{(z+2i)^4}\right] \\
&= \frac{1}{2} \lim_{z \to 2i} \left[\frac{12}{(z+2i)^5}\right] \\
&= -\frac{3i}{512}.
\end{aligned}$$

Therefore, $\int_{-\infty}^{\infty} \frac{1}{(x^2+4)^3} \, dx = 2\pi i \left(\frac{-3i}{512}\right) = \frac{3\pi}{256}.$

Exercises for Section 8.3 (Selected answers or hints are on page 450.)

Use residues to evaluate the following integrals:

1. $\int_{-\infty}^{\infty} \frac{x^2}{(x^2+16)^2}\, dx.$

2. $\int_{-\infty}^{\infty} \frac{1}{x^2+16}\, dx.$

3. $\int_{-\infty}^{\infty} \frac{x}{(x^2+9)^2}\, dx.$

4. $\int_{-\infty}^{\infty} \frac{x+3}{(x^2+9)^2}\, dx.$

5. $\int_{-\infty}^{\infty} \frac{2x^2+3}{(x^2+9)^2}\, dx.$

6. $\int_{-\infty}^{\infty} \frac{1}{x^4+4}\, dx.$

7. $\int_{-\infty}^{\infty} \frac{x^2}{x^4+4}\, dx.$

8. $\int_{-\infty}^{\infty} \frac{x^2}{(x^2+4)^3}\, dx.$

9. $\int_{-\infty}^{\infty} \frac{1}{(x^2+1)^2(x^2+4)}\, dx.$

10. $\int_{-\infty}^{\infty} \frac{x+2}{(x^2+4)(x^2+9)}\, dx.$

11. $\int_{-\infty}^{\infty} \frac{3x^2+2}{(x^2+4)(x^2+9)}\, dx.$

12. $\int_{-\infty}^{\infty} \frac{1}{x^6+1}\, dx.$

13. $\int_{-\infty}^{\infty} \frac{x^4}{x^6+1}\, dx.$

14. $\int_{-\infty}^{\infty} \frac{1}{(x^2+a^2)(x^2+b^2)}\, dx$, where $a > 0$ and $b > 0$.

15. $\int_{-\infty}^{\infty} \frac{x^2}{(x^2+a^2)^3}\, dx$, where $a > 0$.

8.4   Improper Integrals of Trigonometric Functions

Let P and Q be polynomials of degree m and n, respectively, where $n \geq m + 1$. We can show (but omit the proof) that if $Q(x) \neq 0$ for all real x, then

$$\text{P.V.} \int_{-\infty}^{\infty} \frac{P(x)}{Q(x)} \cos x\, dx \quad \text{and} \quad \text{P.V.} \int_{-\infty}^{\infty} \frac{P(x)}{Q(x)} \sin x\, dx$$

are convergent improper integrals. You may encounter integrals of this type in the study of Fourier transforms and Fourier integrals. We now show how to evaluate them.

Particularly important is our use of the identities

$$\cos(\alpha x) = \text{Re}[\exp(i\alpha x)] \quad \text{and} \quad \sin(\alpha x) = \text{Im}[\exp(i\alpha x)],$$

where α is a positive real number. The crucial step in the proof of the following Theorem wouldn't hold if we were to use $\cos(\alpha z)$ and $\sin(\alpha z)$ instead of $\exp(i\alpha z)$, as you will see when you get to Lemma 8.1.

Theorem 8.4. *Let P and Q be polynomials with real coefficients of degree m and n, respectively, where $n \geq m + 1$ and $Q(x) \neq 0$, for all real x. If $\alpha > 0$ and*

$$f(z) = \frac{\exp(i\alpha z)P(z)}{Q(z)}. \tag{8.12}$$

then

$$P.V. \int_{-\infty}^{\infty} \frac{P(x)}{Q(x)} \cos(\alpha x)\, dx = -2\pi \sum_{j=1}^{k} \mathrm{Im}(\mathrm{Res}[f, z_j]), \quad and \tag{8.13}$$

$$P.V. \int_{-\infty}^{\infty} \frac{P(x)}{Q(x)} \sin(\alpha x)\, dx = 2\pi \sum_{j=1}^{k} \mathrm{Re}(\mathrm{Res}[f, z_j]), \tag{8.14}$$

where z_1, z_2, ..., z_{k-1}, z_k are the poles of f that lie in the upper half-plane, and $\mathrm{Re}(\mathrm{Res}[f, z_j])$ and $\mathrm{Im}(\mathrm{Res}[f, z_j])$ are the real and imaginary parts of $\mathrm{Res}[f, z_j]$, respectively.

The proof of Theorem 8.4 is similar to the proof of Theorem 8.3. Before turning to the proof, we illustrate how to use Theorem 8.4

Example 8.17. Evaluate P.V. $\int_{-\infty}^{\infty} \frac{x \sin x}{x^2 + 4}\, dx$.

Solution:

The function f in Equation (8.12) is $f(z) = \frac{\exp(iz)z}{z^2+4}$, which has a simple pole at the point $2i$ in the upper half-plane. Calculating the residue yields

$$\mathrm{Res}[f, 2i] = \lim_{z \to 2i} \frac{\exp(iz)z}{z + 2i} = \frac{2ie^{-2}}{4i} = \frac{1}{2e^2}.$$

Using Equation (8.14) gives

$$P.V. \int_{-\infty}^{\infty} \frac{x \sin x}{x^2 + 4}\, dx = 2\pi \mathrm{Re}(\mathrm{Res}[f, 2i]) = \frac{\pi}{e^2}.$$

Example 8.18. Evaluate P.V. $\int_{-\infty}^{\infty} \frac{\cos x\, dx}{x^4 + 4}$.

Solution:

The function f in Equation (8.12) is $f(z) = \frac{\exp(iz)}{z^4 + 4}$, which has simple poles at the points $z_1 = 1 + i$ and $z_2 = -1 + i$ in the upper half-plane. We get the residues with the aid of L'Hôpital's rule:

$$\begin{aligned}
\mathrm{Res}[f, 1 + i] &= \lim_{z \to 1+i} \frac{(z - 1 - i)\exp(iz)}{z^4 + 4} \\
&= \lim_{z \to 1+i} \frac{[1 + i(z - 1 - i)]\exp(iz)}{4z^3} \\
&= \frac{\exp(-1 + i)}{4(1 + i)^3} \\
&= \frac{\sin 1 - \cos 1 - i(\cos 1 + \sin 1)}{16e}.
\end{aligned}$$

Similarly,

$$\mathrm{Res}[f, -1 + i] = \frac{\cos 1 - \sin 1 - i(\cos 1 + \sin 1)}{16e}.$$

Using Equation (8.13), we get

$$\int_{-\infty}^{\infty} \frac{\cos x}{x^4 + 4}\, dx = -2\pi\left[\text{Im}(\text{Res}[f, 1 + i]) + \text{Im}(\text{Res}[f, -1 + i])\right]$$

$$= \frac{\pi(\cos 1 + \sin 1)}{4e}.$$

We are almost ready to give the proof of Theorem 8.4, but first we need one preliminary result.

Lemma 8.1 (Jordan's lemma). *Suppose that P and Q are polynomials of degree m and n, respectively, where $n \geq m + 1$. If C_R is the upper semicircle $z = Re^{i\theta}$, for $0 \leq \theta \leq \pi$, then*

$$\lim_{R \to \infty} \int_{C_R} \frac{\exp(iz)P(z)}{Q(z)}\, dz = 0.$$

Proof. From $n \geq m + 1$, it follows that $\left|\frac{P(z)}{Q(z)}\right| \to 0$ as $|z| \to \infty$. Therefore, for any $\varepsilon > 0$, there exists $R_\varepsilon > 0$ such that

$$\left|\frac{P(z)}{Q(z)}\right| < \frac{\varepsilon}{\pi} \tag{8.15}$$

whenever $|z| \geq R_\varepsilon$. Using the ML inequality (Theorem 6.3) together with Inequality 8.15, we get

$$\left|\int_{C_R} \frac{\exp(iz)P(z)}{Q(z)}\, dz\right| \leq \int_{C_R} \frac{\varepsilon}{\pi}|e^{iz}|\,|dz|. \tag{8.16}$$

provided $R \geq R_\varepsilon$. The parametrization of C_R leads to the equation

$$|dz| = R\, d\theta, \quad \text{and} \quad |e^{iz}| = e^{-y} = e^{-R\sin\theta}. \tag{8.17}$$

Using the trigonometric identity $\sin(\pi - \theta) = \sin\theta$ and Equations (8.17), we express the integral on the right side of Inequality 8.16 as

$$\int_{C_R} \frac{\varepsilon}{\pi}|e^{iz}|\,|dz| = \frac{\varepsilon}{\pi}\int_0^\pi e^{-R\sin\theta}R\, d\theta = \frac{2\varepsilon}{\pi}\int_0^{\pi/2} e^{-R\sin\theta}R\, d\theta. \tag{8.18}$$

On the interval $0 \leq \theta \leq \pi/2$ we can use the inequality

$$0 \leq \frac{2\theta}{\pi} \leq \sin\theta.$$

We combine this inequality with Inequality 8.16 and Equation 8.18 to conclude that, for $R > R_\varepsilon$,

$$\left|\int_{C_R} \frac{\exp(iz)P(z)}{Q(z)}\, dz\right| \leq \frac{2\varepsilon}{\pi}\int_0^{\pi/2} e^{-\frac{2R\theta}{\pi}}R\, d\theta$$

$$= -\varepsilon e^{-\frac{2R\theta}{\pi}}\Big|_{\theta=0}^{\theta=\pi/2}$$

$$= \varepsilon(1 - e^{-R})$$

$$< \varepsilon.$$

Because $\varepsilon > 0$ is arbitrary, the proof of Jordan's lemma is complete. $\qquad\square$

We now turn to the proof of our main Theorem.

Proof of Theorem 8.4.

Let C be the contour that consists of the segment $-R \leq x \leq R$ of the real axis together with the upper semicircle C_R parametrized by $z = Re^{i\theta}$, for $0 \leq \theta \leq \pi$. Using properties of integrals, we have

$$\int_{-R}^{R} \frac{\exp(i\alpha x)P(x)}{Q(x)}\, dx = \int_{C} \frac{\exp(i\alpha z)P(z)}{Q(z)}\, dz - \int_{C_R} \frac{\exp(i\alpha z)P(z)}{Q(z)}\, dz.$$

If R is sufficiently large, all the poles z_1, z_2, ..., z_k of f will lie inside C, and we can use the residue theorem to obtain

$$\int_{-R}^{R} \frac{\exp(i\alpha x)P(x)}{Q(x)}\, dx = 2\pi i \sum_{j=1}^{k} \operatorname{Res}[f, z_j] - \int_{C_R} \frac{\exp(i\alpha z)P(z)}{Q(z)}\, dz. \qquad (8.19)$$

Since α is a positive real number, the change of variables $\zeta = \alpha z$ shows that the conclusion of Jordan's lemma holds for the integrand $\frac{\exp(i\alpha z)P(z)}{Q(z)}$. Hence we let $R \to \infty$ in Equation (8.19) to obtain

$$\text{P.V.} \int_{-\infty}^{\infty} \frac{[\cos(\alpha x) + i\sin(\alpha x)]P(x)}{Q(x)}\, dx = 2\pi i \sum_{j=1}^{k} \operatorname{Res}[f, z_j]$$

$$= -2\pi \sum_{j=1}^{k} \operatorname{Im}(\operatorname{Res}[f, z_j])$$

$$+ 2\pi i \sum_{j=1}^{k} \operatorname{Re}(\operatorname{Res}[f, z_j]).$$

Equating the real and imaginary parts of this equation gives us Equations (8.13) and (8.14), which completes the proof. $\qquad \square$

Exercises for Section 8.4 (Selected answers or hints are on page 451.)

Use residues to find the Cauchy principal value of the following:

1. $\int_{-\infty}^{\infty} \frac{\cos x}{x^2+9}\, dx,$ and $\int_{-\infty}^{\infty} \frac{\sin x}{x^2+9}\, dx.$

2. $\int_{-\infty}^{\infty} \frac{x\cos x}{x^2+9}\, dx,$ and $\int_{-\infty}^{\infty} \frac{x\sin x}{x^2+9}\, dx.$

3. $\int_{-\infty}^{\infty} \frac{x\sin x}{(x^2+4)^2}\, dx.$

4. $\int_{-\infty}^{\infty} \frac{\cos x}{(x^2+4)^2}\, dx.$

5. $\int_{-\infty}^{\infty} \frac{\cos x}{(x^2+4)(x^2+9)}\, dx.$

6. $\int_{-\infty}^{\infty} \frac{\cos x}{(x^2+1)(x^2+4)}\, dx.$

7. $\int_{-\infty}^{\infty} \frac{\cos x}{x^2-2x+5}\, dx.$

8. $\int_{-\infty}^{\infty} \frac{\cos x}{x^2-4x+5}\, dx.$

9. $\int_{-\infty}^{\infty} \frac{x\sin x}{x^4+4}\, dx.$

262

10. $\int_{-\infty}^{\infty} \frac{x^3 \sin x}{x^4+4}\, dx$.

11. $\int_{-\infty}^{\infty} \frac{\cos 2x}{x^2+2x+2}\, dx$.

12. $\int_{-\infty}^{\infty} \frac{x^3 \sin 2x}{x^4+4}\, dx$.

13. Why do you need to use the exponential function when evaluating improper integrals involving the sine and cosine functions?

8.5   Indented Contour Integrals

If f is continuous on the interval $b < x \leq c$, but discontinuous at b, then the improper integral of f over $[b, c]$ is defined by

$$\int_b^c f(x)\, dx = \lim_{r \to b^+} \int_r^c f(x)\, dx,$$

provided the limit exists. Similarly, if f is continuous on the interval $a \leq x < b$, but discontinuous at b, then the improper integral of f over $[a, b]$ is defined by

$$\int_a^b f(x)\, dx = \lim_{R \to b^-} \int_a^R f(x)\, dx,$$

provided the limit exists. For example,

$$\int_0^9 \frac{1}{2\sqrt{x}}\, dx = \lim_{r \to 0^+} \int_r^9 \frac{1}{2\sqrt{x}}\, dx = \lim_{r \to 0^+} \left(\sqrt{x} \right) \Big|_{x=r}^{x=9} = 3 - \lim_{r \to 0^+} \sqrt{r} = 3.$$

If f is continuous for all values of x in the interval $[a, c]$, except at the value $x = b$, where $a < b < c$, then the Cauchy principal value of f over $[a, c]$ is defined by

$$\text{P.V.} \int_a^c f(x)\, dx = \lim_{r \to 0^+} \left[\int_a^{b-r} f(x)\, dx + \int_{b+r}^c f(x)\, dx \right],$$

provided the limit exists.

Example 8.19.

$$\text{P.V.}\,. \int_{-1}^8 \frac{1}{x^{\frac{1}{3}}}\, dx = \lim_{r \to 0^+} \left[\int_{-1}^{-r} \frac{1}{x^{\frac{1}{3}}}\, dx + \int_r^8 \frac{1}{x^{\frac{1}{3}}}\, dx \right].$$

Evaluating the integrals and computing limits gives

$$\lim_{r \to 0^+} \left[\frac{3}{2} r^{\frac{2}{3}} - \frac{3}{2} + 6 - \frac{3}{2} r^{\frac{2}{3}} \right] = \frac{9}{2}.$$

In this section we show how to use residues to evaluate the Cauchy principal value of the integral of f over $(-\infty, \infty)$ when the integrand f has simple poles on the x-axis. We state our main results and then look at some examples before giving proofs.

Theorem 8.5. *Let* $f(z) = \frac{P(z)}{Q(z)}$, *where P and Q are polynomials with real coefficients of degree m and n, respectively, and $n \geq m + 2$. If Q has simple zeros at the points $t_1, t_2, \ldots, t_l$ on the x axis, then*

$$\text{P.V.} \int_{-\infty}^{\infty} \frac{P(x)}{Q(x)}\, dx = 2\pi i \sum_{j=1}^k \text{Res}[f, z_j] + \pi i \sum_{j=1}^l \text{Res}[f, t_j], \tag{8.20}$$

where $z_1, z_2, \ldots, z_k$ are the poles of f that lie in the upper half-plane.

Theorem 8.6. *Let P and Q be polynomials of degree m and n, respectively, where $n \geq m + 1$, and let Q have simple zeros at the points t_1, t_2, ..., t_l on the x-axis. If α is a positive real number and if $f(z) = \frac{\exp(i\alpha z) P(z)}{Q(z)}$, then*

$$P.V. \int_{-\infty}^{\infty} \frac{P(x)}{Q(x)} \cos \alpha x \, dx = -2\pi \sum_{j=1}^{k} \text{Im}(\text{Res}[f, z_j]) - \pi \sum_{j=1}^{l} \text{Im}(\text{Res}[f, t_j]) \qquad (8.21)$$

and

$$P.V. \int_{-\infty}^{\infty} \frac{P(x)}{Q(x)} \sin \alpha x \, dx = 2\pi \sum_{j=1}^{k} \text{Re}(\text{Res}[f, z_j]) + \pi \sum_{j=1}^{l} \text{Re}(\text{Res}[f, t_j]) \qquad (8.22)$$

where z_1, z_2, ..., z_k are the poles of f that lie in the upper half-plane.

Remark 8.2. *The formulas in these theorems give the Cauchy principal value of the integral, which pays special attention to the manner in which any limits are taken. They are similar to those in Sections 8.3 and 8.4, except here we add one-half the value of each residue at the points t_1, t_2, ..., t_l on the x-axis.*

Example 8.20. Evaluate P.V. $\int_{-\infty}^{\infty} \frac{x}{x^3-8} \, dx$ by using complex analysis.

Solution:

The integrand

$$f(z) = \frac{z}{z^3 - 8} = \frac{z}{(z-2)(z+1+i\sqrt{3})(z+1-i\sqrt{3})}$$

has simple poles at the points $t_1 = 2$ on the x-axis and $z_1 = -1 + i\sqrt{3}$ in the upper half-plane. By Theorem 8.5,

$$P.V. \int_{-\infty}^{\infty} \frac{x}{x^3-8} \, dx = 2\pi i \text{Res}[f, z_1] + \pi i \text{Res}[f, t_1]$$

$$= 2\pi i \frac{-1 - i\sqrt{3}}{12} + \pi i \frac{1}{6} = \frac{\pi\sqrt{3}}{6}.$$

Example 8.21. Evaluate P.V. $\int_{-\infty}^{\infty} \frac{t \, dx}{t^3-8}$ with a computer algebra system.

Solution:

Computer algebra systems such as Mathematica or MAPLE give the indefinite integral

$$\int \frac{t}{t^3-8} \, dt = \frac{\text{Arctan}\left(\frac{1+t}{\sqrt{3}}\right)}{2\sqrt{3}} + \frac{\text{Log}(t-2)}{6} + \frac{\text{Log}(t^2+2t+4)}{12} = g(t).$$

However, for real numbers, we should write the second term as $\frac{\text{Log}[(t-2)^2]}{12}$ and use the equivalent formula:

$$g(t) = \frac{\text{Arctan}\left(\frac{1+t}{\sqrt{3}}\right)}{2\sqrt{3}} + \frac{\text{Log}[(t-2)^2]}{12} + \frac{\text{Log}(t^2+2t+4)}{12}.$$

This antiderivative has the property $\lim_{t \to 2} g(t) = -\infty$, as Figure 8.5 shows.

We also compute

$$\lim_{t \to \infty} g(t) = \frac{\pi\sqrt{3}}{12} \quad \text{and} \quad \lim_{t \to -\infty} g(t) = -\frac{\pi\sqrt{3}}{12}.$$

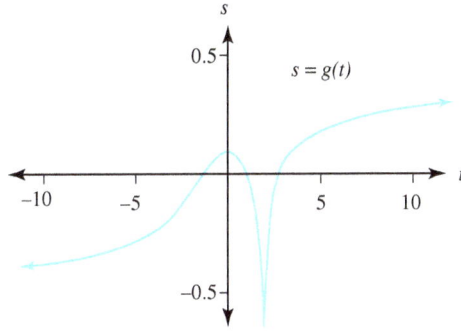

Figure 8.5: Graph of $s = g(t) = \int \frac{t}{t^3-8}\, dt$

and the Cauchy principal limit at $t = 2$ as $r \to 0$ is

$$\lim_{r\to 0^+} [g(2+r) - g(2-r)] = 0.$$

Therefore the Cauchy principal value of the improper integral is

$$\text{P.V. } \int_{-\infty}^{\infty} \frac{t}{t^3-8}\, dt = \lim_{r\to 0^+} \left[\int_{-\infty}^{2-r} \frac{t}{t^3-8}\, dt + \int_{2+r}^{\infty} \frac{t}{t^3-8}\, dt \right]$$

$$= \left(\lim_{r\to 0^+} g(2-r) - \lim_{t\to -\infty} g(t) \right) + \left(\lim_{t\to\infty} g(t) - \lim_{r\to 0^+} g(2+r) \right)$$

$$= \left(\lim_{r\to 0^+} g(2-r) - \lim_{r\to 0^+} g(2+r) \right) + \left(\lim_{t\to\infty} g(t) - \lim_{t\to -\infty} g(t) \right)$$

$$= 0 + \frac{\pi\sqrt{3}}{12} + \frac{\pi\sqrt{3}}{12}$$

$$= \frac{\pi\sqrt{3}}{6}.$$

Example 8.22. Evaluate P.V. $\displaystyle\int_{-\infty}^{\infty} \frac{\sin x}{(x-1)(x^2+4)}\, dx.$

Solution:

The integrand $f(z) = \frac{\exp(iz)}{(z-1)(z^2+4)}$ has simple poles at the points $t_1 = 1$ on the x-axis and $z_1 = 2i$ in the upper half-plane. By Theorem 8.6,

$$\text{P.V. } \int_{-\infty}^{\infty} \frac{\sin x}{(x-1)(x^2+4)}\, dx = 2\pi\text{Re}\left(\text{Res}[f, z_1]\right) + \pi\text{Re}\left(\text{Res}[f, t_1]\right)$$

$$= 2\pi\text{Re}\left(\frac{-2+i}{20e^2}\right) + \pi\text{Re}\left(\frac{\cos 1 + i\sin 1}{5}\right)$$

$$= \frac{\pi}{5}\left(\cos 1 - \frac{1}{e^2}\right).$$

We are almost ready for the proofs of Theorems 8.5 and 8.6. First, we need the following lemma.

Lemma 8.2. *Suppose that f has a simple pole at the point t_0 on the x-axis. If C_r is the contour $C_r : z = t_0 + re^{i\theta}$, for $0 \le \theta \le \pi$, then*

$$\lim_{r\to 0} \int_{C_r} f(z)\, dz = i\pi\text{Res}[f, t_0].$$

265

Proof. The Laurent series for f at $z = t_0$ has the form

$$f(z) = \frac{\text{Res}[f, t_0]}{z - t_0} + g(z), \tag{8.23}$$

where g is analytic at $z = t_0$. Using the parametrization of C_r and Equation (8.23), we get

$$\int_{C_r} f(z)\, dz = \text{Res}[f, t_0] \int_0^\pi \frac{ire^{i\theta}}{re^{i\theta}}\, d\theta + ir \int_0^\pi g(t_0 + re^{i\theta})e^{i\theta}\, d\theta$$

$$= i\pi \text{Res}[f, t_0] + ir \int_0^\pi g(t_0 + re^{i\theta})e^{i\theta} d\theta. \tag{8.24}$$

Since g is continuous at t_0, there is an $M > 0$ so that $|g(t_0 + re^{i\theta})| \le M$, and

$$\left| \lim_{r \to 0} ir \int_0^\pi g(t_0 + re^{i\theta})e^{i\theta} d\theta \right| \le \lim_{r \to 0} r \int_0^\pi M\, d\theta = \lim_{r \to 0} r\pi M = 0.$$

Combining this inequality with Equation (8.24) gives the desired result. $\qquad \square$

We are now ready to prove Theorems 8.5 and 8.6:

Proof of Theorems 8.5 and 8.6.

Since f has only a finite number of poles, we can choose r small enough that the semicircles

$$C_j : z = t_j + re^{i\theta}, \quad \text{for} \quad 0 \le \theta \le \pi \quad \text{and} \quad j = 1, 2, \ldots, l$$

are disjoint and the poles z_1, z_2, $\ldots$, z_k of f in the upper half-plane lie above them, as shown in Figure 8.6.

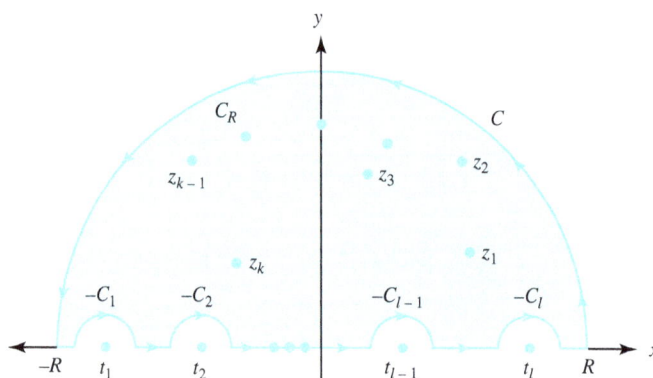

Figure 8.6: The poles t_1, t_2, $\ldots$, t_l of f that lie on the x- axis and the poles z_1, z_2, $\ldots$ z_k that lie above the semicircles C_1, C_2, $\ldots$, C_l

Let R be large enough so that the poles of f in the upper half-plane lie under the semicircle $C_R : z = Re^{i\theta}$, for $0 \le \theta \le \pi$, and the poles of f on the x-axis lie in the interval $-R \le x \le R$. Let C be the simple closed positively oriented contour that consists of C_R and $-C_1$, $-C_2$, $\ldots$, $-C_l$ and the segments of the real axis that lie between the semicircles shown in Figure 8.6. The residue theorem gives $\int_C f(z)\, dz = 2\pi i \sum_{j=1}^{k} \text{Res}[f, z_j]$, which we rewrite as

$$\int_{I_R} f(x)\, dx = 2\pi i \sum_{j=1}^{k} \text{Res}[f, z_j] + \sum_{j=1}^{l} \int_{C_j} f(z)\, dz - \int_{C_R} f(z)\, dz, \tag{8.25}$$

266

where I_R is the portion of the interval $-R \le x \le R$ that lies outside the intervals $(t_j - r, t_j + r)$ for $j = 1, 2, \ldots, l$. Using the same techniques that we used in Theorems 8.3 and 8.4 yields

$$\lim_{R \to \infty} \int_{C_R} f(z)\, dz = 0. \tag{8.26}$$

If we let $R \to \infty$ and $r \to 0$ in Equation (8.25) and use the results of Equation (8.26) and Lemma 8.2, we obtain

$$\text{P.V.} \int_{-\infty}^{\infty} f(x)\, dx = 2\pi i \sum_{j=1}^{k} \text{Res}[f, z_j] + \pi i \sum_{j=1}^{l} \text{Res}[f, t_j]. \tag{8.27}$$

If f is the function given in Theorem 8.5, then Equation (8.27) becomes Equation (8.20). If f is the function given in Theorem 8.6, then equating the real and imaginary parts of Equation (8.27) results in Equations (8.21) and (8.22), respectively, which completes the proof. $\square$

Exercises for Section 8.5 (Selected answers or hints are on page 451.)

Use residues to compute the following integrals:

1. P.V. $\int_{-\infty}^{\infty} \frac{1}{x(x-1)(x-2)}\, dx$.

2. P.V. $\int_{-\infty}^{\infty} \frac{1}{x^3+x}\, dx$.

3. P.V. $\int_{-\infty}^{\infty} \frac{x}{x^3+1}\, dx$.

4. P.V. $\int_{-\infty}^{\infty} \frac{1}{x^3+1}\, dx$.

5. P.V. $\int_{-\infty}^{\infty} \frac{x^2}{x^4-1}\, dx$.

6. P.V. $\int_{-\infty}^{\infty} \frac{x^4}{x^6-1}\, dx$.

7. P.V. $\int_{-\infty}^{\infty} \frac{\sin x}{x}\, dx$.

8. P.V. $\int_{-\infty}^{\infty} \frac{\cos x}{x^2-x}\, dx$.

9. P.V. $\int_{-\infty}^{\infty} \frac{\sin x}{x(\pi^2-x^2)}\, dx$.

10. P.V. $\int_{-\infty}^{\infty} \frac{\cos x}{\pi^2-4x^2}\, dx$.

11. P.V. $\int_{-\infty}^{\infty} \frac{\sin x}{x(x^2+1)}\, dx$.

12. P.V. $\int_{-\infty}^{\infty} \frac{x \cos x}{x^2+3x+2}\, dx$.

13. P.V. $\int_{-\infty}^{\infty} \frac{\sin x}{x(1-x^2)}\, dx$.

14. P.V. $\int_{-\infty}^{\infty} \frac{\cos x}{a^2-x^2}\, dx$.

15. P.V. $\int_{-\infty}^{\infty} \frac{\sin^2 x}{x^2}\, dx$. *Hint:* Use trigonometric identity $\sin^2 x = \frac{1}{2} - \frac{1}{2}\cos 2x$.

16. P.V. $\int_{0}^{\infty} \frac{1}{x^3+1}\, dx$. *Hint:* Use the contour $C = L_1 + C_R - L_2$ shown in Figure 8.7

17. P.V. $\int_{0}^{\infty} \frac{x}{x^3+1}\, dx$. *Hint:* Use the contour $C = L_1 + C_R - L_2$ shown in Figure 8.7

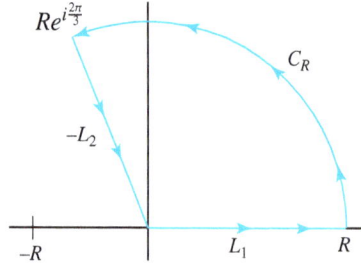

Figure 8.7: The contour $C = L_1 + C_r - L_2$ for Exercises 16 and 17

8.6   Integrands with Branch Points

We now show how to evaluate certain improper real integrals involving the integrand $x^\alpha \frac{P(x)}{Q(x)}$. The complex function z^α is multivalued, so we must first specify the branch to be used.

Let α be a real number with $0 < \alpha < 1$. In this section we use the branch of z^α corresponding to the branch of the logarithm $\log_0$ (see Equation (5.20)) as follows:

$$z^\alpha = e^{\alpha[\log_0(z)]} = e^{\alpha(\ln|z| + i\arg_0 z)} = e^{\alpha(\ln r + i\theta)} = r^\alpha(\cos\alpha\theta + i\sin\alpha\theta). \qquad (8.28)$$

where $z = re^{i\theta} \neq 0$ and $0 < \theta \le 2\pi$. Note that this is not the traditional principal branch of z^a and that, as defined, the function z^a is analytic in the domain $\{re^{i\theta} : r > 0,\ 0 < \theta < 2\pi\}$.

Theorem 8.7. *Let P and Q be polynomials of degree m and n, respectively, where $n \ge m + 2$. If $Q(x) \neq 0$, for $x > 0$, Q has a zero of order at most 1 at the origin, and $f(z) = \frac{z^\alpha P(z)}{Q(z)}$, where $0 < \alpha < 1$, then*

$$P.V. \int_0^\infty \frac{x^\alpha P(x)}{Q(x)}\, dx = \frac{2\pi i}{1 - e^{i\alpha 2\pi}} \sum_{j=1}^{k} \text{Res}[f, z_j],$$

where z_1, z_2, ..., z_k are the nonzero poles of $\frac{P}{Q}$.

Proof. Let C denote the simple closed positively oriented contour that consists of the portions of the circles $C_r(0)$ and $C_R(0)$ and the horizontal segments joining them, per Figure 8.8.

We select a small value of r and a large value of R so that the nonzero poles z_1, z_2, ..., z_k of $\frac{P}{Q}$ lie inside C. Using the residue theorem, we write

$$\int_C f(z)\, dz = 2\pi i \sum_{j=1}^{k} \text{Res}[f, z_j]. \qquad (8.29)$$

If we let $r \to 0$ in Equation (8.29), the integrand $f(z)$ on the upper horizontal line of Figure 8.8 approaches $\frac{x^\alpha P(x)}{Q(x)}$, where x is a real number; however, because of the branch we chose for z^α (see Equation 8.28), the integrand $f(z)$ on the lower horizontal line approaches $\frac{x^\alpha e^{i\alpha 2\pi} P(x)}{Q}(x)$. Therefore

$$\lim_{r \to 0} \int_C f(z)\, dz = \int_0^R \frac{x^\alpha P(x)}{Q(x)}\, dx + \int_R^0 \frac{x^\alpha e^{i\alpha 2\pi} P(x)}{Q(x)}\, dx + \int_{C_R^+(0)} f(z)\, dz. \qquad (8.30)$$

It is here that we need the function Q to have a zero of order at most 1 at the origin. Otherwise, the first two integrals on the right side of Equation (8.30) would not necessarily converge.

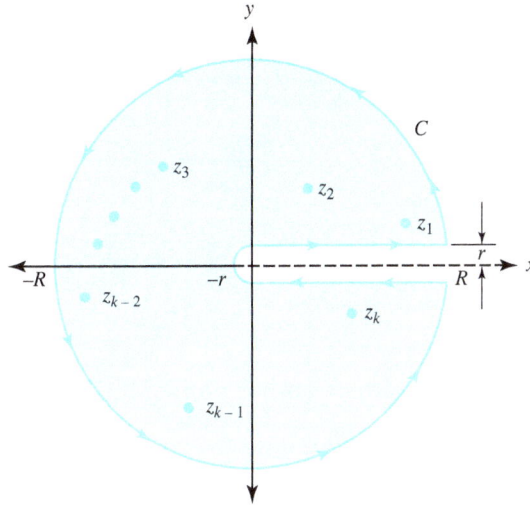

Figure 8.8: Contour C encloses the nonzero poles $z_1, z_2, \ldots, z_k$ of $\frac{P}{Q}$

Combining this result with Equation (8.29) gives

$$\int_0^R \frac{x^\alpha P(x)}{Q(x)} \, dx - \int_0^R \frac{x^\alpha e^{i\alpha 2\pi} P(x)}{Q(x)} \, dx = 2\pi i \sum_{j=1}^k \text{Res}[f, z_j] - \int_{C_R^+(0)} f(z) \, dz,$$

so

$$\left(\int_0^R \frac{x^a P(x)}{Q(x)} \, dx \right) (1 - e^{ia2\pi}) = 2\pi i \sum_{j=1}^k Res[f, z_j] - \int_{C_R^+(0)} f(z) \, dz,$$

which we rewrite as

$$\int_0^R \frac{x^\alpha P(x)}{Q(x)} \, dx = \frac{2\pi i}{1 - e^{i\alpha 2\pi}} \sum_{j=1}^k \text{Res}[f, z_j] - \frac{1}{1 - e^{i\alpha 2\pi}} \int_{C_R^+(0)} f(z) \, dz. \qquad (8.31)$$

Using the ML inequality (Theorem 6.3) gives

$$\lim_{R \to \infty} \int_{C_R^+(0)} f(z) \, dz = 0. \qquad (8.32)$$

The argument is essentially the same as that used to establish Equation (8.11), and we omit the details. If we combine Equations (8.31) and (8.32) and let $R \to \infty$, we arrive at the desired result. □

Example 8.23. Evaluate P.V. $\int_0^\infty \frac{x^\alpha}{x(x+1)} \, dx$, where $0 < a < 1$.

Solution:

The function $f(z) = \frac{z^a}{z(z+1)}$ has a nonzero pole at the point -1, and the denominator has a zero of order at most 1 (in fact, exactly 1) at the origin. Using Theorem 8.7, we compute

269

$$\int_0^\infty \frac{x^a}{x(x+1)}\,dx = \frac{2\pi i}{1 - e^{ia2\pi}}\mathrm{Res}[f, -1]$$

$$= \frac{2\pi i}{1 - e^{ia2\pi}}\left(\frac{e^{ia\pi}}{-1}\right)$$

$$= \frac{\pi}{\frac{e^{ia\pi} - e^{-ia\pi}}{2i}}$$

$$= \frac{\pi}{\sin a\pi}.$$

We can apply the preceding ideas to other multivalued functions.

Example 8.24. Evaluate P.V. $\int_0^\infty \frac{\ln x}{x^2 + a^2}\,dx$, where $a > 0$.

Solution:

We use the function $f(z) = \frac{\log_{-\frac{\pi}{2}} z}{z^2 + a^2}$. Recall that

$$\log_{-\frac{\pi}{2}} z = \ln|z| + i\arg_{-\frac{\pi}{2}} z = \ln r + i\theta,$$

where $z = re^{i\theta} \neq 0$ and $-\frac{\pi}{2} < \theta \leq \frac{3\pi}{2}$. The path C of integration will consist of the segments $[-R, -r]$ and $[r, R]$ of the x-axis together with the upper semicircles $C_r : z = re^{i\theta}$ and $C_R : z = Re^{i\theta}$, for $0 \leq \theta \leq \pi$, as shown in Figure 8.9.

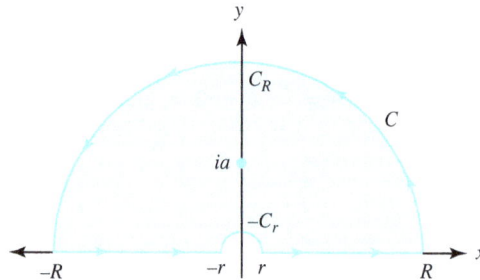

Figure 8.9: The contour C for the integrand $f(z) = \frac{\log_{-\frac{\pi}{2}} z}{z^2 + a^2}$

We chose the branch $\log_{-\frac{\pi}{2}}$ because it is analytic on C and its interior, hence so is the function f. This choice enables us to apply the residue theorem properly (see the hypotheses of Theorem 8.1), and we get

$$\int_C f(z)\,dz = 2\pi i\,\mathrm{Res}[f, ai] = \frac{\pi \ln a}{a} + i\frac{\pi^2}{2a}.$$

Keeping in mind the branch of logarithm that we're using, we then have

$$\int_C f(z)\,dz = \int_{-R}^{-r} f(x)\,dx + \int_{-C_r} f(z)\,dz + \int_r^R f(x)\,dx + \int_{C_R} f(z)\,dz$$

$$= \int_{-R}^{-r} \frac{\ln|x| + i\pi}{x^2 + a^2}\,dx + \int_{-C_r} f(z)\,dz$$

$$+ \int_r^R \frac{\ln x}{x^2 + a^2}\,dx + \int_{C_R} f(z)\,dz$$

$$= \frac{\pi \ln a}{a} + i\frac{\pi^2}{2a}. \tag{8.33}$$

If $R^2 > a^2$, then by the ML inequality (Theorem 6.3)

$$\left| \int_{C_R} f(z)\, dz \right| = \left| \int_0^\pi \frac{\ln R + i\theta}{R^2 e^{i2\theta} + a^2} i R e^{i\theta} d\theta \right|$$
$$\leq \frac{R(\ln R + \pi)\pi}{R^2 - a^2},$$

and L'Hôpital's rule yields $\lim\limits_{R\to\infty} \int_{C_R} f(z) r\, dz = 0$. Engaging in a similar computation shows that $\lim\limits_{r\to 0^+} \int_{C_r} f(z)\, dz = 0$. We use these results when we take limits Equations (8.33) to get

$$\text{P.V.} \left(\int_{-\infty}^0 \frac{\ln|x| + i\pi}{x^2 + a^2}\, dx + \int_0^\infty \frac{\ln x}{x^2 + a^2}\, dx \right) = \frac{\pi \ln a}{a} + i\frac{\pi^2}{2a}.$$

Equating the real parts in this equation gives

$$\text{P.V.} \int_0^\infty \frac{\ln x}{x^2 + a^2}\, dx = \frac{\pi \ln a}{2a}.$$

Remark 8.3. *The theory of this section is not purely esoteric. Many applications of contour integrals surface in government and industry worldwide. Several years ago, for example, a briefing was given at the Korean Institute for Defense Analysis (KIDA) in which a sophisticated problem was analyzed by means of a contour integral whose path of integration was virtually identical to that given in Figure 8.8.*

Exercises for Section 8.6 (Selected answers or hints are on page 451.)

Use residues to compute the following integrals.

1. P.V. $\int_0^\infty \frac{1}{x^{\frac{2}{3}}(1+x)}\, dx$.

2. P.V. $\int_0^\infty \frac{1}{x^{\frac{1}{2}}(1+x)}\, dx$.

3. P.V. $\int_0^\infty \frac{x^{\frac{1}{2}}}{(1+x)^2}\, dx$.

4. P.V. $\int_0^\infty \frac{x^{\frac{1}{2}}}{1+x^2}\, dx$.

5. P.V. $\int_0^\infty \frac{\ln(x^2+1)}{x^2+1}\, dx$.   *Hint:* Use the integrand $f(z) = \frac{\log(z+i)}{z^2+1}$.

6. P.V. $\int_0^\infty \frac{\ln x}{(1+x^2)^2}\, dx$.

7. P.V. $\int_0^\infty \frac{(\ln x)^2}{x^2+1}\, dx$.

8. P.V. $\int_0^\infty \frac{x^{\frac{1}{2}} \ln x}{x^2+1}\, dx$.

9. $\int_0^\infty \frac{\ln x}{x^2+2^2}\, dx$.

10. Carry out the following computations:

    (a) For $f(z) = \frac{z^{\frac{1}{3}}}{z^3(z+1)}$, show that $\text{Res}[f, -1] = -\frac{1}{2} - \frac{\sqrt{3}}{2}i$.

271

(b) Use part (a) and $\alpha = \frac{1}{3}$ to verify the computation $\frac{2\pi i}{1 - e^{i\alpha 2\pi}} \text{Res}[f, -1] = \frac{2\sqrt{3}}{3}\pi$.

(c) Can you conclude that P.V. $\int_0^\infty \frac{x^{\frac{1}{3}}}{x^3(x+1)} dx = \frac{2\sqrt{3}}{3}\pi$? Justify your answer.

11. Carry out the following computations:

   (a) For $f(z) = \frac{z^{\frac{4}{3}}}{z+1}$, show that $\text{Res}[f, -1] = -\frac{1}{2} - \frac{\sqrt{3}}{2}i$.

   (b) Use part (a) and $\alpha = \frac{4}{3}$ to verify the computation $\frac{2\pi i}{1 - e^{i\alpha 2\pi}} \text{Res}[f, -1] = \frac{2\sqrt{3}}{3}\pi$.

   (c) Can you conclude that P.V. $\int_0^\infty \frac{x^{\frac{4}{3}}}{x+1} dx = \frac{2\sqrt{3}}{3}\pi$? Justify your answer.

12. P.V. $\int_0^\infty \frac{1}{x^{\frac{1}{2}}(x+1)^2} dx$.

13. P.V. $\int_0^\infty \frac{1}{x^{\frac{1}{2}}(1+x^2)} dx$.

14. P.V. $\int_0^\infty \frac{x^{\frac{1}{3}}}{(x+1)^2} dx$.

15. P.V. $\int_0^\infty \frac{x^{\frac{1}{3}}}{x^2+1} dx$.

16. P.V. $\int_0^\infty \frac{x^{\frac{1}{3}} \ln x}{x^2+1} dx$ and P.V. $\int_0^\infty \frac{x^{\frac{1}{3}}}{x^2+1} dx$.

   Hint: Use the complex integrand $f(z) = \frac{z^{\frac{1}{3}} \text{Log} z}{z^2+1}$.

17. P.V. $\int_0^\infty \frac{\ln(1+x)}{x^{1+a}} dx$, where $0 < a < 1$.

18. P.V. $\int_0^\infty \frac{\ln x}{(x+a)^2} dx$, where $a > 0$.

19. P.V. $\int_{-\infty}^\infty \frac{\sin x}{x} dx$.   *Hint*: Use the integrand $f(z) = \frac{\exp(iz)}{z}$ and the contour C in Figure 8.8. Let $r \to 0$ and $R \to \infty$.

20. P.V. $\int_{-\infty}^\infty \frac{\sin^2 x}{x^2} dx$.   *Hint*: Use the integrand $f(z) = \frac{1 - \exp(i2z)}{z^2}$ and the contour C in Figure 8.8. Let $r \to 0$ and $R \to \infty$.

21. The Fresnel integrals $\int_0^\infty \cos(x^2) dx$ and $\int_0^\infty \sin(x^2) dx$ are important in the study of optics. Use the integrand $f(z) = \exp(-z^2)$ and the contour C shown in Figure 8.10, and let $R \to \infty$ to get the value of these integrals. Use the fact from calculus that $\int_0^\infty e^{-x^2} dx = \sqrt{\frac{\pi}{2}}$.

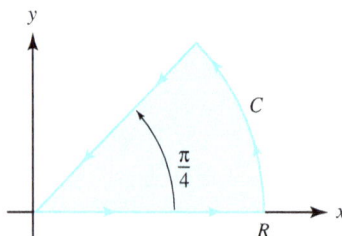

Figure 8.10: For Exercise 21

8.7 The Argument Principle and Rouché's Theorem

We now derive two results based on Cauchy's residue theorem. They have important practical applications and pertain only to functions all of whose isolated singularities are poles.

Definition 8.3 (Meromorphic function). *A function f is said to be **meromorphic** in a domain D provided the only singularities of f are isolated poles and removable singularities.*

We make three important observations relating to this definition.

- Analytic functions are a special case of meromorphic functions.

- Rational functions $f(z) = \frac{P(z)}{Q(z)}$, where $P(z)$ and $Q(z)$ are polynomials, are meromorphic in the entire complex plane.

- By definition, meromorphic functions have no essential singularities.

Suppose that f is analytic at each point on a simple closed contour C and f is meromorphic in the domain that is the interior of C. We assert without proof that Theorem 7.13 can be extended to meromorphic functions so that f has at most a finite number of zeros that lie inside C. Since the function $g(z) = \frac{1}{f(z)}$ is also meromorphic, it can have only a finite number of zeros inside C, and so f can have at most a finite number of poles that lie inside C.

The following theorem, known as the argument principle, is useful in determining the number of zeros and poles that a function has.

Theorem 8.8 (argument principle). *Suppose that f is meromorphic in the simply connected domain D and that C is a simple closed positively oriented contour in D such that f has no zeros or poles for $z \in C$. Then*

$$\frac{1}{2\pi i} \int_C \frac{f'(z)}{f(z)}\, dz = Z_f - P_f, \tag{8.34}$$

where Z_f is the number of zeros of f that lie inside C and P_f is the number of poles of f that lie inside C.

Proof. Let $a_1, a_2, \ldots, a_{Z_f}$ be the zeros of f inside C counted according to multiplicity and let $b_1, b_2, \ldots, b_{P_f}$ be the poles of f inside C counted according to multiplicity. Then $f(z)$ has the representation

$$f(z) = \frac{(z - a_1)(z - a_2) \cdots (z - a_{Z_f})}{(z - b_1)(z - b_2) \cdots (z - b_{P_f})} g(z),$$

where g is analytic and nonzero on C and inside C. A standard calculation shows that

$$\frac{f'(z)}{f(z)} = \frac{1}{(z - a_1)} + \frac{1}{(z - a_2)} + \cdots + \frac{1}{(z - a_{Z_f})}$$

$$- \frac{1}{(z - b_1)} - \frac{1}{(z - b_2)} - \cdots - \frac{1}{(z - b_{P_f})} + \frac{g'(z)}{g(z)}. \tag{8.35}$$

Corollary 6.1 then gives

$$\int_C \frac{dz}{(z - a_j)} = 2\pi i, \quad \text{for} \quad j = 1, 2, \ldots, Z_f, \quad \text{and}$$

$$\int_C \frac{dz}{(z - b_k)} = 2\pi i, \quad \text{for} \quad k = 1, 2, \ldots, P_f.$$

The function $\frac{g'(z)}{g(z)}$ is analytic inside and on C, so the Cauchy-Goursat theorem gives $\int_C \frac{g'(z)}{g(z)}\,dz = 0$. These facts lead to the conclusion of our theorem if we integrate both sides of Equation (8.35) over C. $\qquad\square$

Corollary 8.1. *Suppose that f is analytic in the simply connected domain D. Let C be a simple closed positively oriented contour in D such that for $z \in C$, $f(z) \neq 0$. Then*

$$\frac{1}{2\pi i} \int_C \frac{f'(z)}{f(z)}\,dz = Z_f,$$

where Z_f is the number of zeros of f that lie inside C.

Remark 8.4. *Certain feedback control systems in engineering must be stable. A test for stability involves the function $G(z) = 1 + F(z)$, where F is a rational function. If G does not have any zeros in the region $\{z : \operatorname{Re}(z) \geq 0\}$, then the system is stable. We determine the number of zeros of G by writing $F(z) = \frac{P(z)}{Q(z)}$, where P and Q are polynomials with no common zero. Then $G(z) = \frac{Q(z)+P(z)}{Q(z)}$, and we can check for the zeros of $Q(z) + P(z)$ by using Theorem 8.8. We select a value R so that $G(z) \neq 0$ for $\{z : |z| > R\}$ and then integrate along the contour consisting of the right half of the circle $C_R(0)$ and the line segment between iR and $-iR$. This method is known as the Nyquist stability criterion.*

Why do we label Theorem 8.8 as the *argument principle*? The answer lies with a fascinating application known as the **winding number**. Recall that a branch of the logarithm function, $\log_\alpha$, is defined by

$$\log_\alpha z = \ln |z| + i \arg_\alpha z = \ln r + i\phi,$$

where $z = re^{i\phi} \neq 0$ and $\alpha < \phi \leq \alpha + 2\pi$. Loosely speaking, suppose that for some branch of the logarithm, the composite function $\log_\alpha(f(z))$ were analytic in a simply connected domain D containing the contour C. This would imply that $\log_\alpha\big(f(z)\big)$ is an antiderivative of the function $\frac{f'(z)}{f(z)}$ for all $z \in D$. Theorems 6.9 and 8.8 would then tell us that, as z winds around the curve C, the quantity $\log_\alpha(f(z)) = \ln |f(z)| + i \arg_\alpha f(z)$ would change by $2\pi i(Z_f - P_f)$. Since $2\pi i(Z_f - P_f)$ is purely imaginary, this result tells us that $\arg_\alpha f(z)$ would change by $2\pi(Z_f - P_f)$ radians. In other words, as z winds around C, the integral $\frac{1}{2\pi i} \int_C \frac{f'(z)}{f(z)}\,dz$ would count how many times the curve $f(C)$ winds around the origin.

Unfortunately, we can't always claim that $\log_\alpha\big(f(z)\big)$ is an antiderivative of the function $\frac{f'(z)}{f(z)}$ for all $z \in D$. If it were, the Cauchy-Goursat theorem would imply that $\frac{1}{2\pi i} \int_C \frac{f'(z)}{f(z)}\,dz = 0$. Nevertheless, the heuristics that we gave—indicating that $\frac{1}{2\pi i} \int_C \frac{f'(z)}{f(z)}\,dz$ counts how many times the curve $f(C)$ winds around the origin—still hold true, as we now demonstrate.

Suppose that $C : z(t) = x(t) + iy(t)$ for $a \leq t \leq b$ is a simple closed contour and that we let $a = t_0 < t_1 < \cdots < t_n = b$ be a partition of the interval $[a, b]$. For $k = 0, 1, \ldots, n$, we let $z_k = z(t_k)$ denote the corresponding points on C, where $z_0 = z_n$. If z^* lies inside C, then the curve $C : z(t)$ winds around z^* once as t goes from a to b, as shown in Figure 8.10.

Now suppose that a function f is analytic at each point on C and meromorphic inside C. Then $f(C)$ is a closed curve in the w plane that passes through the points $w_k = f(z_k)$, for $k = 0, 1, \ldots, n$, where $w_0 = w_n$. We can choose subintervals $[t_{k-1}, t_k]$ small enough so that, on the portion of $f(C)$ between w_{k-1} and w_k, we can define a continuous branch of the logarithm

$$\log_{\alpha_k} w = \ln |w| + i \arg_{\alpha_k} w = \ln \rho + i\phi,$$

where $w = \rho e^{i\phi}$ and $\alpha_k < \phi < \alpha_k + 2\pi$, as shown in Figure 8.11. Then

$$\log_{\alpha_k} f(z_k) - \log_{\alpha_k} f(z_{k-1}) = \ln \rho_k - \ln \rho_{k-1} + i\Delta\phi_k,$$

where $\Delta\phi_k = \phi_k - \phi_{k-1}$ measures in radians the amount that the portion of the curve $f(C)$ between w_k and w_{k-1} winds around the origin. With small enough subintervals $[t_{k-1}, t_k]$, the angles α_{k-1} and α_k might be different, but the values $\arg_{\alpha_{k-1}} w_{k-1}$ and $\arg_{\alpha_k} w_{k-1}$ will be the same, so that $\log_{\alpha_{k-1}} w_{k-1} = \log_{\alpha_k} w_{k-1}$.

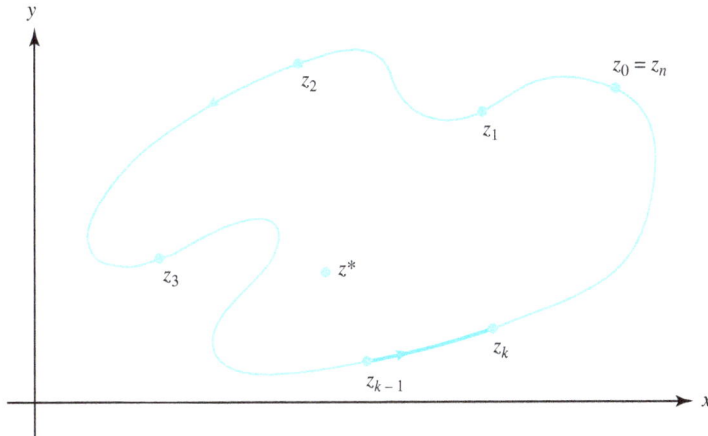

Figure 8.11: The points z_k on the contour C that winds around z^*

We can now show why $\int_C \frac{f'(z)}{f(z)} dz$ counts the number of times that $f(C)$ winds around the origin. We parametrize $C : z(t)$, for $a \le t \le b$, and choose the appropriate branches of $\log_{\alpha_k} w$, giving

$$\int_C \frac{f'(z)}{f(z)} dz = \sum_{k=1}^{n} \int_{t_{k-1}}^{t_k} \frac{f'(z(t))}{f(z(t))} z'(t)\, dt$$

$$= \sum_{k=1}^{n} \left(\log_{\alpha_k} \left[f(z(t_k)) \right] - \log_{\alpha_k} \left[f(z(t_{k-1})) \right] \right)$$

$$= \sum_{k=1}^{n} (\log_{\alpha_k} w_k - \log_{\alpha_k} w_{k-1}).$$

which we rewrite as

$$\int_C \frac{f'(z)}{f(z)} dz = \sum_{k=1}^{n} [\ln \rho_k - \ln \rho_{k-1}] + i \sum_{k=1}^{n} \Delta\phi_k. \qquad (8.36)$$

When we use the fact that $\rho_0 = \rho_n$, the first summation in Equation (8.36) vanishes. The summation of the quantities $\Delta\phi_k$ expresses the accumulated radian measure of $f(C)$ around the origin. Therefore, when we divide both sides of Equation (8.36) by $2\pi i$, its right side becomes an integer (by Theorem 8.8) that must count the number of times $f(C)$ winds around the origin.

Example 8.25. The image of the circle $C_2(0)$ under $f(z) = z^2 + z$ is the curve given by $\{(x, y) = (4\cos 2t + 2\cos t, 4\sin 2t + 2\sin t) : 0 < t < 2\pi\}$ shown in Figure 8.12.

275

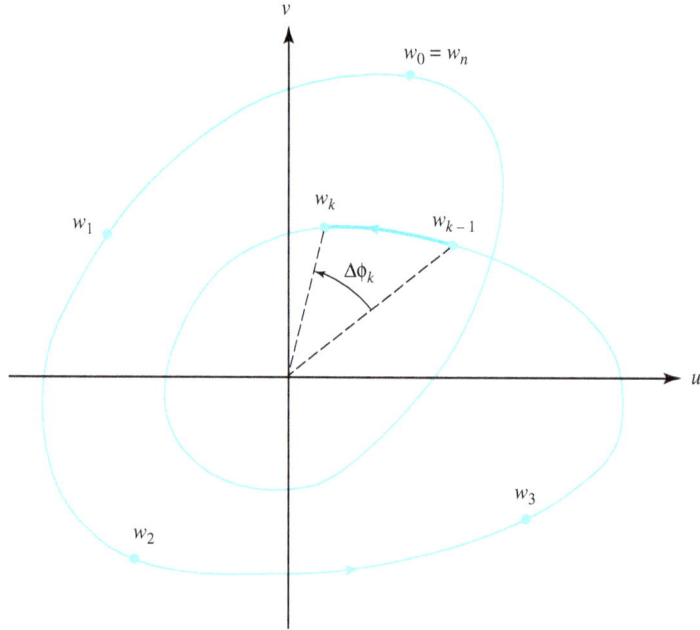

Figure 8.12: The image curve $f\big(C_2(0)\big)$ under $f(z) = z^2 + z$

Note that the image curve $f\big(C_2(0)\big)$ winds twice around the origin. We check this by computing $\frac{1}{2\pi i} \int_{C_2^+(0)} \frac{f'(z)}{f(z)}\, dz = \frac{1}{2\pi i} \int_{C_2^+(0)} \frac{2z+1}{z^2+z} dz$. The residues of the integrand are at 0 and -1. Thus

$$\frac{1}{2\pi i} \int_{C_2^+(0)} \frac{2z+1}{z^2+z} dz = \mathrm{Res}\left[\frac{2z+1}{z^2+z}, 0\right] + \mathrm{Res}\left[\frac{2z+1}{z^2+z}, -1\right]$$
$$= 1 + 1$$
$$= 2.$$

Finally, we note that if $g(z) = f(z) - a$, then $g'(z) = f(z)$, and thus we can generalize what we've just said to compute how many times the curve $f(C)$ winds around the point a. Theorem 8.9 summarizes our discussion.

Theorem 8.9 (winding numbers). *Suppose that f is meromorphic in the simply connected domain D. If C is a simple closed positively oriented contour in D such that for $z \in C$, $f(z) \neq 0$ and $f(z) \neq \infty$, then*

$$W\big(f(C), a\big) = \frac{1}{2\pi i} \int_C \frac{f'(z)}{f(z) - a}\, dz,$$

*known as the **winding number of $f(C)$ about a**, counts the number of times the curve $f(C)$ winds around the point a. If $a = 0$, the integral counts the number of times the curve $f(C)$ winds around the origin.*

Remark 8.5. *Letting $f(z) = z$ in Theorem 8.9 gives*

$$W(C, a) = \frac{1}{2\pi i} \int_C \frac{1}{z - a}\, dz = \begin{cases} 1, & \text{if } a \text{ is inside } C, \quad \text{or} \\ 0, & \text{if } a \text{ is outside } C, \end{cases}$$

which counts the number of times the curve C winds around the point a. If C is not a simple closed curve, but crosses itself perhaps several times, we can show (but omit the proof) that $W(C, a)$ still gives the number of times the curve C winds around the point a. Thus winding number *is indeed an appropriate term*.

We close this section with a result that will help us gain information about the location of the zeros and poles of meromorphic functions.

Theorem 8.10 (Rouché's theorem). *Suppose that f and g are meromorphic functions defined in the simply connected domain D, that C is a simply closed contour in D, and that f and g have no zeros or poles for $z \in C$. If the strict inequality $|f(z) + g(z)| < |f(z)| + |g(z)|$ holds for all $z \in C$, then $Z_f - P_f = Z_g - P_g$.*

Proof. Because g has no zeros or poles on C, we may legitimately divide both sides of the inequality $|f(z) + g(z)| < |f(z)| + |g(z)|$ by $|g(z)|$ to get

$$\left| \frac{f(z)}{g(z)} + 1 \right| < \left| \frac{f(z)}{g(z)} \right| + 1, \quad \text{for all} \quad z \in C. \tag{8.37}$$

For $z \in C$, $\frac{f(z)}{g(z)}$ cannot possibly be zero or any positive real number, as that would contradict Inequality 8.37. This means that C^*, the image of the curve C under the mapping $\frac{f}{g}$, does not contain the interval $[0, \infty)$, and so the function defined by

$$w(z) = \log_0\left(\frac{f(z)}{g(z)}\right) = \ln\left| \frac{f(z)}{g(z)} \right| + i \arg_0\left(\frac{f(z)}{g(z)}\right) = \ln r + i\phi,$$

where $\frac{f(z)}{g(z)} = re^{i\phi} \neq 0$ and $0 < \phi \leq 2\pi$, is analytic in a simply connected domain D^* that contains C^*. We calculate

$$w'(z) = \frac{f'(z)}{f(z)} - \frac{g'(z)}{g(z)},$$

so $w(z) = \log_0(\frac{f(z)}{g(z)})$ is an antiderivative of $\frac{f'(z)}{f(z)} - \frac{g'(z)}{g(z)}$, for all $z \in D^*$. Since C^* is a closed curve in D^*, Theorem 6.9 gives $\int_{C^*} \left(\frac{f'(z)}{f(z)} - \frac{g'(z)}{g(z)} \right) dz = 0$. According to Theorem 8.8 then,

$$\frac{1}{2\pi i} \int_{C^*} \frac{f'(z)}{f(z)} \, dz - \frac{1}{2\pi i} \int_{C^*} \frac{g'(z)}{g(z)} \, dz = (Z_f - P_f) - (Z_g - P_g) = 0,$$

which completes the proof. $\qquad\square$

Corollary 8.2. *Suppose that f and g are analytic functions defined in the simply connected domain D, that C is a simple closed contour in D, and that f and g have no zeros for $z \in C$. If the strict inequality $|f(z) + g(z)| < |f(z)| + |g(z)|$ holds for all $z \in C$, then $Z_f = Z_g$.*

Remark 8.6. *Theorem 8.10 is usually stated with the requirement that f and g satisfy the condition $|f(z) + g(z)| < |g(z)|$, for $z \in C$. The improved theorem that we gave was discovered by Irving Glicksberg (see the* American Mathematical Monthly, *83 (1976), pp. 186-187). The weaker version is adequate for most purposes, however, as the following examples illustrate.*

Example 8.26. Show that all four zeros of the polynomial $g(z) = z^4 - 7z - 1$ lie in the disk $D_2(0) = \{z : |z| < 2\}$.

Solution:

Let $f(z) = -z^4$. Then $f(z) + g(z) = -7z - 1$, and at points on the circle $C_2(0) = \{z : |z| = 2\}$ we have the relation

$$|f(z) + g(z)| \leq |-7z| + |-1| = 7(2) + 1 < 16 = |f(z)|.$$

Of course, if $|f(z) + g(z)| < |f(z)|$, then as we indicated in Remark 8.5 we certainly have $|f(z) + g(z)| < |f(z)| + |g(z)|$, so that the conditions for applying Corollary 8.2 are satisfied on the circle $C_2(0)$. The function f has a zero of order 4 at the origin, so g must have four zeros inside $D_2(0)$.

Example 8.27. Show that the polynomial $g(z) = z^4 - 7z - 1$ has one zero in the disk $D_1(0)$.

Solution:

Let $f(z) = 7z + 1$, then $f(z) + g(z) = z^4$. At points on the circle $C_1(0) = \{z : |z| = 1\}$ we have the relation
$$|f(z) + g(z)| = |z^4| = 1 < 6 = |7| - |1| \leq |7z + 1| = |f(z)|.$$

The function f has one zero at the point $-\frac{1}{7}$ in the disk $D_1(0)$, and the hypotheses of Corollary 8.2 hold on the circle $C_1(0)$. Therefore g has one zero inside $D_1(0)$.

Exercises for Section 8.7 (Selected answers or hints are on page 452.)

1. Let $f(z) = z^5 - z$. Find the number of times the image $f(C)$ winds around the origin if

   (a) $C = C_{\frac{1}{2}}(0)$.

   (b) C is the rectangle with vertices $\pm\frac{1}{2} \pm 3i$.

   (c) $C = C_2(0)$.

   (d) $C = C_{1.25}(i)$.

2. Show that four of the five roots of the equation $z^5 + 15z + 1 = 0$ belong to the annulus $A(0, \frac{3}{2}, 2) = \{z : \frac{3}{2} < |z| < 2\}$.

3. Let $g(z) = z^5 + 4z - 15$.

   (a) Show that there are no zeros in $D_1(0)$.

   (b) Show that there are five zeros in $D_2(0)$.
   Hint: Consider $f(z) = -z^5$.
   Remark: A factorization of the polynomial using numerical approximations for the coefficients is

   $$(z - 1.546)(z^2 - 1.340z + 2.857)(z^2 + 2.885z + 3.397).$$

4. Let $g(z) = z^3 + 9z + 27$.

   (a) Show that there are no zeros in $D_2(0)$.

   (b) Show that there are three zeros in $D_4(0)$.
   Remark: A factorization of the polynomial using numerical approximations for the coefficients is
   $$(z + 2.047)(z^2 - 2.047z + 13.19).$$

5. Let $g(z) = z^5 + 6z^2 + 2z + 1$.

   (a) Show that there are two zeros in $D_1(0)$.

   (b) Show that there are five zeros in $D_2(0)$.

6. Let $g(z) = z^6 - 5z^4 + 10$.

   (a) Show that there are no zeros in $|z| < 1$.

   (b) Show that there are four zeros in $|z| < 2$.

   (c) Show that there are six zeros in $|z| < 3$.

7. Let $g(z) = 3z^3 - 2iz^2 + iz - 7$.

   (a) Show that there are no zeros in $|z| < 1$.

   (b) Show that there are three zeros in $|z| < 2$.

8. Use Rouché's theorem to prove the fundamental theorem of algebra. *Hint*: For the polynomial $g(z) = a_0 + a_1 z + \cdots + a_{n-1} z^{n-1} + a_n z^n$, let $f(z) = -a_n z^n$. Show that, for points z on the circle $C_R(0)$,

$$\left| \frac{f(z) + g(z)}{f(z)} \right| < \frac{|a_0| + |a_1| + \cdots + |a_{n-1}|}{|a_n| R},$$

and conclude that the right side of this inequality is less than 1 when R is large.

9. Suppose that $h(z)$ is analytic and nonzero and $|h(z)| < 1$ for $z \in D_1(0)$. Prove that the function $g(z) = h(z) - z^n$ has n zeros inside the unit circle $C_1(0)$.

10. Suppose that $f(z)$ is analytic inside and on the simple closed contour C. If $f(z)$ is a one-to-one function at points z on C, then prove that $f(z)$ is one-to-one inside C. *Hint*: Consider the image of C.

Chapter 9

Conformal Mapping

Overview

The terminology "conformal mapping" should have a familiar sound. In 1569 the Flemish cartographer Gerardus Mercator (1512–1594) devised a cylindrical map projection that preserves angles. The Mercator projection is still used today for world maps. Another map projection known to the ancient Greeks is the stereographic projection. It is also conformal (*i.e.*, angle preserving), and we introduced it in Chapter 2 when we defined the Riemann sphere. In complex analysis a function preserves angles if and only if it is analytic or anti-analytic (*i.e.*, the conjugate of an analytic function). A significant result, known as Riemann mapping theorem, states that any simply connected domain (other than the entire complex plane) can be mapped conformally onto the unit disk.

9.1 Basic Properties of Conformal Mappings

Let f be an analytic function in the domain D and let z_0 be a point in D. If $f'(z_0) \neq 0$, then we can express f in the form

$$f(z) = f(z_0) + f'(z_0)(z - z_0) + \eta(z)(z - z_0), \tag{9.1}$$

where $\eta(z) \to 0$ as $z \to z_0$. If z is near z_0, then the transformation $w = f(z)$ has the **linear approximation**

$$S(z) = A + B(z - z_0) = Bz + A - Bz_0,$$

where $A = f(z_0)$ and $B = f'(z_0)$. Because $\eta(z) \to 0$ when $z \to z_0$, for points near z_0 the transformation $w = f(z)$ has an effect much like the linear mapping $w = S(z)$. The effect of the linear mapping S is a rotation of the plane through the angle $\alpha = \operatorname{Arg} f'(z_0)$, followed by a magnification by the factor $|f'(z_0)|$, followed by a rigid translation by the vector $A - Bz_0$. Consequently, the mapping $w = S(z)$ preserves the angles at the point z_0. We now show that the mapping $w = f(z)$ also preserves angles at z_0.

Let $C : z(t) = x(t) + iy(t)$, $-1 \leq t \leq 1$ denote a smooth curve that passes through the point $z(0) = z_0$. A vector $\mathbf{T}$ tangent to C at the point z_0 is given by

$$\mathbf{T} = z'(0),$$

where the complex number $z'(0)$ is expressed as a vector.

The angle of inclination of **T** with respect to the positive x-axis is

$$\beta = \text{Arg } z'(0).$$

The image of C under the mapping $w = f(z)$ is the curve K given by the formula $K : w(t) = u\big(x(t), y(t)\big) + iv\big(x(t), y(t)\big)$. We can use the chain rule to show that a vector **T*** tangent to K at the point $w_0 = f(z_0)$ is given by

$$\mathbf{T}^* = w'(0) = f'(z_0)z'(0).$$

The angle of inclination of **T*** with respect to the positive u-axis is

$$\gamma = \text{Arg } f'(z_0) + \text{Arg} z'(0) = \alpha + \beta, \tag{9.2}$$

where $\alpha = \text{Arg } f'(z_0)$. Therefore the effect of the transformation $w = f(z)$ is to rotate the angle of inclination of the tangent vector **T** at z_0 through the angle $\alpha = \text{Arg } f'(z_0)$ to obtain the angle of inclination of the tangent vector **T***at w_0. This situation is illustrated in Figure 9.1.

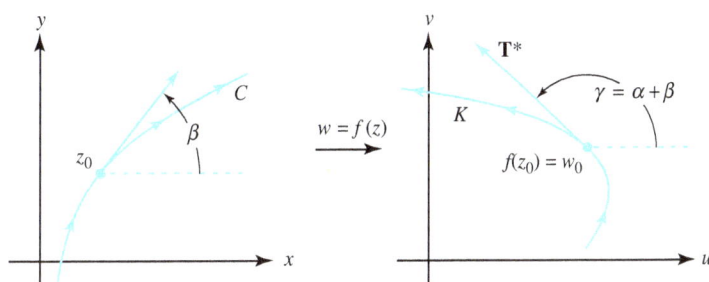

Figure 9.1: Tangents at the points z_0 and w_0, where $f'(z_0) \neq 0$

A mapping $w = f(z)$ is said to be angle preserving, or *conformal* at z_0, if it preserves angles between oriented curves in magnitude as well as in orientation. Theorem 9.1 shows where a mapping by an analytic function is conformal.

Theorem 9.1. *Let f be an analytic function in the domain D, and let z_0 be a point in D. If $f'(z_0) \neq 0$, then f is conformal at z_0.*

Proof. We let C_1 and C_2 be two smooth curves passing through z_0 with tangents given by $\mathbf{T}_1$ and $\mathbf{T}_2$, respectively. We let β_1 and β_2 denote the angles of inclination of $\mathbf{T}_1$ and $\mathbf{T}_2$, respectively.

The image curves K_1 and K_2 that pass through the point $w_0 = f(z_0)$ have tangents denoted $\mathbf{T}_1^*$ and $\mathbf{T}_2^*$, respectively. From Equation (9.2), the angles of inclination γ_1 and γ_2 of $\mathbf{T}_1^*$ and $\mathbf{T}_2^*$ are related to β_1 and β_2 by the equations

$$\gamma_1 = \alpha + \beta_1, \quad \text{and} \quad \gamma_2 = \alpha + \beta_2, \tag{9.3}$$

where $\alpha = \text{Arg } f'(z_0)$. From Equations (9.3) we conclude that

$$\gamma_2 - \gamma_1 = \beta_2 - \beta_1.$$

$\square$

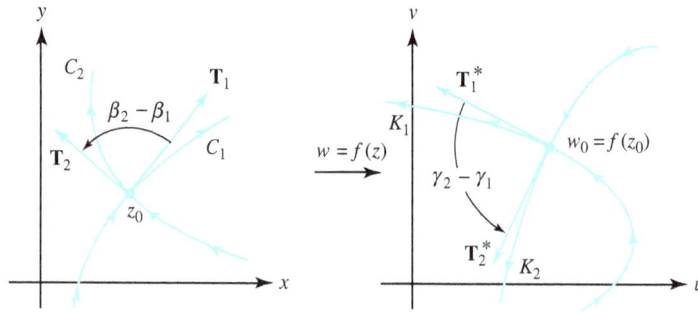

Figure 9.2: Angle preservation when $f'(z_0) \neq 0$

The analytic mapping $w = f(z)$ is conformal at the point z_0, where $f'(z_0) \neq 0$ because the angle $\gamma_2 - \gamma_1$ from K_1 to K_2 is the same in magnitude and orientation as the angle $\beta_2 - \beta_1$ from C_1 to C_2. This situation is shown in Figure 9.2.

Example 9.1. Show that the mapping $w = f(z) = \cos z$ is conformal at the points $z_1 = i$, $z_2 = 1$, and $z_3 = \pi + i$, and determine the angle of rotation given by $\alpha = \text{Arg} f'(z)$ at the given points.

Solution:

Because $f'(z) = -\sin z$, we conclude that the mapping $w = \cos z$ is conformal at all points except $z = n\pi$, where n is an integer. Calculation reveals that

$$
f'(i) = -\sin(i) = -i\sinh 1,
$$
$$
f'(1) = -\sin 1, \quad \text{and}
$$
$$
f'(\pi + i) = -\sin(\pi + i) = i\sinh 1.
$$

Therefore the angles of rotation are given, respectively, by

$$
\alpha_1 = \text{Arg}\big(f'(i)\big) = -\frac{\pi}{2},
$$
$$
\alpha_2 = \text{Arg}\big(f'(1)\big) = \pi, \quad \text{and}
$$
$$
\alpha_3 = \text{Arg}\big(f'(\pi + i)\big) = \frac{\pi}{2}.
$$

Let f be a nonconstant analytic function. If $f'(z_0) = 0$, then z_0 is called a *critical point* of f, and the mapping $w = f(z)$ is not conformal at z_0. Theorem 9.2 shows what happens at a critical point.

Theorem 9.2. *Let f be analytic at z_0. If $f'(z_0) = 0, \ldots, f^{(k-1)}(z_0) = 0$ and $f^{(k)}(z_0) \neq 0$, then the mapping $w = f(z)$ magnifies angles at the vertex z_0 by a factor k.*

Proof. Since f is analytic at z_0, it has a Taylor series expansion. Because $a_n = \dfrac{f^{(n)}(z_0)}{n!} = 0$, for $n = 1, 2, \ldots, k-1$, the series representation for f is

$$
f(z) = f(z_0) + a_k(z - z_0)^k + a_{k+1}(z - z_0)^{k+1} + \cdots . \tag{9.4}
$$

From Equation (9.4) we conclude that

$$
f(z) - f(z_0) = (z - z_0)^k g(z), \tag{9.5}
$$

282

where g is analytic at z_0 and $g(z_0) = a_k \neq 0$. Consequently, if $w = f(z)$ and $w_0 = f(z_0)$, then using Equation (9.5) we obtain

$$\arg(w - w_0) = \text{Arg}[f(z) - f(z_0)] = k\text{Arg}(z - z_0) + \text{Arg}[g(z)]. \qquad (9.6)$$

If C is a smooth curve that passes through z_0 and $z \to z_0$ along C, then $w \to w_0$ along the image curve K. The angle of inclination of the tangents $\mathbf{T}$ to C and $\mathbf{T}^*$ to K, respectively, are then given by the following limits:

$$\beta = \lim_{z \to z_0} \text{Arg}(z - z_0) \quad \text{and} \quad \gamma = \lim {}_- w \to w_0 \ \text{Arg}(w - w_0). \qquad (9.7)$$

From Equations (9.6) and (9.7) it follows that

$$\gamma = \lim_{z \to z_0} (k\text{Arg}(z - z_0) + \text{Arg}[g(z)]) = k\beta + \delta, \qquad (9.8)$$

where $\delta = \text{Arg}[g(z_0)] = \text{Arg}(a_k)$.

If C_1 and C_2 are two smooth curves that pass through z_0, and K_1 and K_2 are their images, then from Equation (9.8) it follows that

$$\Delta\gamma = \gamma_2 - \gamma_1 = k(\beta_2 - \beta_1) = k\Delta\beta.$$

That is, the angle $\Delta\gamma$ from K_1 to K_2 is k times as large as the angle $\Delta\beta$ from C_1 to C_2. Therefore angles at the vertex z_0 are magnified by the factor k. This situation is shown in Figure 9.3. $\qquad \square$

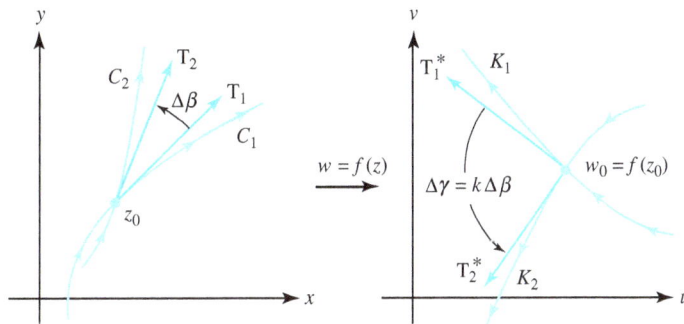

Figure 9.3: The analytic mapping $w = f(z)$ at point z_0 expands angles by a factor of k when $f'(z_0) = 0, \ldots, f^{(k-1)}(z_0) = 0$, and $f^{(k)}(z_0) \neq 0$

Example 9.2. Show that the mapping $w = f(z) = z^2$ maps the unit square $S = \{x + iy : 0 < x < 1, \ 0 < y < 1\}$ onto the region in the upper half-plane $\text{Im}(w) > 0$, which lies under the parabolas

$$u = 1 - \frac{1}{4}v^2, \quad \text{and} \quad u = -1 + \frac{1}{4}v^2,$$

as shown in Figure 9.4.

Solution:

The derivative is $f'(z) = 2z$, and we conclude that the mapping $w = z^2$ is conformal for all $z \neq 0$. Note that the right angles at the vertices $z_1 = 1$, $z_2 = 1 + i$, and $z_3 = i$ are mapped onto right angles at the vertices $w_1 = 1$, $w_2 = 2i$, and $w_3 = -1$, respectively. At the point $z_0 = 0$, we have $f'(0) = 0$ and $f''(0) \neq 0$. Hence angles at the vertex $z_0 = 0$ are magnified by the factor $k = 2$. In particular, the right angle at $z_0 = 0$ is mapped onto the straight angle at $w_0 = 0$.

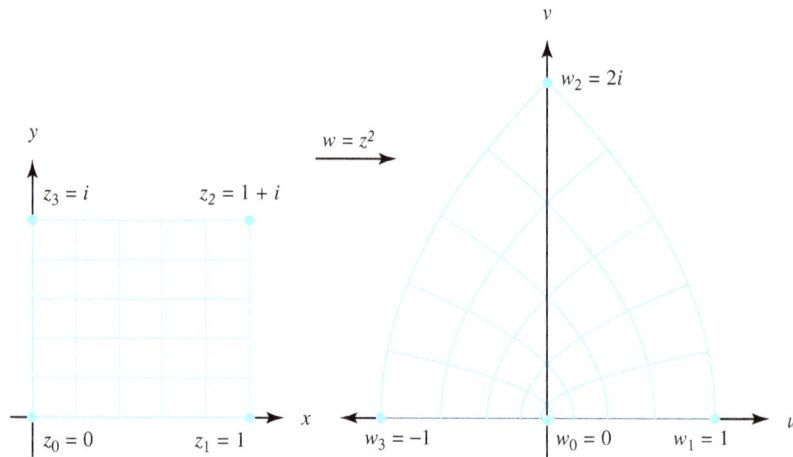

Figure 9.4: The mapping $w = z^2$

Another property of a conformal mapping $w = f(z)$ is obtained by considering the modulus of $f'(z_0)$. If z_1 is near z_0, we can use Equation (9.1) and neglect the term $\eta(z_1)(z_1 - z_0)$. We then have the approximation

$$w_1 - w_0 = f(z_1) - f(z_0) \approx f'(z_0)(z_1 - z_0). \tag{9.9}$$

From Equation (9.9), the distance $|w_1 - w_0|$ between the images of the points z_1 and z_0 is given approximately by $|f'(z_0)||z_1 - z_0|$. Therefore we say that the transformation $w = f(z)$ changes small distances near z_0 by the *scale factor* $|f'(z_0)|$. For example, the scale factor of the transformation $w = f(z) = z^2$ near the point $z_0 = 1 + i$ is $|f'(1+i)| = |2(1+i)| = 2\sqrt{2}$.

We also need to say a few things about the inverse transformation $z = g(w)$ of a conformal mapping $w = f(z)$ near a point z_0, where $f'(z_0) \neq 0$. A complete justification of the following assertions relies on theorems studied in advanced calculus.[1] We express the mapping $w = f(z)$ in the coordinate form

$$u = u(x, y), \quad \text{and} \quad v = v(x, y). \tag{9.10}$$

The mapping in Equations (9.10) represents a transformation from the xy plane into the uv plane, and the *Jacobian determinant*, $J(x, y)$, is defined by

$$J(x, y) = \begin{vmatrix} u_x(x, y) & u_y(x, y) \\ v_x(x, y) & v_y(x, y). \end{vmatrix} \tag{9.11}$$

The transformation in Equations (9.10) has a local inverse, provided $J(x, y) \neq 0$. Expanding Equation (9.11) and using the Cauchy-Riemann equations, we obtain

$$\begin{aligned} J(x_0, y_0) &= u_x(x_0, y_0)v_y(x_0, y_0) - v_x(x_0, y_0)u_y(x_0, y_0) \\ &= u_x^2(x_0, y_0) + v_x^2(x_0, y_0) \\ &= |f'(z_0)|^2 \neq 0. \end{aligned} \tag{9.12}$$

[1]See, for instance, R. Creighton Buck, *Advanced Calculus*, 3rd ed. (New York, McGraw-Hill), pp. 358—361, 1978.

284

Consequently, Equations (9.11) and (9.12) imply that a local inverse $z = g(w)$ exists in a neighborhood of the point w_0. The derivative of g at w_0 is given by the familiar expression

$$\begin{aligned} g'(w_0) &= \lim_{w \to w_0} \frac{g(w) - g(w_0)}{w - w_0} \\ &= \lim_{z \to z_0} \frac{z - z_0}{f(z) - f(z_0)} \\ &= \frac{1}{f'(z_0)} = \frac{1}{f'(g(w_0))}. \end{aligned}$$

Exercises for Section 9.1 (Selected answers or hints are on page 452.)

1. State where the following mappings are conformal.

   (a) $w = \exp z$.
   (b) $w = \sin z$.
   (c) $w = z^2 + 2z$.
   (d) $w = \exp(z^2 + 1)$.
   (e) $w = \dfrac{1}{z}$.
   (f) $w = \dfrac{z+1}{z-1}$.

2. Find the angle of rotation $\alpha = \operatorname{Arg}(f'(z))$ and the scale factor $|f'(z)|$ of the mapping $w = f(z)$ at the indicated points.

   (a) $w = \dfrac{1}{z}$ at the points 1, $1 + i$, and i.
   (b) $w = \ln r + i\theta$, where $-\pi < \theta \le \pi$ at the points 1, $1 + i$, i, and -1.
   (c) $w = r^{\frac{1}{2}} \cos \frac{\theta}{2} + i r^{\frac{1}{2}} \sin \frac{\theta}{2}$, where $-\pi < \theta < \pi$, at the points i, 1, $-i$, and $3 + 4i$.
   (d) $w = \sin z$ at the points $\frac{\pi}{2} + i$, 0, and $-\frac{\pi}{2} + i$.

3. Consider the mapping $w = z^2$. If $a \ne 0$ and $b \ne 0$, show that the lines $x = a$ and $y = b$ are mapped onto orthogonal parabolas.

4. Consider the mapping $w = z^{\frac{1}{2}}$, where $z^{\frac{1}{2}}$ denotes the principal branch of the square root function. If $a > 0$ and $b > 0$, show that the lines $x = a$ and $y = b$ are mapped onto orthogonal curves.

5. Consider the mapping $w = \exp z$. Show that the lines $x = a$ and $y = b$ are mapped onto orthogonal curves.

6. For $w = \sin z$ show that the line segment $-\frac{\pi}{2} < x < \frac{\pi}{2}$, $y = 0$, and the vertical line $x = a$, where $|a| < \frac{\pi}{2}$, are mapped onto orthogonal curves.

7. Consider the mapping $w = \operatorname{Log} z$, where $\operatorname{Log} z$ denotes the principal branch of the logarithm function. Show that the positive x-axis and the vertical line $x = 1$ are mapped onto orthogonal curves.

8. If f is analytic at z_0 and $f'(z_0) \ne 0$, show that the function $g(z) = \overline{f(z)}$ preserves the magnitude, but reverses the sense, of angles at z_0.

9. If $w = f(z)$ is a mapping, where $f(z)$ is not analytic, then what behavior would you expect regarding the angles between curves?

9.2   Bilinear Transformations

Another important class of elementary mappings was studied by Augustus Ferdinand Möbius (1790–1868). These mappings are conveniently expressed as the quotient of two linear expressions. They arise naturally in mapping problems involving the function $\text{Arctan}(z)$. In this section, we show how they are used to map a disk one-to-one and onto a half-plane.

If we let a, b, c, and d denote four complex constants with the restriction that $ad \neq bc$, then the function

$$w = S(z) = \frac{az + b}{cz + d} \tag{9.13}$$

is called a **bilinear transformation**, a **Möbius transformation**, or a **linear fractional transformation**. If the expression for S in Equation (9.13) is multiplied by the quantity $cz + d$, then the resulting expression has the bilinear form $cwz - az + dw - b = 0$. We collect terms involving z and write $z(cw - a) = -dw + b$. Then, for values of $w \neq \frac{a}{c}$, the inverse transformation is given by

$$z = S^{-1}(w) = \frac{-dw + b}{cw - a}. \tag{9.14}$$

We can extend S and S^{-1} to mappings in the extended complex plane. The value $S(\infty)$ should equal the limit of $S(z)$ as $z \to \infty$. Therefore we define

$$S(\infty) = \lim_{z \to \infty} S(z) = \lim_{z \to \infty} \frac{a + \frac{b}{z}}{c + \frac{d}{z}} = \frac{a}{c},$$

and the inverse is $S^{-1}(\frac{a}{c}) = \infty$. Similarly, the value $S^{-1}(\infty)$ is obtained by

$$S^{-1}(\infty) = \lim_{w \to \infty} S^{-1}(w) = \lim_{w \to \infty} \frac{-d + \frac{b}{w}}{c - \frac{a}{w}} = \frac{-d}{c},$$

and the inverse is $S(\frac{-d}{c}) = \infty$. With these extensions we conclude that the transformation $w = S(z)$ is a one-to-one mapping of the extended complex z plane onto the extended complex w plane.

We now show that a bilinear transformation carries the class of circles and lines onto itself. If S is an arbitrary bilinear transformation given by Equation (9.13) and $c = 0$, then S reduces to a linear transformation, which carries lines onto lines and circles onto circles. If $c \neq 0$, then we can write S in the form

$$S(z) = \frac{a(cz + d) + bc - ad}{c(cz + d)} = \frac{a}{c} + \left(\frac{bc - ad}{c} \right) \left(\frac{1}{cz + d} \right). \tag{9.15}$$

The condition $ad \neq bc$ precludes the possibility that S reduces to a constant. Equation (9.15) indicates that S can be considered as a composition of functions. It is a linear mapping $\xi = cz + d$, followed by the reciprocal transformation $Z = \frac{1}{\xi}$, followed by $w = \frac{a}{c} + \frac{bc - ad}{c} Z$. In Chapter 2 we showed that each function in this composition maps the class of circles and lines onto itself; it follows that the bilinear transformation S has this property. A half-plane can be considered to be a family of parallel lines and a disk as a family of circles. Therefore we conclude that a bilinear transformation maps the class of half-planes and disks onto itself. Example 9.3 illustrates this idea.

Example 9.3. Show that $w = S(z) = \frac{i(1-z)}{1+z}$ maps the unit disk $|z| < 1$ one-to-one and onto the upper half-plane $\text{Im}(w) > 0$.

Solution:

We first consider the unit circle $C : |z| = 1$, which forms the boundary of the disk and find its image in the w plane. If we write $S(z) = \frac{-iz+i}{z+1}$, then we see that $a = -i$, $b = ic = 1$, and $d = 1$. Using Equation (9.14), we find that the inverse is given by

$$z = S^{-1}(w) = \frac{-dw + b}{cw - a} = \frac{-w + i}{w + i}. \tag{9.16}$$

If $|z| = 1$, then Equation (9.16) implies that the images of points on the unit circle satisfy the equation

$$|w + i| = |-w + i|. \tag{9.17}$$

Squaring both sides of Equation (9.17), we obtain $u^2 + (1 + v)^2 = u^2 + (1 - v)^2$, which can be simplified to yield $v = 0$, which is the equation of the u-axis in the w plane.

The circle C divides the z plane into two portions, and its image is the u-axis, which divides the w plane into two portions. The image of the point $z = 0$ is $w = S(0) = i$, so we expect that the interior of the circle C is mapped onto the portion of the w plane that lies above the u-axis. To show that this outcome is true, we let $|z| < 1$. Then Equation (9.16) implies that the image values must satisfy the inequality $|-w + i| < |w + i|$, which we write as

$$d_1 = |w - i| < |w - (-i)| = d_2.$$

If we interpret d_1 as the distance from w to i and d_2 as the distance from w to $-i$, then a geometric argument shows that the image point w must lie in the upper half-plane $\text{Im}(w) > 0$, as shown in Figure 9.5. As S is one-to-one and onto in the extended complex plane, it follows that S maps the disk onto the half-plane.

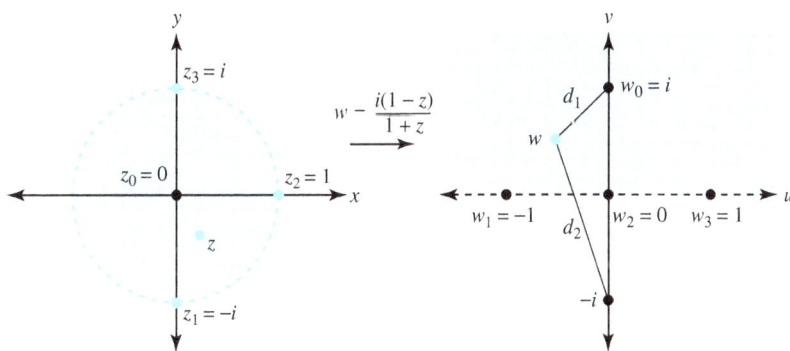

Figure 9.5: The image of $|z| < 1$ under $w = \frac{i(1-z)}{1+z}$

The general formula for a bilinear transformation (Equation (9.13)) appears to involve four independent coefficients: a, b, c, and d. But as $S(z) \neq K$, either $a \neq 0$ or $c \neq 0$, we can express the transformation with three unknown coefficients and write either

$$S(z) = \frac{z + \frac{b}{a}}{\frac{cz}{a} + \frac{d}{a}}, \quad \text{or} \quad S(z) = \frac{\frac{az}{c} + \frac{b}{c}}{z + \frac{d}{c}},$$

respectively. Doing so permits us to determine a unique a bilinear transformation if three distinct image values $S(z_1) = w_1$, $S(z_2) = w_2$, and $S(z_3) = w_3$ are specified. To determine such a mapping, we can conveniently use an implicit formula involving z and w.

Theorem 9.3 (The implicit formula). *There exists a unique bilinear transformation that maps three distinct points, z_1, z_2, and z_3, onto three distinct points, w_1, w_2, and w_3, respectively. An implicit formula for the mapping is given by*

$$\left(\frac{z - z_1}{z - z_3}\right)\left(\frac{z_2 - z_3}{z_2 - z_1}\right) = \left(\frac{w - w_1}{w - w_3}\right)\left(\frac{w_2 - w_3}{w_2 - w_1}\right). \tag{9.18}$$

Proof. We algebraically manipulate Equation (9.18) and solve for w in terms of z. The result is an expression for w that has the form of Equation (9.13), where the coefficients a, b, c, and d involve various combinations of the values z_1, z_2, z_3, w_1, w_2, and w_3. The details are left as an exercise.

If we set $z = z_1$ and $w = w_1$ in Equation (9.18), then both sides of the equation are zero, showing that w_1 is the image of z_1. If we set $z = z_2$ and $w = w_2$ in Equation (9.18), then both sides of the equation take on the value 1. Hence w_2 is the image of z_2. Taking reciprocals, we write Equation (9.18) in the form

$$\left(\frac{z - z_3}{z - z_1}\right)\left(\frac{z_2 - z_1}{z_2 - z_3}\right) = \left(\frac{w - w_3}{w - w_1}\right)\left(\frac{w_2 - w_1}{w_2 - w_3}\right). \tag{9.19}$$

If we set $z = z_3$ and $w = w_3$ in Equation (9.19), then both sides of the equation are zero. Therefore w_3 is the image of z_3, and we have shown that the transformation has the required properties. □

Example 9.4. Construct the bilinear transformation $w = S(z)$ that maps the points $z_1 = -i$, $z_2 = 1$, and $z_3 = i$ onto the points $w_1 = -1$, $w_2 = 0$, and $w_3 = 1$, respectively.

Solution:

We use the implicit formula (Equation (9.18)) and write

$$\left(\frac{z + i}{z - i}\right)\left(\frac{1 - i}{1 + i}\right) = \left(\frac{w + 1}{w - 1}\right)\left(\frac{0 - 1}{0 + 1}\right) = \frac{w + 1}{-w + 1}.$$

Expanding this equation, we obtain

$$(1 + i)zw + (1 - i)w + (1 + i)z + (1 - i)$$
$$= (-1 + i)zw + (-1 - i)w + (1 - i)z + (1 + i).$$

Then, collecting terms involving w and zw on the left results in

$$2w + 2zw = 2i - 2iz,$$

from which we obtain $w(1 + z) = i(1 - z)$. Therefore the desired bilinear transformation is

$$w = S(z) = \frac{i(1 - z)}{1 + z}.$$

Example 9.5. Find the bilinear transformation $w = S(z)$ that maps the points $z_1 = -2$, $z_2 = -1 - i$, and $z_3 = 0$ onto $w_1 = -1$, $w_2 = 0$, and $w_3 = 1$, respectively.

Solution:

Again, we use the implicit formula and write

$$\left(\frac{z - (-2)}{z - 0}\right)\left(\frac{-1 - i - 0}{-1 - i - (-2)}\right) = \left(\frac{w - (-1)}{w - 1}\right)\left(\frac{0 - 1}{0 - (-1)}\right).$$

Using the fact that $\frac{-1-i}{1-i} = \frac{1}{i}$, we rewrite this equation as

$$\frac{z + 2}{iz} = \frac{1 + w}{1 - w}.$$

We now expand the equation and obtain $z + 2 - zw - 2w = iz + izw$, which can be solved for w in terms of z, giving the desired solution

$$w = S(z) = \frac{(1 - i)z + 2}{(1 + i)z + 2}.$$

We let D be a region in the z plane that is bounded by either a circle or a straight line C. We further let z_1, z_2, and z_3 be three distinct points that lie on C and have the property that an observer moving along C from z_1 to z_3 through z_2 finds the region D to be on the left. If C is a circle and D is the interior of C, then we say that C is positively oriented. Conversely, the ordered triple (z_1, z_2, z_3) uniquely determines a region that lies to the left of C.

We let G be a region in the w plane that is bounded by either a circle of a straight line K. We further let w_1, w_2, and w_3 be three distinct points that lie on K such that an observer moving along K from w_1 to w_3 through w_2 finds the region G to be on the left. Because a bilinear transformation is a conformal mapping that maps the class of circles and straight lines onto itself, we can use the implicit formula to construct a bilinear transformation $w = S(z)$ that is a one-to-one mapping of D onto G.

Example 9.6. Show that the mapping

$$w = S(z) = \frac{(1 - i)z + 2}{(1 + i)z + 2}$$

maps the disk $D : |z + 1| < 1$ onto the upper half-plane $\text{Im}(w) > 0$.

Solution:

For convenience, we choose the ordered triple $z_1 = -2$, $z_2 = -1 - i$, and $z_3 = 0$, which gives the circle $C : |z + 1| = 1$ a positive orientation and the disk D a left orientation. From Example 9.5, the corresponding image points are

$$w_1 = S(z_1) = -1, \quad w_2 = S(z_2) = 0, \quad \text{and} \quad w_3 = S(z_3) = 1.$$

Because the ordered triple of points w_1, w_2, and w_3 lie on the u-axis, it follows that the image of circle C is the u-axis. The points w_1, w_2, and w_3 give the upper half-plane $G : \text{Im}(w) > 0$ a left orientation. Therefore $w = S(z)$ maps the disk D onto the upper half-plane G. To check our work, we choose a point z_0 that lies in D and find the half-plane in which its image, w_0, lies. The choice $z_0 = -1$ yields $w_0 = S(-1) = i$. Hence the upper half-plane is the correct image. This situation is illustrated in Figure 9.6.

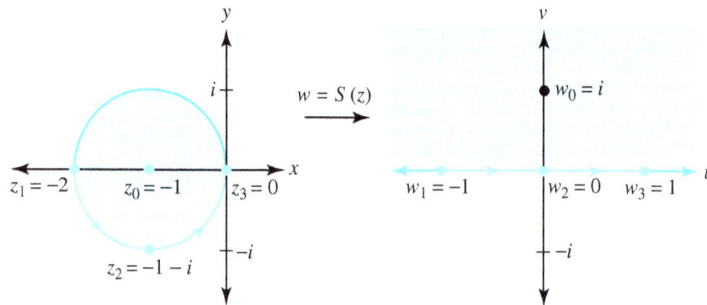

Figure 9.6: The bilinear mapping $w = S(z) = (1-i)z + 2](1+i)z + 2]$

Corollary 9.1. *(The implicit formula with a point at infinity) In Equation 9.18, the point at infinity can be introduced as one of the prescribed points in either the z plane or the w plane.*

Proof:

Case 1 If $z_3 = \infty$, then we can write $\frac{z_2 - z_3}{z - z_3} = \frac{z_2 - \infty}{z - \infty} = 1$ and substitute this expression into Equation (9.18) to obtain

$$\frac{z - z_1}{z_2 - z_1} = \left(\frac{w - w_1}{w - w_3}\right)\left(\frac{w_2 - w_3}{w_2 - w_1}\right).$$

Case 2 If $w_3 = \infty$, then we can write $\frac{w_2 - w_3}{w - w_3} = \frac{w_2 - \infty}{w - \infty} = 1$ and substitute this expression into Equation (9.18) to obtain

$$\left(\frac{z - z_1}{z - z_3}\right)\left(\frac{z_2 - z_3}{z_2 - z_1}\right) = \frac{w - w_1}{w_2 - w_1}. \tag{9.20}$$

Equation (9.20) is sometimes used to map the crescent-shaped region that lies between the tangent circles onto an infinite strip.

Example 9.7. Find the bilinear transformation that maps the crescent-shaped region that lies inside the disk $|z - 2| < 2$ and outside the circle $|z - 1| = 1$ onto a horizontal strip.

Solution:

For convenience we choose $z_1 = 4$, $z_2 = 2 + 2i$, and $z_3 = 0$ and the image values $w_1 = 0$, $w_2 = 1$, and $w_3 = \infty$, respectively. The ordered triple z_1, z_2, and z_3 gives the circle $|z - 2| = 2$ a positive orientation and the disk $|z - 2| < 2$ has a left orientation. The image points w_1, w_2, and w_3 all lie on the extended u-axis, and they determine a left orientation for the upper half-plane $\text{Im}(w) > 0$. Therefore we can use the second implicit formula given in Equation (9.20) to write

$$\left(\frac{z - 4}{z - 0}\right)\left(\frac{2 + 2i - 0}{2 + 2i - 4}\right) = \frac{w - 0}{1 - 0},$$

which determines a mapping of the disk $|z - 2| < 2$ onto the upper half-plane $\text{Im}(w) > 0$. We simplify the preceding equation to obtain the desired solution:

$$w = S(z) = \frac{-iz + 4i}{z}.$$

A straightforward calculation shows that the points $z_4 = 1 - i$, $z_5 = 2$, and $z_6 = 1 + i$ are mapped onto the points

$$w_4 = S(1 - i) = -2 + i, \quad w_5 = S(2) =, \quad \text{and} \quad w_6 = S(1 + i) = 2 + i,$$

290

respectively. The points w_4, w_5, and w_6 lie on the horizontal line $\text{Im}(w) = 1$ in the upper half-plane. Therefore the crescent-shaped region is mapped onto the horizontal strip $0 < \text{Im}(w) < 1$, as shown in Figure 9.7.

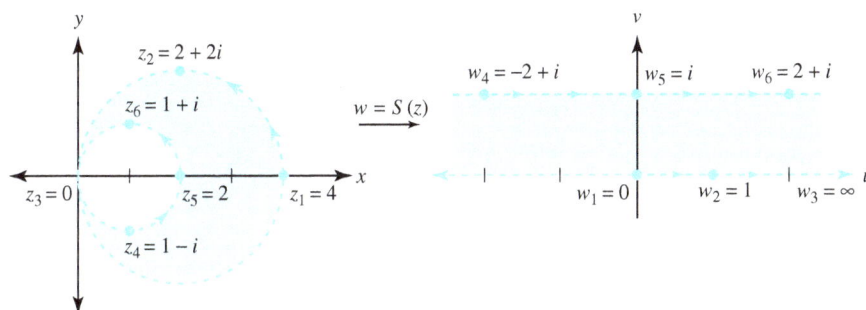

Figure 9.7: The mapping $w = S(z) = \frac{-iz+4i}{z}$

9.2.1 Lines of Flux

In electronics, images of certain lines represent lines of electric flux, which comprise the trajectory of an electron placed in an electrical field. Consider the bilinear transformation

$$w = S(z) = \frac{z}{z-a} \quad \text{and} \quad z = S^{-1}(w) = \frac{aw}{w-1}.$$

The half rays $\{\text{Arg}(w) = c\}$, where c is a constant, that meet at the origin $w = 0$ represent the lines of electric flux produced by a source located at $w = 0$ (and a sink at $w = \infty$). The preimage of this family of lines is a family of circles that pass through the points $z = 0$ and $z = a$. We visualize these circles as the lines of electric flux from one point charge to another. The limiting case as $a \to 0$ is called a dipole. The graphs for $a = 1$, $a = 0.5$, and $a = 0.1$ are shown in Figure 9.8.

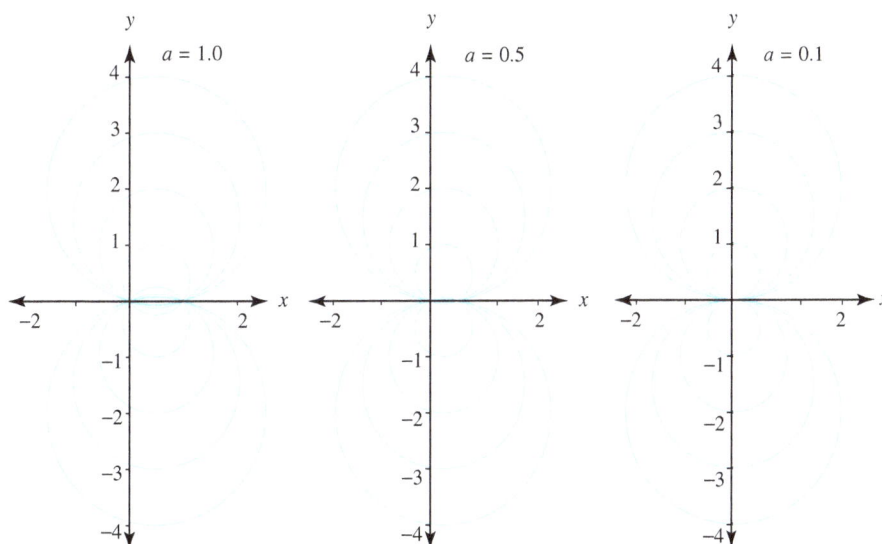

Figure 9.8: Images of $\text{Arg}(w) = c$ under the mapping $z = \frac{aw}{w-1}$

Exercises for Section 9.2 (Selected answers or hints are on page 453.)

1. If $w = S(z) = \frac{(1-i)z+2}{(1+i)z+2}$, find $S^{-1}(w)$.

2. If $w = S(z) = \frac{i+z}{i-z}$, find $S^{-1}(w)$.

3. Find the image of the right half-plane $\mathrm{Re}(z) > 0$ under $w = \frac{i(1-z)}{1+z}$.

4. Show that the bilinear transformation $w = \frac{i(1-z)}{1+z}$ maps the portion of the disk $|z| < 1$ that lies in the upper half-plane $\mathrm{Im}(z) > 0$ onto the first quadrant $u > 0$, $v > 0$.

5. Find the image of the upper half-plane $\mathrm{Im}(z) > 0$ under the transformation
$$w = \frac{(1-i)z+2}{(1+i)z+2}.$$

6. Find the bilinear transformation $w = S(z)$ that maps the points $z_1 = 0$, $z_2 = i$, and $z_3 = -i$ onto $w_1 = -1$, $w_2 = 1$, and $w_3 = 0$, respectively.

7. Find the bilinear transformation $w = S(z)$ that maps the points $z_1 = -i$, $z_2 = 0$, and $z_3 = i$ onto $w_1 = -1$, $w_2 = i$, and $w_3 = 1$, respectively.

8. Find the bilinear transformation $w = S(z)$ that maps the points $z_1 = 0$, $z_2 = 1$, and $z_3 = 2$ onto $w_1 = 0$, $w_2 = 1$, and $w_3 = \infty$, respectively.

9. Find the bilinear transformation $w = S(z)$ that maps the points $z_1 = 1$, $z_2 = i$, and $z_3 = -1$ onto $w_1 = 0$, $w_2 = 1$, and $w_3 = \infty$, respectively.

10. Show that the transformation $w = \frac{i+z}{i-z}$ maps the unit disk $|z| < 1$ onto the right half-plane $\mathrm{Re}(w) > 0$.

11. Find the image of the lower half-plane $\mathrm{Im}(z) < 0$ under $w = \frac{i+z}{i-z}$.

12. If $S_1(z) = \frac{z-2}{z+1}$ and $S_2(z) = \frac{z}{z+3}$, find $S_1(S_2(z))$ and $S_2(S_1(z))$.

13. Find the image of the quadrant $x > 0$, $y > 0$ under $w = \frac{z-1}{z+1}$.

14. Show that Equation (9.18) can be written in the form of Equation (9.13).

15. Find the image of the horizontal strip $0 < y < 2$ under $w = \frac{z}{z-i}$.

16. Show that the bilinear transformation $w = S(z) = \frac{az+b}{cz+d}$ is conformal at all points $z \neq \frac{-d}{c}$.

17. A *fixed point* of a mapping $w = f(z)$ is a point z_0 such that $f(z_0) = z_0$. Show that a bilinear transformation can have at most two fixed points.

18. Find the fixed points of

    (a) $w = \frac{z-1}{z+1}$.

    (b) $w = \frac{4z+3}{2z-1}$.

9.3 Mappings Involving Elementary Functions

In Section 5.1 we showed that the function $w = f(z) = \exp z$ is a one-to-one mapping of the fundamental period strip $-\pi < y \leq \pi$ in the z plane onto the w plane with the point $w = 0$ deleted. Because $f'(z) \neq 0$, the mapping $w = \exp z$ is a conformal mapping at each point z in the complex plane. The family of horizontal lines $y = c$ for $-\pi < c \leq \pi$ and the segments $x = a$ for $-\pi < y \leq \pi$ form an orthogonal grid in the fundamental period strip. Their images under the mapping $w = \exp z$ are the rays $\rho > 0$ and $\phi = c$ and the circles $|w| = e^a$, respectively. These images form an orthogonal curvilinear grid in the w plane, as shown in Figure 9.9. If $-\pi < c < d \leq \pi$, then the rectangle $R = \{x + iy : a < x < b, \ c < y < d\}$ is mapped one-to-one and onto the region $G = \{\rho e^{i\phi} : e^a < \rho < e^b, \ c < \phi < d\}$. The inverse mapping is the principal branch of the logarithm $z = \text{Log } w$.

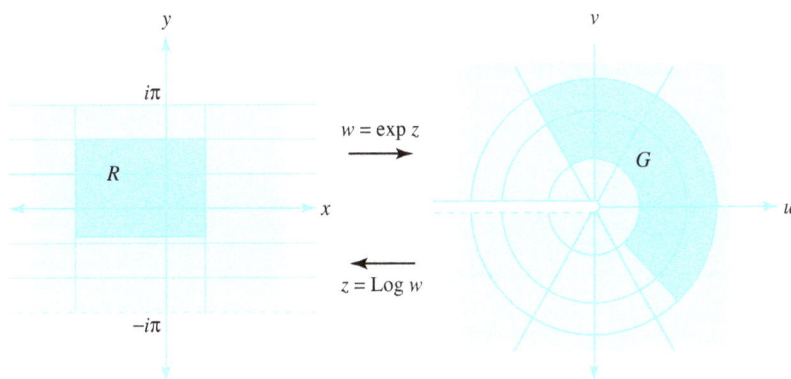

Figure 9.9: The conformal mapping $w = \exp z$

In this section we show how compositions of conformal transformations are used to construct mappings with specified characteristics.

Example 9.8. Show that the transformation $w = f(z) = \frac{e^z - i}{e^z + i}$ is a one-to-one conformal mapping of the horizontal strip $0 < y < \pi$ onto the disk $|w < 1|$. Furthermore, the x-axis is mapped onto the lower semicircle bounding the disk, and the line $y = \pi$ is mapped onto the upper semicircle.

Solution:

The function f is the composition of $Z = \exp z$ followed by $w = \frac{Z-i}{Z+i}$. The transformation $Z = \exp z$ maps the horizontal strip $0 < y < \pi$ onto the upper half plane $\text{Im}(Z) > 0$; the x-axis is mapped on to the positive x-axis; and the line $y = \pi$ is mapped onto the negative x-axis. Then the bilinear transformation $w = \frac{Z-i}{Z+i}$ maps the upper half plane $\text{Im}(Z) > 0$ onto the disk $|w| < 1$, the positive x-axis is mapped onto the lower semicircle; and the negative x-axis onto the upper semicircle. Figures 9.10 and 9.11 illustrate the composite mappings.

Example 9.9. Show that the transformation $w = f(z) = \text{Log}(\frac{1+z}{1-z})$ is a one-to-one conformal mapping of the unit disk $|z| < 1$ onto the horizontal strip $|v| < \frac{\pi}{2}$. Furthermore, the upper semicircle of the disk is mapped onto the line $v = \frac{\pi}{2}$ and the lower semicircle onto $v = -\frac{\pi}{2}$.

Solution:

The function $w = f(z)$ is the composition of the bilinear transformation $Z = \frac{1+z}{1-z}$ followed by the logarithmic mapping $w = \text{Log} z$. The image of the disk $|z| < 1$ under the bilinear mapping $Z = \frac{1+z}{1-z}$ is the right half-plane $\text{Re}(Z) > 0$; the upper semicircle is mapped onto the positive y-axis; and the lower semicircle is mapped onto the negative y-axis. The logarithmic function

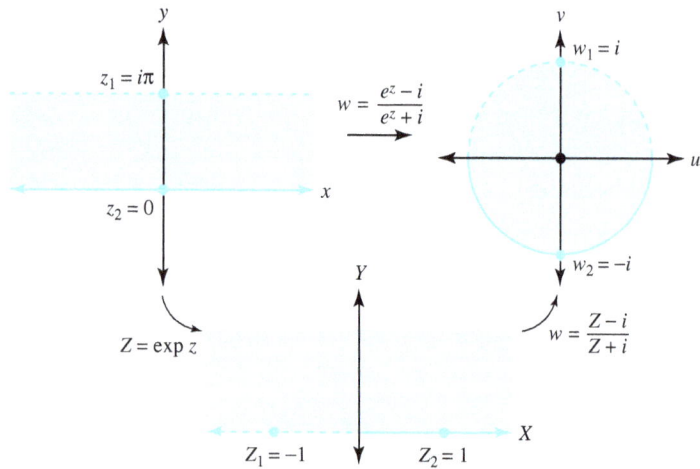

Figure 9.10: The composite transformation $w = \frac{e^z - i}{e^z + i}$

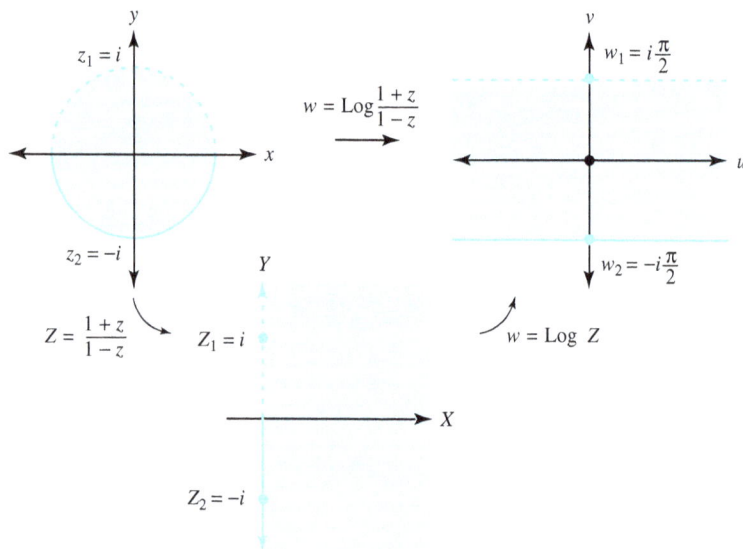

Figure 9.11: The composite transformation $w = \text{Log}(\frac{1+z}{1-z})$

$w = \text{Log}\, Z$ then maps the right half-plane onto the horizontal strip, the image of the positive y-axis is the line $v = \frac{\pi}{2}$, and the image of the negative y-axis is the line $v = \frac{\pi}{2}$. Figures 9.11 and 9.12 illustrate the composite mappings.

Example 9.10. Show that the transformation $w = f(z) = (\frac{1+z}{1-z})^2$ is a one-to-one conformal mapping of the portion of the disk $|z| < 1$ that lies in the upper half-plane $\text{Im}(z) > 0$ onto the upper half-plane $\text{Im}(w) > 0$. Furthermore, show that the image of the semicircular portion of the boundary is mapped onto the negative u-axis, and the segment $-1 < x < 1$, $y = 0$ is mapped onto the positive u-axis.

Solution:

The function $w = f(z)$ is the composition of the bilinear transformation $Z = \frac{1+z}{1-z}$ followed by the mapping $w = Z^2$. The image of the half-disk under the bilinear mapping $Z = \frac{1+z}{1-z}$ is the first quadrant $X > 0$, $Y > 0$; the image of the segment $y = 0$, $-1 < x < 1$, is the positive x-axis; and the image of the semicircle is the positive y-axis. The mapping $w = Z^2$ then maps the first quadrant in the Z plane onto the upper half-plane $\text{Im}(w) > 0$, as shown in Figure 9.12.

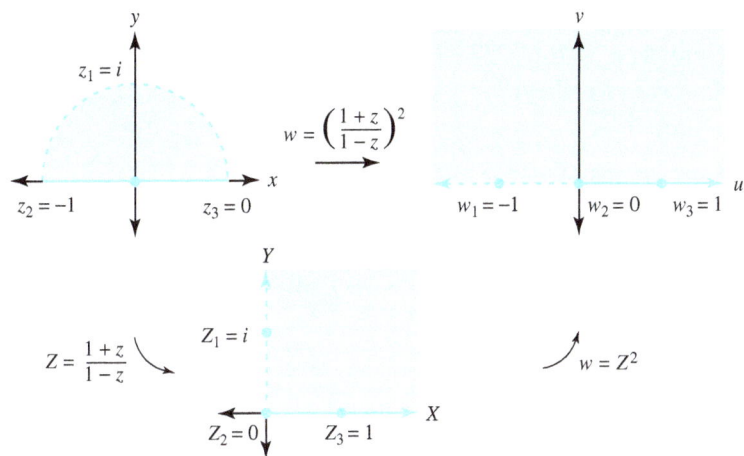

Figure 9.12: The composite transformation $w = (\frac{1+z}{1-z})^2$

Example 9.11. Consider the function $w = f(z) = (z^2 - 1)^{\frac{1}{2}}$, which is the composition of the functions $Z = z^2 - 1$ and $w = Z^{\frac{1}{2}}$, where the branch of the square root is given by the formula $Z^{\frac{1}{2}} = R^{\frac{1}{2}}(\cos\frac{\varphi}{2} + i\sin\frac{\varphi}{2})$, for $0 \leq \varphi < 2\pi$. Show that the transformation $w = f(z)$ maps the upper half-plane $\text{Im}(z) > 0$ one-to-one and onto the upper half-plane $\text{Im}(w) > 0$ slit along the segment $u = 0$, $0 < v \leq 1$.

Solution:

The function $Z = z^2 - 1$ maps the upper half-plane $\text{Im}(z) > 0$ one-to-one and onto the Z-plane slit along the ray $y = 0$, $x \geq -1$. Then the function $w = Z^{\frac{1}{2}}$ maps the slit plane onto the slit half-plane, as shown in Figure 9.13.

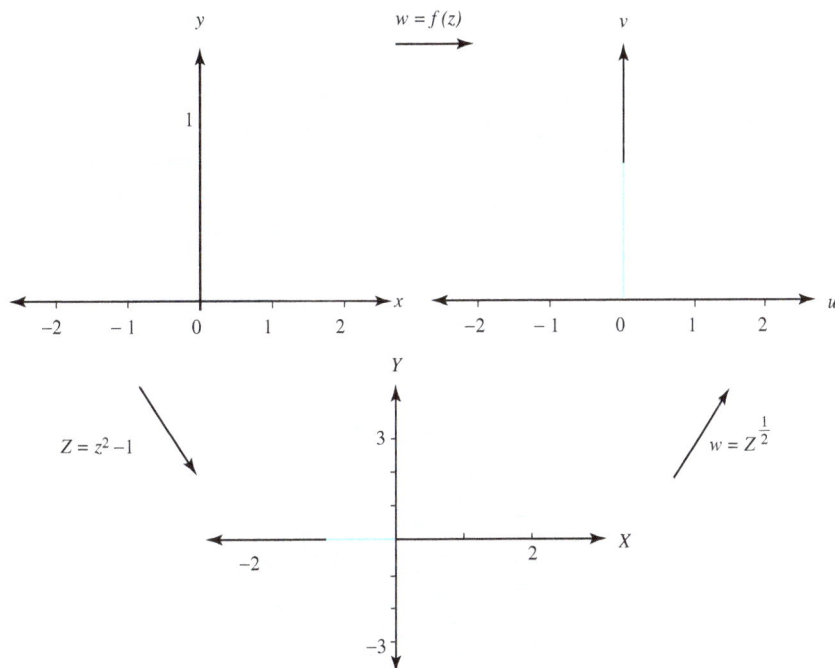

Figure 9.13: The composite transformation $w = f(z) = (z^2 - 1)^{\frac{1}{2}}$ and the intermediate steps $Z = z^2 - 1$ and $w = Z^{\frac{1}{2}}$

Remark 9.1. *The images of the horizontal lines $y = b$ are curves in the w plane that bend around the segment from 0 to i. The curves represent the streamlines of a fluid flowing across the w plane. We discuss fluid flows in more detail in Section 10.7.*

9.3.1 The Mapping $w = (z^2 - 1)^{\frac{1}{2}}$

The double-valued function $f(z) = (z^2 - 1)^{\frac{1}{2}}$ has a branch that is continuous for values of z distant from the origin. This feature is spurred by our desire for the approximation $(z^2 - 1)^{\frac{1}{2}} \approx z$ to hold for values of z distant from the origin. We begin by expressing $(z^2 - 1)^{\frac{1}{2}}$ as

$$w = f_1(z) = (z - 1)^{\frac{1}{2}}(z + 1)^{\frac{1}{2}}, \qquad (9.21)$$

where the principal branch of the square root function is used in both factors. We claim that the mapping $w = f_1(z)$ is a one-to-one conformal mapping from the domain set D_1, consisting of the z plane slit along the segment $-1 \leq x \leq 1$, $y = 0$, onto the range set H_1, consisting of the w plane slit along the segment $u = 0$, $-1 \leq v \leq 1$. To verify this claim, we investigate the two formulas on the right side of Equation (9.21) and express them in the form

$$(z - 1)^{\frac{1}{2}} = \sqrt{r_1}\, e^{i\frac{\theta_1}{2}},$$

where $r_1 = |z - 1|$ and $\theta_1 = \text{Arg}(z - 1)$, and

$$(z + 1)^{\frac{1}{2}} = \sqrt{r_2}\, e^{i\frac{\theta_2}{2}},$$

where $r_2 = |z + 1|$ and $\theta_2 = \text{Arg}(z + 1)$.

The discontinuities of $\text{Arg}(z - 1)$ and $\text{Arg}(z + 1)$ are points on the real axis such that $x \leq 1$ and $x \leq -1$, respectively. We now show that $f_1(z)$ is continuous on the ray $x < -1$, $y = 0$.

We let $z_0 = x_0 + iy_0$ denote a point on the ray $x < -1$, $y = 0$ and then obtain the following limits as z approaches z_0 from the upper and lower half-planes, respectively:

$$\lim_{z \to z_0,\ \text{Im}(z) > 0} f_1(z) = \left(\lim_{r_1 \to |x_0 - 1|,\ \theta_1 \to \pi} \sqrt{r_1}\, e^{i\frac{\theta_1}{2}} \right) \left(\lim_{r_2 \to |x_0 + 1|,\ \theta_2 \to \pi} \sqrt{r_2}\, e^{i\frac{\theta_2}{2}} \right)$$
$$= \left(\sqrt{|x_0 - 1|}(i) \right) \left(\sqrt{|x_0 + 1|}(i) \right)$$
$$= -\sqrt{|x_0^2 - 1|},$$

and

$$\lim_{z \to z_0,\ \text{Im}(z) < 0} f_1(z) = \left(\lim_{r_1 \to |x_0 - 1|,\ \theta_1 \to -\pi} \sqrt{r_1}\, e^{i\frac{\theta_1}{2}} \right) \left(\lim_{r_2 \to |x_0 + 1|,\ \theta_2 \to -\pi} \sqrt{r_2}\, e^{i\frac{\theta_2}{2}} \right)$$
$$= \left(\sqrt{|x_0 - 1|}(-i) \right) \left(\sqrt{|x_0 + 1|}(-i) \right)$$
$$= -\sqrt{|x_0^2 - 1|}.$$

Both limits agree with the value of $f_1(z_0)$, so it follows that $f_1(z)$ is continuous along the ray $x < -1$, $y = 0$.

We can easily find the inverse mapping and express it similarly:

$$z = g_1(w) = (w^2 + 1)^{\frac{1}{2}} = (w + i)^{\frac{1}{2}}(w - i)^{\frac{1}{2}},$$

where the branches of the square root function are given by

$$(w+i)^{\frac{1}{2}} = \sqrt{\rho_1}\, e^{i\frac{\phi_1}{2}},$$

where $\rho_1 = |w+i|$, $\phi_1 = \arg_{-\frac{\pi}{2}}(w+i)$, and $-\frac{\pi}{2} < \arg_{-\frac{\pi}{2}}(w+i) < \frac{3\pi}{2}$, and

$$(w-i)^{\frac{1}{2}} = \sqrt{\rho_2}\, e^{i\frac{\phi_2}{2}},$$

where $\rho_2 = |w-i|$, $\phi_2 = \arg_{-\frac{\pi}{2}}(w-i)$, and $-\frac{\pi}{2} < \arg_{-\frac{\pi}{2}}(w-i) < \frac{3\pi}{2}$.

A similar argument shows that $g_1(w)$ is continuous for all w except those points that lie on the segment $u = 0$, $-1 \le v \le 1$. Verification that

$$g_1\big(f_1(z)\big) = z, \quad \text{and} \quad f_1\big(g_1(w)\big) = w$$

hold for z in D_1 and w in H_1, respectively, is straightforward. Therefore we conclude that $w = f_1(z)$ is a one-to-one mapping from D_1 onto H_1. Verifying that $f_1(z)$ is also analytic on the ray $x < -1$, $y = 0$, is tedious. We leave it as a challenging exercise.

9.3.2 The Riemann Surface for $w = (z^2 - 1)^{\frac{1}{2}}$

Using the other branch of the square root, we find that $w = f_2(z) = -f_1(z)$ is a one-to-one conformal mapping from the domain set D_2, consisting of the z plane slit along the segment $-1 \le x \le 1$, $y = 0$, onto the range set H_2, consisting of the w plane slit along the segment $u = 0$, $-1 \le v \le 1$. Figure 9.14 shows the sets D_1 and H_1 for $f_1(z)$ and D_2 and H_2 for $f_2(z)$.

We obtain the Riemann surface for $w = (z^2 - 1)^{\frac{1}{2}}$ by gluing the edges of D_1 and D_2 together and the edges of H_1 and H_2 together. In the domain set, we glue edges A to a, B to b, C to c, and D to d. In the image set, we glue edges A' to a', B' to b', C' to c', and D' to d'. The result is a Riemann domain surface and Riemann image surface for the mapping, as illustrated in Figures 9.15(a) and 9.15(b), respectively.

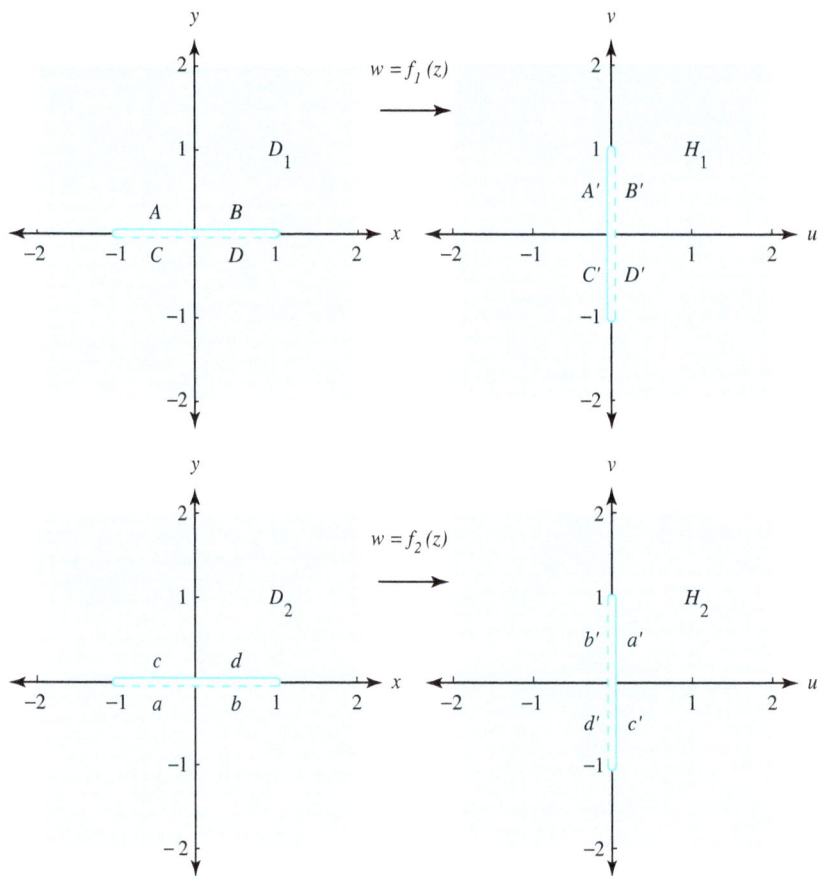

Figure 9.14: The mappings $w = f_1(z)$ and $w = f_2(z)$

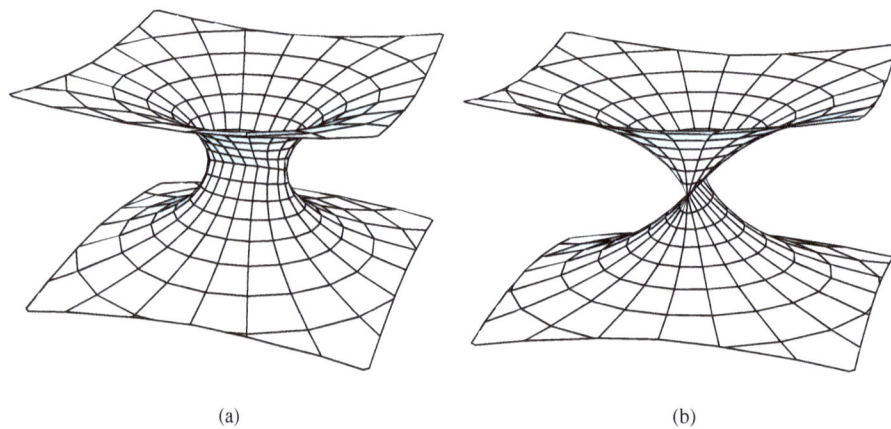

(a)   (b)

Figure 9.15: The Riemann surfaces for the mapping $w = (z^2 - 1)^{\frac{1}{2}}$

Exercises for Section 9.3 (Selected answers or hints are on page 453.)

1. Find the image of the semi-infinite strip $0 < x < \frac{\pi}{2}$, $y > 0$, under the transformation $w = \exp(iz)$.

2. Find the image of the rectangle $0 < x < \ln 2$, $0 < y < \frac{\pi}{2}$, under the transformation $w = \exp z$.

3. Find the image of the first quadrant $x > 0$, $y > 0$, under $w = \frac{2}{\pi} \operatorname{Log} z$.

4. Find the image of the annulus $1 < |z| < e$ under $w = \operatorname{Log} z$.

5. Show that the multivalued function $w = \log z$ maps the annulus $1 < |z| < e$ onto the vertical strip $0 < \operatorname{Re}(w) < 1$.

6. Show that $w = \frac{2-z^2}{z^2}$ maps the portion of the right half-plane $\operatorname{Re}(z) > 0$ that lies to the right of the hyperbola $x^2 - y^2 = 1$ onto the unit disk $|w| < 1$.

7. What is the image the horizontal strip $-\pi < \operatorname{Im}(z) < 0$ when mapped by $w = \frac{e^z - i}{e^z + i}$?

8. Show that $w = \frac{e^z - 1}{e^z + 1}$ maps the horizontal strip $|y| < \frac{\pi}{2}$ onto the unit disk $|w| < 1$.

9. Find the image of the upper half-plane $\operatorname{Im}(z) > 0$ under $w = \operatorname{Log}(\frac{1+z}{1-z})$.

10. Find the image of the portion of the upper half-plane $\operatorname{Im}(z) > 0$ that lies outside the circle $|z| = 1$ under the transformation $w = \operatorname{Log}(\frac{1+z}{1-z})$.

11. Show that the function $w = \frac{(1+z)^2}{(1-z)^2}$ maps the portion of the disk $|z| < 1$ that lies in the first quadrant onto the portion of the upper half plane $\operatorname{Im}(w) > 0$ that lies outside the unit disk.

12. Find the image of the upper half-plane $\operatorname{Im}(z) > 0$ under the transformation $w = \operatorname{Log}(1 - z^2)$.

13. Find the branch of $w = (z^2 + 1)^{\frac{1}{2}}$ that maps the right half-plane $\operatorname{Re}(z) > 0$ onto the right half-plane $\operatorname{Re}(w) > 0$ slit along the segment $0 < u \le 1$, $v = 0$.

14. Show that the transformation $w = \frac{z^2 - 1}{z^2 + 1}$ maps the portion of the first quadrant $x > 0$, $y > 0$, that lies outside the circle $|z| = 1$ onto the first quadrant $u > 0$, $v > 0$.

15. Find the image of the sector $r > 0$, $0 < \theta < \frac{\pi}{4}$, under $w = \frac{i - z^4}{i + z^4}$.

16. Show that the function $f_1(z)$ in Equation (9.21) is analytic on the ray $x < -1$, $y = 0$.

9.4 Mapping by Trigonometric Functions

The trigonometric functions can be expressed with compositions that involve the exponential function followed by a bilinear function. We can find images of certain regions by following the shapes of successive images in the composite mapping. We begin with the tangent function, and make use of the following figure.

Example 9.12. Show that the transformation $w = \tan z$ is a one-to-one conformal mapping of the vertical strip $|x| < \frac{\pi}{4}$ onto the unit disk $|w| < 1$.

Solution:

$$w = \tan z = \frac{1}{i} \frac{e^{iz} - e^{-iz}}{e^{iz} + e^{-iz}} = \frac{-ie^{i2z} + i}{e^{i2z} + 1}.$$

Then, mapping $w = \tan z$ can be considered to be the composition

$$w = \frac{-iZ + i}{Z + 1}, \quad \text{and} \quad Z = e^{i2z}.$$

299

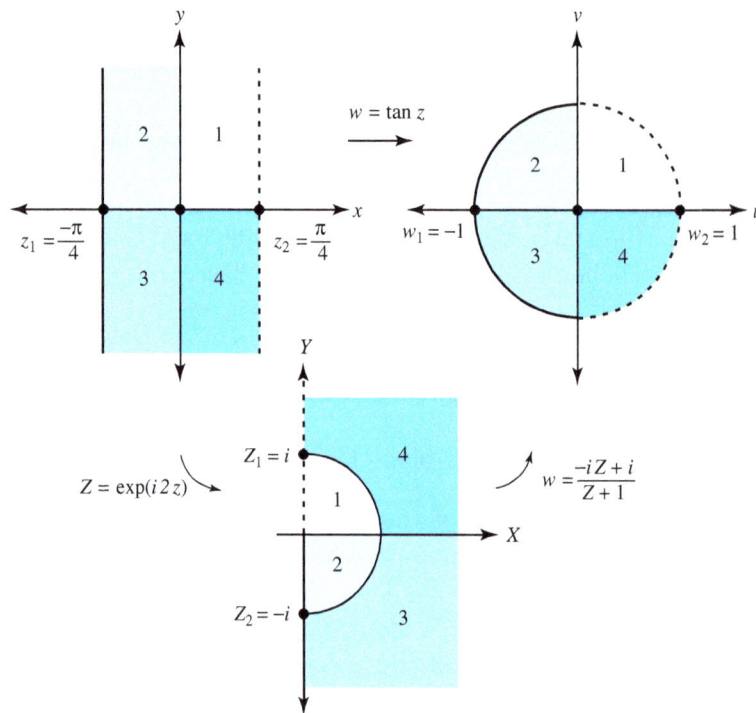

Figure 9.16: The composite transformation $w = \tan z$

The function $Z = \exp(i2z)$ maps the vertical strip $|x| < \frac{\pi}{4}$ one-to-one and onto the right half-plane $\text{Re}(Z) > 0$. Then the bilinear transformation given by $w = \frac{-iZ+i}{Z+1}$ maps the half-plane one-to-one and onto the disk, as shown in Figure 9.16.

Example 9.13. Show that the transformation $w = f(z) = \sin z$ is a one-to-one conformal mapping of the vertical strip $|x| < \frac{\pi}{2}$ onto the w plane slit along the rays $u \leq -1$, $v = 0$, and $u \geq 1$, $v = 0$.

Solution:

Because $f'(z) = \cos z \neq 0$ for values of z satisfying the inequality $-\frac{\pi}{2} < \text{Re}(z) < \frac{\pi}{2}$, it follows that $w = \sin z$ is a conformal mapping. Using Equation 5, we write

$$u + iv = \sin z = \sin x \cosh y + i \cos z \sinh y.$$

If $|a| < \frac{\pi}{2}$, then the image of the vertical line $x = a$ is the curve in the w plane given by the parametric equations

$$u = \sin a \cosh y, \quad \text{and} \quad v = \cos a \sinh y,$$

for $-\infty < y < \infty$. Next, we rewrite these equations as

$$\cosh y = \frac{u}{\sin a}, \quad \text{and} \quad \sinh y = \frac{v}{\cos a}.$$

We now eliminate y from these equations by squaring and using the hyperbolic identity $\cosh^2 y - \sinh^2 y = 1$. The result is the single equation

$$\frac{u^2}{\sin^2 a} - \frac{v^2}{\cos^2 a} = 1. \tag{9.22}$$

300

The curve given by Equation (9.22) is identified as a hyperbola in the uv plane that has foci at the points $(\pm 1, 0)$. Therefore the vertical line $x = a$ is mapped one-to-one onto the branch of the hyperbola given by Equation (9.22) that passes through the point $(\sin a, 0)$. If $0 < a < \frac{\pi}{2}$, then it is the right branch; if $-\frac{\pi}{2} < a < 0$, it is the left branch. The image of the y-axis, which is the line $x = 0$, is the v-axis. The images of several vertical lines are shown in Figure 9.17(a).

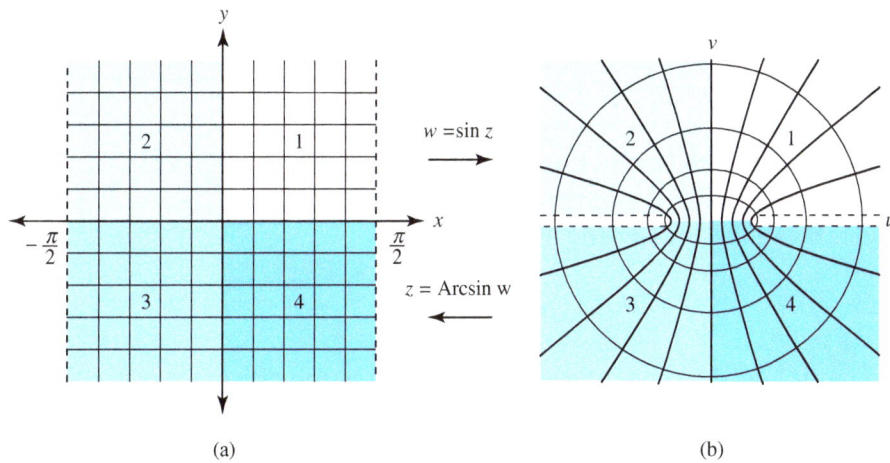

Figure 9.17: The transformation $w = \sin z$

The image of the horizontal segment $-\frac{\pi}{2} < x < \frac{\pi}{2}$, $y = b$ is the curve in the w plane given by the parametric equations

$$u = \sin x \cosh b, \quad \text{and} \quad v = \cos x \sinh b$$

for $-\frac{\pi}{2} < x < \frac{\pi}{2}$. We rewrite them as

$$\sin x = \frac{u}{\cosh b}, \quad \text{and} \quad \cos x = \frac{v}{\sinh b}.$$

We now eliminate x from the equations by squaring and using the trigonometric identity $\sin^2 x + \cos^2 x = 1$. The result is the single equation

$$\frac{u^2}{\cosh^2 b} + \frac{v^2}{\sinh^2 b} = 1. \tag{9.23}$$

The curve given by Equation (9.23) is identified as an ellipse in the uv plane that passes through the points $(\pm \cosh b, 0)$ and $(0, \pm \sinh b)$ and has foci at the points $(\pm 1, 0)$. Therefore, if $b > 0$, then $v = \cos x \sinh b > 0$, and the image of the horizontal segment is the portion of the ellipse given by Equation (9.23) that lies in the upper half-plane $\text{Im}(w) > 0$. If $b < 0$, then it is the portion that lies in the lower half-plane. Figure 9.17(b) shows the images of several segments.

9.4.1 The Complex Arcsine Function

We now develop explicit formulas for the real and imaginary parts of the principal value of the arcsine function $w = f(z) = \text{Arcsin} z$. We use this mapping to solve problems involving steady temperatures and ideal fluid flow in Section 9.7. The mapping is found by solving the equation

$$x + iy = \sin w = \sin u \cosh v + i \cos u \sinh v \tag{9.24}$$

for u and v expressed as functions of x and y. To solve for u, we first equate the real and imaginary parts of Equation (9.24) and obtain the equations

$$\cosh v = \frac{x}{\sin u}, \quad \text{and} \quad \sinh v = \frac{y}{\cos u}.$$

Then we eliminate v from these equations and obtain the single equation

$$\frac{x^2}{\sin^2 u} - \frac{y^2}{\cos^2 u} = 1.$$

If we treat u as a constant, this equation represents a hyperbola in the xy plane, the foci occur at the points $(\pm 1, 0)$, and the transverse axis is given by $2 \sin u$. Therefore a point (x, y) on the hyperbola must satisfy the equation

$$2 \sin u = \sqrt{(x+1)^2 + y^2} - \sqrt{(x-1)^2 + y^2}.$$

The quantity on the right side of this equation is the difference of the distances from (x, y) to $(-1, 0)$ and from (x, y) to $(1, 0)$. We now solve the equation for u to obtain the real part:

$$u(x, y) = \operatorname{Arcsin}\left[\frac{\sqrt{(x+1)^2 + y^2} - \sqrt{(x-1)^2 + y^2}}{2} \right]. \tag{9.25}$$

The principal branch of the real function $\operatorname{Arcsin}(t)$ is used in Equation (9.25), where the range values satisfy the inequality $-\frac{\pi}{2} < \operatorname{Arcsin}(t) < \frac{\pi}{2}$.

Similarly, we can start with Equation (9.24) and obtain the equations

$$\sin u = \frac{x}{\cosh v}, \quad \text{and} \quad \cos u = \frac{y}{\sinh v}.$$

We then eliminate u from these equations and obtain the single equation

$$\frac{x^2}{\cosh^2 v} + \frac{y^2}{\sinh^2 v} = 1.$$

If we treat v as a constant, then this equation represents an ellipse in the xy plane, the foci occur at the points $(\pm 1, 0)$, and the major axis has length $2 \cosh v$. Therefore a point (x, y) on this ellipse must satisfy the equation

$$2 \cosh v = \sqrt{(x+1)^2 + y^2} + \sqrt{(x-1)^2 + y^2}.$$

The quantity on the right side of this equation is the sum of the distances from (x, y) to $(-1, 0)$ and from (x, y) to $(1, 0)$.

The function $z = \sin w$ maps points in the upper half (lower half) of the vertical strip $\frac{-\pi}{2} < u < \frac{\pi}{2}$ onto the upper half-plane (lower half-plane), respectively. Hence we can solve the preceding equation and obtain v as a function of x and y:

$$v(x, y) = (\operatorname{sign} y)\operatorname{Arccosh}\left[\frac{\sqrt{(x+1)^2 + y^2} + \sqrt{(x-1)^2 + y^2}}{2} \right], \tag{9.26}$$

where $\operatorname{sign} y = 1$, if $y \geq 0$, and $\operatorname{sign} y = -1$, if $y < 0$. In Equation (9.26) we use the real function given by $\operatorname{Arccosh}(t) = \ln(t + \sqrt{t^2 - 1})$, with $t \geq 1$.

Thus, the mapping $w = \text{Arcsin}\, z$ is a one-to-one conformal mapping of the z plane cut along the rays $x \leq -1$, $y = 0$, and $x \geq 1$, $y = 0$, onto the vertical strip $-\frac{\pi}{2} \leq u \leq \frac{\pi}{2}$ in the w-plane, which can be construed from Figure 9.17 if we interchange the roles of the z and w planes. The image of the square $0 \leq x \leq 4$, $0 \leq y \leq 4$, under $w = \text{Arcsin}\, z$, is shown in Figure 9.18. We obtained it by plotting the two families of curves $\{(u(c,t),\ v(c,t)) : 0 \leq t \leq 4\}$ and $\{(u(t,c),\ v(t,c)) : 0 \leq t \leq 4\}$, where $c = \frac{k}{5}$, $k = 0, 1, \ldots, 20$. The formulas in Equations (9.25) and (9.26) are also convenient for evaluating $\text{Arcsin}(z)$, as shown in Example 9.14

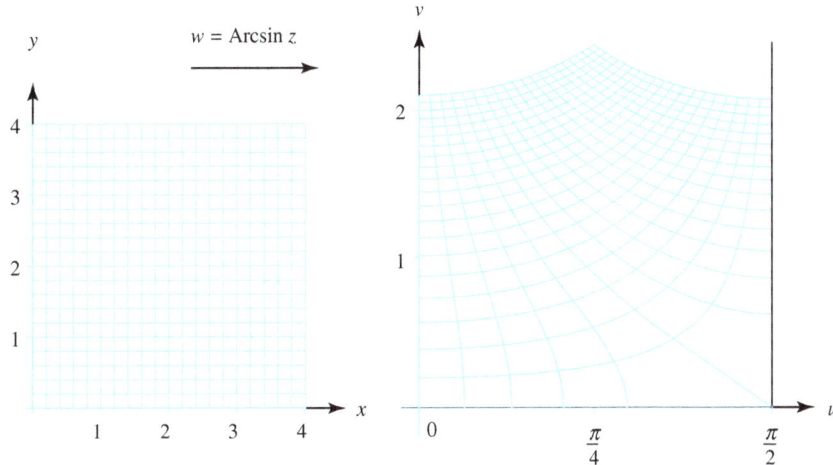

Figure 9.18: The mapping $w = \text{Arcsin}\, z$

Example 9.14. Find the principal value $\text{Arcsin}(1 + i)$.

Solution:

Using Formulas 9.25 and 9.26, we get

$$u(1,1) = \text{Arcsin}\left(\frac{\sqrt{5} - 1}{2}\right) \approx 0.666239432, \quad \text{and}$$

$$v(1,1) = \text{Arccosh}\left(\frac{\sqrt{5} + 1}{2}\right) \approx 1.061275062.$$

Therefore, we have

$$\text{Arcsin}(1 + i) \approx 0.666239432 + i1.061275062.$$

Is there any reason to assume that there exists a conformal mapping for some specified domain D onto another domain G? Our final theorem concerning the existence of conformal mappings is attributed to Riemann and is presented in Lars V. Ahlfors, *Complex Analysis* (New York: McGraw-Hill) Chapter 6, 1966.

Theorem 9.4 (Riemann mapping theorem). *If D is any simply connected domain in the plane (other than the entire plane itself), then there exists a one-to-one conformal mapping $w = f(z)$ that maps D onto the unit disk $|w| < 1$.*

Exercises for Section 9.4 (Selected answers or hints are on page 453.)

1. Find the image of the semi-infinite strip $-\frac{\pi}{4} < x < 0$, $y > 0$, under the mapping $w = \tan z$.

2. Find the image of the vertical strip $0 < \text{Re}(z) < \frac{\pi}{2}$ under the mapping $w = \tan z$.

3. Find the image of the vertical line $x = \frac{\pi}{4}$ under the transformation $w = \sin z$.

4. Find the image of the horizontal line $y = 1$ under the transformation $w = \sin z$.

5. Find the image of the rectangle $R = \{x + iy : 0 < x < \frac{\pi}{4}, \, 0 < y < 1\}$ under the transformation $w = \sin z$.

6. Find the image of the semi-infinite strip $-\frac{\pi}{2} < x < 0$, $y > 0$, under the mapping $w = \sin z$.

7. (a) Find $\lim\limits_{y \to +\infty} \text{Arg}[\sin(\frac{\pi}{6} + iy)]$.

   (b) Find $\lim\limits_{y \to +\infty} \text{Arg}[\sin(-\frac{2\pi}{3} + iy)]$.

8. Use Equations (9.25) and (9.26) to find

   (a) $\text{Arcsin}(2 + 2i)$.
   (b) $\text{Arcsin}(-2 + i)$.
   (c) $\text{Arcsin}(1 - 3i)$.
   (d) $\text{Arcsin}(-4 - i)$.

9. Show that $w = \sin z$ maps the rectangle $R = \{x + iy : -\frac{\pi}{2} < x < \frac{\pi}{2}, 0 < y < b\}$ one-to-one and onto the portion of the upper half-plane $\text{Im}(w) > 0$ that lies inside the ellipse

$$\frac{u^2}{\cosh^2 b} + \frac{v^2}{\sinh^2 b} = 1.$$

10. Find the image of the vertical strip $-\frac{\pi}{2} < x < 0$ under the mapping $w = \cos z$.

11. Find the image of the horizontal strip $0 < \text{Im}(z) < \frac{\pi}{2}$ under $w = \sinh z$.

12. Find the image of the right half-plane $\text{Re}(z) > 0$ under the mapping

$$w = \text{Arctan}(z) = \frac{i}{2} \text{Log}\left(\frac{i + z}{i - z}\right).$$

13. Find the image of the first quadrant $x > 0$, $y > 0$, under $w = \text{Arcsin} \, z$.

14. Find the image of the first quadrant $x > 0$, $y > 0$, under $w = \text{Arcsin}(z^2)$.

15. Show that the transformation $w = \sin^2 z$ is a one-to-one conformal mapping of the semi-infinite strip $0 < x < \frac{\pi}{2}, y > 0$, onto the upper half plane $\text{Im}(w) > 0$.

16. Find the image of the semi-infinite strip $|x| < \frac{\pi}{2}$, $y > 0$, under the mapping $w = \text{Log}(\sin z)$.

Chapter 10

Applications of Harmonic Functions

Overview

A wide variety of problems in engineering and physics involve harmonic functions, which are the real or imaginary part of an analytic function. The standard applications are two dimensional steady state temperatures, electrostatics, fluid flow and complex potentials. The techniques of conformal mapping and integral representation can be used to construct a harmonic function with prescribed boundary values. Noteworthy methods include Poisson's integral formulae; the Joukowski transformation; and Schwarz-Christoffel transformation. Modern computer software is capable of implementing these complex analysis methods.

10.1 Preliminaries

In most applications involving harmonic functions, a harmonic function that takes on prescribed values along certain contours must be found. In presenting the material in this chapter, we assume that you are familiar with the material covered in Sections 2.4, 3.3, 5.1, and 5.2. If you aren't, please review it before proceeding.

Example 10.1. Find the function $u(x, y)$ that is harmonic in the vertical strip $a \leq \mathrm{Re}(z) \leq b$ and takes on the boundary values

$$u(a, y) = U_1 \quad \text{and} \quad u(b, y) = U_2$$

along the vertical lines $x = a$ and $x = b$, respectively.

Solution:

Intuition suggests that we should seek a solution that takes on constant values along the vertical lines of the form $x = x_0$ and that $u(x, y)$ be a function of x alone; that is,

$$u(x, y) = P(x), \quad \text{for} \quad a \leq x \leq b, \quad \text{and for all} \quad y.$$

Laplace's equation, $u_{xx}(x, y) + u_{yy}(x, y) = 0$, implies that $P''(x) = 0$, which implies $P(x) = mx + c$, where m and c are constants. The stated boundary conditions $u(a, y) = P(a) = U_1$ and $u(b, y) = P(b) = U_2$ lead to the solution

$$u(x, y) = U_1 + \frac{U_2 - U_1}{b - a}(x - a).$$

The level curves $u(x, y) = $ constant are vertical lines as indicated in Figure 10.1.

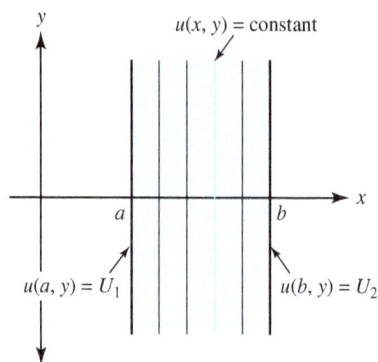

Figure 10.1: Level curves of $u(x, y) = U_1 + \frac{U_2 - U_1}{b-a}(x - a)$

Example 10.2. Find the function $\Psi(x, y)$ that is harmonic in the sector $0 < \text{Arg}\, z < \alpha$, where $\alpha \leq \pi$, and takes on the boundary values

$$\Psi(x, 0) = C_1 \quad \text{for} \quad x > 0, \text{ and}$$
$$\Psi(x, y) = C_2 \quad \text{at points on the ray} \quad r > 0, \ \theta = \alpha.$$

Solution:

Recalling that the function $\text{Arg}(z)$ is harmonic and takes on constant values along rays emanating from the origin, we see that a solution has the form

$$\Psi(x, y) = a + b\text{Arg}(z),$$

where a and b are constants. The boundary conditions lead to

$$\Psi(x, y) = C_1 + \frac{C_2 - C_1}{\alpha}\text{Arg}(z).$$

The situation is shown in Figure 10.2.

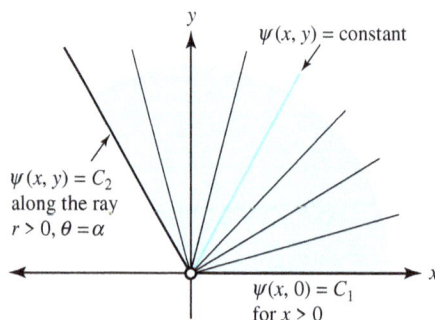

Figure 10.2: Level curves of $\Psi(x, y) = C_1 + (C_2 - C_1)\frac{1}{\alpha}\text{Arg}(z)$

Example 10.3. Find the function $\Phi(x, y)$ that is harmonic in the annulus $1 < |z| < R$ and takes on the boundary values

$$\Phi(x, y) = K_1, \quad \text{when} \quad |z| = 1, \quad \text{and}$$
$$\Phi(x, y) = K_2, \quad \text{when} \quad |z| = R.$$

306

Solution:

This problem is a companion to the one in Example 10.2. Here we use the fact that $\ln |z|$ is a harmonic function, for all $z \neq 0$. The solution is

$$\Phi(x, y) = K_1 + \frac{K_2 - K_1}{\ln R} \ln |z|,$$

and the level curves $\Phi(x, y) = $ constant are concentric circles, as illustrated in Figure 10.3.

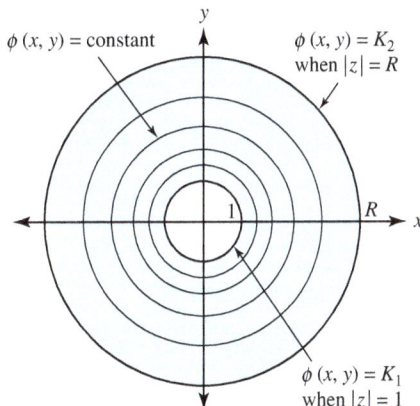

Figure 10.3: Level curves of $\Phi(x, y) = K_1 + \frac{K_2 - K_1}{\ln R} \ln |z|$

10.2   The Dirichlet Problem

Theorem 10.1. *Let* $\Phi(u, v)$ *be harmonic in a domain* G *in the* w *plane. Then* Φ *satisfies Laplace's equation*

$$\Phi_{uu}(u, v) + \Phi_{vv}(u, v) = 0 \tag{10.1}$$

at each point $w = u + iv$ *in* G. *If* $w = f(z) = u(x, y) + iv(x, y)$ *is a conformal mapping from a domain* D *in the* z *plane onto* G, *then the composition*

$$\phi(x, y) = \Phi\big(u(x, y), v(x, y)\big) \tag{10.2}$$

is harmonic in D, *and* ϕ *satisfies Laplace's equation*

$$\phi_{xx}(x, y) + \phi_{yy}(x, y) = 0 \tag{10.3}$$

at each point $z = x + iy$ *in* D.

Proof. Equations 10.1 and 10.3 are Laplace's equations for the harmonic functions Φ and ϕ, respectively (see Section 3.3). A direct proof that the function ϕ in Equation 10.2 is harmonic would involve a tedious calculation of the partial derivatives ϕ_{xx} and ϕ_{yy}. An easier proof involves the use of a complex variable technique. We assume that there is a harmonic conjugate $\Psi(u, v)$ so that the function

$$g(w) = \Phi(u, v) + i \, \Psi(u, v)$$

is analytic in a neighborhood of the point $w_0 = f(z_0)$. Then the composition $h(z) = g\big(f(z)\big)$ is analytic in a neighborhood of z_0 and can be written

$$h(z) = \Phi\big(u(x, y), v(x, y)\big) + i\Psi\big(u(x, y), v(x, y)\big).$$

Recall (Theorem 3.8) that the real part of the analytic function $h(z)$ is harmonic, so $\Phi\big(u(x, y), v(x, y)\big)$ is harmonic in a neighborhood of z_0. $\qquad\square$

307

Example 10.4. Show that $\phi(x, y) = \text{Arctan}(\frac{2x}{x^2+y^2-1})$ is harmonic in the disk $|z| < 1$.

Solution:

A straightforward use of the techniques in Chapter 9 will show that

$$f(z) = \frac{i+z}{i-z} = \frac{1-x^2-y^2}{x^2+(y-1)^2} - \frac{i2x}{x^2+(y-1)^2}$$

is a conformal mapping of the disk $|z| < 1$ onto the right half-plane $\text{Re}(w) > 0$. The results from Exercise 7b, Section 5.2, show that the function

$$\Phi(u, v) = \text{Arctan}\left(\frac{v}{u}\right) = \text{Arg}(u + iv)$$

is harmonic in the right half-plane $\text{Re}(w) > 0$. Taking the real and imaginary parts of $f(z)$, we write

$$u(x, y) = \frac{1-x^2-y^2}{x^2+(y-1)^2} \quad \text{and} \quad v(x, y) = \frac{-2x}{x^2+(y-1)^2}.$$

Substituting these equations into the formula for $\Phi(u, v)$ and using Equation 10.2 reveals that $\phi(x, y) = \text{Arctan}\left(\frac{v(x,y)}{u(x,y)}\right) = \text{Arctan}(\frac{2x}{x^2+y^2-1})$ is harmonic for $|z| < 1$.

Let D be a domain whose boundary is made up of piecewise smooth contours joined end to end. The **Dirichlet problem** is to find a function ϕ that is harmonic in D such that ϕ takes on prescribed values at points on the boundary. Let's first look at this problem in the upper half-plane.

Example 10.5. Show that the function

$$\Phi(u, v) = \frac{1}{\pi}\text{Arctan}\left(\frac{v}{u - u_0}\right) = \frac{1}{\pi}\text{Arg}(w - u_0) \tag{10.4}$$

is harmonic in the upper half-plane $\text{Im}(w) > 0$ and takes on the boundary values

$$\Phi(u, 0) = 0 \quad \text{for} \quad u > u_0, \quad \text{and}$$
$$\Phi(u, 0) = 1 \quad \text{for} \quad u < u_0.$$

Solution:

The function

$$g(w) = \frac{1}{\pi}\text{Log}(w - u_0) = \frac{1}{\pi}\ln|w - u_0| + \frac{i}{\pi}\text{Arg}(w - u_0)$$

is analytic in the upper half-plane $\text{Im}(w) > 0$, and its imaginary part is the harmonic function $\frac{1}{\pi}\text{Arg}(w - u_0)$.

Remark 10.1. *Let t be a real number. The convention $\text{Arctan}(\pm\infty) = \pm\frac{\pi}{2}$ allows $\text{Arctan}(t)$ to denote the branch of the inverse tangent with range in $-\frac{\pi}{2} < \text{Arctan}(t) < \frac{\pi}{2}$. With that understanding we can write the solution given in Equation 10.4 as $\Phi(u, v) = \frac{1}{\pi}\text{Arctan}(\frac{v}{u-u_0})$.*

Theorem 10.2 (*N-value Dirichlet problem for the upper half-plane*)*. Let $u_1 < u_2 < \cdots < u_{N-1}$ denote $N - 1$ real constants. The function*

$$\Phi(u, v) = a_{N-1} + \frac{1}{\pi}\sum_{k=1}^{N-1}(a_{k-1} - a_k)\text{Arg}(w - u_k) \tag{10.5}$$

$$= a_{N-1} + \frac{1}{\pi}\sum_{k=1}^{N-1}(a_{k-1} - a_k)\text{Arctan}\left(\frac{v}{u - u_k}\right)$$

is harmonic in the upper half-plane $\text{Im}(w) > 0$ *and takes on the boundary values*

$$\Phi(u, 0) = a_0, \quad for \quad u < u_1;$$

$$\Phi(u, 0) = a_k, \quad for \quad u_k < u < u_{k+1}, \quad for \quad k = 1, 2, \ldots, N-2;$$

$$\Phi(u, 0) = a_{N-1}, \quad for \quad u > u_{N-1}.$$

The situation is illustrated in Figure 10.4.

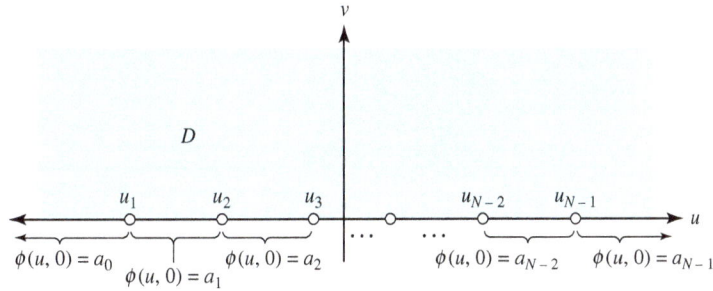

Figure 10.4: The boundary conditions for the harmonic function $\Phi(u, v)$

Proof. Each term in the sum in Equation 10.5 is harmonic, so it follows that Φ is harmonic for $\text{Im}(w) > 0$. To show that Φ has the prescribed boundary conditions, we fix j and let $u_j < u < u_{j+1}$. Using Example 10.5, we get

$$\frac{1}{\pi}\text{Arg}(u - u_k) = 0, \quad if \quad k \le j, \quad and$$

$$\frac{1}{\pi}\text{Arg}(u - u_k) = 1, \quad if \quad k > j.$$

Substituting these equations into Equation 10.5 gives

$$\Phi(u, 0) = a_{N-1} + \sum_{k=1}^{j}(a_{k-1} - a_k)(0) + \sum_{k-j+1}^{N-1}(a_{k-1} - a_k)(1)$$

$$= a_{N-1} + (a_{N-2} - a_{N-1}) + \cdots + (a_{j+1} - a_{j+2}) + (a_j - a_{j+1})$$

$$= a_j, \quad for \quad u_j < u < u_{j+1}.$$

You can verify that the boundary conditions are correct for $u < u_1$ and $u > u_{N-1}$ to complete the proof. $\square$

Example 10.6. Find the function $\phi(x, y)$ that is harmonic in the upper half-plane $\text{Im}(z) > 0$ and takes on the boundary values indicated in Figure 10.5.

Solution:
This is a four-value Dirichlet problem in the upper half-plane defined by $\text{Im}(z) > 0$. For the z plane, the solution in Equation 10.5 becomes

$$\phi(x, y) = a_3 + \frac{1}{\pi}\sum_{k=1}^{3}(a_{k-1} - a_k)\text{Arg}(z - x_k).$$

309

Here we have $a_0 = 4$, $a_1 = 1$, $a_2 = 3$, and $a_3 = 2$ and $x_1 = -1$, $x_2 = 0$, and $x_3 = 1$, which we substitute into equation for ϕ to obtain

$$\phi(x, y) = 2 + \frac{4-1}{\pi}\text{Arg}(z+1) + \frac{1-3}{\pi}\text{Arg}(z-0) + \frac{3-2}{\pi}\text{Arg}(z-1)$$

$$= 2 + \frac{3}{\pi}\text{Arctan}\left(\frac{y}{x+1}\right) - \frac{2}{\pi}\text{Arctan}\left(\frac{y}{x}\right) + \frac{1}{\pi}\text{Arctan}\left(\frac{y}{x-1}\right).$$

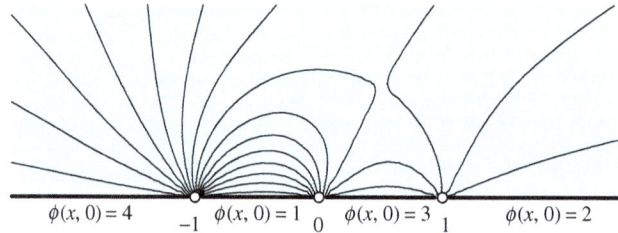

$\phi(x, 0) = 4 \qquad \phi(x, 0) = 1 \qquad \phi(x, 0) = 3 \qquad \phi(x, 0) = 2$
$\qquad\qquad -1 \qquad\qquad 0 \qquad\qquad 1$

Figure 10.5: The boundary values for the Dirichlet problem

Example 10.7. Find the function $\phi(x, y)$ that is harmonic in the upper half-plane $\text{Re}(z) > 0$ and takes on the boundary values

$$\phi(x, 0) = 1, \quad \text{for} \quad |x| < 1;$$
$$\phi(x, 0) = 0, \quad \text{for} \quad |x| > 1.$$

Solution:

This three-value Dirichlet problem has $a_0 = 0$, $a_1 = 1$, $a_2 = 0$, $x_1 = -1$, and $x_2 = 1$. Applying Equation 10.5 yields

$$\phi(x, y) = 0 + \frac{0-1}{\pi}\text{Arg}(z+1) + \frac{1-0}{\pi}\text{Arg}(z-1)$$

$$= -\frac{1}{\pi}\text{Arctan}\left(\frac{y}{x+1}\right) + \frac{1}{\pi}\text{Arctan}\left(\frac{y}{x-1}\right).$$

A three-dimensional graph of $u = \phi(x, y)$ is shown in Figure 10.6.

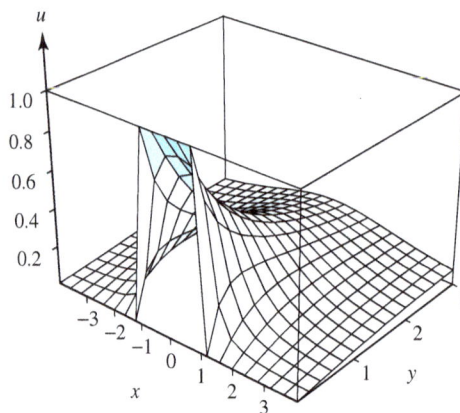

Figure 10.6: $\phi(x, y)$ with boundary values $\phi(x, 0) = 1$ for $|x| < 1$, and $\phi(x, 0) = 0$ for $|x| > 1$

We now state the N-value Dirichlet problem for a simply connected domain. We let D be a simply connected domain bounded by the simple closed contour C and let $z_1, z_2, \ldots, z_N$ denote N points that lie along C in this specified order as C is traversed in the positive direction (counterclockwise). Then we let C_k denote the portion of C that lies strictly between z_k and z_{k+1}, for $k = 1, 2, \ldots, N-1$, and let C_N denote the portion that lies strictly between z_N and z_1. Finally, we let $a_1, a_2, \ldots, a_N$ be real constants. We want to find a function $\phi(x, y)$ that is harmonic in D and continuous on $D \cup C_1 \cup C_2 \cup \cdots \cup C_N$ that takes on the boundary values

$$\phi(x, y) = a_1, \quad \text{for} \quad z = x + iy \quad \text{on} \quad C_1; \tag{10.6}$$
$$\phi(x, y) = a_2, \quad \text{for} \quad z = x + iy \quad \text{on} \quad C_2;$$
$$\vdots$$
$$\phi(x, y) = a_N, \quad \text{for} \quad z = x + iy \quad \text{on} \quad C_N.$$

The situation is illustrated in Figure 10.7.

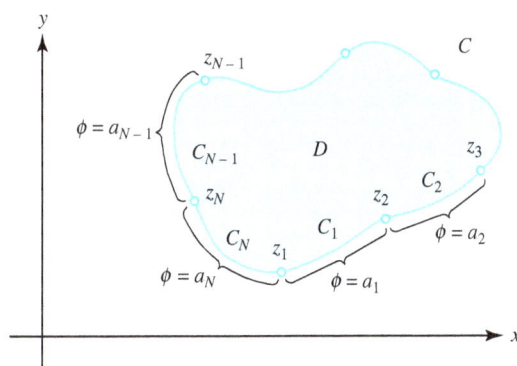

Figure 10.7: The boundary values for $\phi(x, y)$ for the Dirichlet problem in the simply connected domain D

One method for finding ϕ is to find a conformal mapping

$$w = f(z) = u(x, y) + iv(x, y) \tag{10.7}$$

of D onto the upper half-plane $\text{Im}(w) > 0$, such that the N points $z_1, z_2, \ldots, z_N$ are mapped onto the points $u_k = f(z_k)$, for $k = 1, 2, \ldots, N-1$, and z_N is mapped onto $u_N = +\infty$ along the u axis in the w plane.

When we use Theorem 10.1, the mapping in Equation 10.7 gives rise to a new N-value Dirichlet problem in the upper half-plane $\text{Im}(w) > 0$ for which Theorem 10.2 gives the solution. If we set $a_0 = a_N$, then the solution to the Dirichlet problem in D with the boundary values from Equation 10.6 is

$$\phi(x, y) = a_{N-1} + \frac{1}{\pi} \sum_{k=1}^{N-1} (a_{k-1} - a_k) \text{Arg}[f(z) - u_k]$$

$$= a_{N-1} + \frac{1}{\pi} \sum_{k=1}^{N-1} (a_{k-1} - a_k) \text{Arctan} \left(\frac{v(x, y)}{u(x, y) - u_k} \right).$$

This method relies on our ability to construct a conformal mapping from D onto the upper half-plane $\text{Im}(w) > 0$. Theorem 9.4 guarantees the existence of such a conformal mapping.

311

Example 10.8. Find a function $\phi(x, y)$ that is harmonic in the unit disk $|z| < 1$ and takes on the boundary values

$$\phi(x, y) = 0, \quad \text{for} \quad x + iy = e^{i\theta}, \quad 0 < \theta < \pi; \tag{10.8}$$

$$\phi(x, y) = 1, \quad \text{for} \quad x + iy = e^{i\theta}, \quad \pi < \theta < 2\pi.$$

Solution:

Example 10.3 showed that the function

$$u + iv = \frac{i(1 - z)}{1 + z} = \frac{2y}{(x + 1)^2 + y^2} + i \frac{1 - x^2 - y^2}{(x + 1)^2 + y^2} \tag{10.9}$$

is a one-to-one conformal mapping of the unit disk $|z| < 1$ onto the upper half-plane $\text{Im}(w) > 0$. Equation 10.9 reveals that the points $z = x + iy$ lying on the upper semicircle $y > 0$, $1 - x^2 - y^2 = 0$ are mapped onto the positive u axis. Similarly, the lower semicircle is mapped onto the negative u axis, as shown in Figure 10.8. The mapping given by Equation 10.9 gives rise to a new Dirichlet problem of finding a harmonic function $\Phi(u, v)$ that has the boundary values

$$\Phi(u, 0) = 0, \quad \text{for} \quad u > 0, \quad \text{and} \quad \Phi(u, 0) = 1, \quad \text{for} \quad u < 0,$$

as shown in Figure 10.8. Using the result of Example 10.5 and the functions u and v from Equation 10.9, we get the solution to Equation 10.8:

$$\phi(x, y) = \frac{1}{\pi} \text{Arctan}\left(\frac{v(x, y)}{u(x, y)}\right) = \frac{1}{\pi} \text{Arctan}\left(\frac{1 - x^2 - y^2}{2y}\right).$$

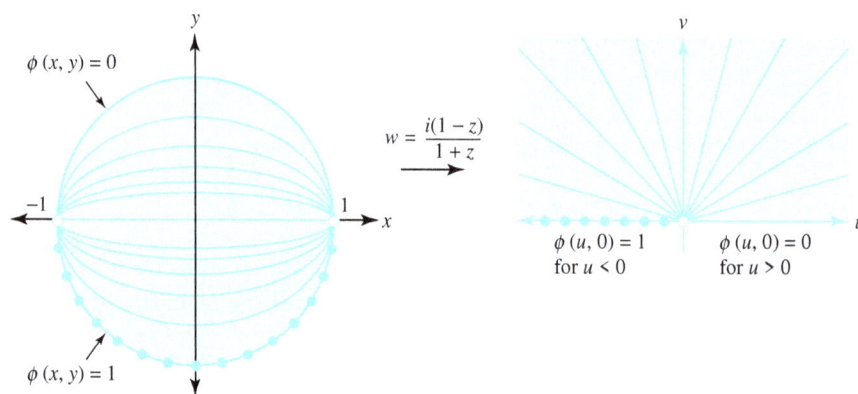

Figure 10.8: The Dirichlet problems for $|z| < 1$ and $\text{Im}(w) > 0$

Example 10.9. Find a function $\phi(x, y)$ that is harmonic in the upper half-disk defined by $H = \{z = x + iy : y > 0, |z| < 1\}$ and takes on the boundary values

$$\phi(x, y) = 0, \quad \text{for} \quad x + iy = e^{i\theta}, \quad 0 < \theta < \pi;$$

$$\phi(x, 0) = 1, \quad \text{for} \quad -1 < x < 1.$$

Solution:

When we use the result of Exercise 4, Section 9.2, the function in Equation 10.9 maps the upper half-disk H onto the first quadrant $Q : u > 0$, $v > 0$. The conformal mapping given in Equation

10.9 maps the points $z = x + iy$ that lie on the segment $y = 0$, $-1 < x < 1$ onto the positive v axis.

Equation 10.9 gives rise to a new Dirichlet problem of finding a harmonic function $\Phi(u, v)$ in Q that has the boundary values

$$\Phi(u, 0) = 0, \quad \text{for} \quad u > 0, \quad \text{and} \quad \Phi(0, v) = 1, \quad \text{for} \quad v > 0,$$

as shown in Figure 10.9. In this case, the method in Example 10.2 can be used to show that $\Phi(u, v)$ is given by

$$\Phi(u, 0) = 0 + \frac{1 - 0}{\frac{\pi}{2}} \text{Arg}(w) = \frac{2}{\pi} \text{Arg}(w) = \frac{2}{\pi} \text{Arctan}\left(\frac{v}{u}\right).$$

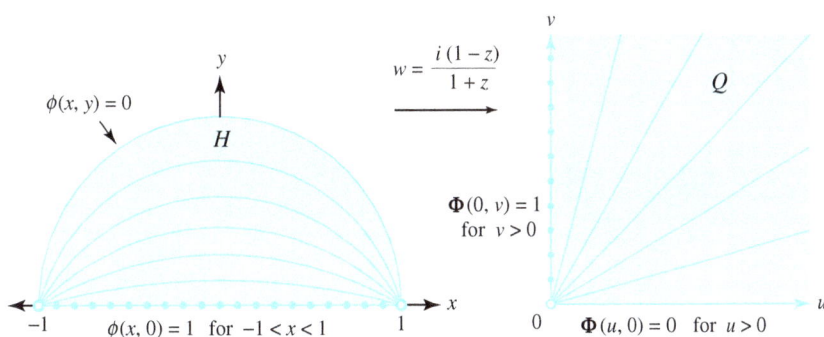

Figure 10.9: The Dirichlet problems for the domains H and Q

Using the functions u and v in Equation 10.9 in the preceding equation, we find the solution of the Dirichlet problem in H:

$$\phi(x, y) = \frac{2}{\pi} \text{Arctan}\left(\frac{v(x, y)}{u(x, y)}\right) = \frac{2}{\pi} \text{Arctan}\left(a \frac{1 - x^2 - y^2}{2y}\right).$$

A three-dimensional graph $u = \phi(x, y)$ in cylindrical coordinates is shown in Figure 10.10.

Example 10.10. Find a function $\phi(x, y)$ that is harmonic in the quarter disk defined by $G = \{z = x + iy : x > 0, |z| < 1\}$ and takes on the boundary values

$$\phi(x, y) = 0, \quad \text{for} \quad x + iy = z = e^{i\theta}, \quad 0 < \theta < \frac{\pi}{2};$$
$$\phi(x, 0) = 1, \quad \text{for} \quad 0 \le x < 1;$$
$$\phi(0, y) = 1, \quad \text{for} \quad 0 \le y < 1.$$

Solution:

The function

$$u + iv = z^2 = x^2 - y^2 + i2xy \tag{10.10}$$

maps the quarter-disk onto the upper half-disk $H = \{w = u + iv : v > 0, |w| < 1\}$. The new Dirichlet problem in H is shown in Figure 10.11. From the result of Example 10.9 the solution $\Phi(u, v)$ in H is

$$\Phi(u, v) = \frac{2}{\pi} \text{Arctan}\left(\frac{1 - u^2 - v^2}{2v}\right). \tag{10.11}$$

313

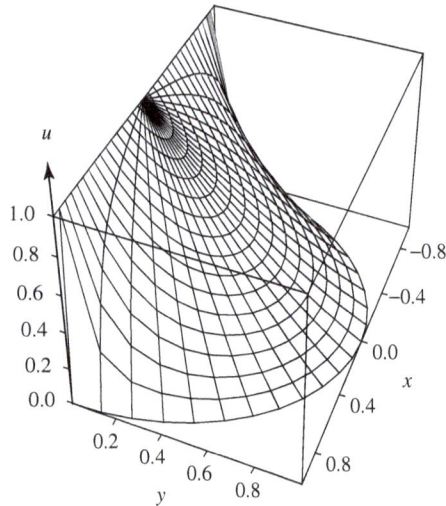

Figure 10.10: The graph $u = \frac{2}{\pi}\text{Arctan}\left(\frac{1-x^2-y^2}{2y}\right) = \frac{2}{\pi}\text{Arctan}\left(\frac{1-r^2}{2r\sin\theta}\right)$

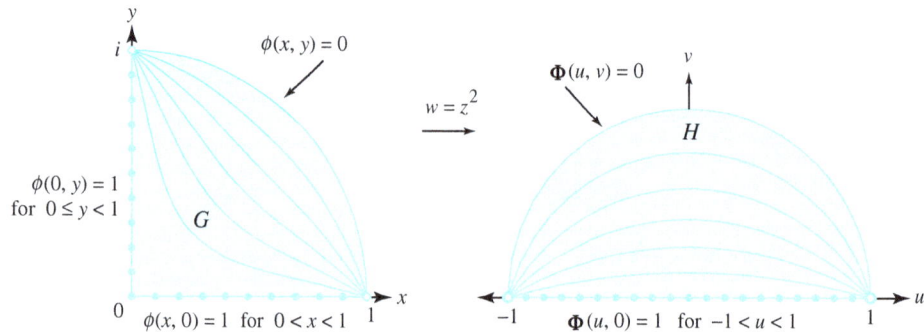

Figure 10.11: The Dirichlet problems for the domains G and H

Using Equation 10.10, we can show that $u^2 + v^2 = (x^2 + y^2)^2$ and $2v = 4xy$, which we use in Equation 10.11 to construct the solution ϕ in G:

$$\phi(x, y) = \frac{2}{\pi}\text{Arctan}\left(\frac{1 - (x^2 + y^2)^2}{4xy}\right).$$

A three-dimensional graph $u = \phi(x, y)$ in cylindrical coordinates is shown in Figure 10.12.

Exercises for Section 10.2 (Selected answers or hints are on page 454.)

1. Find the function $\phi(x, y)$ that is harmonic in the horizontal strip $1 < \text{Im}(z) < 2$ and has the boundary values

$$\phi(x, 1) = 6 \quad \text{for all} \quad x, \quad \text{and} \quad \phi(x, 2) = -3 \quad \text{for all} \quad x.$$

2. Find the function $\phi(x, y)$ that is harmonic in the sector $0 < \text{Arg}(z) < \frac{\pi}{3}$ and has the boundary values

$$\phi(x, y) = 2 \quad \text{for} \quad \text{Arg}(z) = \frac{\pi}{3}, \quad \text{and} \quad \phi(x, 0) = 1 \quad \text{for} \quad x > 0.$$

314

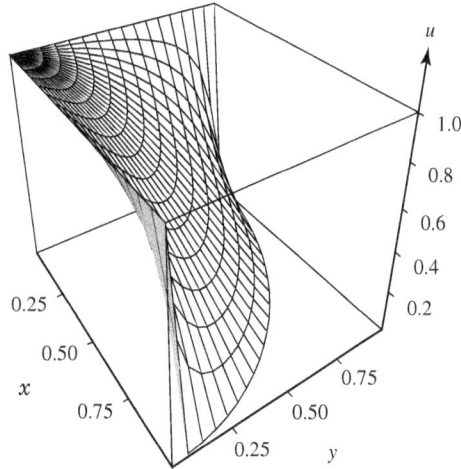

Figure 10.12: The graph $u = \frac{2}{\pi} \text{Arctan} \left(\frac{1-(x^2+y^2)^2}{4xy} \right) = \frac{2}{\pi} \text{Arctan} \left(\frac{1-r^4}{4r^2 \cos\theta \sin\theta} \right)$

3. Find the function $\phi(x, y)$ that is harmonic in the annulus $1 < |z| < 2$ and has the boundary values

$$\phi(x, y) = 5 \quad \text{when} \quad |z| = 1, \quad \text{and} \quad \phi(x, y) = 8 \quad \text{when} \quad |z| = 2.$$

4. Find the function $\phi(x, y)$ that is harmonic in the upper half-plane $\text{Im}(z) > 0$ and has the boundary values

$$\phi(x, 0) = 0 \quad \text{for} \quad -1 < x < 1, \quad \text{and} \quad \phi(x, 0) = 1 \quad \text{for} \quad |x| > 1.$$

5. Find the function $\phi(x, y)$ that is harmonic in the upper half-plane $\text{Im}(z) > 0$ and has the boundary values

$$\phi(x, 0) = 3 \quad \text{for} \quad x < -3, \quad \text{and} \quad \phi(x, 0) = 7 \quad \text{for} \quad -3 < x < -1;$$
$$\phi(x, 0) = 1 \quad \text{for} \quad -1 < x < 2, \quad \text{and} \quad \phi(x, 0) = 4 \quad \text{for} \quad x > 2.$$

6. Find the function $\phi(x, y)$ that is harmonic in the first quadrant $x > 0$, $y > 0$ and has the boundary values

$$\phi(0, y) = 0 \quad \text{for} \quad y > 1, \quad \text{and} \quad \phi(0, y) = 1 \quad \text{for} \quad 0 < y < 1;$$
$$\phi(x, 0) = 1 \quad \text{for} \quad 0 \leq x < 1, \quad \text{and} \quad \phi(x, 0) = 0 \quad \text{for} \quad x > 1.$$

7. Find the function $\phi(x, y)$ that is harmonic in the unit disk $|z| < 1$ and has the boundary values

$$\phi(x, y) = 0 \quad \text{for} \quad x + iy = z = e^{i\theta}, \quad 0 < \theta < \pi;$$
$$\phi(x, y) = 5 \quad \text{for} \quad x + iy = z = e^{i\theta}, \quad \pi < \theta < 2\pi.$$

8. Find the function $\phi(x, y)$ that is harmonic in the unit disk $|z| < 1$ and has the boundary values

$$\phi(x, y) = 8 \quad \text{for} \quad x + iy = z = e^{i\theta}, \quad 0 < \theta < \pi;$$
$$\phi(x, y) = 4 \quad \text{for} \quad x + iy = z = e^{i\theta}, \quad \pi < \theta < 2\pi.$$

315

9. Find the function $\phi(x, y)$ that is harmonic in the upper half-disk $y > 0$, $|z| < 1$ and has the boundary values

$$\phi(x, y) = 5 \quad \text{for} \quad x + iy = z = e^{i\theta}, \quad 0 < \theta < \pi;$$
$$\phi(x, 0) = -5 \quad \text{for} \quad -1 < x < 1.$$

10. Find the function $\phi(x, y)$ that is harmonic in the portion of the upper half-plane $\text{Im}(z) > 0$ that lies outside the circle $|z| = 1$ and has the boundary values

$$\phi(x, y) = 1 \quad \text{for} \quad x + iy = z = e^{i\theta}, \quad 0 < \theta < \pi;$$
$$\phi(x, 0) = 0 \quad \text{for} \quad |x| > 1.$$

Hint: Use the mapping $w = -\frac{1}{z}$ and the result of Example 10.9.

11. Find the function $\phi(x, y)$ that is harmonic in the quarter-disk

$$x > 0, \, y > 0, \, |z| < 1$$

and has the boundary values

$$\phi(x, y) = 3, \quad \text{for} \quad x + iy = z = e^{i\theta}, \quad 0 < \theta < \frac{\pi}{2};$$
$$\phi(x, 0) = -3 \quad \text{for} \quad 0 \le x < 1;$$
$$\phi(0, y) = -3 \quad \text{for} \quad 0 < y < 1.$$

12. Find the function $\phi(x, y)$ that is harmonic in the unit disk $|z| < 1$ and has the boundary values

$$\phi(x, y) = 1 \quad \text{for} \quad x + iy = z = e^{i\theta}, \quad -\frac{\pi}{2} < \theta < \frac{\pi}{2};$$
$$\phi(x, y) = 0 \quad \text{for} \quad x + iy = z = e^{i\theta}, \quad \frac{\pi}{2} < \theta < \frac{3\pi}{2}.$$

10.3   Poisson's Integral Formula

The Dirichlet problem for the upper half-plane $\text{Im}(z) > 0$ is to find a function $\phi(x, y)$ that is harmonic in the upper half-plane and has the boundary values $\phi(x, 0) = U(x)$, where $U(x)$ is a real-valued function of the real variable x.

Theorem 10.3 (Poisson's integral formula). *Let $U(t)$ be a real-valued function that is piecewise continuous and bounded for all real t. The function*

$$\phi(x, y) = \frac{y}{\pi} \int_{-\infty}^{\infty} \frac{U(t)}{(x - t)^2 + y^2} \, dt \tag{10.12}$$

is harmonic in the upper half-plane $\text{Im}(z) > 0$ and has the boundary values

$$\phi(x, 0) = U(x)$$

wherever U is continuous.

Proof. Equation 10.12 is easy to determine from the results of Theorem 10.2 regarding the Dirichlet problem. Let $t_1 < t_2 < \cdots < t_N$ denote N points that lie along the x axis. Let $t_0^* < t_1^* < \cdots < t_N^*$ be $N+1$ points chosen so that $t_{k-1}^* < t_k < t_k^*$, for $k = 1, 2, \ldots, N$, and that $U(t)$ is continuous at each value t_k^*. Then according to Theorem 10.2, the function

$$\Phi(x, y) = U(t_N^*) + \frac{1}{\pi} \sum_{k=1}^{N} \left[U(t_{k-1}^*) - U(t_k^*) \right] \operatorname{Arg}(z - t_k) \tag{10.13}$$

is harmonic in the upper half-plane and takes on the boundary values

$$llll\Phi(x, 0) = U(t_0^*), \quad \text{for} \quad x < t_1,$$
$$\Phi(x, 0) = U(t_k^*), \quad \text{for} \quad t_k < x < t_{k+1}, \quad \text{and}$$
$$\Phi(x, 0) = U(t_N^*), \quad \text{for} \quad x > t_N,$$

as shown in Figure 10.13.

We use properties of the argument of a complex number (see Section 1.4) to write Equation 10.13 in the form

$$\Phi(x, y) = \frac{1}{\pi} U(t_0^*) \operatorname{Arg}(z - t_1) + \frac{1}{\pi} \sum_{k=1}^{N-1} U(t_k^*) \operatorname{Arg}\left(\frac{z - t_{k+1}}{z - t_k} \right)$$
$$+ \frac{1}{\pi} U(t_N^*) [\pi - \operatorname{Arg}(z - t_N)].$$

Hence the value Φ is given by the weighted mean

$$\Phi(x, y) = \frac{1}{\pi} \sum_{k=0}^{N} U(t_k^*) \Delta \theta_k, \tag{10.14}$$

where the angles $\Delta \theta_k$, for $k = 0, 1, \ldots, N$, sum to π, and are also shown in Figure 10.13.

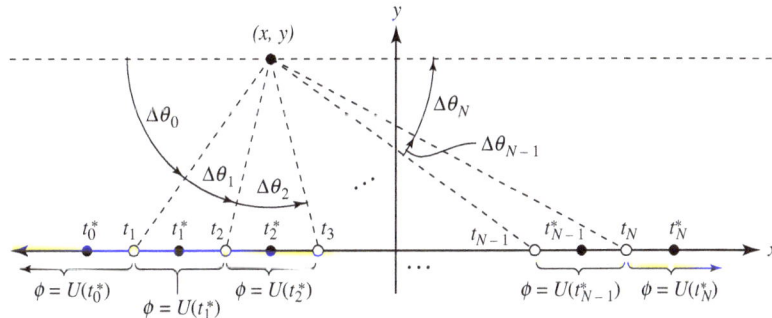

Figure 10.13: Boundary values for Φ

Using the substitutions

$$\theta = \operatorname{Arg}(z - t) = \operatorname{Arctan}\left(\frac{y}{x - t} \right) \quad \text{and} \quad d\theta = \frac{y}{(x - t)^2 + y^2} \, dt, \tag{10.15}$$

we write Equation 10.14 as

$$\Phi(x, y) = \frac{y}{\pi} \sum_{k=0}^{N} \frac{U(t_k^*) \, \Delta t_k}{(x - t_k^*)^2 + y^2}.$$

317

The limit of this Riemann sum becomes the improper integral

$$\phi(x, y) = \frac{y}{\pi} \int_{-\infty}^{\infty} \frac{U(t)}{(x - t)^2 + y^2} \, dt.$$

□

Example 10.11. Find the function $\phi(x, y)$ that is harmonic in the upper half-plane $\mathrm{Im}(z) > 0$ and has the boundary values

$$\phi(x, 0) = 1, \quad \text{for} \quad -1 < x < 1, \quad \text{and} \quad \phi(x, 0) = 0, \quad \text{for} \quad |x| > 1.$$

Solution:

Using Equation 10.12, we obtain

$$\phi(x, y) = \frac{y}{\pi} \int_{-1}^{1} \frac{1}{(x - t)^2 + y^2} \, dt = \frac{1}{\pi} \int_{-1}^{1} \frac{y}{(x - t)^2 + y^2} \, dt.$$

Using the antiderivative in Equation 10.15, we write this solution as

$$\phi(x, y) = \frac{1}{\pi} \mathrm{Arctan} \left(\frac{y}{x - t} \right) \Big|_{t=-1}^{t=1}$$

$$= \frac{1}{\pi} \mathrm{Arctan} \left(\frac{y}{x - 1} \right) - \frac{1}{\pi} \mathrm{Arctan} \left(\frac{y}{x + 1} \right).$$

Example 10.12. Find the function $\phi(x, y)$ that is harmonic in the upper half-plane $\mathrm{Im}(z) > 0$ and has the boundary values

$$\phi(x, 0) = x, \quad \text{for} \quad -1 < x < 1, \quad \text{and} \quad \phi(x, 0) = 0, \quad \text{for} \quad |x| > 1.$$

Solution:

Using Equation 10.12, we obtain

$$\phi(x, y) = \frac{y}{\pi} \int_{-1}^{1} \frac{t}{(x - t)^2 + y^2} \, dt$$

$$= \frac{y}{\pi} \int_{-1}^{1} \frac{(x - t)(-1)}{(x - t)^2 + y^2} \, dt + \frac{x}{\pi} \int_{-1}^{1} \frac{y}{(x - t)^2 + y^2} \, dt.$$

Using calculus techniques and Equations 10.15, we write the solution as

$$\phi(x, y) = \frac{y}{2\pi} \ln \left[\frac{(x - 1)^2 + y^2}{(x + 1)^2 + y^2} \right] + \frac{x}{\pi} \mathrm{Arctan} \left(\frac{y}{x - 1} \right) - \frac{x}{\pi} \mathrm{Arctan} \left(\frac{y}{x + 1} \right).$$

The function $\phi(x, y)$ is continuous in the upper half-plane, and on the boundary $\phi(x, 0)$, it has discontinuities at $x = \pm 1$ on the real axis. The graph in Figure 10.14 shows this phenomenon.

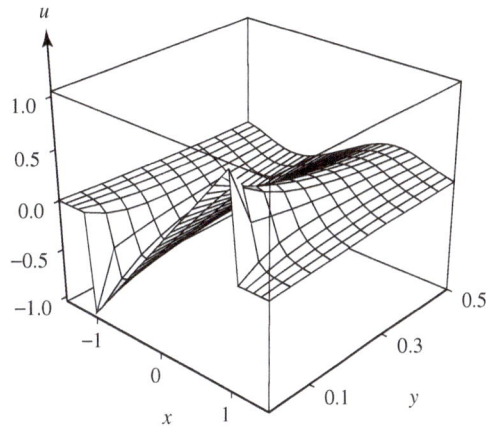

Figure 10.14: $\phi(x, y)$ with boundary values $\phi(x, 0) = x$ for $|x| < 1$, and $\phi(x, 0) = 0$ for $|x| > 1$

Example 10.13. Find $\phi(x, y)$ that is harmonic in the upper half-plane $\text{Im}(z) > 0$ and that has the boundary values

$$\phi(x, 0) = x \quad \text{for} \quad |x| < 1, \quad \phi(x, 0) = -1 \quad \text{for} \quad x < -1, \quad \text{and} \quad \phi(x, 0) = 1 \quad \text{for} \quad x > 1.$$

***Solution*:**

Using techniques from Section 10.2, we find that the function

$$v(x, y) = 1 - \frac{1}{\pi} \text{Arctan}\left(\frac{y}{x+1}\right) - \frac{1}{\pi} \text{Arctan}\left(\frac{y}{x-1}\right)$$

is harmonic in the upper half-plane and has the following boundary values:

$$v(x, 0) = 0 \quad \text{for} \quad |x| < 1, \quad v(x, 0) = -1 \quad \text{for} \quad x < -1, \quad \text{and} \quad v(x, 0) = 1 \quad \text{for} \quad x > 1.$$

This function can be added to the one in Example 10.12 to obtain the desired result:

$$\phi(x, y) = 1 + \frac{y}{2\pi} \ln\left[\frac{(x-1)^2 + y^2}{(x+1)^2 + y^2}\right] + \frac{x-1}{\pi} \text{Arctan}\left(\frac{y}{x-1}\right) - \frac{x+1}{\pi} \text{Arctan}\left(\frac{y}{x+1}\right).$$

Figure 10.15 shows the graph of $\phi(x, y)$.

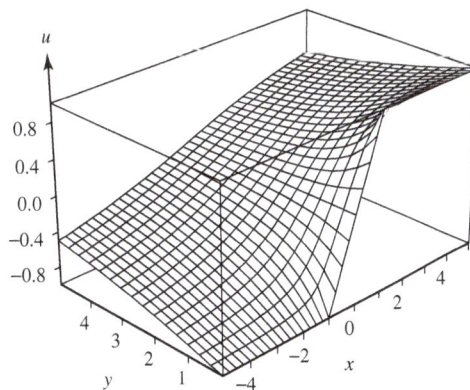

Figure 10.15: The graph of $\phi(x, y)$ with designated boundary values

319

Exercises for Section 10.3 (Selected answers or hints are on page 454.)

1. Use Poisson's integral formula to find the harmonic function $\phi(x, y)$ in the upper half-plane that takes on the boundary values

$$\phi(t, 0) = U(t) = 0, \quad \text{for} \quad t < 0;$$
$$\phi(t, 0) = U(t) = t, \quad \text{for} \quad 0 < t < 1;$$
$$\phi(t, 0) = U(t) = 0, \quad \text{for} \quad 1 < t.$$

2. Use Poisson's integral formula to find the harmonic function $\phi(x, y)$ in the upper half-plane that takes on the boundary values

$$\phi(t, 0) = U(t) = 0, \quad \text{for} \quad t < 0;$$
$$\phi(t, 0) = U(t) = t, \quad \text{for} \quad 0 < t < 1;$$
$$\phi(t, 0) = U(t) = 1, \quad \text{for} \quad 1 < t.$$

3. Use Poisson's integral formula for the upper half-plane to conclude

$$\phi(x, y) = e^{-y} \cos x = \frac{y}{\pi} \int_{-\infty}^{\infty} \frac{\cos t}{(x - t)^2 + y^2} \, dt.$$

4. Use Poisson's integral formula for the upper half-plane to conclude

$$\phi(x, y) = e^{-y} \sin x = \frac{y}{\pi} \int_{-\infty}^{\infty} \frac{\sin t}{(x - t)^2 + y^2} \, dt.$$

5. Show that the function $\phi(x, y)$ given by Poisson's integral formula is harmonic by applying Leibniz's rule, which permits you to write

$$\left(\frac{\partial^2}{\partial x^2} + \frac{\partial^2}{\partial y^2} \right) \phi(x, y) = \frac{1}{\pi} \int_{-\infty}^{\infty} U(t) \left[\left(\frac{\partial^2}{\partial x^2} + \frac{\partial^2}{\partial y^2} \right) \frac{y}{(x - t)^2 + y^2} \right] dt.$$

6. Let $U(t)$ be a real-valued function that satisfies the conditions for Poisson's integral formula for the upper half-plane. If $U(t)$ is an even function so that $U(-t) = U(t)$, then show that the harmonic function $\phi(x, y)$ has the property $\phi(-x, y) = \phi(x, y)$.

7. Let $U(t)$ be a real-valued function that satisfies the conditions for Poisson's integral formula for the upper half-plane. If $U(t)$ is an odd function so that for all t $U(-t) = -U(t)$, then show that the harmonic function $\phi(x, y)$ has the property $\phi(-x, y) = -\phi(x, y)$.

10.4   Two-Dimensional Mathematical Models

We now consider problems involving steady state heat flow, electrostatics, and ideal fluid flow that can be solved with conformal mapping techniques. Conformal mapping transforms a region in which the problem is posed to one in which the solution is easy to obtain. As our solutions involve only two independent variables, x and y, we first mention a basic assumption needed for the validity of the model.

The physical problems we just mentioned are real-world applications and involve solutions in three-dimensional Cartesian space. Such problems generally would involve the Laplacian in three variables and the divergence and curl of three-dimensional vector functions. Since complex analysis involves only x and y, we consider the special case in which the solution does not vary with the coordinate along the axis perpendicular to the xy plane. For steady state heat flow and electrostatics, this assumption means that the temperature, T, or the potential, V, varies only with x and y. Thus for the flow of ideal fluids, the fluid motion is the same in any plane that is parallel to the z plane. Curves drawn in the z plane are to be interpreted as cross sections that correspond to infinite cylinders perpendicular to the z plane. An infinite cylinder is the limiting case of a "long" physical cylinder, so the mathematical model that we present is valid provided the three-dimensional problem involves a physical cylinder long enough that the effects at the ends can be reasonably neglected.

In Sections 10.1 and 10.2, we showed how to obtain solutions $\phi(x, y)$ for harmonic functions. For applications, we need to consider the family of level curves

$$\{\phi(x, y) = K_1 : K_1 \text{ is a real constant }\} \tag{10.16}$$

and the conjugate harmonic function $\psi(x, y)$ and its family of level curves

$$\{\psi(x, y) = K_2 : K_2 \text{ is a real constant }\}. \tag{10.17}$$

For convenience, we introduce the term **complex potential** for the analytic function

$$F(z) = \phi(x, y) + i\psi(x, y).$$

We use Theorem 10.4, regarding the orthogonality of the families of level curves (Equations 10.16 and 10.17), to develop ideas concerning the physical applications that we will consider.

Theorem 10.4 (Orthogonal families of level curves). *Suppose $\phi(x, y)$ is harmonic in a domain D, $\psi(x, y)$ is its harmonic conjugate, and $F(z) = \phi(x, y) + i\psi(x, y)$ is the complex potential. Then the two families of level curves given in Equations 10.16 and 10.17, respectively, are orthogonal in the sense that if (a, b) is a point common to the two curves $\phi(x, y) = K_1$ and $\psi(x, y) = K_2$, and if $F,'(a + ib) \neq 0$, then these two curves intersect orthogonally.*

Proof. Since $\phi(x, y) = K_1$ is an implicit equation of a plane curve, the gradient vector grad ϕ, evaluated at (a, b), is perpendicular to the curve at (a, b). This vector is given by

$$\mathbf{N}_1 = \phi_x(a, b) + i\phi_y(a, b).$$

Similarly, the vector $\mathbf{N}_2$ defined by

$$\mathbf{N}_2 = \psi_x(a, b) + i\psi_y(a, b)$$

is orthogonal to the curve $\psi(x, y) = K_2$ at (a, b). Using the Cauchy-Riemann equations, $\phi_x = \psi_y$ and $\phi_y = -\psi_x$, we have

$$\begin{aligned}
\mathbf{N}_1 \cdot \mathbf{N}_2 &= \phi_x(a, b)\psi_x(a, b) + \phi_y(a, b)\psi_y(a, b) \tag{10.18} \\
&= -\phi_x(a, b)\phi_y(a, b) + \phi_y(a, b)\phi_x(a, b) \\
&= 0.
\end{aligned}$$

In addition, $F,'(a + ib) \neq 0$, so we have

$$\phi_x(a, b) + i\psi_x(a, b) \neq 0.$$

The Cauchy-Riemann equations and the facts $\phi_x(a, b) \neq 0$ and $\psi_x(a, b) \neq 0$ imply that both $\mathbf{N}_1$ and $\mathbf{N}_2$ are nonzero. Therefore Equation 10.18 implies that $\mathbf{N}_1$ is perpendicular to $\mathbf{N}_2$, and hence the curves are orthogonal. $\qquad \square$

The complex potential $F(z) = \phi(x, y) + i\psi(x, y)$ has many physical interpretations. Suppose, for example, that we have solved a problem in steady state temperatures. Then we can obtain the solution to a similar problem with the same boundary conditions in electrostatics by interpreting the isothermals as equipotential curves and the heat flow lines as flux lines. This implies that heat flow and electrostatics correspond directly.

Or suppose that we have solved a fluid flow problem. Then we can obtain a solution to an analogous problem in heat flow by interpreting the equipotentials as isothermals and streamlines as heat flow lines. Various interpretations of the families of level curves given in Equations 10.16 and 10.17 and correspondences between families are summarized in Table 10.1

Physical Phenomenon	$\phi(x, y) = constant$	$\psi(x, y) = constant$
Heat flow	Isothermals	Heat flow lines
Electrostatics	Equipotential curves	Flux lines
Fluid flow	Equipotentials	Streamlines
Gravitational field	Gravitational potential	Lines of force
Magnetism	Potential	Lines of force
Diffusion	Concentration	Lines of flow
Elasticity	Strain function	Stress lines
Current flow	Potential	Lines of flow

Table 10.1: Interpretations for Level Curves

10.5   Steady State Temperatures

In the theory of heat conduction, an assumption is made that heat flows in the direction of decreasing temperature. Another assumption is that the time rate at which heat flows across a surface area is proportional to the component of the temperature gradient in the direction perpendicular to the surface area. If the temperature $T(x, y)$ does not depend on time, then the heat flow at the point (x, y) is given by the vector

$$\mathbf{V}(x, y) = -K\nabla T(x, y) = -K[T_x(x, y) + iT_y(x, y)],$$

where K is the thermal conductivity of the medium and is assumed to be constant. If Δz denotes a straight-line segment of length Δs, then the amount of heat flowing across the segment per unit of time is

$$\mathbf{V} \bullet \mathbf{N} \, \Delta s, \tag{10.19}$$

where $\mathbf{N}$ is the unit vector perpendicular to the segment.

If we assume that no thermal energy is created or destroyed within the region, then the net amount of heat flowing through any small rectangle with sides of length Δx and Δy is identically zero (see Figure 10.16(a)). This observation leads to the conclusion that $T(x, y)$ is a harmonic function. The following heuristic argument is often used to suggest that $T(x, y)$

satisfies Laplace's equation. Using Expression (10.19), we find that the amount of heat flowing out of the right edge of the rectangle in Figure 10.16(a) is approximately

$$\mathbf{V} \bullet \mathbf{N}_1 \, \Delta s_1 = -K[T_x(x + \Delta x, y) + iT_y(x + \Delta x, y)] \bullet (1 + 0i)\Delta y \qquad (10.20)$$
$$= -KT_x(x + \Delta x, y)\Delta y,$$

and the amount of heat flowing out of the left edge is

$$\mathbf{V} \bullet \mathbf{N}_2 \, \Delta s_2 = -K[T_x(x, y) + iT_y(x, y)] \bullet (-1 + 0i)\Delta y \qquad (10.21)$$
$$= KT_x(x, y)\Delta y.$$

If we add the contributions in Equations (10.20) and (10.21) we get

$$-K\left[\frac{T_x(x + \Delta x, y) - T_x(x, y)}{\Delta x}\right]\Delta x\Delta y \approx -KT_{xx}(x, y)\Delta x\Delta y. \qquad (10.22)$$

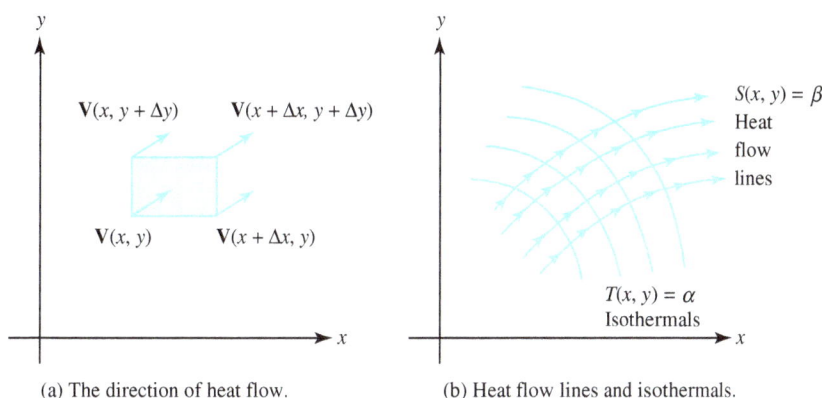

(a) The direction of heat flow. (b) Heat flow lines and isothermals.

Figure 10.16: Steady state temperatures

Similarly, the contribution for the amount of heat flowing out of the top and bottom edges is

$$-K\left[\frac{T_y(x, y + \Delta y) - T_y(x, y)}{\Delta y}\right]\Delta y\Delta y \approx -KT_{yy}(x, y)\Delta x\Delta y. \qquad (10.23)$$

Adding the quantities in Equations (10.22) and (10.23), we find that the net heat flowing out of the rectangle is approximated by the equation

$$-K[T_{xx}(x, y) + T_{yy}(x, y)]\Delta x\Delta y = 0,$$

which implies that $T(x, y)$ satisfies Laplace's equation and is a harmonic function.

If the domain in which $T(x, y)$ is defined is simply connected, then a conjugate harmonic function $S(x, y)$ exists, and

$$F(z) = T(x, y) + iS(x, y)$$

is an analytic function. The curves $T(x, y) = K_1$ are called **isothermals** and are lines connecting points of the same temperature. The curves $S(x, y) = K_2$ are called **heat flow lines**, and we can visualize the heat flowing along these curves from points of higher temperature to points of lower temperature. The situation is illustrated in Figure 10.16(b).

Boundary value problems for steady state temperatures are realizations of the Dirichlet problem where the value of the harmonic function $T(x, y)$ is interpreted as the temperature at the point (x, y).

Example 10.14. Suppose that two parallel planes are perpendicular to the z plane and pass through the horizontal lines $y = a$ and $y = b$ and that the temperature is held constant at the values $T(x, a) = T_1$ and $T(x, b) = T_2$, respectively, on these planes. Show that T is given by

$$T(x, y) = T_1 + \frac{T_2 - T_1}{b - a}(y - a).$$

Solution:

A reasonable assumption is that the temperature at all points on the plane passing through the line $y = y_0$ is constant. Hence $T(x, y) = t(y)$, where $t(y)$ is a function of y alone. Laplace's equation implies that $t,''(y) = 0$, and an argument similar to that in Example 10.1 will show that the solution $T(x, y)$ has the form given in the preceding equation.

The isothermals $T(x, y) = \alpha$ are easily seen to be horizontal lines. The conjugate harmonic function is

$$S(x, y) = \frac{T_1 - T_2}{b - a} x,$$

and the heat flow lines $S(x, y) = \beta$ are vertical segments between the horizontal lines. If $T_1 > T_2$, then the heat flows along these segments from the plane through $y = a$ to the plane through $y = b$, as illustrated in Figure 10.17.

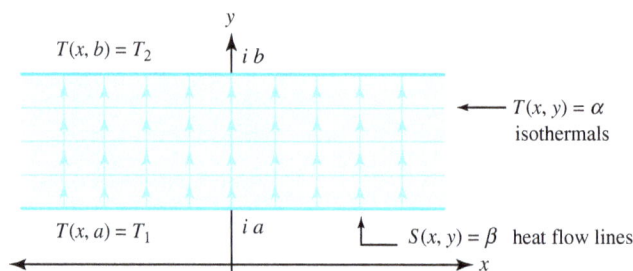

Figure 10.17: The temperature between parallel planes where $T_1 > T_2$

Example 10.15. Find the temperature $T(x, y)$ at each point in the upper half-plane $\text{Im}(z) > 0$ if the temperature along the x axis satisfies

$$T(x, 0) = T_1 \quad \text{for} \quad x > 0, \quad \text{and} \quad T(x, 0) = T_2 \quad \text{for} \quad x < 0.$$

Solution:

Since $T(x, y)$ is a harmonic function, this problem is an example of a Dirichlet problem. From Example 11.2, it follows that the solution is

$$T(x, y) = T_1 + \frac{T_2 - T_1}{\pi} \text{Arg}(z).$$

The isotherms $T(x, y) = \alpha$ are rays emanating from the origin. The conjugate harmonic function is $S(x, y) = \frac{1}{\pi}(T_1 - T_2) \ln |z|$, and the heat flow lines $S(x, y) = \beta$ are semicircles centered at the origin. If $T_1 > T_2$, then the heat flows counterclockwise along the semicircles, as shown in Figure 10.18.

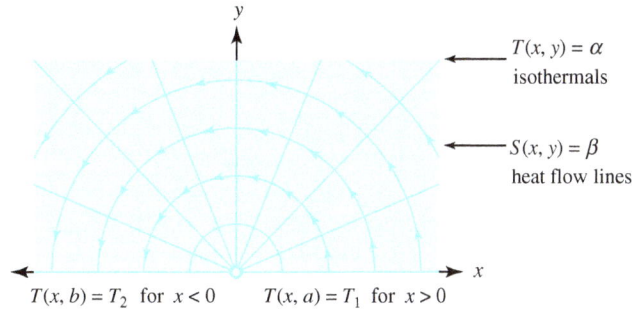

Figure 10.18: The temperature $T(x, y)$ in the upper half-plane where $T_1 > T_2$

Example 10.16. Find the temperature $T(x, y)$ at each point in the upper half-disk $H = \{z : \text{Im}(z) > 0, |z| < 1\}$ if the temperatures at points on the boundary satisfy

$$T(x, y) = 100 \quad \text{for} \quad x + iy = z = e^{i\theta}, \quad 0 < \theta < \pi;$$
$$T(x, 0) = 50 \quad \text{for} \quad -1 < x < 1.$$

Solution:

As discussed in Example 10.9, the function

$$u + iv = \frac{i(1 - z)}{1 + z} = \frac{2y}{(x + 1)^2 + y^2} + i\frac{1 - x^2 - y^2}{(x + 1)^2 + y^2} \tag{10.24}$$

is a one-to-one conformal mapping of the half-disk H onto the first quadrant $Q : u > 0, v > 0$. The conformal map given by Equation 10.24 gives rise to a new problem of finding the temperature $T^*(u, v)$ that satisfies the boundary conditions

$$T^*(u, 0) = 100 \quad \text{for} \quad u > 0, \quad \text{and} \quad T^*(0, v) = 50 \quad \text{for} \quad v > 0.$$

If we use Example 11.2, the harmonic function $T^*(u, v)$ is given by

$$T^*(u, v) = 100 + \frac{50 - 100}{\frac{\pi}{2}} \text{Arg}(w) = 100 - \frac{100}{\pi} \text{Arctan}\left(\frac{v}{u}\right). \tag{10.25}$$

Substituting the expressions for u and v from Equation (10.24) into Equation (10.25) yields the desired solution:

$$T(x, y) = 100 - \frac{100}{\pi} \text{Arctan}\left(\frac{1 - x^2 - y^2}{2y}\right).$$

The isothermals $T(x, y) = $ constant are circles that pass through the points ± 1, as shown in Figure 10.19.

10.5.1 An Insulated Segment on the Boundary

We now turn to the problem of finding the steady state temperature function $T(x, y)$ inside the simply connected domain D whose boundary consists of three adjacent curves C_1, C_2, and C_3, where $T(x, y) = T_1$ along C_1, and $T(x, y) = T_2$ along C_2. The region is insulated along C_3, so that fact that no heat flows across C_3 implies that

$$\mathbf{V}(x, y) \bullet \mathbf{N}(x, y) = -K\mathbf{N}(x, y) \bullet \nabla T(x, y) = 0,$$

325

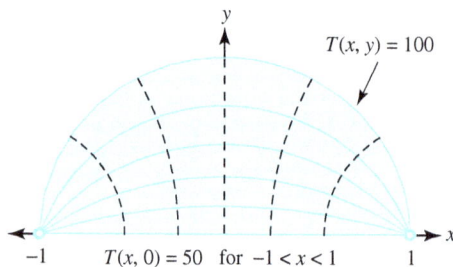

Figure 10.19: The temperature $T(x,y)$ in a half-disk

where $\mathbf{N}(x,y)$ is perpendicular to C_3. Thus the direction of heat flow must be parallel to this portion of the boundary. In other words, C_3 must be part of a heat flow line $S(x,y) = \text{constant}$, and the isothermals $T(x,y) = \text{constant}$ intersect C_3 orthogonally.

We can solve this problem by finding a conformal mapping

$$w = f(z) = u(x,y) + iv(x,y) \tag{10.26}$$

from D onto the semi-infinite strip $G : 0 < u < 1$, $v > 0$ so that the image of the curve C_1 is the ray $u = 0$; the image of the curve C_2 is the ray given by $u = 1$, $v > 0$; and the thermally insulated curve C_3 is mapped onto the thermally insulated segment $0 < u < 1$ of the u axis, as shown in Figure 10.20.

The new problem in G is to find the steady state temperature function $T^*(u,v)$ so that along the rays, we have the boundary values

$$T^*(0,v) = T_1 \quad \text{for} \quad v > 0, \quad \text{and} \quad T^*(1,v) = T_2 \quad \text{for} \quad v > 0. \tag{10.27}$$

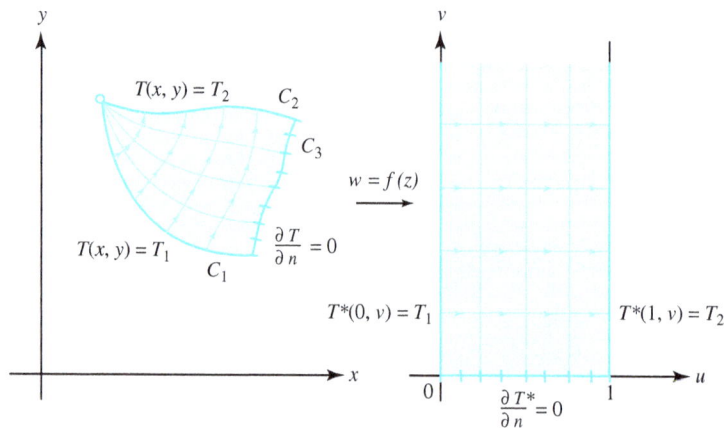

Figure 10.20: Steady state temperatures with one boundary portion insulated

The condition that a segment of the boundary is insulated can be expressed mathematically by saying that the normal derivative of $T^*(u,v)$ is zero. That is,

$$\frac{\partial T^*}{\partial n} = T_v^*(u,0) = 0, \tag{10.28}$$

where n is a coordinate measured perpendicular to the segment. We can easily verify that the function

$$T^*(u,v) = T_1 + (T_2 - T_1)u$$

326

satisfies the conditions stated in Equations 10.27 and 10.28 for region G. Therefore, using Equation 10.26, we find that the solution in D is

$$T(x,y) = T_1 + (T_2 - T_1)u(x,y).$$

The isothermals $T(x,y) = $ constant and their images under $w = f(z)$ are also illustrated in Figure 10.20.

Example 10.17. Find the steady state temperature $T(x,y)$ for the domain D consisting of the upper half-plane $\text{Im}(z) > 0$, where $T(x,y)$ has the boundary conditions

$$T(x,0) = 1 \quad \text{for} \quad x > 1, \quad \text{and} \quad T(x,0) = -1 \quad \text{for} \quad x < -1;$$

$$\frac{\partial T}{\partial n} = T_y(x,0) = 0 \quad \text{for} \quad -1 < x < 1,$$

where (again) n is a coordinate measured perpendicular to the segment.

Solution:

The mapping $w = \text{Arcsin}(z)$ conformally maps D onto the strip $v > 0$, $-\frac{\pi}{2} < u < \frac{\pi}{2}$, where the new problem is to find the steady state temperature $T^*(u,v)$ with boundary conditions

$$T^*\left(\frac{\pi}{2}, v\right) = 1 \quad \text{for} \quad v > 0, \quad \text{and} \quad T^*\left(-\frac{\pi}{2}, v\right) = -1 \quad \text{for} \quad v > 0;$$

$$\frac{\partial T^*}{\partial n} = T_v^*(u,0) = 0 \quad \text{for} \quad -\frac{\pi}{2} < u < \frac{\pi}{2}.$$

Using the result of Example 10.1, we can easily obtain the solution $T^*(u,v) = \frac{2}{\pi}u$.

Therefore the solution in D is $T(x,y) = \frac{2}{\pi}\text{Re}\big(\text{Arcsin}(z)\big)$.

If an explicit solution is required, we can use Formula (9.25) to obtain

$$T(x,y) = \frac{2}{\pi}\text{Arcsin}\frac{\sqrt{(x+1)^2 + y^2} - \sqrt{(x-1)^2 + y^2}}{2},$$

where $-\frac{\pi}{2} < \text{Arcsin}(t) < \frac{\pi}{2}$; see Figure 10.21.

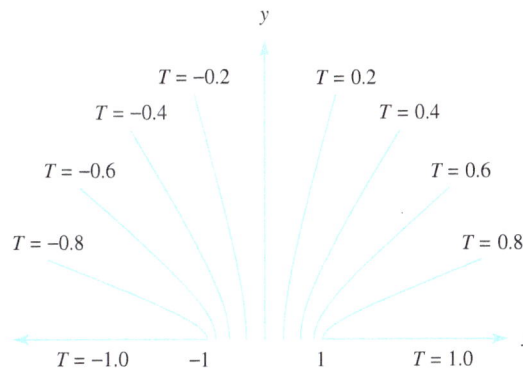

Figure 10.21: The temperature $T(x,y)$ with $T_y(x,0) = 0$ for $-1 < x < 1$, and boundary values $T(x,0) = -1$ for $x < -1$ and $T(x,0) = 1$, for $x > 1$

Exercises for Section 10.5 (Selected answers or hints are on page 454.)

1. Show that $H(x, y, z) = \frac{1}{\sqrt{x^2+y^2+z^2}}$ satisfies Laplace's equation $H_{xx} + H_{yy} + H_{zz} = 0$ in three-dimensional Cartesian space, but that $h(x, y) = \frac{1}{\sqrt{x^2+y^2}}$ does not satisfy equation $h_{xx} + h_{yy} = 0$ in two-dimensional Cartesian space.

2. Find the temperature function $T(x, y)$ in the infinite strip bounded by the lines $y = -x$ and $y = 1 - x$ that satisfies the following boundary values (shown in Figure 10.22).

$$T(x, -x) = 25 \quad \text{for all} \quad x;$$
$$T(x, 1-x) = 75 \quad \text{for all} \quad x.$$

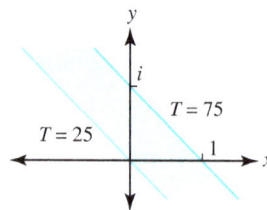

Figure 10.22: For Exercise 2

3. Find the temperature function $T(x, y)$ in the first quadrant $x > 0$, $y > 0$ that satisfies the following boundary values (shown in Figure 10.23).

$$T(x, 0) = 10 \quad \text{for} \quad x > 1,$$
$$T(x, 0) = 20 \quad \text{for} \quad 0 < x < 1,$$
$$T(0, y) = 20 \quad \text{for} \quad 0 \le y < 1,$$
$$T(0, y) = 10 \quad \text{for} \quad y < 1.$$

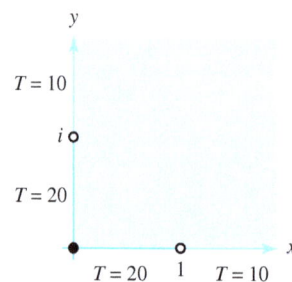

Figure 10.23: For Exercise 3

4. Find the temperature function $T(x, y)$ inside the unit disk $|z| < 1$ that satisfies the following boundary values (shown in Figure 10.24).
   Hint: Use $w = \frac{i(1-z)}{1+z}$.

$$T(x, y) = 20 \quad \text{for} \quad x + iy = z = e^{i\theta}, \quad 0 < \theta < \frac{\pi}{2};$$
$$T(x, y) = 60 \quad \text{for} \quad x + iy = z = e^{i\theta}, \quad \frac{\pi}{2} < \theta < 2\pi.$$

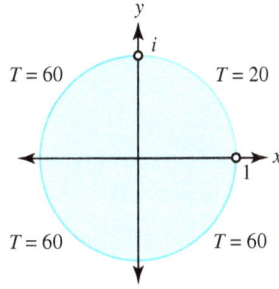

Figure 10.24: For Exercise 4

5. Find the temperature function $T(x, y)$ in the semi-infinite strip $-\frac{\pi}{2} < x < \frac{\pi}{2}$, $y > 0$ that satisfies the following boundary values (shown in Figure 10.25):

$$T\left(\frac{\pi}{2}, y\right) = 100 \quad \text{for} \quad y > 0;$$

$$T(x, y) = 0 \quad \text{for} \quad -\frac{\pi}{2} < x < \frac{\pi}{2};$$

$$T\left(-\frac{\pi}{2}, y\right) = 100 \quad \text{for} \quad y > 0.$$

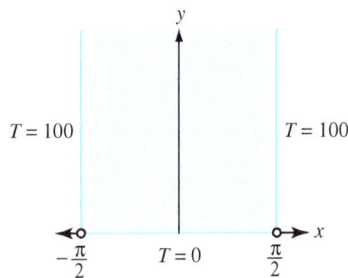

Figure 10.25: For Exercise 5

6. Find the temperature function $T(x, y)$ in the domain $r > 1$, $0 < \theta < \pi$ that satisfies the following boundary values (shown in Figure 10.26). *Hint:* $w = \frac{i(1-z)}{1+z}$.

$$T(x, 0) = 0 \quad \text{for} \quad x > 1;$$
$$T(x, 0) = 0 \quad \text{for} \quad x < -1;$$
$$T(x, y) = 100 \quad \text{if} \quad z = e^{i\theta}, \ 0 < \theta < \pi.$$

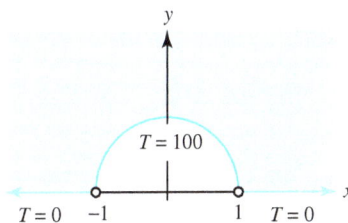

Figure 10.26: For Exercise 6

329

7. Find the temperature function $T(x, y)$ in the domain

$$1 < r < 2,\ 0 < \theta < \frac{\pi}{2}$$

that satisfies the following boundary conditions (shown in Figure 10.27):

$$T(x, y) = 0 \quad \text{for} \quad x + iy = z = e^{i\theta},\ 0 < \theta < \frac{\pi}{2};$$

$$T(x, y) = 50 \quad \text{for} \quad x + iy = z = 2e^{i\theta},\ 0 < \theta < \frac{\pi}{2};$$

$$\frac{\partial T}{\partial n} = T_y(x, 0) = 0 \quad \text{for} \quad 1 < x < 2;$$

$$\frac{\partial T}{\partial n} = T_x(0, y) = 0 \quad \text{for} \quad 1 < y < 2.$$

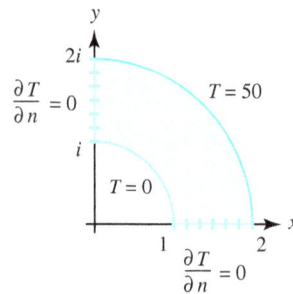

Figure 10.27: For Exercise 7

8. Find the temperature function $T(x, y)$ in the domain $0 < r < 1$, $0 < \text{Arg}(z) < \alpha$ that satisfies the following boundary conditions (shown in Figure 10.28). Use $w = \text{Log}(z)$.

$$T(x, 0) = 100 \quad \text{for} \quad 0 < x < 1;$$

$$T(x, y) = 50 \quad \text{for} \quad x + iy = z = re^{i\alpha},\ 0 < r < 1;$$

$$\frac{\partial T}{\partial n} = 0 \quad \text{for} \quad x + iy = z = e^{i\theta},\ 0 < \theta < \alpha.$$

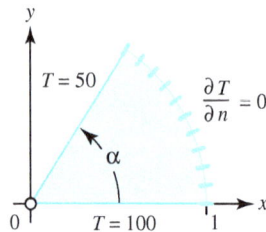

Figure 10.28: For Exercise 8

9. Find the temperature function $T(x, y)$ in the first quadrant $x > 0$, $y > 0$ that satisfies the

330

following boundary conditions (shown in Figure 10.29).

$$T(x,0) = 100 \quad \text{for} \quad x > 1;$$
$$T(0,y) = -50 \quad \text{for} \quad y > 1;$$
$$\frac{\partial T}{\partial n} = T_y(x,0) = 0 \quad \text{for} \quad 0 < x < 1;$$
$$\frac{\partial T}{\partial n} = T_x(0,y) = 0 \quad \text{for} \quad 0 < y < 1.$$

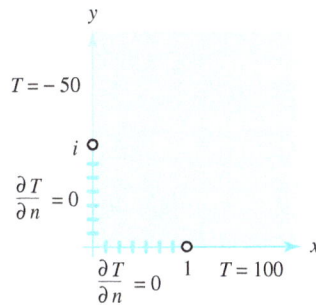

Figure 10.29: For Exercise 9

10. Find the temperature function $T(x,y)$ in the infinite strip $0 < y < \pi$ that satisfies the following boundary conditions (shown in Figure 10.30). *Hint*: Use $w = e^z$.

$$T(x,0) = 50 \quad \text{for} \quad x > 0;$$
$$T(x,\pi) = -50 \quad \text{for} \quad x > 0;$$
$$\frac{\partial T}{\partial n} = T_y(x,0) = 0 \quad \text{for} \quad x < 0;$$
$$\frac{\partial T}{\partial n} = T_y(x,\pi) = 0 \quad \text{for} \quad x < 0.$$

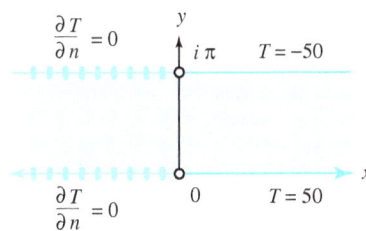

Figure 10.30: For Exercise 10

11. Find the temperature function $T(x,y)$ in the upper half-plane $\text{Im}(z) > 0$ that satisfies the following boundary conditions (shown in Figure 10.31).

$$T(x,0) = 100 \quad \text{for} \quad 0 < x < 1;$$
$$T(x,0) = -100 \quad \text{for} \quad -1 < x < 0;$$
$$\frac{\partial T}{\partial n} = T_y(x,0) = 0 \quad \text{for} \quad x > 1;$$
$$\frac{\partial T}{\partial n} = T_y(x,0) = 0 \quad \text{for} \quad x < -1.$$

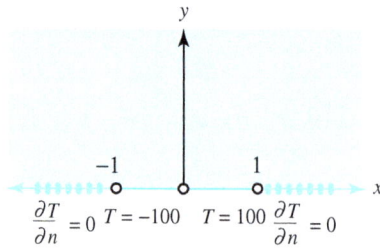

Figure 10.31: For Exercise 11

12. Find the temperature function $T(x, y)$ in the first quadrant $x > 0$, $y > 0$ that satisfies the following boundary conditions (shown in Figure 10.32).

$$T(x, 0) = 50 \quad \text{for} \quad x > 0;$$
$$T(0, y) = -50 \quad \text{for} \quad y > 1;$$
$$\frac{\partial T}{\partial n} = T_x(0, y) = 0 \quad \text{for} \quad 0 < y < 1.$$

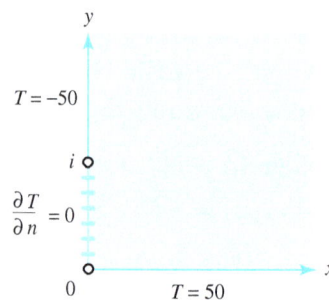

Figure 10.32: For Exercise 12

13. For the temperature function

$$T(x, y) = 100 - \frac{100}{\pi} \text{Arctan} \left(\frac{1 - x^2 - y^2}{2y} \right)$$

in the upper half-disk $\{z : |z| < 1, \text{Im}(z) > 0\}$, show that the isothermals $T(x, y) = \alpha$ are portions of circles that pass through the points $+1$ and -1, as illustrated in Figure 10.33.

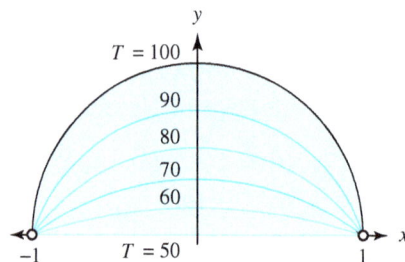

Figure 10.33: For Exercise 13

332

14. For the temperature function

$$T(x, y) = \frac{300}{\pi} \mathrm{Re}\big(\mathrm{Arcsin}(x + iy)\big)$$

in the upper half-plane $\mathrm{Im}(z) > 0$, show that the isothermals $T(x, y) = \alpha$ are portions of hyperbolas that have foci at the points ± 1, as illustrated in Figure 10.34.

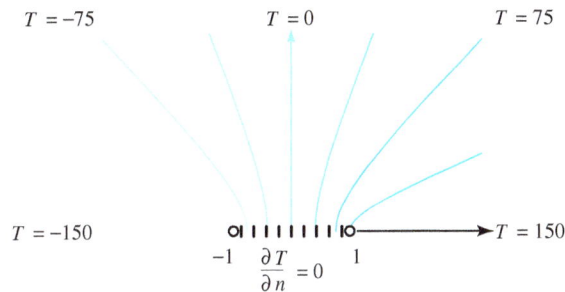

Figure 10.34: For Exercise 14

15. Find the temperature function in the portion of the upper half-plane $\mathrm{Im}(z) > 0$ that lies inside the ellipse

$$\frac{x^2}{\cosh^2 2} + \frac{y^2}{\sinh^2 2} = 1$$

and satisfies the following boundary conditions (shown in Figure 10.35). *Hint*: Use $w = \mathrm{Arcsin}(z)$.

$$T(x, y) = 80 \quad \text{for} \quad (x, y) \quad \text{on the ellipse,}$$
$$T(x, 0) = 40 \quad \text{for} \quad -1 < x < 1,$$
$$\frac{\partial T}{\partial n} = T_y(x, 0) = 0 \quad \text{when} \quad 1 < |x| < \cosh 2.$$

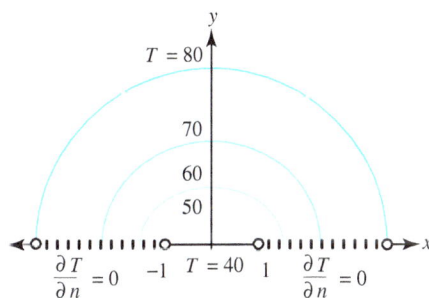

Figure 10.35: For Exercise 15

10.6 Two-Dimensional Electrostatics

A two-dimensional electrostatic field is produced by a system of charged wires, plates, and cylindrical conductors that are perpendicular to the z plane. The wires, plates, and cylinders are assumed to be long enough so that the effects at the ends can be neglected, as mentioned in Section 10.4. This assumption results in an electric field $\mathbf{E}(x, y)$ that can be interpreted as the force acting on a unit positive charge placed at the point (x, y). In the study of electrostatics, the vector field $\mathbf{E}(x, y)$ is shown to be *conservative* and is derivable from a function $\phi(x, y)$, called the **electrostatic potential**, expressed as

$$\mathbf{E}(x, y) = -\nabla\phi(x, y) = -\phi_x(x, y) - i\phi_y(x, y).$$

If we make the additional assumption that there are no charges within the domain D, then Gauss's law for electrostatic fields implies that the line integral of the outward normal component of $\mathbf{E}(x, y)$ taken around any small rectangle lying inside D is identically zero. A heuristic argument similar to the one we used for steady state temperatures, with $T(x, y)$ replaced by $\phi(x, y)$, will show that the value of the line integral is

$$-[\phi_{xx}(x, y) + \phi_{yy}(x, y)]\Delta x \Delta y.$$

This quantity equals zero, so we conclude that $\phi(x, y)$ is a harmonic function. If we designate $\psi(x, y)$ as the harmonic conjugate, then

$$F(z) = \phi(x, y) + i\psi(x, y)$$

is the complex potential (not to be confused with the electrostatic potential).

The curves $\phi(x, y) = K_1$ are called the **equipotential curves**, and the curves $\psi(x, y) = K_2$ are called the **lines of flux**. If a small test charge is allowed to move under the influence of the field $\mathbf{E}(x, y)$, then it will travel along a line of flux. Boundary value problems for the potential function $\phi(x, y)$ are mathematically the same as those for steady state heat flow, and they are realizations of the Dirichlet problem where the harmonic function is $\phi(x, y)$.

Example 10.18. Consider two parallel conducting planes that pass perpendicular to the z plane through the lines $x = a$ and $x = b$, which are kept amt the potentials U_1 and U_2, respectively. Then, according to the result of Example 10.1, the electrical potential is

$$\phi(x, y) = U_1 + \frac{U_2 - U_1}{b - a}(x - a).$$

Example 10.19. Find the electrical potential $\phi(x, y)$ in the region between two infinite coaxial cylinders $r = a$ and $r = b$, which are kept at the potentials U_1 and U_2, respectively.

Solution:

The function $w = \log z = \ln|z| + i\arg z$ maps the annular region between the circles $r = a$ and $r = b$ onto the infinite strip $\ln a < u < \ln b$ in the w plane, as shown in Figure 10.36. The potential $\Phi(u, v)$ in the infinite strip has the boundary values

$$\Phi(\ln a, v) = U_1, \quad \text{and} \quad \Phi(\ln b, v) = U_2 \quad \text{for all} \quad v.$$

The result of Example 10.18 gives the electrical potential $\Phi(u, v)$:

$$\Phi(u, v) = U_1 + \frac{U_2 - U_1}{\ln b - \ln a}(u - \ln a).$$

We can use the fact that $u = \ln|z|$ to conclude that the potential $\phi(x, y)$ is

$$\phi(x, y) = U_1 + \frac{U_2 - U_1}{\ln b - \ln a}(\ln|z| - \ln a).$$

The equipotentials $\phi(x, y) = $ constant are concentric circles centered on the origin, and the lines of flux are portions of rays emanating from the origin. If $U_2 < U_1$, then the situation is as illustrated in Figure 10.36.

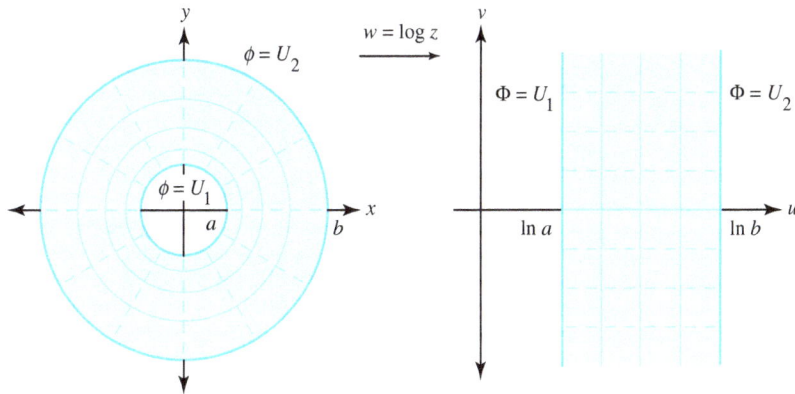

Figure 10.36: The electrical field in a coaxial cylinder, where $U_2 < U_1$

Example 10.20. Find the electrical potential $\phi(x, y)$ produced by two charged half-planes that are perpendicular to the z plane and pass through the rays $x < -1$, $y = 0$ and $x > 1$, $y = 0$, where the planes are kept at the fixed potentials

$$\phi(x, 0) = -300 \quad \text{for} \quad x < -1, \quad \text{and} \quad \phi(x, 0) = 300 \quad \text{for} \quad x > 1.$$

Solution:

The result of Example 9.13 shows that the function $w = \text{Arcsin}(z)$ is a conformal mapping of the z plane slit along the two rays $x < -1$, $y = 0$ and $x > 1$, $y = 0$ onto the vertical strip $-\frac{\pi}{2} < u < \frac{\pi}{2}$. Thus, the problem reduces to finding the potential $\Phi(u, v)$ that satisfies the boundary values

$$\Phi\left(-\frac{\pi}{2}, v\right) = -300 \quad \text{and} \quad \Phi\left(\frac{\pi}{2}, v\right) = 300 \quad \text{for all} \quad v.$$

From Example 10.1,

$$\Phi(u, v) = \frac{600}{\pi} u.$$

As in the discussion of Example 10.17, the solution in the z plane is

$$\phi(x, y) = \frac{600}{\pi} \text{Re}\big(\text{Arcsin}(z)\big)$$

$$= \frac{600}{\pi} \text{Arcsin}\left(\frac{\sqrt{(x+1)^2 + y^2} - \sqrt{(x-1)^2 + y^2}}{2}\right).$$

Several equipotential curves are shown in Figure 10.37.

Example 10.21. Find the electrical potential $\phi(x, y)$ in the disk $D : |z| < 1$ that satisfies the boundary values

$$\phi(x, y) = 80 \quad \text{for} \quad x + iy = z \quad \text{on} \quad C_1 = \left\{ z = e^{i\theta} : 0 < \theta < \frac{\pi}{2} \right\};$$

$$\phi(x, y) = 0 \quad \text{for} \quad x + iy = z \quad \text{on} \quad C_2 = \left\{ z = e^{i\theta} : \frac{\pi}{2} < \theta < 2\pi \right\}.$$

Solution:

335

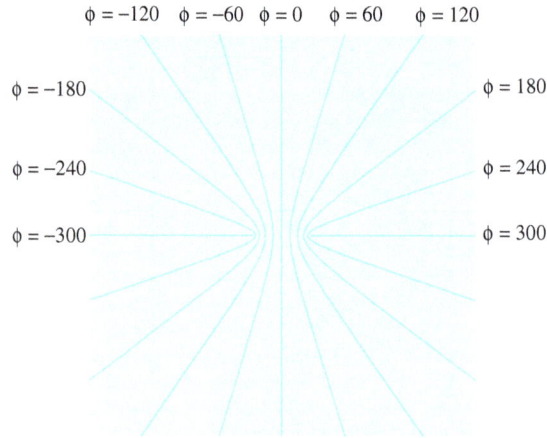

$\phi = -120$ $\phi = -60$ $\phi = 0$ $\phi = 60$ $\phi = 120$

$\phi = -180$ $\phi = 180$

$\phi = -240$ $\phi = 240$

$\phi = -300$ $\phi = 300$

Figure 10.37: Equipotentials of two charged half-planes perpendicular to the complex plane

The mapping $w = S(z) = \frac{(1-i)(z-i)}{z-1}$ is a one-to-one conformal mapping of D onto the upper half-plane $\operatorname{Im}(w) > 0$ with the property that C_1 is mapped onto the negative u axis and C_2 is mapped onto the positive u axis. The potential $\Phi(u, v)$ in the upper half-plane that satisfies the new boundary values

$$\Phi(u, 0) = 80 \quad \text{for} \quad u < 0, \quad \text{and} \quad \Phi(u, 0) = 0 \quad \text{for} \quad u > 0$$

is given by

$$\Phi(u, v) = \frac{80}{\pi}\operatorname{Arg}(w) = \frac{80}{\pi}\operatorname{Arctan}\left(\frac{v}{u}\right). \tag{10.29}$$

A straightforward calculation shows that

$$u + iv = S(z) = \frac{(x-1)^2 + (y-1)^2 - 1 + i(1 - x^2 - y^2)}{(x-1)^2 + y^2}.$$

We substitute the real and imaginary parts, u and v from this equation, into Equation (10.29) to obtain the desired solution:

$$\phi(x, y) = \frac{80}{\pi}\operatorname{Arctan}\left(\frac{1 - x^2 - y^2}{(x-1)^2 + (y-1)^2 - 1}\right).$$

The level curve $\Phi(u, v) = \alpha$ in the upper half-plane is a ray emanating from the origin, and the preimage $\phi(x, y) = \alpha$ in the unit disk is an arc of a circle that passes through the points 1 and i. Several level curves are illustrated in Figure 10.38.

Exercises for Section 10.6 (Selected answers or hints are on page 455.)

1. Find the electrostatic potential $\phi(x, y)$ between the two coaxial cylinders $r = 1$ and $r = 2$ that has the boundary values shown in Figure 10.39:

$$\phi(x, y) = 100 \quad \text{when} \quad |z| = 1,$$
$$\phi(x, y) = 200 \quad \text{when} \quad |z| = 2.$$

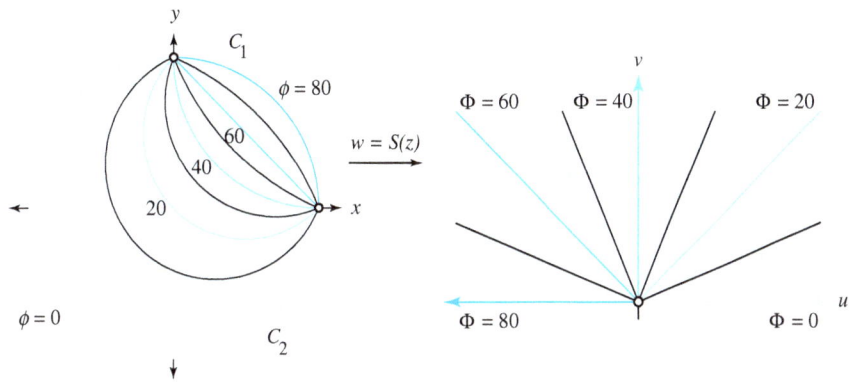

Figure 10.38: The potentials ϕ and Φ

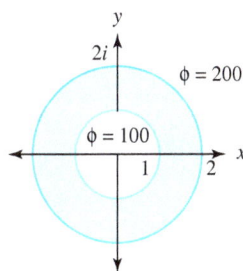

Figure 10.39: For Exercise 1

2. Find the electrostatic potential $\phi(x, y)$ in the upper half-plane $\text{Im}(z) > 0$ that satisfies the boundary values shown in Figure 10.40:

$$
\begin{aligned}
\phi(x, 0) &= 100 \quad \text{for} \quad x > 1; \\
\phi(x, 0) &= 0 \quad \text{for} \quad -1 < x < 1; \\
\phi(x, 0) &= -100 \quad \text{for} \quad x < -1.
\end{aligned}
$$

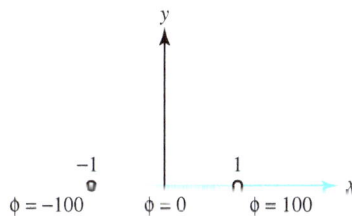

Figure 10.40: For Exercise 2

3. Find the electrostatic potential $\phi(x, y)$ in the crescent-shaped region that lies inside the disk $|z - 2| < 2$ and outside the circle $|z - 1| = 1$ that satisfies the following boundary values (shown in Figure 10.41).

$$
\begin{aligned}
\phi(x, y) &= 100 \quad \text{for} \quad |z - 2| = 2, \quad z \neq 0, \\
\phi(x, y) &= 50 \quad \text{for} \quad |z - 1| = 1, \quad z \neq 0.
\end{aligned}
$$

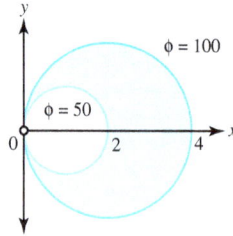

Figure 10.41: For Exercise 3

4. Find the electrostatic potential $\phi(x, y)$ in the semi-infinite strip $-\frac{\pi}{2} < x < \frac{\pi}{2}$, $y > 0$ that has the boundary values shown in Figure 10.42:

$$\phi\left(\frac{\pi}{2}, y\right) = 0 \quad \text{for} \quad y > 0;$$

$$\phi(x, 0) = 50 \quad \text{for} \quad -\frac{\pi}{2} < x < \frac{\pi}{2};$$

$$\phi\left(-\frac{\pi}{2}, y\right) = 100 \quad \text{for} \quad y > 0.$$

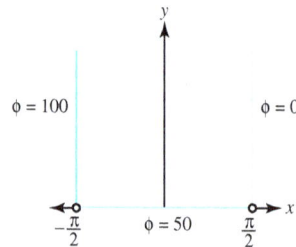

Figure 10.42: For Exercise 4

5. Find the electrostatic potential $\phi(x, y)$ in the domain D in the half-plane $\mathrm{Re}(z) > 0$ that lies to the left of the hyperbola $2x^2 - 2y^2 = 1$ and satisfies the following boundary values (shown in Figure 10.43).

$$\phi(0, y) = 50 \quad \text{for all} \quad y;$$

$$\phi(x, y) = 100 \quad \text{when} \quad 2x^2 - 2y^2 = 1.$$

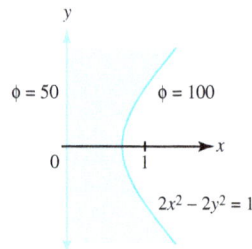

Figure 10.43: For Exercise 5

6. Find the electrostatic potential $\phi(x, y)$ in the infinite strip $0 < x < \frac{\pi}{2}$ that satisfies the

following boundary values (shown in Figure 10.44). *Hint:* Use $w = \sin z$.

$$\phi(0, y) = 100 \quad \text{for} \quad y > 0;$$
$$\phi\left(\frac{\pi}{2}, y\right) = 0 \quad \text{for all} \quad y;$$
$$\phi(0, y) = -100 \quad \text{for} \quad y < 0.$$

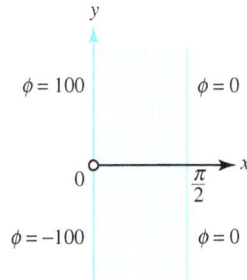

Figure 10.44: For Exercise 6

7. Consider the conformal mapping $w = S(z) = \frac{2z-6}{z+3}$.

   (a) Show that $S(z)$ maps the domain D that is the portion of the right half-plane $\text{Re}(z) > 0$ that lies exterior to the circle $|z - 5| = 4$ onto the annulus $1 < |w| < 2$.

   (b) Find the electrostatic potential $\phi(x, y)$ in the domain D that satisfies the boundary values shown in Figure 10.45:

   $$\phi(0, y) = 100 \quad \text{for all} \quad y;$$
   $$\phi(x, y) = 200 \quad \text{when} \quad |z - 5| = 4.$$

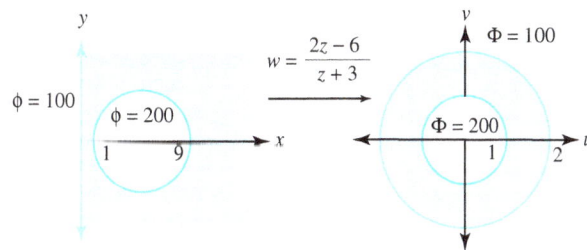

Figure 10.45: For Exercise 7b

8. Consider the conformal mapping $w = S(z) = \frac{z-10}{2z-5}$.

   (a) Show that $S(z)$ maps the domain D that is the portion of the disk $|z| < 5$ that lies outside the circle $|z - 2| = 2$ onto the annulus defined by $1 < |w| < 2$.

   (b) Find the electrostatic potential $\phi(x, y)$ in the domain D that satisfies the boundary values shown in Figure 10.46.

   $$\phi(x, y) = 100 \quad \text{when} \quad |z| = 5;$$
   $$\phi(x, y) = 200 \quad \text{when} \quad |z - 2| = 2.$$

339

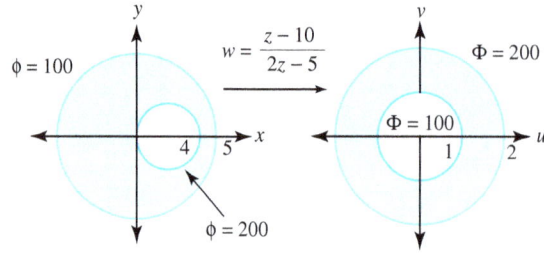

Figure 10.46: For Exercise 8b

10.7   Two-Dimensional Fluid Flow

Suppose that a fluid flows over the complex plane and that the velocity at the point $z = x + iy$ is given by the velocity vector

$$\mathbf{V}(x, y) = p(x, y) + iq(x, y). \tag{10.30}$$

We also require that the velocity does not depend on time and the components $p(x, y)$ and $q(x, y)$ have continuous partial derivatives. The divergence of the vector field in Equation 10.30 is given by

$$\text{div}\mathbf{V}(x, y) = p_x(x, y) + q_y(x, y)$$

and is a measure of the extent to which the velocity field diverges near the point. We consider only fluid flows for which the divergence is zero. This condition is more precisely characterized by the requirement that the net flow through any simply closed contour be identically zero.

If we consider the flow out of the small rectangle shown in Figure 10.47, then the rate of outward flow equals the line integral of the exterior normal component of $\mathbf{V}(x, y)$ taken over the sides of the rectangle. The exterior normal component is given by $-q$ on the bottom edge, p on the right edge, q on the top edge, and $-p$ on the left edge. Integrating and setting the resulting net flow to zero yields

$$\int_y^{y+\Delta y} [p(x + \Delta x, t) - p(x, t)]\, dt + \int_x^{x+\Delta x} [q(t, y + \Delta y) - q(t, y)]\, dt = 0. \tag{10.31}$$

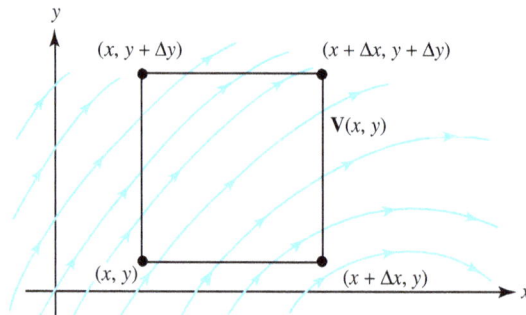

Figure 10.47: A two-dimensional vector field

Both p and q are continuously differentiable, so the mean value theorem implies that

$$p(x + \Delta x, t) - p(x, t) = p_x(x_1, t)\Delta x, \quad \text{and} \tag{10.32}$$
$$q(t, y + \Delta y) - q(t, y) = q_y(t, y_2)\Delta y,$$

340

where $x < x_1 < x + \Delta x$ and $y < y_2 < y + \Delta y$. Substitution of the expressions displayed in Equation (10.32) into Equation (10.31) and subsequently dividing through by $\Delta x \, \Delta y$ results in

$$\frac{1}{\Delta y} \int_y^{y+\Delta y} p_x(x_1, t) \, dt + \frac{1}{\Delta x} \int_x^{x+\Delta x} q_y(t, y_2) \, dt = 0.$$

We can use the mean value theorem for integrals with this equation to show that

$$p_x(x_1, y_1) + q_y(x_2, y_2) = 0,$$

where $y < y_1 < y + \Delta y$ and $x < x_2 < x + \Delta x$. Letting $\Delta x \to 0$ and $\Delta y \to 0$ in this equation yields

$$p_x(x, y) + q_y(x, y) = 0, \tag{10.33}$$

which is called the *equation of continuity*.

The curl of the vector field in Equation 10.30 has magnitude

$$\left| \mathbf{curl} \, \mathbf{V}(x, y) \right| = q_x(x, y) - p_y(x, y)$$

and is an indication of how the field swirls in the vicinity of a point. Imagine that a "fluid element" at the point (x, y) is suddenly frozen and then moves freely in the fluid. The fluid element will rotate with an angular velocity given by

$$\frac{1}{2} q_y(x, y) - \frac{1}{2} p_x(x, y) = \frac{1}{2} \left| \mathbf{curl} \, \mathbf{V}(x, y) \right|.$$

We consider only fluid flows for which the curl is zero. Such fluid flows are called *irrotational*. This condition is more precisely characterized by requiring that the line integral of the tangential component of $\mathbf{V}(x, y)$ along any simply closed contour be identically zero. If we consider the rectangle in Figure 10.47, then the tangential component is given by p on the bottom edge, q on the right edge, $-p$ on the top edge, and $-q$ on the left edge. Integrating and equating the resulting *circulation* integral to zero yields

$$\int_y^{y+\Delta y} [q(x + \Delta x, t) - q(x, t)] \, dt - \int_x^{x+\Delta x} [p(t, y + \Delta y) - p(t, y)] \, dt = 0.$$

As before, we apply the mean value theorem and divide through by $\Delta x \, \Delta y$, and obtain the equation

$$\frac{1}{\Delta y} \int_y^{y+\Delta y} q_x(x_1, t) \, dt - \frac{1}{\Delta x} \int_x^{x+\Delta x} p_y(t, y_2) \, dt = 0.$$

We can use the mean value for integrals with this equation to deduce that $q_x(x_1, y_1) - p_y(x_2, y_2) = 0$. Letting $\Delta x \to 0$ and $\Delta y \to 0$ yields

$$q_x(x, y) - p_y(x, y) = 0.$$

Equation (10.33) and this equation show that the function $f(z) = p(x, y) - iq(x, y)$ satisfies the Cauchy-Riemann equations and is an analytic function. If we let $F(z)$ denote the antiderivative of $f(z)$, then

$$F(z) = \phi(x, y) + i\psi(x, y), \tag{10.34}$$

which is the **complex potential** of the flow and has the property

$$\overline{F'(z)} = \phi_x(x, y) - i\psi_x(x, y) = p(x, y) + iq(x, y) = \mathbf{V}(x, y).$$

Since $\phi_x = p$ and $\phi_y = q$, we also have

$$\nabla\phi(x,y) = p(x,y) + iq(x,y) = \mathbf{V}(x,y),$$

so $\phi(x,y)$ is the **velocity potential** for the flow, and the curves

$$\phi(x,y) = K_1$$

are called **equipotentials**. The function $\psi(x,y)$ is called the **stream function**. The curves

$$\psi(x,y) = K_2$$

are called **streamlines** and describe the paths of the fluid particles. To demonstrate this result, we implicitly differentiate $\psi(x,y) = K_2$ and find that the slope of a vector tangent is given by

$$\frac{dy}{dx} = -\frac{\psi_x(x,y)}{\psi_y(x,y)}.$$

Using the fact that $\psi_y = \phi_x$ and this equation, we find that the tangent vector to the curve is

$$\mathbf{T} = \phi_x(x,y) - i\psi_x(x,y) = p(x,y) + iq(x,y) = \mathbf{V}(x,y).$$

The salient idea of the preceding discussion is the conclusion that, if

$$F(z) = \phi(x,y) + i\psi(x,y) \tag{10.35}$$

is an analytic function, then the family of curves

$$\{\psi(x,y) = K_2\}$$

represents the streamlines of a fluid flow.

The boundary condition for an ideal fluid flow is that $\mathbf{V}$ should be parallel to the boundary curve containing the fluid (the fluid flows parallel to the walls of a containing vessel). In other words, if Equation 10.35 is the complex potential for the flow, then the boundary curve must be given by $\psi(x,y) = K$ for some constant K; that is, the boundary curve must be a streamline.

Theorem 10.5 (Invariance of flow). *Let*

$$F_1(w) = \Phi(u,v) + i\Psi(u,v)$$

denote the complex potential for a fluid flow in a domain G in the w plane, where the velocity is

$$\mathbf{V}_1(u,v) = \overline{F_1{}'(w)}.$$

If the function $w = S(z) = u(x,y) + iv(x,y)$ is a one-to-one conformal mapping from a domain D in the z-plane onto G, then the composite function

$$F_2(z) = F_1(S(z)) = \Phi\big(u(x,y),\, v(x,y)\big) + i\Psi\big(u(x,y),\, v(x,y)\big)$$

is the complex potential for a fluid flow in D, where the velocity is

$$\mathbf{V}_2(x,y) = \overline{F_2{}'(z)}.$$

The situation is shown in Figure 10.48.

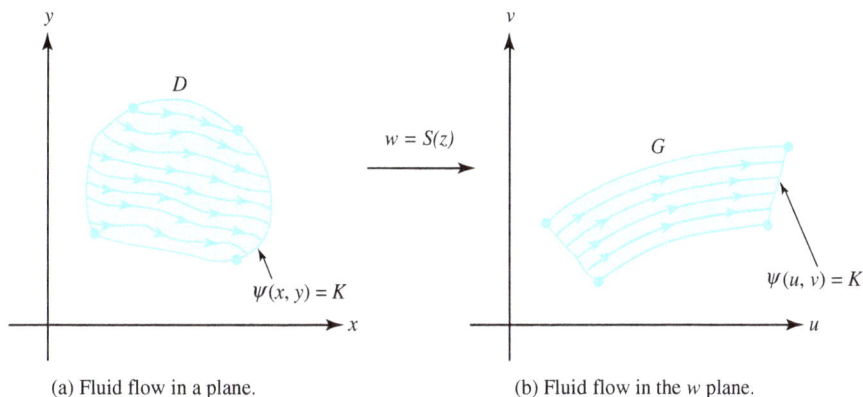

(a) Fluid flow in a plane. (b) Fluid flow in the w plane.

Figure 10.48: The image of a fluid flow under a conformal mapping

Proof. From Equation (10.34), $F_1(w)$ is an analytic function. Since the composition of analytic functions is analytic, $F_2(z)$ is the required complex potential for an ideal fluid flow in D. □

We note that the functions

$$\phi(x, y) = \Phi\big(u(x, y),\, v(x, y)\big) \quad \text{and} \quad \psi(x, y) = \Psi\big(u(x, y),\, v(x, y)\big)$$

are the new velocity potential and stream function, respectively, for the flow in D. A streamline or natural boundary curve

$$\psi(x, y) = K$$

in the z plane is mapped onto a streamline or natural boundary curve

$$\Psi(u, v) = K$$

in the w plane by the transformation $w = S(z)$. One method for finding a flow inside a domain D in the z plane is to conformally map D onto a domain G in the w plane in which the flow is known.

For an ideal fluid with uniform density ρ, the fluid pressure $P(x, y)$ and speed $|\mathbf{V}(x, y)|$ are related by the following special case of Bernoulli's equation:

$$\frac{P(x, y)}{\rho} + \frac{1}{2}\big|\mathbf{V}(x, y)\big| = \text{ constant.}$$

Note that the pressure is greatest when the speed is least.

Example 10.22. The complex potential $F(z) = (a + ib)z$ has the velocity potential and stream function of

$$\phi(x, y) = ax - by \quad \text{and} \quad \psi(x, y) = bx + ay,$$

respectively, and gives rise to the fluid flow defined in the entire complex plane that has a uniform parallel velocity of

$$\mathbf{V}(x, y) = \overline{F'(z)} = a - ib.$$

The streamlines are parallel lines given by the equation $bx + ay = \text{constant}$ and are inclined at an angle $\alpha = -\text{Arctan}(\frac{b}{a})$, as indicated in Figure 10.49.

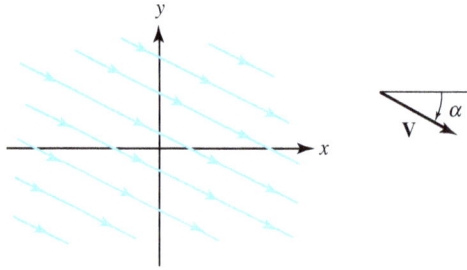

Figure 10.49: A uniform parallel flow

Example 10.23. Consider the complex potential $F(z) = \frac{A}{2}z^2$, where A is a positive real number. The velocity potential and stream function are given by

$$\phi(x, y) = \frac{A}{2}(x^2 - y^2) \quad \text{and} \quad \psi(x, y) = Axy,$$

respectively. The streamlines

$$\psi(x, y) = \text{constant}$$

form a family of hyperbolas with asymptotes along the coordinate axes. The velocity vector $\mathbf{V} = A\overline{z}$ indicates that in the upper half-plane $\text{Im}(z) > 0$, the fluid flows down along the streamlines and spreads out along the x axis, as against a wall, as depicted in Figure 10.50.

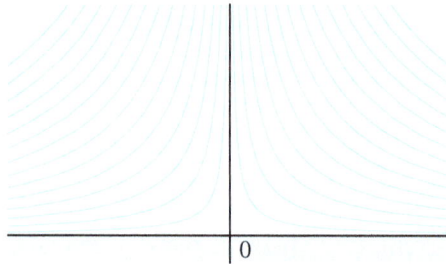

Figure 10.50: The fluid flow with complex potential $F(z) = \frac{A}{2}z^2$

Example 10.24. Find the complex potential for an ideal fluid flowing from left to right across the complex plane and around the unit circle $|z| = 1$.

Solution:

We use the fact that the conformal mapping $w = S(z) = z + \frac{1}{z})$ maps the domain $D = \{z : |z| < 1\}$ one-to-one and onto the w plane slit along the segment $-2 \le u \le 2$, $v = 0$. The complex potential for a uniform horizontal flow parallel to this slit in the w plane is

$$F_1(w) = Aw,$$

where A is a positive real number. The stream function for the flow in the w plane is $\psi(u, v) = Av$ so that the slit lies along the streamline $\Psi(u, v) = 0$.

The composite function $F_2(z) = F_1(S(z))$ determines the fluid flow in the domain D, where the complex potential is

$$F_2(z) = A\left(z + \frac{1}{z}\right),$$

344

where $A > 0$. We can use polar coordinates to express $F_2(z)$ as

$$F_2(z) = F_2(re^{i\theta}) = A\left(r + \frac{1}{r}\right)\cos\theta + iA\left(r - \frac{1}{r}\right)\sin\theta.$$

The streamline $\psi(r, \theta) = A(r - \frac{1}{r})\sin\theta = 0$ consists of the rays

$$r > 1, \ \theta = 0 \quad \text{and} \quad r > 1, \ \theta = \pi$$

along the x axis and the curve $r - \frac{1}{r} = 0$, which is the unit circle $r = 1$. Thus the unit circle can be considered as a boundary curve for the fluid flow.

The approximation $F_2(z) = A(z + \frac{1}{z}) \approx Az$ is valid for large values of z, so we can approximate the flow with a uniform horizontal flow having speed $|\mathbf{V}| = A$ at points that are distant from the origin. The streamlines

$$\psi(x, y) = \text{ constant}$$

and their images

$$\Psi(u, v) = \text{ constant}$$

under the mapping $w = S(z) = z + \frac{1}{z}$ are illustrated in Figure 10.51.

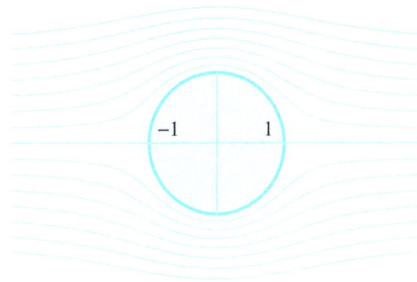

Figure 10.51: Fluid flow around a circle

Example 10.25. Find the complex potential for an ideal fluid flowing from left to right across the complex plane and around the segment from $-i$ to i.

Solution:

We use the conformal mapping

$$w = S(z) = (z^2 + 1)^{\frac{1}{2}} = (z + i)^{\frac{1}{2}}(z - i)^{\frac{1}{2}}$$

where the branch of the square root of $Z = z \pm i$ in each factor is $Z^{\frac{1}{2}} = R^{\frac{1}{2}}e^{i\theta/2}$, where $R = |Z|$, and $\theta = \arg_{-\frac{\pi}{2}}(Z)$, $-\frac{\pi}{2} < \theta \leq \frac{3\pi}{2}$. The function given by $w = S(z)$ is a one-to-one conformal mapping of the domain D consisting of the z plane slit along the segment $x = 0, -1 \leq y \leq 1$ onto the domain G consisting of the w plane slit along the segment $-1 \leq u \leq 1, v = 0$. The complex potential for a uniform horizontal flow parallel to the slit in the w plane is given by $F_1(w) = Aw$, where for convenience we choose $A = 1$ and where the slit lies along the streamline $\Psi(u, v) = c = 0$. The composite function

$$F_2(z) = F_1\big(S(z)\big) = A(z^2 + 1)^{\frac{1}{2}}$$

345

is the complex potential for a fluid flow in the domain D. The streamlines given by $\psi(x, y) = c$ for the flow in D are obtained by finding the preimage of the streamline $\Psi(u, v) = c$ in G given by the parametric equations

$$v = c \quad \text{and} \quad u = t, \quad \text{for} \quad -\infty < t < \infty.$$

The corresponding streamline in D is found by solving the equation

$$t + ic = (z^2 + 1)^{\frac{1}{2}}$$

for x and y in terms of t. Squaring both sides of this equation yields

$$t^2 - c^2 - 1 + i2ct = x^2 - y^2 + i2xy.$$

Equating the real and imaginary parts leads to the system of equations

$$x^2 - y^2 = t^2 - c^2 - 1 \quad \text{and} \quad xy = ct.$$

Eliminating t in the last two equations gives $c^2 = (x^2 + c^2)(y^2 - c^2)$, and we can solve for y in terms of x to get

$$y = c\sqrt{\frac{1 + c^2 + x^2}{c^2 + x^2}}$$

for streamlines in D. For large values of x, this streamline approaches the asymptote $y = c$ and approximates a horizontal flow, as shown in Figure 10.52.

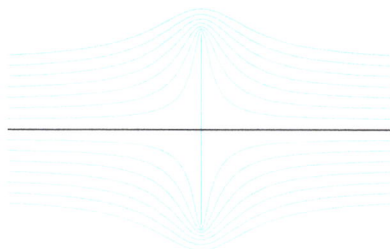

Figure 10.52: Flow around a segment

Exercises for Section 10.7 (Selected answers or hints are on page 455.)

1. Consider the ideal fluid flow for the complex potential $F(z) = A(z + \frac{1}{z})$, where A is a positive real number.

   (a) Show that the velocity vector at the point $(1, \theta)$, $z = re^{i\theta}$ on the unit circle is given by $\mathbf{V}(1, \theta) = A(1 - \cos 2\theta - i \sin 2\theta)$.

   (b) Show that the velocity vector $\mathbf{V}(1, \theta)$ is tangent to the unit circle $|z| = 1$ at all points except -1 and $+1$.
   Hint: Show that $\mathbf{V} \cdot \mathbf{P} = 0$, where $\mathbf{P} = \cos \theta + i \sin \theta$.

   (c) Show that the speed at the point $(1, \theta)$ on the unit circle is given by $|\mathbf{V}| = 2A|\sin \theta|$ and that the speed attains the maximum of $2A$ at the points $\pm i$ and is zero at the points ± 1. Where is the pressure the greatest?

346

2. Show that the complex potential $F(z) = ze^{-i\alpha} + \frac{e^{i\alpha}}{z}$ determines the ideal fluid flow around the unit circle $|z| = 1$, where the velocity at points distant from the origin is given approximately by $\mathbf{V} \approx e^{i\alpha}$; that is, the direction of the flow for large values of z is inclined at an angle α with the x axis, as shown in Figure 10.53.

3. Consider the ideal fluid flow in the channel bounded by the hyperbolas $xy = 1$ and $xy = 4$ in the first quadrant, where the complex potential is given by $F(z) = \frac{A}{2}z^2$ and A is a positive real number.

   (a) Find the speed at each point, and find the point on the boundary at which the speed attains a minimum value.

   (b) Where is the pressure greatest?

4. Show that the stream function is given by $\psi(r, \theta) = Ar^3 \sin 3\theta$ for an ideal fluid flow around the angular region $0 < \theta < \frac{\pi}{3}$ indicated in Figure 10.54. Sketch several streamlines of the flow. *Hint:* Use the conformal mapping $w = z^3$.

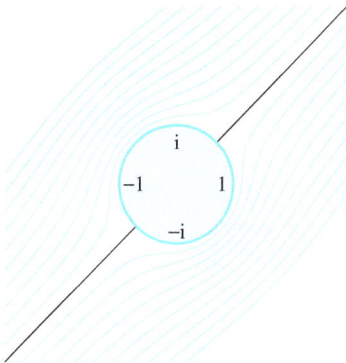

Figure 10.53: For Exercise 2

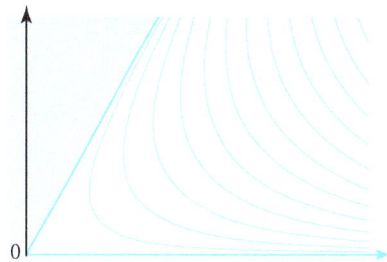

Figure 10.54: For Exercise 4

5. Consider the ideal fluid flow, where the complex potential is

$$F(z) = Az^{\frac{3}{2}} = Ar^{\frac{3}{2}}\left(\cos\frac{3\theta}{2} + i\sin\frac{3\theta}{2}\right), \quad \text{for} \quad 0 \le \theta \le 2\pi.$$

   (a) Find the stream function $\psi(r, \theta)$.

   (b) Sketch streamlines of the flow in the region $0 < \theta < \frac{4\pi}{3}$ indicated in Figure 10.55.

6. Consider the complex potential $F(z) = A(z^2 + \frac{1}{z^2})$

   (a) Let $A > 0$. Show that $F(z)$ determines an ideal fluid flow around the domain $r > 1$, $0 < \theta < \frac{\pi}{2}$ indicated in Figure 10.56, which shows the flow around a circle in the first quadrant. *Hint:* Use the conformal mapping $w = z^2$.

   (b) Show that the speed at the point $z = e^{i\theta}$ on the quarter-circle $r = 1$, $0 < \theta < \frac{\pi}{2}$ is given by $\mathbf{V} = 4A|\sin 2\theta|$.

   (c) Determine the stream function for the flow and sketch several streamlines.

7. Show that $F(z) = \sin z$ is the complex potential for the ideal fluid flow inside the semi-infinite strip $-\frac{\pi}{2} < x < \frac{\pi}{2}$, $y > 0$ indicated in Figure 10.57. Find the stream function.

347

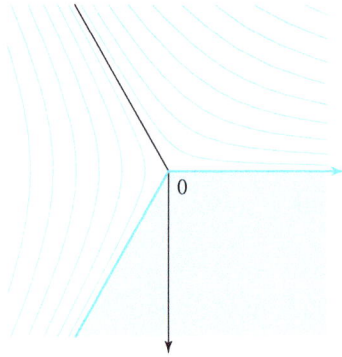

Figure 10.55: For Exercise 5b

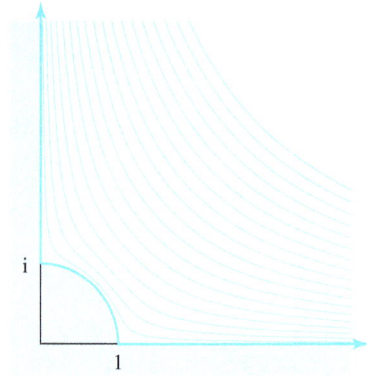

Figure 10.56: For Exercise 6a

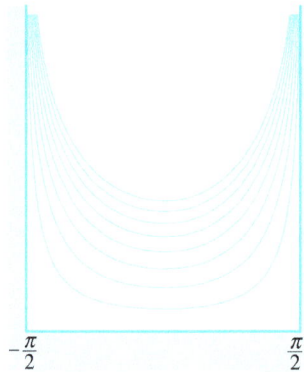

Figure 10.57: For Exercise 7

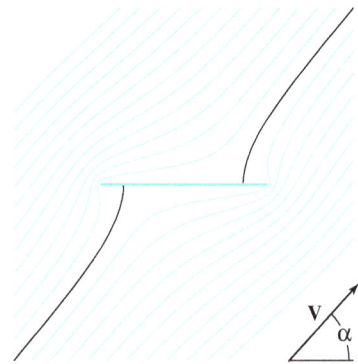

Figure 10.58: For Exercise 8

8. Let $w = S(z) = \frac{1}{2}[z + (z^2 - 4)^{\frac{1}{2}}]$ denote the branch of the inverse of $z = w + \frac{1}{w}$ that is a one-to-one mapping of the z plane slit along the segment $-2 \le x \le 2$, $y = 0$ onto the domain $|w| > 1$. Use the complex potential $F_2(w) = we^{-i\alpha} + \frac{1}{w}e^{i\alpha}$ in the w plane to show that the complex potential $F_1(z) = z\cos\alpha - i(z^2 - 4)^{\frac{1}{2}}\sin\alpha$ determines the ideal fluid flow around the segment $-2 \le x \le 2$, $y = 0$, where the velocity at points distant from the origin is given by $\mathbf{V} \approx e^{i\alpha}$, as shown in Figure 10.58.

9. Consider the complex potential $F(z) = -i\,\mathrm{Arcsin}(z)$

   (a) Show that $F(z)$ determines the ideal fluid flow through the aperture from -1 to $+1$, as indicated in Figure 10.59.

   (b) Show that the streamline $\psi(x, y) = c$ for the flow is a portion of the hyperbola $\frac{x^2}{\sin^2 c} - \frac{y^2}{\cos^2 c} = 1$.

10.8   The Joukowski Airfoil

The Russian scientist N. E. Joukowski studied the function

$$J(z) = z + \frac{1}{z}.$$

He showed that the image of a circle passing through $z_1 = 1$ and containing the point $z_2 = -1$ is mapped onto a curve shaped like the cross section of an airplane wing. We call this curve the **Joukowski airfoil**. If the streamlines for a flow around the circle are known, then their

348

Figure 10.59: For Exercise 9a

images under the mapping $w = J(z)$ will be streamlines for a flow around the Joukowski airfoil, as shown in Figure 10.60.

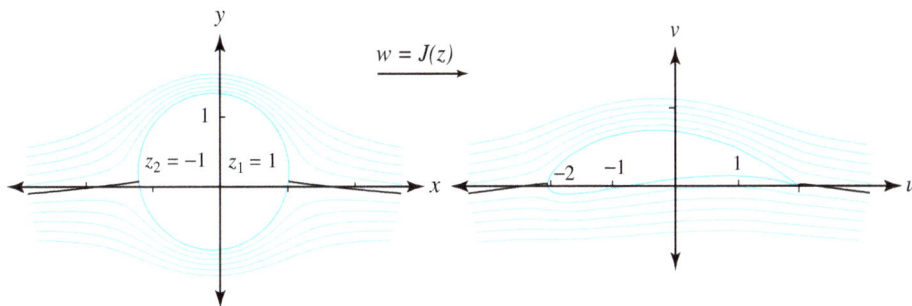

Figure 10.60: Image of a fluid flow under $w = J(z) = z + \frac{1}{z}$

The mapping $w = J(z)$ is two-to-one, because $J(z) = J(\frac{1}{z})$, for $z \neq 0$. The region $|z| > 1$ is mapped one-to-one onto the w plane slit along the portion of the real axis $-2 \leq u \leq 2$. To visualize this mapping, we investigate the implicit form, which we obtain by using the substitutions

$$w - 2 = z - 2 + \frac{1}{z} = \frac{z^2 - 2z + 1}{z} = \frac{(z-1)^2}{z}, \quad \text{and}$$

$$w + 2 = z + 2 + \frac{1}{z} = \frac{z^2 + 2z + 1}{z} = \frac{(z+1)^2}{z}.$$

Forming the quotient of these two quantities results in the relationship

$$\frac{w-2}{w+2} = \left(\frac{z-1}{z+1}\right)^2.$$

The inverse of $T(w) = \frac{w-2}{w+2}$ is $S_3(z) = \frac{2+2z}{1-z}$. If we use the notation $S_1(z) = \frac{z-1}{z+1}$ and $S_2(z) = z^2$, then we can express $J(z)$ as the composition of S_1, S_2, and S_3:

$$w = J(z) = S_3\Big(S_2\big(S_1(z)\big)\Big). \tag{10.36}$$

We can easily show that $w = J(z) = z + \frac{1}{z}$ maps the four points $z_1 = -i$, $z_2 = 1$, $z_3 = i$, and $z_4 = -1$ onto $w_1 = 0$, $w_2 = 2$, $w_3 = 0$, and $w_4 = -2$, respectively. However, the composition

349

functions in Equation (10.36) must be considered in order to visualize the geometry involved. First, the bilinear transformation $Z = S_1(z)$ maps the region $|z| > 1$ onto the right half-plane $\text{Re}(Z) > 0$, and the points $z_1 = -i$, $z_2 = 1$, $z_3 = i$, and $z_4 = -1$ are mapped onto $Z_1 = -i$, $Z_2 = 0$, $Z_3 = i$, and $Z_4 = i\infty$, respectively. Second, the function $W = S_2(Z)$ maps the right half plane onto the W plane slit along its negative real axis, and the points $Z_1 = -i$, $Z_2 = 0$, $Z_3 = i$, and $Z_4 = i\infty$, are mapped onto $W_1 = -1$, $W_2 = 0$, $W_3 = -1$, and $W_4 = -\infty$, respectively. Then the bilinear transformation $w = S_3(W)$ maps the latter region onto the w plane slit along the portion of the real axis $-2 \le u \le 2$, and the points $W_1 = -1$, $W_2 = 0$, $W_3 = -1$, and $W_4 = -\infty$ are mapped onto $w_1 = 0$, $w_2 = 2$, $w_3 = 0$, and $w_4 = -2$, respectively. These three compositions are shown in Figure 10.61.

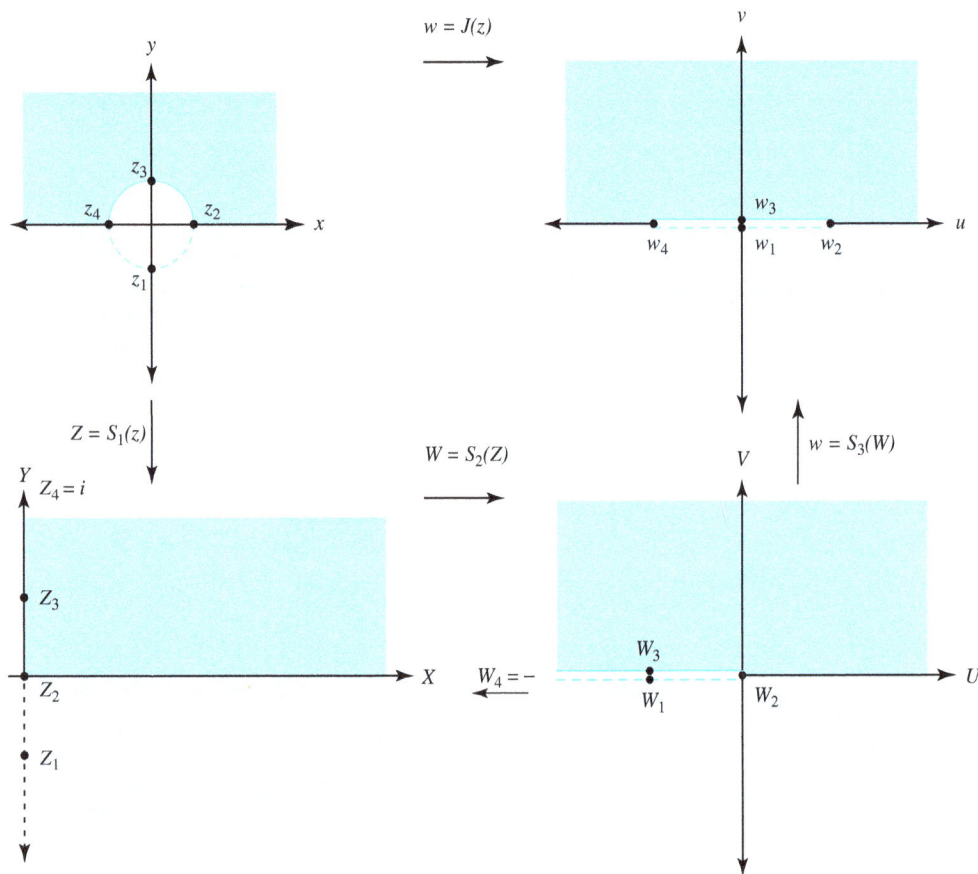

Figure 10.61: The composition mappings for $J(z) = S_3\Big(S_2\big(S_1(z)\big)\Big)$

The circle C_0 with center $c_0 = ia$ on the imaginary axis passes through the points $z_2 = 1$ and $z_4 = -1$ and has radius $r_0 = \sqrt{1 + a^2}$. With the restriction that $0 < a < 1$, then this circle intersects the x axis at the point z_2 with angle $\alpha_0 = \frac{\pi}{2} - \text{Arctan}(a)$, with $\frac{\pi}{4} < \alpha_0 < \frac{\pi}{2}$. We want to track the image of C_0 in the Z, W, and w planes. First, the image of this circle C_0 under $Z = S_1(z)$ is the line L_0 that passes through the origin and is inclined at the angle α_0. Second, the function $W = S_2(Z)$ maps the line L_0 onto the ray R_0 inclined at the angle $2\alpha_0$. Finally, the transformation given by $w = S_3(W)$ maps the ray R_0 onto the arc of the circle A_0 that passes through the points $w_2 = 2$ and $w_4 = -2$ and intersects the x axis at w_2 with angle $2\alpha_0$, where $\frac{\pi}{2} < 2\alpha_0 < \pi$. The restriction on the angle α_0, and hence $2\alpha_0$, is necessary in order for the arc A_0 to have a low profile. The arc A_0 lies in the center of the Joukowski airfoil and is shown in Figure 10.62.

350

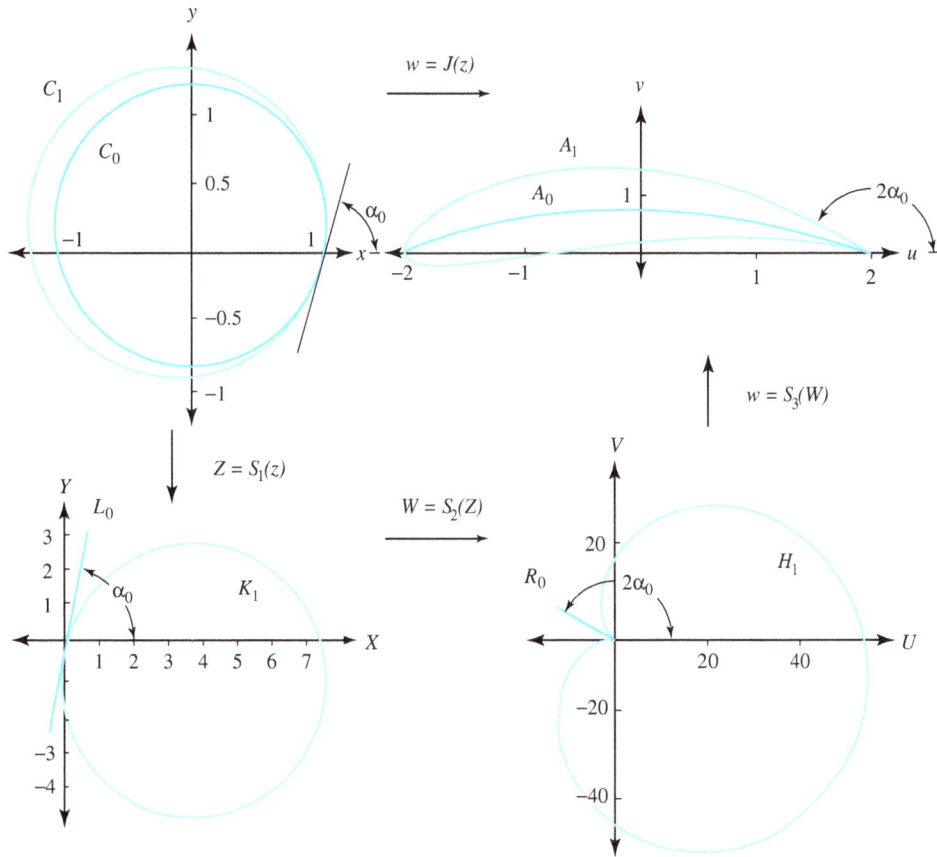

Figure 10.62: The images of the circles C_0 and C_1 under the mapping $J(z) = S_3\big(S_2\big(S_1(z)\big)\big)$

If we let b be fixed, $0 < b < 1$, then the larger circle C_1 with center given by $c_1 = -h + i(1+h)b$ on the imaginary axis will pass through the points $z_2 = 1$ and $z_4 = -1 - 2h$ and have radius $r_1 = (1+h)\sqrt{1+b^2}$. The circle C_1 also intersects the x axis at the point z_2 at the angle α_0. The image of circle C_1 under $Z = S_1(z)$ is the circle K_1, which is tangent to L_0 at the origin. The function $W = S_2(Z)$ maps the circle K_1 onto the cardioid H_1. Finally, $w = S_3(W)$ maps the cardioid H_1 onto the Joukowski airfoil A_1 that passes through the point $w_2 = 2$ and surrounds the point $w_4 = -2$, as shown in Figure 10.62. An observer traversing C_1 counterclockwise will traverse the image curves K_1 and H_1 clockwise but will traverse A_1 counterclockwise. Thus the points z_4, Z_4, W_4, and w_4 will always be to the observer's left.

Now we are ready to visualize the flow around the Joukowski airfoil. We start with the fluid flow around a circle (see Figure 10.51). This flow is adjusted with a linear transformation $z^* = az + b$ so that it flows horizontally around the circle C_1, as shown in Figure 10.63. Then the mapping $w = J(z^*)$ creates a flow around the Joukowski airfoil, per Figure 10.64.

10.8.1 Flow with Circulation

The function $F(z) = sz + \frac{s}{z} + \frac{k}{2\pi i}\mathrm{Log}(z)$, where $s > 0$ and k is real, is the complex potential for a uniform horizontal flow past the unit circle $|z| = 1$, with circulation strength k and velocity at infinity $V_\infty = s$. For illustrative purposes, we let $s = 1$ and use the substitution $a = \frac{-k}{2\pi}$.

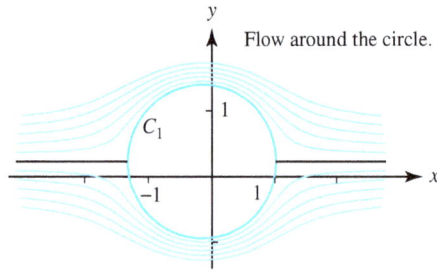

Figure 10.63: Flow around the circle

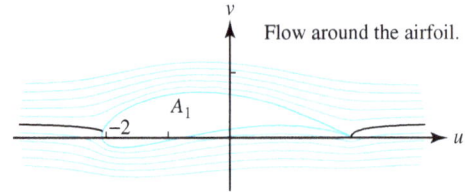

Figure 10.64: Flow around the Joukowski airfoil

Now the complex potential has the form

$$F(z) = z + \frac{1}{z} + ai\mathrm{Log}(z), \tag{10.37}$$

and the corresponding velocity function is

$$V(x,y) = \overline{F,'(z)} = 1 - (\overline{z})^{-2} - ai(\overline{z})^{-1}.$$

We can express the complex potential in $F = \phi + i\psi$ form:

$$F(z) = re^{i\theta} + \frac{1}{r}e^{i\theta} + ia(\ln r + i\theta)$$

$$= \left(r + \frac{1}{r}\right)\cos\theta - a\theta + i\left[\left(r - \frac{1}{r}\right)\sin\theta + a\ln r\right].$$

For the flow given by $\psi = c$, where c is a constant, we have

$$\psi(r\cos\theta, r\sin\theta) = \left(r - \frac{1}{r}\right)\sin\theta + a\ln r = c. \quad \text{(Streamlines)}.$$

Setting $r = 1$ in this equation, we get $\psi(\cos\theta, \sin\theta) = 0$ for all θ, so the unit circle is a natural boundary curve for the flow.

Points at which the flow has zero velocity are called *stagnation points*. To find them we solve $F,'(z) = 0$: For the function in Equation 10.37 we have $1 - \frac{1}{z^2} + \frac{a}{z} = 0$. Multiplying through by z^2 and rearranging terms gives $z^2 + aiz - 1 = 0$. Now we invoke the quadratic equation to obtain

$$z = \frac{-ai \pm \sqrt{4 - z^2}}{2} \quad \text{(Stagnation point(s))}.$$

If $0 \leq |a| < 2$, then there are two stagnation points on the unit circle $|z| = 1$. If $a = 2$, then there is one stagnation point on the unit circle. If $|a| > 2$, then the stagnation point lies outside the unit circle. We are mostly interested in the case with two stagnation points. When $a = 0$, the two stagnation points are $z = \pm 1$, which is the flow discussed in Example 10.25. The cases $a = 1$, $a = \sqrt{3}$, $a = 2$, and $a = 2.2$ are shown in Figure 10.65.

We are now ready to combine the preceding ideas. For illustrative purposes, we consider a C_1 circle with center $c_0 = -0.15 + 0.23i$ that passes through the points $z_2 = 1$ and $z_4 = -1.3$ and has radius $r_0 = 0.23\sqrt{\frac{13}{2}}$. We use the linear transformation $Z = S(z) = -0.15 + 0.23i + r_0 z$ to map the flow with circulation $k = -0.52p$ (or $a = 0.26$) around $|z| = 1$ onto the flow around the circle C_1, as shown in Figure 10.66.

Flow with circulation $a = 1$

Flow with circulation $a = 3^{\frac{1}{2}}$

Flow with circulation $a = 2$

Flow with circulation $a = 2.2$

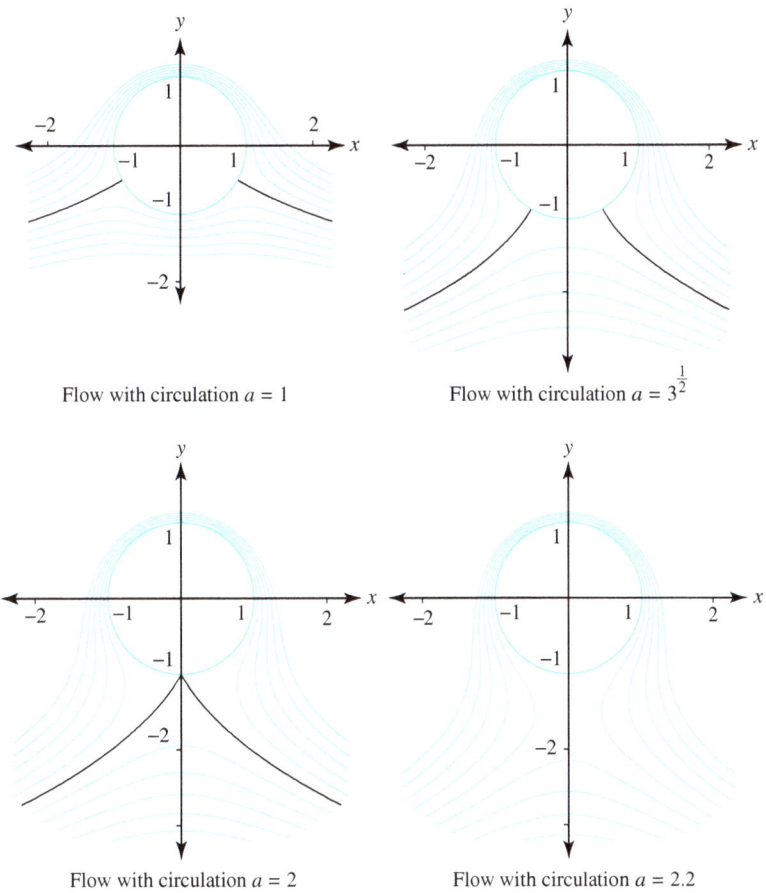

Figure 10.65: Flows past the unit circle with circulation a

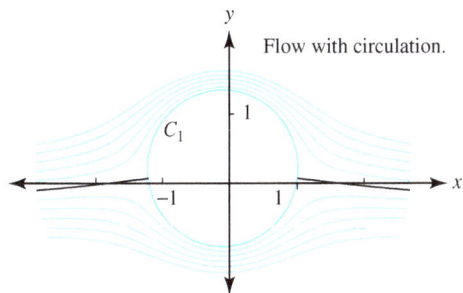

Flow with circulation.

Figure 10.66: Flow around C_1

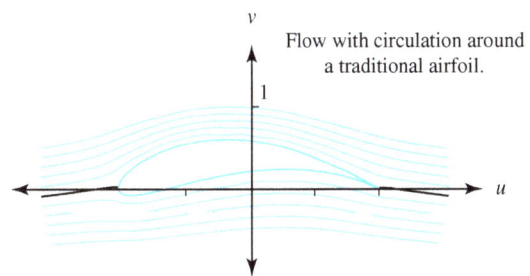

Flow with circulation around a traditional airfoil.

Figure 10.67: Flow around a Joukowski airfoil

Then we use the mapping $w = J(Z) = Z + \frac{1}{Z}$ to map this flow around the Joukowski airfoil, as shown in Figure 10.67 and compare it to the flows shown in Figures 10.63 and 10.64. If the second transformation in the composition given by $w = J(z) = S_3\Big(S_2\big(S_1(z)\big)\Big)$ is modified to be $S_2(z) = z^{1.925}$, then the image of the flow shown in Figure 10.66 will be the flow around the modified airfoil shown in Figure 10.68. The advantage of this latter airfoil is that the sides of its tailing edge form an angle of 0.15π radians, or $27°$, which is more realistic than the angle of $0°$ of the traditional Joukowski airfoil.

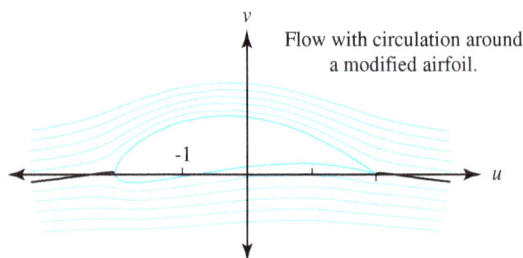

Figure 10.68: Flow with circulation around a modified Joukowski airfoil

Exercises for Section 10.8 (Selected answers or hints are on page 455.)

1. Find the inverse of the Joukowski transformation.

2. Consider the Joukowski transformation $w = z + \frac{1}{z}$.

   (a) Show that the circles $C_r = t\{|z| = r : r > 1\}$ are mapped onto the ellipses

   $$\frac{u^2}{\left(r + \frac{1}{r}\right)^2} + \frac{v^2}{\left(r - \frac{1}{r}\right)} = 1.$$

   (b) Show that the ray $r > 0, \theta = \alpha$ is mapped onto a branch of the hyperbola

   $$\frac{u^2}{4\cos^2\alpha} - \frac{v^2}{4\sin^2\alpha} = 1.$$

3. Let C_0 be a circle that passes through the points 1 and -1 and has center $c_0 = ia$.

   (a) Find the equation of the circle C_0.
   (b) Show that the image of the circle C_0 under $w = \frac{z-1}{z+1}$ is a line L_0 that passes through the origin.
   (c) Show that the line L_0 is inclined at the angle $\alpha_0 = \frac{\pi}{2} - \text{Arctan}(a)$.

4. Show that a line through the origin is mapped onto a ray by the mapping $w = z^2$.

5. Let R_0 be a ray through the origin inclined at an angle β_0.

   (a) Show that the image of the ray R_0 under $w = \frac{2+2z}{1-z}$ is an arc A_0 of a circle that passes through 2 and -2.
   (b) Show that the arc A_0 is inclined at the angle β_0.

6. Show that a circle passing through the origin is mapped onto a cardioid by $w = z^2$. Show that the cusp in the cardioid forms an angle of $0°$.

7. Let H_1 be a cardioid whose cusp is at the origin. The image of H_1 under $w = \frac{2+2z}{1-z}$ will be a Joukowski airfoil. Show that trailing edge forms an angle of $0°$.

8. Consider the modified Joukowski airfoil when $W = S_2(Z) = Z^{1.925}$ is used to map the Z plane onto the W plane. Refer to Figure 10.69 and discuss why the angle of the trailing edge of the modified Joukowski airfoil A_1 forms an angle of 0.15π radians.
   Hint: The image of the circle C_0 is the line L_0, then two rays $R_{0.1}$ and $R_{0.2}$, and then two arcs $A_{0,1}$ and $A_{0,2}$ in the respective Z, W, and w planes. The image of the circle C_1 is the circle K_1, then the "cardioid like" curve H_1, then the modified Joukowski airfoil A_1.

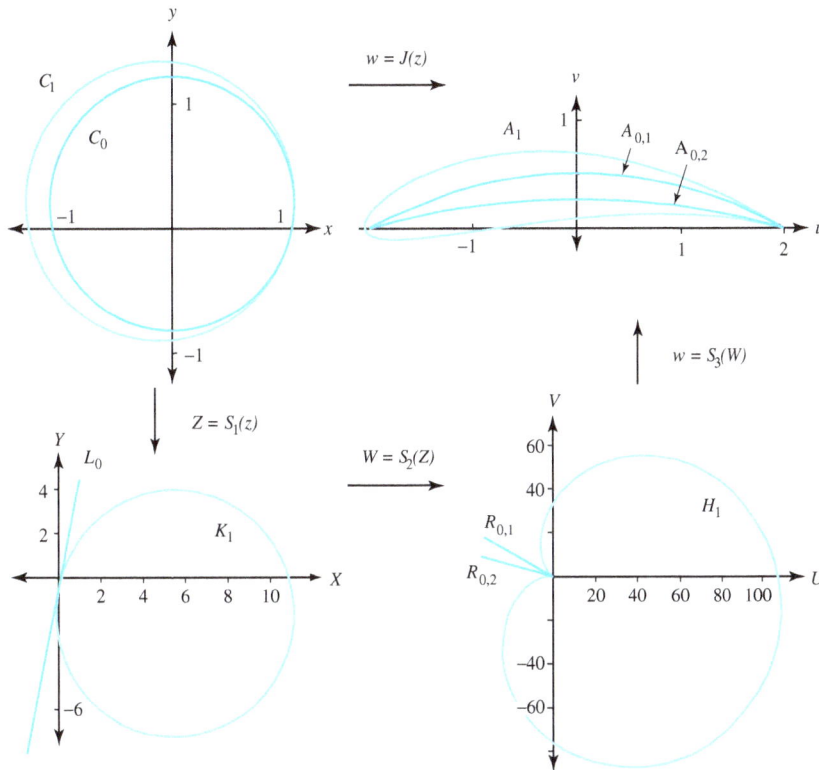

Figure 10.69: For Exercise 8

10.9   The Schwarz-Christoffel Transformation

To proceed further, we must review the rotational effect of a conformal mapping $w = f(z)$ at a point z_0. If the contour C has the parameterization $z(t) = x(t) + iy(t)$, then a vector τ tangent to C at the point z_0 is

$$\tau = z'(t_0) = x'(t_0) + iy'(t_0).$$

The image of C is a contour K given by $w = u\big(x(t), y(t)\big) + iv\big(x(t), y(t)\big)$, and a vector $\mathbf{T}$ tangent to K at the point $w_0 = f(z_0)$ is

$$\mathbf{T} = w'(z_0) = f'(z_0)z'(t_0).$$

If the angle of inclination of τ is $\beta = \mathrm{Arg}(z)'(t_0)$, then the angle of inclination of $\mathbf{T}$ is

$$\mathrm{Arg}(\mathbf{T}) - \mathrm{Arg}[f'(z_0)z'(t_0)] = \mathrm{Arg}\big(f'(z_0)\big) + \beta.$$

Hence the angle of inclination of the tangent τ to C at z_0 is rotated through the angle $\mathrm{Arg}\big(f'(z_0)\big)$ to obtain the angle of inclination of the tangent $\mathbf{T}$ to K at the point w_0.

Many applications involving conformal mappings require the construction of a one-to-one conformal mapping from the upper half-plane $\mathrm{Im}(z) > 0$ onto a domain G in the w plane where the boundary consists of straight-line segments. Let's consider the case where G is the interior of a polygon P with vertices $w_1, w_2, \ldots, w_n$ specified in the positive sense (counterclockwise). We want to find a function $w = f(z)$ with the property

$$w_k = f(x_k) \quad \text{for} \quad k = 1, 2, \ldots, n - 1, \quad \text{and} \tag{10.38}$$
$$w_n = f(\infty) \quad \text{where} \quad x_1 < x_2 < \cdots < x_{n-1} < \infty.$$

355

Two German mathematicians Herman Amandus Schwarz (1843–1921) and Elwin Bruno Christoffel (1929–1900) independently discovered a method for finding f, which we present as Theorem 10.6.

Theorem 10.6 (Schwarz-Christoffel). *Let P be a polygon in the w plane with vertices $w_1, w_2, \ldots, w_n$ and exterior angles α_k, where $-\pi < \alpha_k < \pi$. There exists a one-to-one conformal mapping $w = f(z)$ from the upper half-plane $\mathrm{Im}(z) > 0$ onto G that satisfies the boundary conditions in Equations 10.38. The derivative $f'(z)$ is*

$$f'(z) = A(z - x_1)^{-\frac{\alpha_1}{\pi}}(z - x_2)^{-\frac{\alpha_2}{\pi}}\cdots(z - x_{n-1})^{-\frac{\alpha_{n-1}}{\pi}}, \tag{10.39}$$

and the function f can be expressed as an indefinite integral

$$f(z) = B + A\int(z - x_1)^{-\frac{\alpha_1}{\pi}}(z - x_2)^{-\frac{\alpha_2}{\pi}}\cdots(z - x_{n-1})^{-\frac{\alpha_{n-1}}{\pi}}\,dz, \tag{10.40}$$

where A and B are suitably chosen constants. Two of the points $\{x_k\}$ may be chosen arbitrarily, and the constants A and B determine the size and position of P.

Proof. The proof relies on finding how much the tangent

$$\tau_j = 1 + 0i$$

(which always points to the right) at the point $(x, 0)$ must be rotated by the mapping $w = f(z)$ so that the line segment $x_{j-1} < x < x_j$ is mapped onto the edge of P that lies between the points $w_{j-1} = f(x_{j-1})$ and $w_j = f(x_j)$. The amount of rotation is determined by $\mathrm{Arg}\,f'(x)$, so Equation (10.39) specifies $f'(z)$ in terms of the values x_j and the amount of rotation α_j that is required at the vertex $f(x_j)$. □

If we let $x_0 = -\infty$ and $x_n = \infty$, then, for values of x that lie in the interval $x_{j-1} < x < x_j$, the amount of rotation is

$$\mathrm{Arg}\,f'(x) = \mathrm{Arg}(A) - \frac{1}{\pi}\Big[\alpha_1\mathrm{Arg}(x - x_1) + \alpha_2\mathrm{Arg}(x - x_2)$$
$$+ \cdots + \alpha_{n-1}\mathrm{Arg}(x - x_{n-1})\Big]$$

Because $\mathrm{Arg}(x - x_k) = 0$ for $1 \le k < j$, and $\mathrm{Arg}(x - x_k) = \pi$ for $j \le k \le n - 1$, we can write this equation as

$$\mathrm{Arg}\big(f'(x)\big) = \mathrm{Arg}(A) - \alpha_j - \alpha_{j+1} - \cdots - \alpha_{n-1}.$$

The angle of inclination of the tangent vector $\mathbf{T}_j$ to the polygon P at the point $w = f(x)$ for $x_{j-1} < x < x_j$ is

$$\gamma_j = \mathrm{Arg}(A) - \alpha_j - \alpha_{j+1} - \cdots - \alpha_{n-1}.$$

The angle of inclination of the tangent vector $\mathbf{T}_{j+1}$ to the polygon P at the point $w = f(x)$, for $x_j < x < x_{j+1}$, is

$$\gamma_{j+1} = \mathrm{Arg}(A) - \alpha_{j+1} - \alpha_{j+2} - \cdots - \alpha_{n-1}.$$

The angle of inclination of the vector tangent to the polygon P jumps abruptly by the amount α_j as the point $w = f(x)$ moves along the side $\widehat{w_{j-1}w_j}$ through the vertex w_j to the side $\widehat{w_jw_{j+1}}$. Therefore the exterior angle to the polygon P at the vertex w_j is given by the angle α_j and satisfies the inequality $-\pi < \alpha_j < \pi$, for $j = 1, 2, \ldots, n-1$. Since the sum of the exterior angles of a polygon equals 2π, we have $\alpha_n = 2\pi - \alpha_1 - \alpha_2 - \cdots - \alpha_{n-1}$ and only $n-1$ angles need to be specified. The case $n = 5$ is illustrated in Figure 10.70.

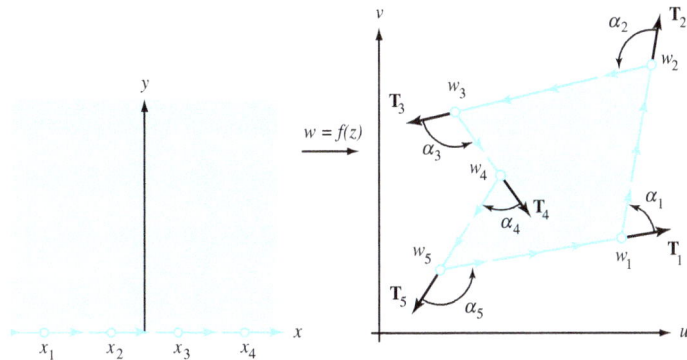

Figure 10.70: A Schwarz-Christoffel mapping with $n = 5$ and $\alpha_1 + \alpha_2 + \cdots + \alpha_4 > \pi$

If the case $\alpha_1 + \alpha_2 + \cdots + \alpha_{n-1} \le \pi$ occurs, then $\alpha_n > \pi$, and the vertices $w_1, w_2, \ldots, w_n$ cannot form a closed polygon. For this case, Equations (10.39) and (10.40) will determine a mapping from the upper half plane $\text{Im}(z) > 0$ onto an infinite region in the w plane, where the vertex w_n is at infinity. The case $n = 5$ is illustrated in Figure 10.71.

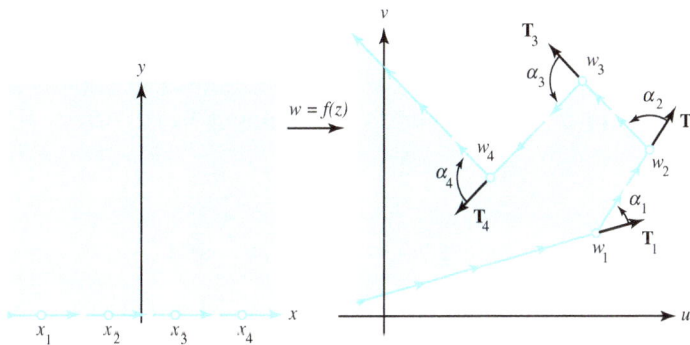

Figure 10.71: A Schwarz-Christoffel mapping with $n = 5$ and $\alpha_1 + \alpha_2 + \cdots + \alpha_4 \le \pi$

Equation (10.40) gives a representation for f in terms of an indefinite integral. Note that these integrals do not represent elementary functions unless the image is an infinite region. Also, the integral will involve a multivalued function, and we must select a specific branch to fit the boundary values specified in the problem. Table 10.2, which appears at the end of this chapter, is useful for our purposes.

Example 10.26. Use the Schwarz-Christoffel formula to verify that the function $w = f(z) = \text{Arcsin}\,z$ maps the upper half-plane $\text{Im}(z) > 0$ onto the semi-infinite strip $-\frac{\pi}{2} < u < \frac{\pi}{2}$, $v > 0$ shown in Figure 10.72.

Solution:

If we choose $x_1 = -1$, $x_2 = 1$, $w_1 = -\frac{\pi}{2}$, and $w_2 = \frac{\pi}{2}$, then $\alpha_1 = \frac{\pi}{2}$ and $\alpha_2 = \frac{\pi}{2}$, and Equation

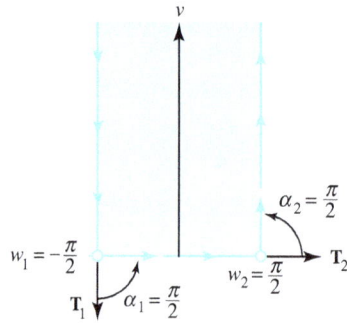

Figure 10.72: The region of interest

(10.39) for $f'(z)$ becomes

$$f'(z) = A(z+1)^{-(\pi/2)/\pi}(z-1)^{-(\pi/2)/\pi} = \frac{A}{(z^2-1)^{\frac{1}{2}}}.$$

Then, using Table 10.2, the indefinite integral becomes

$$f(z) = i(A)\mathrm{Arcsin}(z) + B.$$

Using the image values $f(-1) = -\frac{\pi}{2}$ and $f(1) = \frac{\pi}{2}$, we obtain the system

$$-\frac{\pi}{2} = A\left(-\frac{\pi}{2}\right) + B \quad \text{and} \quad \frac{\pi}{2} = A\left(i\frac{\pi}{2}\right) + B.$$

Solving gives $B = 0$ and $A = -i$. Hence, the required function is

$$f(z) = \mathrm{Arcsin}(z).$$

Example 10.27. Verify that $w = f(z) = (z^2 - 1)^{\frac{1}{2}}$ maps the upper half-plane $\mathrm{Im}(z) > 0$ onto the upper half-plane $\mathrm{Im}(w) > 0$ slit along the segment from 0 to i. (Use the principal square root throughout.)

Solution:

If we choose $x_1 = -1, x_2 = 0, x_3 = 1, w_1 = -d, w_2 = i$, and $w_3 = d$, then the formula

$$g'(z) = A(z+1)^{-\frac{\alpha_1}{\pi}}\left(z^{-\frac{\alpha_2}{\pi}}\right)(z-1)^{-\frac{\alpha_3}{\pi}}$$

will determine a mapping $w = g(z)$ from the upper half-plane $\mathrm{Im}(z) > 0$ onto the portion of the upper half-plane $\mathrm{Im}(w) > 0$ that lies outside the triangle with vertices $\pm d, i$ as indicated in Figure 10.73(a). If we let $d \to 0$, then $w_1 \to 0$, $w_3 \to 0$, $\alpha_1 \to \frac{\pi}{2}$, $\alpha_2 \to -\pi$, and $\alpha_3 \to \frac{\pi}{2}$. The limiting formula for the derivative $g'(z)$ becomes

$$f'(z) = A(z+1)^{-\frac{1}{2}}(z)(z-1)^{-\frac{1}{2}},$$

which will determine a mapping $w = f(z)$ from the upper half-plane $\mathrm{Im}(z) > 0$ onto the upper half-plane $\mathrm{Im}(w) > 0$ slit from 0 to i as indicated in Figure 10.73(b).

An easy computation reveals that $f(z)$ is given by

$$f(z) = A\int \frac{z}{(z^2-1)^{\frac{1}{2}}}\, dz = A(z^2-1)^{\frac{1}{2}} + B,$$

and the boundary values $f(\pm 1) = 0$ and $f(0) = i$ lead to the solution

$$f(z) = (z^2 - 1)^{\frac{1}{2}}.$$

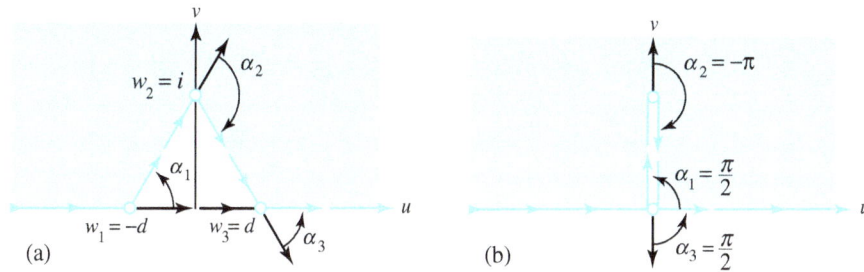

Figure 10.73: The regions of interest

Example 10.28. Show that the function

$$w = f(z) = \frac{1}{\pi}\mathrm{Arcsin}(z) + \frac{i}{\pi}\mathrm{Arcsin}\left(\frac{1}{z} + \frac{1+i}{2}\right)$$

maps the upper half-plane $\mathrm{Im}(z) > 0$ onto the right angle channel in the first quadrant, which is bounded by the coordinate axes and the rays $x \geq 1$, $y = 1$ and $y \geq 1$, $x = 1$, as depicted in Figure 10.74(b).

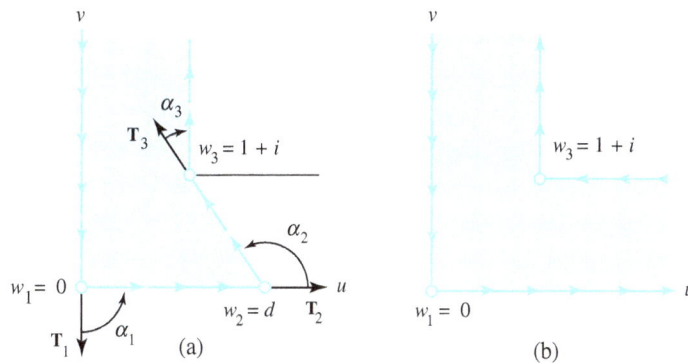

Figure 10.74: The regions of interest

Solution:

If we choose $x_1 = -1, x_2 = 0, x_3 = 1, w_1 = 0, w_2 = d$, and $w_3 - 1 + i$, then the formula

$$g'(z) = A_1(z+1)^{-\frac{\alpha_1}{\pi}}\left(z^{-\frac{\alpha_2}{\pi}}\right)(z-1)^{-\frac{\alpha_3}{\pi}}$$

will determine a mapping $w - g(z)$ of the upper half-plane onto the domain that is indicated in Figure 10.74(a). With $\alpha_1 = \frac{\pi}{2}$, we let $d \to \infty$, then $\alpha_2 \to \pi$ and $\alpha_3 \to -\frac{\pi}{2}$, and the limiting formula for the derivative $g'(z)$ becomes

$$f'(z) = A_1(z+1)^{-(\pi/2)/\pi}(z)^{-(\pi)/\pi}(z-1)^{-(-\pi/2)/\pi}$$

$$= A_1\frac{1}{z}\frac{(z-1)^{\frac{1}{2}}}{(z+1)^{\frac{1}{2}}}$$

$$= A_1\frac{z-1}{z(z^2-1)^{\frac{1}{2}}}$$

$$= A\frac{z-1}{z(1-z^2)^{\frac{1}{2}}},$$

359

where $A = -iA_1$, which will determine a mapping $w = f(z)$ from the upper half plane onto the channel as indicated in Figure 10.74b. Using Table 11.2, we obtain

$$f(z) = A \left[\int \frac{1}{(1 - z^2)^{\frac{1}{2}}} \, dz - i \int \frac{1}{z(z^2 - 1)^{\frac{1}{2}}} \, dz \right]$$
$$= A \left[\mathrm{Arcsin}(z) + i\mathrm{Arcsin}\left(\frac{1}{z}\right) \right] + B.$$

If we use the principal branch of the inverse sine function, then the boundary values $f(-1) = 0$ and $f(1) = 1 + i$ lead to the system

$$A\left(-\frac{\pi}{2} - i\frac{\pi}{2}\right) + B = 0, \quad \text{and} \quad A\left(\frac{\pi}{2} + i\frac{\pi}{2}\right) + B = 1 + i,$$

which we can solve to obtain $A = \frac{1}{\pi}$ and $B = \frac{1+i}{2}$. Hence the required solution is

$$w = f(z) = \frac{1}{\pi}\mathrm{Arcsin}(z) + \frac{i}{\pi}\mathrm{Arcsin}\left(\frac{1}{z}\right) + \frac{1+i}{2}.$$

Exercises for Section 10.9 (Selected answers or hints are on page 456.)

1. Let a and K be real constants with $0 < K < 2$. Use the Schwarz-Christoffel formula to show that the function $w = f(z) = (z - a)^K$ maps the upper half-plane $\mathrm{Im}(z) > 0$ onto the sector $0 < \arg_0 w < K\pi$ shown in Figure 10.75.

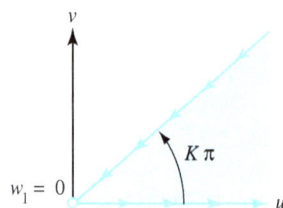

Figure 10.75: For Exercise 1

2. Let a be a real constant. Use the Schwarz-Christoffel formula to show that the function $w = f(z) = \mathrm{Log}(z - a)$ maps the upper half-plane $\mathrm{Im}(z) > 0$ onto the infinite strip $0 < v < \pi$ shown in Figure 10.76.
   Hint: Set $x_1 = a - 1$ $x_2 = a$, $w_1 = i\pi$, and $w_2 = -d$. Then let $d \to \infty$.

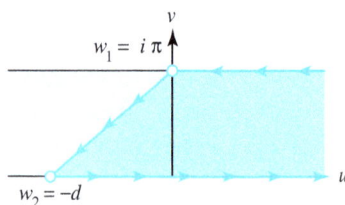

Figure 10.76: For Exercise 2

3. For the remaining exercises, construct the derivative $f'(z)$ and use the Schwarz-Christoffel formula, Equation 10.40, and techniques of integration to complete the required tasks.

(a) Show that $w = f(z) = \frac{1}{\pi}(z^2 - 1)^{\frac{1}{2}} + \frac{1}{\pi}\text{Log}\left[z + (z^2 - 1)^{\frac{1}{2}}\right] - i$ maps the upper half-plane onto the domain indicated in Figure 10.77.
Hint: Set $x_1 = -1$, $x_2 = 1$, $w_1 = 0$, and $w_2 = -i$.

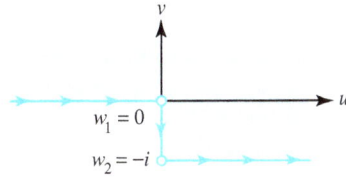

Figure 10.77: For Exercise 3a

(b) Show that $w = f(z) = \frac{2}{\pi}(z^2 - 1)^{\frac{1}{2}} + \frac{2}{\pi}\text{Arcsin}\frac{1}{z}$ maps the upper half-plane onto the domain indicated in Figure 10.78.
Hint: Set $x_1 = w_1 = -1$, $x_2 = 0$, $x_3 = w_3 = 1$, and $w_2 = -id$. Then let $d \to \infty$.

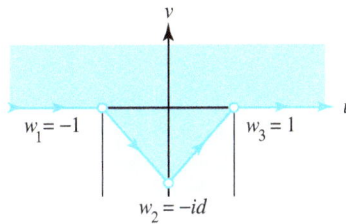

Figure 10.78: For Exercise 3b

(c) Show that $w = f(z) = \frac{1}{2}\text{Log}(z^2 - 1) = \text{Log}\left[(z^2 - 1)^{\frac{1}{2}}\right]$ maps the upper half-plane $\text{Im}(z) > 0$ onto the infinite strip $0 < v < \pi$ slit along the ray $u \le 0, v = \frac{\pi}{2}$, per Figure 10.79.
Hint: Set $x_1 = -1$, $x_2 = 0$, $x_3 = 1$, $w_1 = i\pi - d$, , $w_2 = \frac{i\pi}{2}$, and $w_3 = -d$. Then let $d \to \infty$.

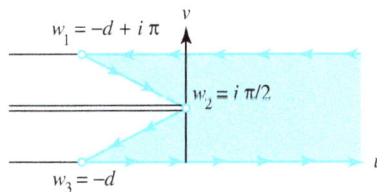

Figure 10.79: For Exercise 3c

(d) Show that $w = f(z) = -\frac{2}{\pi}z(1 - z^2)^{\frac{1}{2}} - \frac{2}{\pi}\text{Arcsin}(z)$ maps the upper half-plane onto the domain indicated in Figure 10.80.
Hint: Set $x_1 = -1$, $x_2 = 1$, $w_1 = 1$, and $w_2 = -1$.

(e) Show that $w = f(z) = z + \text{Log}(z)$ maps the upper half-plane $\text{Im}(z) > 0$ onto the upper half-plane $\text{Im}(w) > 0$ slit along the ray $u \le -1$, $v = \pi$, as shown in Figure 10.81. *Hint*: Set $x_1 = -1$, $x_2 = 0$, $w_1 = -1 + i\pi$, and $w_2 = -d$. Then Let $d \to \infty$.

(f) Show that $w = f(z) = i\pi + 2(z + 1)^{\frac{1}{2}} + \text{Log}\left[\frac{1 - (z+1)^{\frac{1}{2}}}{1 + (z+1)^{\frac{1}{2}}}\right]$ maps the upper half-plane onto the domain indicated in Figure 10.82.
Hint: Set $x_1 = -1$, $x_2 = 0$, $w_1 = i\pi$, and $w_2 = -d$. Then let $d \to \infty$.

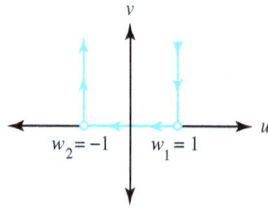

Figure 10.80: For Exercise 3d

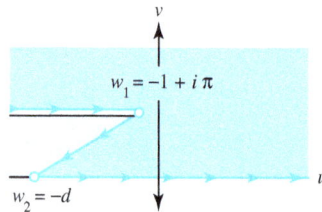

Figure 10.81: For Exercise 3e

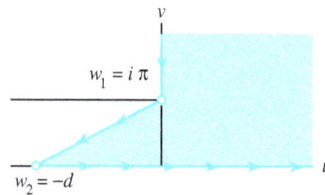

Figure 10.82: For Exercise 3f

(g) Show that $w = f(z) = (z-1)^\alpha \left[\frac{1+\alpha z}{1-\alpha}\right]^{1-\alpha}$ maps the upper half-plane $\mathrm{Im}(z) > 0$ onto the upper half-plane $\mathrm{Im}(w) > 0$ slit along the segment from 0 to $e^{i\alpha\pi}$, per Figure 10.83.

Hint: Show that $f'(z) = A \left[\frac{z+(1-\alpha)}{\alpha}\right]^{-\alpha} (z)(z-1)^{\alpha-1}$.

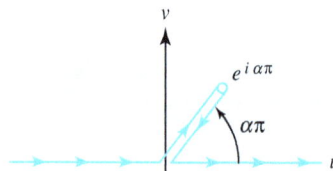

Figure 10.83: For Exercise 3g

(h) Show that $w = f(z) = 4(z+1)^{\frac{1}{4}} + \log \left[\frac{(z+1)^{\frac{1}{4}}-1}{(z+1)^{\frac{1}{4}}+1}\right] + i \log \left[\frac{i-(z+1)^{\frac{1}{4}}}{i+(z+1)^{\frac{1}{4}}}\right]$ maps the upper half-plane onto the domain indicated in Figure 10.84. *Hint:* Set $z_1 = -1$, $z_2 = 0$, $w_1 = i\pi$, and $w_2 = -d$. Then let $d \to \infty$. Use the change of variable $z + 1 = s^4$ in the resulting integral.

(i) Show that $w = f(z) = -\frac{i}{z} z^{\frac{1}{2}}(z-3)$ maps the upper half-plane onto the domain indicated in Figure 10.85. *Hint:* Set $x_1 = 0$, $x_2 = 1$, $w_1 = -d$, and $w_2 = i$ and let $d \to 0$.

(j) Show that $w = f(z) = \int \frac{1}{(1-z^2)^{\frac{3}{4}}} dz$ maps the upper half-plane $\mathrm{Im}(z) > 0$ onto a

362

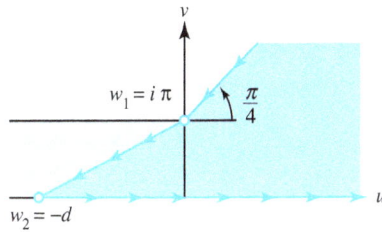

Figure 10.84: For Exercise 3h

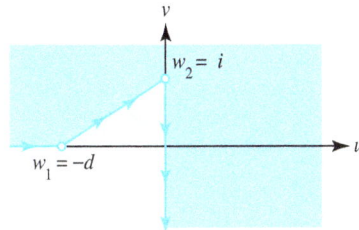

Figure 10.85: For Exercise 3i

right triangle with angles $\frac{\pi}{2}$, $\frac{\pi}{4}$, and $\frac{\pi}{4}$.

(k) Show that $w = f(z) = \int \dfrac{1}{(1-z^2)^{\frac{2}{3}}}\, dz$ maps the upper half-plane onto an equilateral triangle.

(l) Show that $w = f(z) = \int \dfrac{1}{(z-z^3)^{\frac{1}{2}}}\, dz$ maps the upper half-plane onto a square.

(m) Show that $w = f(z) = 2(z+1)^{\frac{1}{2}} - \mathrm{Log}\left[\dfrac{1-(z+1)^{\frac{1}{2}}}{1+(z+1)^{\frac{1}{2}}}\right]$ maps the upper half-plane $\mathrm{Im}(z) > 0$ onto the domain indicated in Figure 10.86.
Hint: Set $x_1 = -1$, $x_2 = 0$, $x_3 = 1$, $w_1 = 0$, $w_2 = d$, and $w_3 = 2\sqrt{2} - 2\ln(\sqrt{2}-1) + i\pi$. Then let $d \to \infty$.

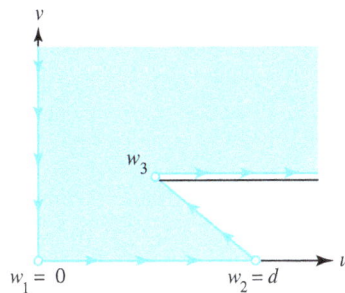

Figure 10.86: For Exercise 3m

10.10   Image of a Fluid Flow

We have already examined several two-dimensional fluid flows and have shown that the image of a flow under a conformal transformation is a flow. The conformal mapping $w = f(z) = u(x,y) + iv(x,y)$, which we obtained by using the Schwarz-Christoffel formula, allows us to find the streamlines for flows in domains in the w plane that are bounded by straight-line segments.

363

The first technique involves finding the image of a fluid flowing horizontally from left to right across the upper half-plane $\text{Im}(z) > 0$. The image of the streamline $-\infty < t < \infty$, $y = c$ is a streamline given by the parametric equations

$$u = u(t, c) \quad \text{and} \quad u = v(t, c), \quad \text{for} \quad -\infty < t < \infty$$

and is oriented in the positive sense (counterclockwise). The streamline $u = u(t, 0)$, $v = (t, 0)$ is considered to be a boundary wall for a containing vessel for the fluid flow.

Example 10.29. Consider the conformal mapping

$$w = f(z) = \frac{1}{\pi}(z^2 - 1)^{\frac{1}{2}} + \frac{1}{\pi}\text{Log}\left[z + (z^2 - 1)^{\frac{1}{2}}\right],$$

which we obtained by using the Schwarz-Christoffel formula. It maps the upper half-plane $\text{Im}(z) > 0$ onto the domain in the w plane that lies above the boundary curve consisting of the rays $u \leq 0$, $v = 1$ and $u \geq 0$, $v = 0$ and the segment $u = 0$, $-1 \leq v \leq 0$. Furthermore, the image of horizontal streamlines in the z plane are curves in the w plane given by the parametric equation

$$w = f(t + ic) = \frac{1}{\pi}(t^2 - c^2 - 1 + i2ct)^{\frac{1}{2}}$$
$$+ \frac{1}{\pi}\text{Log}\left[t + ic + (t^2 - c^2 - 1 + i2ct)^{\frac{1}{2}}\right],$$

for $-\infty < t < \infty$. The new flow is that of a step in the bed of a deep stream and is illustrated in Figure 10.87(a). The function $w = f(z)$ is also defined for values of z in the lower half-plane, and the images of horizontal streamlines that lie above or below the x axis are mapped onto streamlines that flow past a long rectangular obstacle, which is illustrated in Figure 10.87(b).

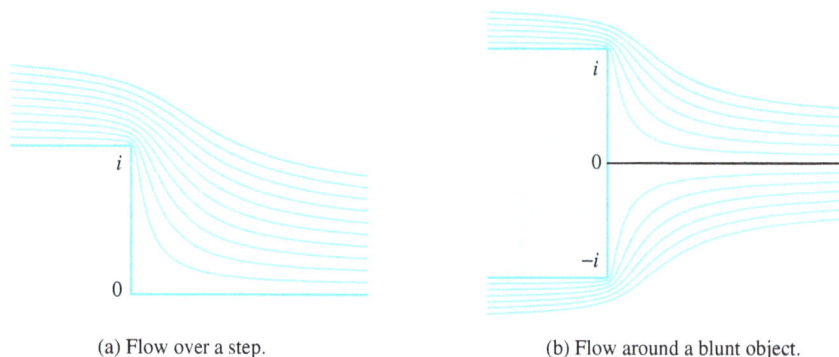

(a) Flow over a step.   (b) Flow around a blunt object.

Figure 10.87: Streamline mappings

Exercises for Section 10.10 (Selected answers or hints are on page 456.)

For Exercises 1-4, use the Schwarz-Christoffel formula to find a conformal mapping $w = f(z)$ that will map the flow in the upper half-plane $\text{Im}(z) > 0$ onto the flows indicated.

1. Use Figure 10.88 to find the flow over the vertical segment from 0 to i.

2. Use Figure 10.89 to find the flow around an infinitely long rectangular barrier.

3. Use Figure 10.90 to find the flow around

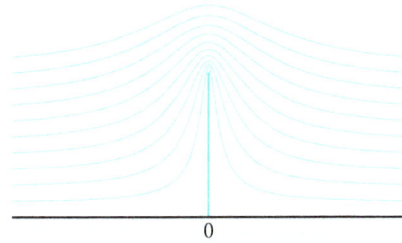

Figure 10.88: For Exercise 1

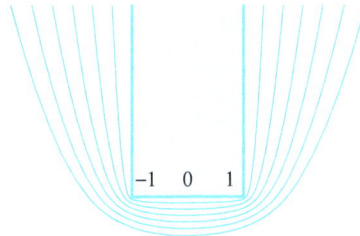

Figure 10.89: For Exercise 2

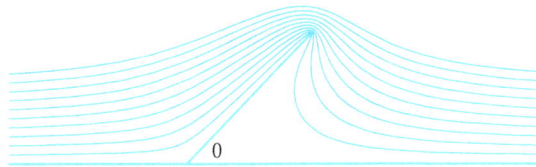

(a) Flow around an inclined segment.

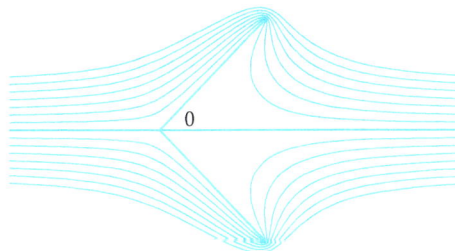

(b) Flow around a V-shape.

Figure 10.90: For Exercise 3

  (a) one inclined segment in the upper half-plane.

  (b) two inclined segments forming a "V" in the plane.

4. Use Figure 10.91 to find the flow over a dam.

5. For flow around an infinitely long rectangular barrier with a pointed "nose," find

  (a) the flow up an inclined step, as shown in Figure 10.92(a).

  (b) the flow around a pointed object, as shown in Figure 10.92(b).

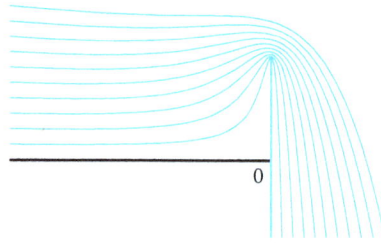

Flow over a dam.

Figure 10.91: For Exercise 4

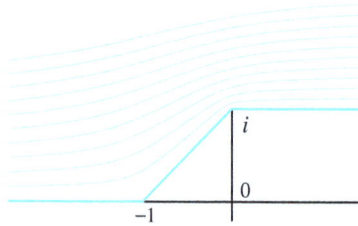

(a) Flow up an inclined step.

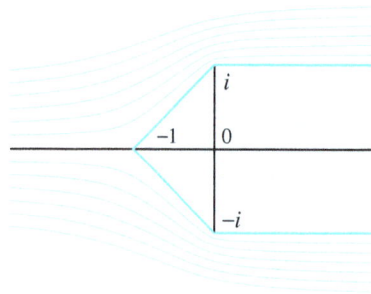

(b) Flow around a pointed object.

Figure 10.92: For Exercise 5

10.11 Sources and Sinks

If the two-dimensional motion of an ideal fluid consists of an outward radial flow from a point and is symmetrical in all directions, then the point is called a *simple source*. A source at the origin can be considered as a line perpendicular to the z plane along which fluid is being emitted. If the rate of emission of volume of fluid per unit length is $2\pi m$, then the origin is said to be a source of strength m, the complex potential for the flow is

$$F(z) = m \log z,$$

and the velocity $\mathbf{V}$ at the point (x, y) is given by

$$\mathbf{V}(x, y) = \overline{F,'(z)} = \frac{m}{\bar{z}}.$$

For fluid flows, a sink is a negative source and is a point of inward radial flow at which the fluid is considered to be absorbed or annihilated. Figure 10.93 illustrates these concepts.

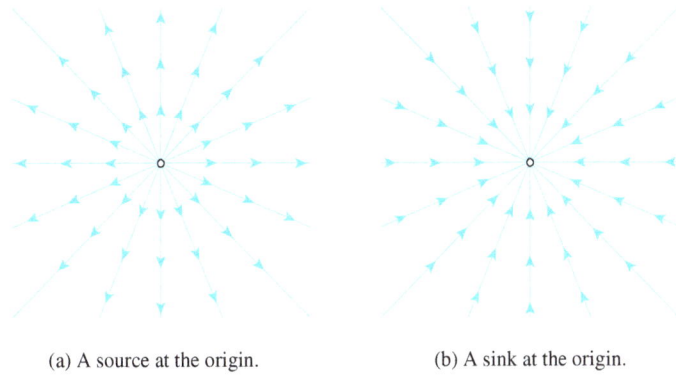

(a) A source at the origin.   (b) A sink at the origin.

Figure 10.93: Sources and sinks for an ideal fluid

10.11.1   Source: A Charged Line

In the case of electrostatics, a source will correspond to a uniformly charged line perpendicular to the z plane at the point z_0. We will show that if the line L is located at $z_0 = 0$ and carries a charge density of $\frac{q}{2}$ coulombs per unit length, then the magnitude of the electrical field is $|\mathbf{E}(x, y)| = \frac{q}{\sqrt{x^2+y^2}}$. Hence, $\mathbf{E}$ is given by

$$\mathbf{E}(x, y) = \frac{qz}{|z|^2} = \frac{q}{\bar{z}}, \tag{10.41}$$

and the complex potential is

$$F(z) = -q \log z, \quad \text{and} \quad \mathbf{E}(x, y) = -\overline{F,'(z)}.$$

A sink for electrostatics is a negatively charged line perpendicular to the z plane. The electric field for electrostatic problems corresponds to the velocity field for fluid flow problems, except that their corresponding potentials differ by a sign change.

To establish Equation (10.41), we start with Coulomb's law, which states that two particles with charges q and Q exert a force on one another with magnitude $\frac{CqQ}{r^2}$, where r is the distance between particles and C is a constant that depends on the scientific units. For simplicity, we assume that $C = 1$ and the test particle at the point z has charge $Q = 1$.

The contribution $\Delta\mathbf{E}_1$ induced by the element of charge $\frac{q\Delta h}{2}$ along the segment of length Δh situated at a height h above the plane has magnitude $|\Delta\mathbf{E}_1|$ given by

$$|\Delta\mathbf{E}_1| = \frac{(\frac{q}{2})\Delta h}{r^2 + h^2}.$$

It has the same magnitude as $\Delta\mathbf{E}_2$ induced by the element $\frac{q\Delta h}{2}$ located a distance $-h$ below the plane. From the vertical symmetry involved their sum, $\Delta\mathbf{E}_2 + \Delta\mathbf{E}_2$, lies parallel to the plane along the ray from the origin, as shown in Figure 10.94.

By the principal of superposition, we add all contributions from the elements of charge along L to obtain $\mathbf{E} = \sum \Delta\mathbf{E}_k$. By vertical symmetry, $\mathbf{E}$ lies parallel to the complex plane along the ray from the origin through the point z. Hence the magnitude of $\mathbf{E}$ is the sum of all components $|\Delta\mathbf{E}|\cos t$ that are parallel to the complex plane, where t is the angle between $\Delta\mathbf{E}$ and the plane. Letting $\Delta h \to 0$ in this summation process produces the definite integral

$$|\mathbf{E}(x, y)| = \int_{-\infty}^{\infty} |\Delta\mathbf{E}| \cos t \, dh = \int_{-\infty}^{\infty} \frac{(\frac{q}{2})\cos t}{r^2 + h^2} \, dh.$$

367

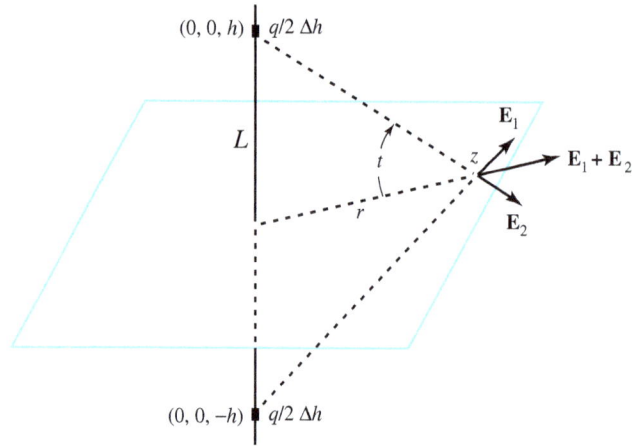

Figure 10.94: Contributions to $\mathbf{E}$ from the elements of charge $\frac{q\Delta h}{2}$ situated at $(0, 0, \pm h)$, above and below the z-plane.

Next, we use the change of variable $h = r \tan t$ and $dh = r \sec^2 t \, dt$ and the trigonometric identity $\sec^2 t = \frac{r^2 + h^2}{r^2}$ to obtain the equivalent integral:

$$|\mathbf{E}(x, y)| = \int_{-\frac{\pi}{2}}^{\frac{\pi}{2}} \frac{\left(\frac{q}{2}\right) \cos t}{r^2 + h^2} \frac{r^2 + h^2}{r} \, dt = \frac{q}{2r} \int_{-\frac{\pi}{2}}^{\frac{\pi}{2}} \cos t \, dt = \frac{q}{r}.$$

Multiplying this magnitude $\frac{q}{r}$ by the unit vector $\frac{z}{|z|}$ establishes Formula (10.41). If $q > 0$, then the field is directed away from $z_0 = 0$ and, if $q < 0$, then it is directed toward $z_0 = 0$. An electrical field located at $z_0 \neq 0$ is given by

$$E(x, y) = \frac{q(z - z_0)}{|z - z_0|^2} = \frac{q}{\overline{z} - \overline{z_0}},$$

and the corresponding complex potential is

$$F(z) = -q \log(z - z_0).$$

Example 10.30 (Source and sink of equal strength). Let a source and sink of unit strength be located at the points $+1$ and -1, respectively. The complex potential for a fluid flowing from the source at $+1$ to the sink at -1 is

$$F(z) = \log(z - 1) - \log(z + 1) = \log\left(\frac{z - 1}{z + 1}\right).$$

The velocity potential and stream function are, respectively

$$\phi(x, y) = \ln\left|\frac{z - 1}{z + 1}\right| \quad \text{and} \quad \psi(x, y) = \arg\left(\frac{z - 1}{z + 1}\right).$$

To solve for the streamline $\psi(x, y) = c$ we start with

$$c = \arg\left(\frac{z - 1}{z + 1}\right) = \arg\left(\frac{x^2 + y^2 - 1 + i2y}{(x + 1)^2 + y^2}\right) = \arctan\left(\frac{2y}{x^2 + y^2 - 1}\right),$$

and obtain the equation $(\tan c)(x^2 + y^2 - 1) = 2y$. A straightforward calculation shows that points on the streamline must satisfy the equation

$$x^2 + (y - \cot c)^2 = 1 + \cot^2 c,$$

(a) Source and sink of equal strength.

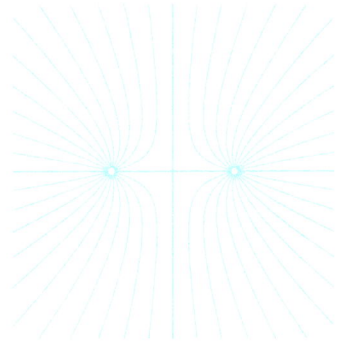

(b) Two sources of equal strength.

Figure 10.95: Fields depicting electrical strength

which is the equation of a circle with center at $(0, \cot c)$ that passes through the points $(\pm 1, 0)$. Several streamlines are indicated in Figure 10.95(a).

Example 10.31 (Two sources of equal strength). Let two sources of unit strength be located at the points ± 1. The resulting complex potential for a fluid flow is

$$F(z) = \log(z - 1) + \log(z + 1) = \log(z^2 - 1).$$

The velocity potential and stream function are, respectively

$$\phi(x, y) = \ln|z^2 - 1| \quad \text{and} \quad \psi(x, y) = \arg(z^2 - 1),$$

To solve for the streamline $\psi(x, y) = c$ we start with

$$c = \arg(z^2 - 1) = \arg(x^2 - y^2 - 1 + i2xy) = \arctan\left(\frac{2xy}{x^2 - y^2 - 1}\right),$$

and obtain the equation $x^2 + 2xy \cot c - y^2 = 1$. If we express this equation in the form

$$\left(x - y \tan \frac{c}{2}\right)\left(x + y \cot \frac{c}{2}\right) = 1, \quad \text{or}$$

$$\left(x \cos \frac{c}{2} - y \sin \frac{c}{2}\right)\left(x \sin \frac{c}{2} + y \cos \frac{c}{2}\right) = \left(\sin \frac{c}{2}\right)\left(\cos \frac{c}{2}\right) = \frac{\sin c}{2},$$

and use the rotation of axes

$$x^* = x \cos\left(-\frac{c}{2}\right) + y \sin\left(-\frac{c}{2}\right), \quad \text{and} \quad y^* = -x \sin\left(-\frac{c}{2}\right) + y \cos\left(-\frac{c}{2}\right),$$

369

then the streamlines must satisfy the equation $x^*y^* = \frac{\sin c}{2}$ and are rectangular hyperbolas with centers at the origin that pass through the points ± 1. Figure 10.95(b) depicts several streamlines.

Let an ideal fluid flow in a domain in the z plane be effected by a source located at the point z_0. Then the flow at points z, which lie in a small neighborhood of the point z_0, is approximated by that of a source with the complex potential

$$\log(z - z_0) + \text{ constant.}$$

If $w = S(z)$ is a conformal mapping and $w_0 = S(z_0)$, then $S(z)$ has a nonzero derivative at z_0, and

$$w - w_0 = (z - z_0)[S'(z_0) + \eta(z)],$$

where $\eta(z) \to 0$ as $z \to z_0$. Taking logarithms yields

$$\log(w - w_0) = \log(z - z_0) + \text{Log}[S'(z_0) + \eta(z)].$$

Because $S'(z_0) \neq 0$, the term $\text{Log}[S'(z_0) + \eta(z)]$ approaches the constant value $\text{Log}[S'(z_0)]$ as $z \to z_0$. Also, because $\log(z - z_0)$ is the complex potential for a source located at the point z_0, the image of a source under a conformal mapping is a source.

We can use the technique of conformal mapping to determine the fluid flow in a domain D in the z plane that is produced by sources and sinks. If we can construct a conformal mapping $w = S(z)$ so that the image of sources, sinks, and boundary curves for the flow in D are mapped onto sources, sinks, and boundary curves in a domain G where the complex potential is known to be $F_1(w)$, then the complex potential in D is given by $F_2(z) = F_1\big(S(z)\big)$.

Example 10.32. Suppose that the lines $x = \pm\frac{\pi}{2}$ are considered as walls of a containing vessel for a fluid flow produced by a single source of unit strength located at the origin. The conformal mapping $w = S(z) = \sin z$ maps the infinite strip bounded by the lines $x = \pm\frac{\pi}{2}$ onto the w plane slit along the boundary rays $u \leq -1$, $v = 0$ and $u \geq 1$, $v = 0$, and the image of the source at $z_0 = 0$ is a source located at $w_0 = 0$. The complex potential

$$F_1(w) = \log w$$

determines a fluid flow in the w plane past the boundary curves $u \leq -1$, $v = 0$ and $u \geq 1$, $v = 0$, which lie along streamlines of the flow. Therefore the complex potential for the fluid flow in the infinite strip in the z plane is

$$F_2(z) = \log(\sin z).$$

Several streamlines for the flow are illustrated in Figure 10.96.

Example 10.33. Suppose that the lines $x = \pm\frac{\pi}{2}$ are considered as walls of a containing vessel for the fluid flow produced by a single source of unit strength located at the point $z_1 = \frac{\pi}{2}$ and a sink of unit strength located at the point $z_2 = -\frac{\pi}{2}$. The conformal mapping $w = S(z) = \sin z$ maps the infinite strip bounded by the lines $x = \pm\frac{\pi}{2}$ onto the w plane slit along the boundary rays $K_1 : u \leq -1$, $v = 0$ and $K_2 : u \geq 1$, $v = 0$. The image of the source at z_1 is a source at $w_1 = 1$, and the image of the sink at z_2 is a sink at $w_2 = -1$. The potential

$$F_1(w) = \log\left(\frac{w - 1}{w + 1}\right)$$

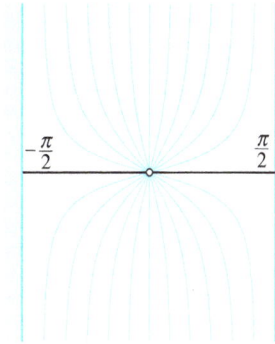

Figure 10.96: A source in the center of a strip

determines a fluid flow in the w plane past the boundary curves K_1 and K_2, which lie along streamlines of the flow. Therefore the complex potential for the fluid flow in the infinite strip in the z plane is

$$F_2(z) = \log\left(\frac{\sin z - 1}{\sin z + 1}\right).$$

Several streamlines for the flow are illustrated in Figure 10.97.

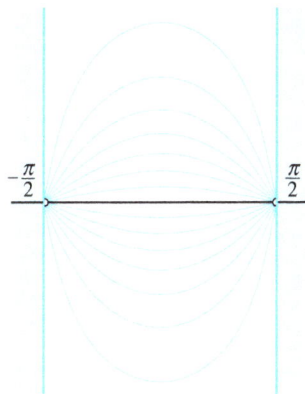

Figure 10.97: A source and a sink on the edges of a strip

We can use the technique of transformation of a source to determine the effluence from a channel extending from infinity. In this case, we construct a conformal mapping $w = S(z)$ from the upper half-plane $\text{Im}(z) > 0$ so that the single source located at $z_0 = 0$ is mapped to the point w_0 at infinity that lies along the channel. The streamlines emanating from $z_0 = 0$ in the upper half-plane are mapped onto streamlines issuing from the channel.

Example 10.34. Consider the conformal mapping

$$w = S(z) = \frac{2}{\pi}(z^2 - 1)^{\frac{1}{2}} + \frac{2}{\pi}\text{Arcsin}\left(\frac{1}{z}\right),$$

which maps the upper half-plane $\text{Im}(z) > 0$ onto the domain consisting of the upper half-plane $\text{Im}(w) > 0$ joined to the channel $-1 \le u \le 1$, $v \le 0$. The point $z_0 = 0$ is mapped onto the point $w_0 = -i\infty$ along the channel. Images of the rays $r > 0$, $\theta = \alpha$ are streamlines issuing from the channel as indicated in Figure 10.98.

371

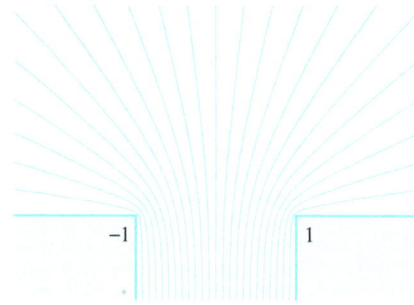

Figure 10.98: Effluence from a channel into a half-plane

Exercises for Section 10.11 (Selected answers or hints are on page 456.)

1. Let the coordinate axes be walls of a containing vessel for a fluid flow in the first quadrant that is produced by a source of unit strength located at $z_1 = 1$ and a sink of unit strength located at $z_2 = i$. Show that $F(z) = \log\left(\frac{z^2-1}{z^2+1}\right)$ is the complex potential for the flow shown in Figure 10.99.

2. Let the coordinate axes be walls of a containing vessel for a fluid flow in the first quadrant that is produced by two sources of equal strength located at the points $z_1 = 1$ and $z_2 = i$. Find the complex potential $F(z)$ for the flow in Figure 10.100

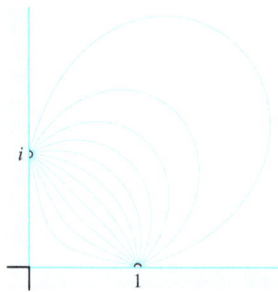

Figure 10.99: For Exercise 1

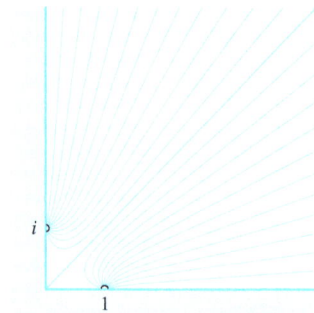

Figure 10.100: For Exercise 2

3. Let the lines $x = 0$ and $x = \frac{\pi}{2}$ form the walls of a containing vessel for a fluid flow in the infinite strip $0 < x < \frac{\pi}{2}$ that is produced by a single source located at the point $z_0 = 0$. Find the complex potential for the flow in Figure 10.101.

4. Let the rays $x = 0$, $y > 0$ and $x = \pi$ $y > 0$ and the segment $y = 0$, $0 < x < \pi$ form the walls of a containing vessel for a fluid flow in the semi-infinite strip $0 < x < \pi$, $y > 0$ that is produced by two sources of equal strength located at the points $z_1 = 0$ and $z_2 = \pi$. Find the complex potential for the flow shown in Figure 10.102
   Hint: Use the fact that $\sin(\frac{\pi}{2} + z) = \sin(\frac{\pi}{2} - z)$.

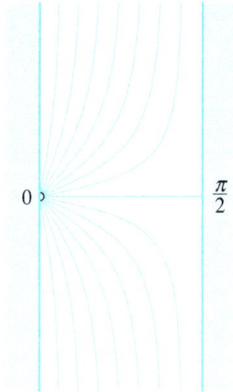

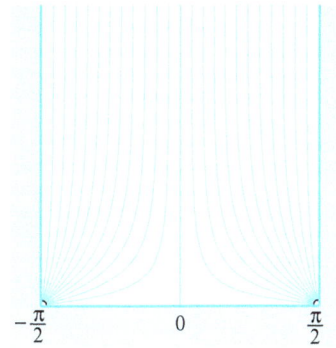

Figure 10.101: For Exercise 3   Figure 10.102: For Exercise 4

5. Let the y axis be considered a wall of a containing vessel for a fluid flow in the right half plane $\text{Re}(z) > 0$ that is produced by a single source located at the point $z_0 = 1$. Find the complex potential for the flow shown in Figure 10.103.

6. The complex potential $F(z) = \frac{1}{z}$ determines an electrostatic field that is referred to as a dipole.

   (a) Show that $F(z) = \lim\limits_{a \to 0} \frac{\log(z) - \log(z-a)}{a}$, and that a dipole is the limiting case of a source and sink.

   (b) Show that the lines of flux of a dipole are circles that pass through the origin, as shown in Figure 10.104.

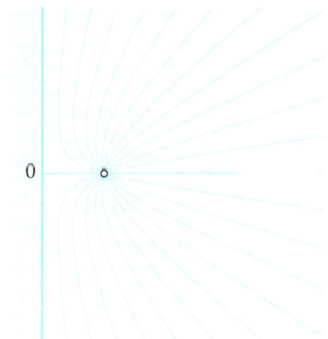

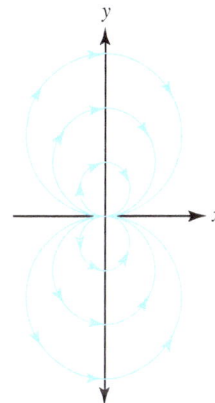

Figure 10.103: For Exercise 5   Figure 10.104: For Exercise 6b

7. Use a Schwarz-Christoffel transformation to find a conformal mapping $w = S(z)$ that will map the flow in the upper half-plane onto the flow from a channel into a quadrant, as indicated in Figure 10.105.

8. Use a Schwarz-Christoffel transformation to find a conformal mapping $w = S(z)$ that will map the flow in the upper half-plane onto the flow from a channel into a sector, as indicated in Figure 10.106.

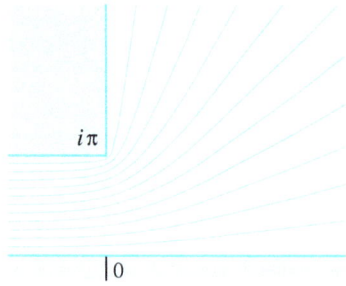

Figure 10.105: For Exercise 7

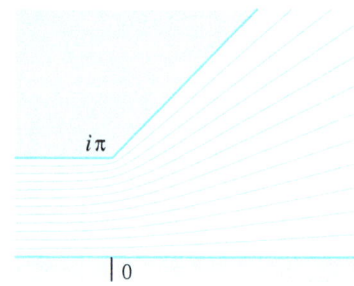

Figure 10.106: For Exercise 8

9. Use a Schwarz-Christoffel transformation to find a conformal mapping $w = S(z)$ that will map the flow in the upper half-plane onto the flow in a right-angled channel indicated in Figure 10.107.

10. Use a Schwarz-Christoffel transformation to find a conformal mapping $w = S(z)$ that will map the flow in the upper half-plane onto the flow from a channel back into a quadrant, as indicated in Figure 10.108, where $w_0 = 2\sqrt{2} - 2\ln(\sqrt{2} - 1) + i\pi$.

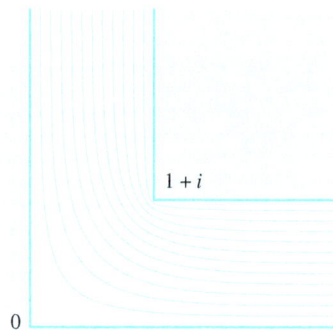

Figure 10.107: For Exercise 9

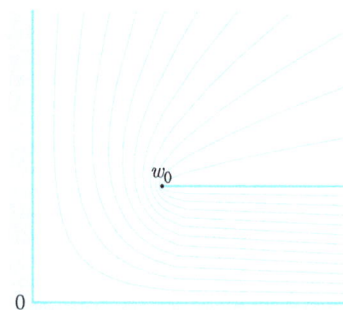

Figure 10.108: For Exercise 10

11. Consider the complex potential $F(z) = w$ given implicitly by $z = w + e^w$.

   (a) Show that $F(z) = w$ determines the ideal fluid flow through an open channel bounded by the rays $y = \pi$, $-\infty < x < -1$, and $y = -\pi$, $-\infty < x < -1$ into the plane.

   (b) Show that the streamline $\psi(x, y) = c$ of the flow is given by the parametric equations

$$x = t + e,' \cos c, \quad \text{and} \quad y = c + e,' \sin c, \quad \text{for} \quad -\infty < t < \infty,$$

   as shown in Figure 10.109.

374

Figure 10.109: For Exercise 11b

Table of Integrals

Integral	Value
$\int \dfrac{1}{(z^2-1)^{\frac{1}{2}}}\,dz$	$i \arcsin z$
$\int \dfrac{1}{(z^2-1)^{\frac{1}{2}}}\,dz$	$\log\left(z + (z^2-1)^{\frac{1}{2}}\right) - \dfrac{i\pi}{2}$
$\int \dfrac{1}{z^2+1}\,dz$	$\arctan z.$
$\int \dfrac{1}{z^2+1}\,dz$	$\dfrac{i}{2}\log\left(\dfrac{i+z}{i-z}\right)$
$\int \dfrac{1}{z(z^2-1)^{\frac{1}{2}}}\,dz$	$-\arcsin\dfrac{1}{z}$
$\int \dfrac{1}{z(z^2-1)^{\frac{1}{2}}}\,dz$	$i\log\left[\dfrac{1}{z} + \left(\dfrac{1}{z^2}-1\right)^{\frac{1}{2}}\right]$
$\int \dfrac{1}{z(z+1)^{\frac{1}{2}}}\,dz$	$-2\operatorname{arctanh}(z+1)^{\frac{1}{2}}$
$\int \dfrac{1}{z(z+1)^{\frac{1}{2}}}\,dz$	$\log\left[\dfrac{1-(z+1)^{\frac{1}{2}}}{1+(z+1)^{\frac{1}{2}}}\right]$
$\int (1-z^2)^{\frac{1}{2}}\,dz$	$\dfrac{1}{2}\left[z(1-z^2)^{\frac{1}{2}} + \arcsin z\right]$
$\int (1-z^2)^{\frac{1}{2}}\,dz$	$\dfrac{i}{2}\left[z(z^2-1)^{\frac{1}{2}} + \log(z + (z^2-1)^{\frac{1}{2}})\right]$

Table 10.2: Indefinite Integrals

Chapter 11

Fourier Series and the Laplace Transform

11.1   Fourier Series

In this chapter we show how Fourier series, the Fourier transform, and the Laplace transform are related to the study of complex analysis. We develop the Fourier series representation of a real-valued function $U(t)$ of the real variable t. Complex Fourier series and Fourier transforms are then discussed. Finally, we develop the Laplace transform and the complex variable technique for finding its inverse. This chapter focuses on applying these ideas to solving problems involving real-valued functions, so many of the theorems throughout are stated without proof.

Let $U(t)$ be a real-valued function that is periodic with period 2π, that is,

$$U(t + 2\pi) = U(t) \quad \text{for all } t.$$

One such function is $s = U(t) = \sin(t - \frac{\pi}{2}) + 0.7 \cos(2t - \pi - \frac{1}{4}) + 1.7$, and its graph is obtained by repeating the portion of the graph in any interval of length 2π, as shown in Figure 11.1.

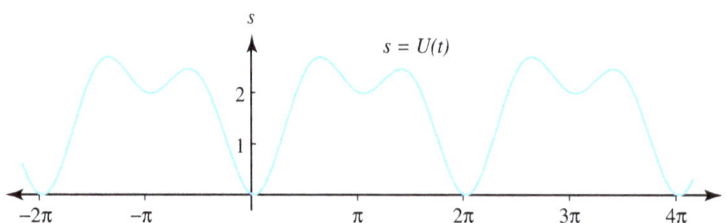

Figure 11.1: A function U with period 2π

Familiar examples of real functions that have period 2π are $\sin nt$ and $\cos nt$, where n is an integer. This raises the question whether any periodic function can be represented by a sum of terms involving $a_n \cos nt$ and $b_n \sin nt$, where a_n and b_n are real constants. As we shall soon see, the answer to this question is often yes.

Definition 11.1 (Piecewise Continuous). *The function U is piecewise continuous on the closed interval $[a, b]$ if there exists values $t_0, t_1, \ldots, t_n$ with $a = t_0 < t_1 < \cdots < t_n = b$ such that U is continuous in each of the open intervals $t_{k-1} < t < t_k (k = 1, 2, \ldots n)$ and has left- and right-hand limits at the values t_k for $k = 0, 1, \ldots, n$.*

We use the symbols $U(a^-)$ and $U(a^+)$ for the left- and right-hand limits, respectively, of a function $U(t)$ as t approaches the point a. The graph of a piecewise continuous function is illustrated in Figure 11.2.

$$U(t) = \begin{cases} \frac{2}{3}(t - \frac{1}{2})^2 + \frac{1}{4} & \text{when } 1 \le t < 2, \\ \frac{5}{2} - (t-2)^2 & \text{when } 2 < t < 3, \\ 1 + \frac{t-3}{4} & \text{when } 3 < t < 4, \\ \frac{6}{5} - (t-5)^3 & \text{when } 4 < t \le 6. \end{cases}$$

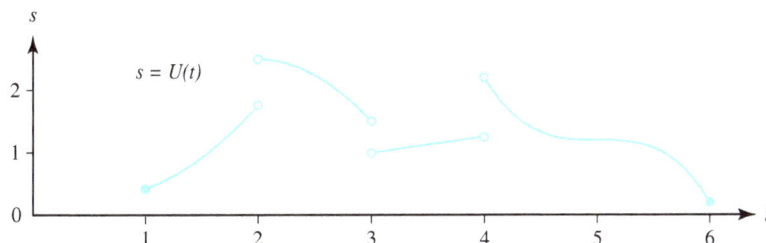

Figure 11.2: A piecewise continuous function U over the interval $[1, 6]$

The left- and right-hand limits at $t_0 = 2$, $t_1 = 3$, and $t_2 = 4$ are easy to determine:

$$\begin{array}{llll} \text{At } t = 2, & U(2^-) = \frac{7}{4} & \text{and} & U(2^+) = \frac{5}{2}. \\ \text{At } t = 3, & U(3^-) = \frac{3}{2} & \text{and} & U(3^+) = 1. \\ \text{At } t = 4, & U(4^-) = \frac{5}{4} & \text{and} & U(4^+) = \frac{11}{5}. \end{array}$$

Definition 11.2 (Fourier Series). *If $U(t)$ is periodic with period 2π and is piecewise continuous on $[-\pi, \pi]$ then the Fourier series $S(t)$ for $U(t)$ is*

$$S(t) = \frac{a_0}{2} + \sum_{n=1}^{\infty}(a_n \cos nt + b_n \sin jt), \tag{11.1}$$

where the coefficients a_n and b_n are given by the so-called Euler's Formulas:

$$a_n - \frac{1}{\pi}\int_{-\pi}^{\pi} U(t)\cos nt\, dt \quad for \quad n = 0, 1, \ldots \tag{11.2}$$

and

$$b_n = \frac{1}{\pi}\int_{-\pi}^{\pi} U(t)\sin nt\, dt \quad for \quad n = 1, 2, \ldots \tag{11.3}$$

The factor $\frac{1}{2}$ in the constant term $\frac{a_0}{2}$ on the right side of Equation (11.1) has been introduced for convenience so that a_0 could be obtained from the general formula in Equation (11.2) by setting $j = 0$. The reasons for this will be explained shortly. The next result discusses convergence of the Fourier series.

Theorem 11.1 (Fourier Expansion). *Assume that $S(t)$ is the Fourier series for $U(t)$. If $U'(t)$ is piecewise continuous on $[-\pi, \pi]$, then $S(t)$ is convergent for all $t \in [-\pi, \pi]$ where $U(t)$ is continuous. If $t = a$ is a point of discontinuity of U, then*

$$S(a) = \frac{U(a^-) + U(a^+)}{2},$$

377

where $U(a^-)$ and $U(a^+)$ denote the left- and right-hand limits, respectively. With this understanding, we have the Fourier expansion:

$$U(t) = \frac{a_0}{2} + \sum_{n=1}^{\infty} (a_n \cos nt + b_n \sin nt). \qquad (11.4)$$

Example 11.1. Show that the function $U(t) = \frac{t}{2}$ for $t \in (-\pi, \pi)$, extended periodically by the equation $U(t + 2\pi) = U(t)$, has the Fourier series expansion

$$U(t) = \sum_{n=1}^{\infty} \frac{(-1)^{n+1}}{n} \sin nt.$$

Solution:

Using Euler's Formulas 11.2 and integration by parts, we obtain

$$a_n = \frac{1}{\pi} \int_{-\pi}^{\pi} \frac{t}{2} \cos nt \, dt = \left(\frac{t \sin nt}{2\pi n} + \frac{\cos nt}{2\pi n^2} \right) \bigg|_{-\pi}^{\pi} = 0 \quad \text{for} \quad n = 1, 2, \ldots,$$

and then using Formula (11.3) we get

$$b_n = \frac{1}{\pi} \int_{-\pi}^{\pi} \frac{t}{2} \sin nt \, dt = \left(\frac{-t \cos nt}{2\pi n} + \frac{\sin nt}{2\pi n^2} \right) \bigg|_{-\pi}^{\pi}$$

$$= \frac{-\cos n\pi}{n} = \frac{(-1)^{n+1}}{n} \quad \text{for} \quad n = 1, 2, \ldots.$$

The coefficient a_0 is computed by the calculation

$$a_0 = \frac{1}{\pi} \int_{-\pi}^{\pi} \frac{t}{2} \, dt = \frac{t^2}{4\pi} \bigg|_{-\pi}^{\pi} = 0.$$

Substituting the coefficients a_j and b_j in Equation (11.1) produces the required solution. The graphs of $U(t)$ and the first three partial sums $S_1(t) = \sin t$, $S_2(t) = \sin t - \frac{1}{2} \sin 2t$, and $S_3(t) = \sin t - \frac{1}{2} \sin 2t + \frac{1}{3} \sin 3t$ are shown in Figure 11.3.

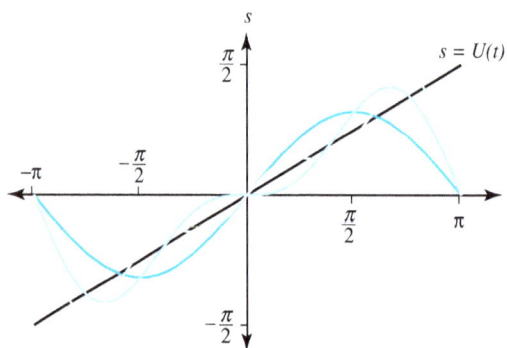

Figure 11.3: $U(t) = \frac{t}{2}$, and approximations $S_1(t)$, $S_2(t)$, and $S_3(t)$

We now state some general properties of Fourier series that are useful for calculating the coefficients. The proofs are left for the reader.

Theorem 11.2. *If $U(t)$ and $V(t)$ have Fourier series representations, then their sum $W(t) = U(t) + V(t)$ has a Fourier series representation, and the Fourier coefficients of W are obtained by adding the corresponding coefficients of U and V.*

378

Theorem 11.3 (Fourier Cosine Series). *Assume that $U(x)$ is an even function. If $U(t)$ has period 2π, and $U(t)$ and $U'(t)$ are piecewise continuous, then the Fourier series for $U(t)$ involves only the cosine terms, i.e., $b_n = 0$ for all n:*

$$U(t) = \frac{a_0}{2} + \sum_{n=1}^{\infty} a_n \cos nt,$$

where

$$a_n = \frac{2}{\pi} \int_0^{\pi} U(t) \cos nt \, dt, \quad \text{for} \quad n = 0, 1, \dots .$$

Theorem 11.4 (Fourier Sine Series). *Assume that $U(t)$ is an odd function. If $U(t)$ has period 2π and if $U(t)$ and $U'(x)$ are piecewise continuous, then the Fourier series for $U(t)$ involves only sine terms, i.e., $a_n = 0$ for all n:*

$$U(t) = \sum_{n=1}^{\infty} b_n \sin nt,$$

where

$$b_n = \frac{2}{\pi} \int_0^{\pi} U(t) \sin nt \, dt \quad \text{for} \quad n = 1, 2, \dots .$$

Theorem 11.5 (Termwise Integration). *If $U(t)$ has a Fourier series representation given in Equation 11.4, then the integral of $U(t)$ has a Fourier series representation which can be obtained by termwise integration of the Fourier Series of $U(t)$, that is,*

$$\int_0^t U(\tau) = \sum_{n=1}^{\infty} \left(\frac{a_n + a_0(-1)^{n+1}}{n} \sin nt - \frac{b_n}{n} \cos nt \right),$$

where we have used the expansion $a_0(\frac{t}{2}) = \sum_{n=1}^{\infty} \frac{a_0(-1)^{n+1}}{n} \sin nt$ from Example 11.1.

Theorem 11.6 (Termwise Differentiation). *If $U'(t)$ has a Fourier series representation, and $U(t)$ is given by Equation (11.4), then*

$$U'(t) = \sum_{n=1}^{\infty} (nb_n \cos nt - na_n \sin jn).$$

Example 11.2. Show that the function $U(t) = |t|$ for $t \in (-\pi, \pi)$, extended periodically the equation $U(t + 2\pi) - U(t)$, has the Fourier series representation

$$U(t) = |t| = \frac{\pi}{2} - \frac{4}{\pi} \sum_{n=1}^{\infty} \frac{1}{(2n-1)^2} \cos[(2n-1)t].$$

Solution:

The function $U(t)$ is an even function, hence we can use Theorem 11.3 to conclude that $b_n = 0$ for all n, and

$$a_n = \frac{2}{\pi} \int_0^{\pi} t \cos nt \, dt = \left(\frac{2t \sin nt}{\pi n} + \frac{2 \cos nt}{\pi n^2} \right)\Big|_0^{\pi}$$

$$= \frac{2 \cos n\pi - 2}{\pi n^2} = \frac{2(-1)^n - 2}{\pi n^2} \quad \text{for} \quad n = 1, 2, \dots .$$

The coefficient a_0 is computed by the calculation:

$$a_0 = \frac{2}{\pi} \int_0^{\pi} t \, dt = \frac{t^2}{\pi}\Big|_0^{\pi} = \pi.$$

Using the a_n and Theorem 11.3 produces the required solution.

11.1.1 Proof of Euler's Formulas

The following intuitive proof will justify the Euler formulas given in Equations (11.2) and (11.3) To determine a_0 we integrate both $U(t)$ and the Fourier series representation in Equation (11.1) from $-\pi$ to π, which results in

$$\int_{-\pi}^{\pi} U(t)\,dt = \int_{-\pi}^{\pi}\left[\frac{a_0}{2} + \sum_{n=1}^{\infty}(a_n\cos nt + b_n\sin nt)\right]dt.$$

Next, we perform integration term by term, and we obtain

$$\int_{-\pi}^{\pi} U(t)\,dt = \frac{a_0}{2}\int_{-\pi}^{\pi} 1\,dt + \sum_{n=1}^{\infty} a_n \int_{-\pi}^{\pi}\cos nt\,dt + \sum_{n=1}^{\infty} b_n \int_{-\pi}^{\pi}\sin nt\,dt.$$

The value of the first integral on the right side of this equation is 2π and all the other integrals are zero. Hence we obtain the desired formula for a_0:

$$a_0 = \frac{1}{\pi}\int_{-\pi}^{\pi} U(t)\,dt.$$

To determine a_m, we let $m > 1$ denote a fixed integer and multiply both $U(t)$ and the Fourier series representation in Equation (11.1) by the term $\cos mt$, and then we integrate and obtain

$$\int_{-\pi}^{\pi} U(t)\cos mt\,dt = \frac{a_0}{2}\int_{-\pi}^{\pi}\cos mt\,dt \tag{11.5}$$

$$+ \sum_{n=1}^{\infty} a_j \int_{-\pi}^{\pi}\cos mt\cos nt\,dt + \sum_{n=1}^{\infty} b_n \int_{-\pi}^{\pi}\cos mt\sin nt\,dt.$$

The value of the first term on the right side of Equation (11.5) is easily seen to be zero:

$$\frac{a_0}{2}\int_{-\pi}^{\pi}\cos mt\,dt = \left.\frac{a_0\sin mt}{2m}\right|_{-\pi}^{\pi} = 0. \tag{11.6}$$

The value of the term involving $\cos mt\cos jt$ is found by using the trigonometric identity:

$$\cos mt\cos nt = \frac{1}{2}\Big(\cos[(m+n)t] + \cos[(m-n)t]\Big).$$

Calculation reveals that if $m \neq n$ (and $m > 0$), then

$$a_n \int_{-\pi}^{\pi}\cos mt\cos nt\,dt = \frac{a_m}{2}\left[\int_{-\pi}^{\pi}\cos[(m+n)t]\,dt + \int_{-\pi}^{\pi}\cos[(m-n)t]dt\right] \tag{11.7}$$

$$= 0.$$

When $m = n$, the value of the integral becomes:

$$a_m \int_{-\pi}^{\pi}\cos^2 mt\,dt = \pi a_m. \tag{11.8}$$

The value of the term on the right side of Equation (11.5) involving the integrand $\cos mt\sin nt$ is found by using the trigonometric identity

$$\cos mt\sin nt = \frac{1}{2}\Big(\sin[(m+n)t] + \sin[(m-n)t]\Big),$$

and for all values of m and n, we obtain

$$b_n \int_{-\pi}^{\pi} \cos mt \sin nt \, dt = \frac{b_m}{2} \left[\int_{-\pi}^{\pi} \sin[(m+n)t] \, dt \right.$$

$$\left. + \int_{-\pi}^{\pi} \sin[(m-n)t] \, dt \right] \qquad (11.9)$$

$$= 0.$$

Therefore, we can use the results of Equations (11.6)–(11.9) in Equation (11.5) to obtain

$$\int_{-\pi}^{\pi} U(t) \cos mt \, dt = \pi a_m \quad \text{for} \quad m = 0, 1, 2, \ldots,$$

and Equation (11.2) is established. An exercise asks you to establish Euler's Formula (11.3) for the coefficients $\{b_n\}$. A complete discussion of the details of the proof of Theorem 11.1 can be found in some advanced texts. See for instance, John W. Dettman, Chapter 8 in *Applied Complex Variables,* The Macmillan Company, New York, 1965.

Exercises for Section 11.1 (Selected answers or hints are on page 457.)

For Exercises 1–2 and 6–11, find the Fourier series representation.

1. $U(t) = \begin{cases} 1 & \text{for} \quad 0 < t < \pi, \\ -1 & \text{for} \quad -\pi < t < 0. \end{cases}$ See Figure 11.4.

2. $V(t) = \begin{cases} \frac{\pi}{2} - t & \text{for} \quad 0 \le t \le \pi. \\ \frac{\pi}{2} + t & \text{for} \quad -\pi < t < 0. \end{cases}$ See Figure 11.5.

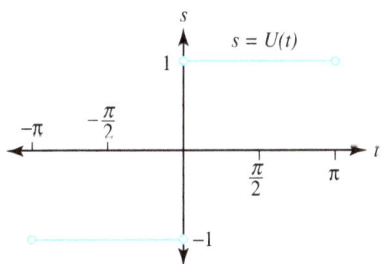

Figure 11.4: Graph of $U(t)$ for Exercise 1   Figure 11.5: Graph of $V(t)$ for Exercise 2

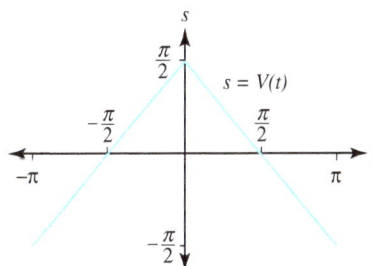

3. For Exercises 1 and 2, verify that $U(t) = -V'(t)$ by termwise differentiation of the Fourier series representation for $V(t)$.

4. For Exercise 1, set $t = \frac{\pi}{2}$ and conclude that $\frac{\pi}{4} = \sum_{j=1}^{\infty} \frac{(-1)^{j-1}}{2j-1}$.

5. For Exercise 2, set $t = 0$ and conclude that $\frac{\pi^2}{8} = \sum_{j=1}^{\infty} \frac{1}{(2j-1)^2}$.

6. $U(t) = \begin{cases} -1 & \text{for } \frac{\pi}{2} < t < \pi, \\ 1 & \text{for } -\frac{\pi}{2} < t < \frac{\pi}{2}, \\ -1 & \text{for } -\pi < t < -\frac{\pi}{2}. \end{cases}$ See Figure 11.6.

7. $U(t) = \begin{cases} \pi - t & \text{for } \frac{\pi}{2} < t < \pi, \\ t & \text{for } -\frac{\pi}{2} < t < \frac{\pi}{2}, \\ -\pi - t & \text{for } -\pi < t < -\frac{\pi}{2}. \end{cases}$ See Figure 11.7.

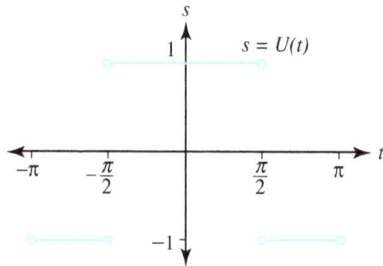

Figure 11.6: Graph of $U(t)$ for Exercise 6

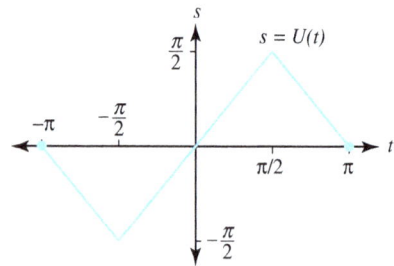

Figure 11.7: Graph of $U(t)$ for Exercise 7

8. $U(t)$, as defined in Figure 11.8.

9. $U(t) = \begin{cases} 1 & \text{for } \frac{\pi}{2} < t < \pi, \\ 0 & \text{for } -\frac{\pi}{2} < t < \frac{\pi}{2}, \\ -1 & \text{for } -\pi < t < \frac{-\pi}{2}. \end{cases}$ See Figure 11.9.

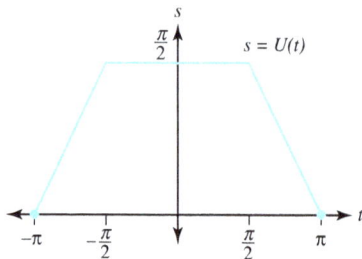

Figure 11.8: Graph of $U(t)$ for Exercise 8

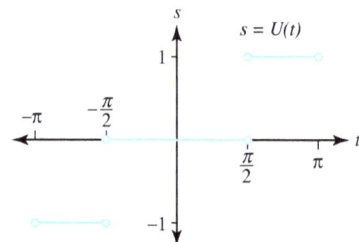

Figure 11.9: Graph of $U(t)$ for Exercise 9

10. $U(t)$, as defined in Figure 11.10.

11. $U(t)$, as defined in Figure 11.11.

12. Establish Euler's Formula (11.3) for the coefficients $\{b_n\}$.

382

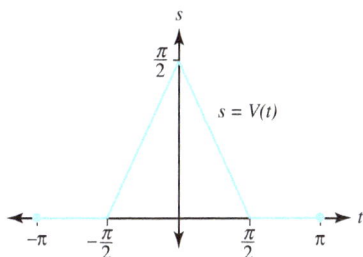

Figure 11.10: Graph of $U(t)$ for Exercise 10   Figure 11.11: Graph of $U(t)$ for Exercise 11

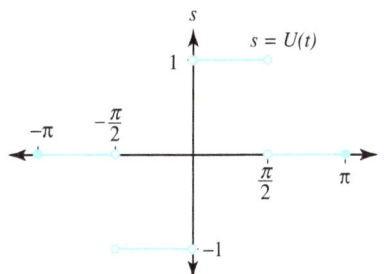

11.2   The Dirichlet Problem for the Unit Disk

The Dirichlet problem for the unit disk $D : |z| < 1$ is to find a real-valued function $u(x,y)$ that is harmonic in the unit disk D and that takes on the boundary values

$$u(\cos\theta, \sin\theta) = U(\theta), \quad \text{for} \quad -\pi < \theta \le \pi \tag{11.10}$$

at points $z = (\cos\theta, \sin\theta)$ on the unit circle, as shown in Figure 11.12.

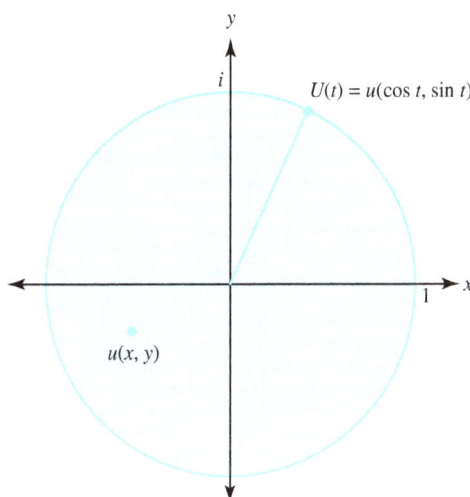

Figure 11.12: The Dirichlet problem for the unit disk $|z| < 1$.

Theorem 11.7 (Dirichet Solution). *If $U(t)$ has period 2π, and has the Fourier series representation*

$$U(t) = \frac{a_0}{2} + \sum_{n=1}^{\infty}(a_n \cos nt + b_n \sin nt),$$

then the solution u to the Dirichlet problem in D is

$$u(r\cos\theta, r\sin\theta) = \frac{a_0}{2} + \sum_{n=1}^{\infty}(a_n r^n \cos n\theta + b_n r^n \sin n\theta), \tag{11.11}$$

where $x + iy = re^{i\theta}$ denotes a complex number in the closed disk $|z| \le 1$.

We will prove this theorem shortly, but note meanwhile that it is easy to see that the series representation in Equation (11.11) for u takes on the boundary values given in Equation (11.10) at points on the unit circle $|z| = 1$. Since each term $r^n \cos n\theta$ and $r^n \sin n\theta$ in series given in Equation (11.11) is harmonic, it is reasonable to conclude that the infinite series representing u will also be harmonic.

Before proving Theorem 11.7 we establish a result that gives an integral representation for a function $u(x, y)$ that is harmonic in a domain containing the closed unit disk. The result is the analog to Poisson's integral formula for the upper half plane.

Theorem 11.8 (Poisson Integral Formula for the Unit Disk). *Let $u(x, y)$ be a function that is harmonic in a simply connected domain that contains the closed unit disk $|z| \leq 1$. If $u(x, y)$ takes on the boundary values*

$$u(\cos \theta, \sin \theta) = U(\theta) \quad for \quad -\pi < \theta \leq \pi,$$

then u has the integral representation

$$u(r \cos \theta, r \sin \theta) = \frac{1}{2\pi} \int_{-\pi}^{\pi} \frac{(1 - r^2) U(t)}{1 + r^2 - 2r \cos(t - \theta)} \, dt \tag{11.12}$$

that is valid for $|z| < 1$.

Proof of Theorem 11.8. Since $u(x, y)$ is harmonic in the simply connected domain, there exists a conjugate harmonic function $v(x, y)$ such that $f(z) = u(x, y) + iv(x, y)$ is analytic. Let C denote the contour consisting of the unit circle; then Cauchy's integral formula

$$f(z) = \frac{1}{2\pi i} \int_C \frac{f(\xi)}{\xi - z} \, d\xi \tag{11.13}$$

expresses the value of $f(z)$ at any point z inside C in terms of the values of $f(\xi)$ at points ξ that lie on the circle C. If we set $z^* = (\bar{z})^{-1}$ then z^* lies outside the unit circle C and the Cauchy-Goursat theorem gives

$$0 = \frac{1}{2\pi i} \int_C \frac{f(\xi)}{\xi - z^*} \, d\xi. \tag{11.14}$$

If we subtract Equation (11.14) from Equation (11.13), then use the parameterization $\xi = e^{it}$, $d\xi = i e^{it} \, dt$ and the substitutions $z = re^{i\theta}$, $z^* = \frac{1}{r} e^{i\theta}$, we get

$$f(z) = \frac{1}{2\pi} \int_{-\pi}^{\pi} \left(\frac{e^{it}}{e^{it} - re^{i\theta}} - \frac{e^{it}}{e^{it} - \frac{1}{r} e^{i\theta}} \right) f(e^{it}) \, dt.$$

The expression inside the parentheses on the right side of this equation can be written as

$$\frac{e^{it}}{e^{it} - re^{i\theta}} - \frac{e^{it}}{e^{it} - \frac{1}{r} e^{i\theta}} = \frac{1}{1 - re^{i(\theta - t)}} + \frac{re^{i(t-\theta)}}{1 - re^{i(t-\theta)}}$$

$$= \frac{1 - r^2}{1 + r^2 - 2r \cos(t - \theta)}, \tag{11.15}$$

and it follows that

$$f(z) = \frac{1}{2\pi} \int_{-\pi}^{\pi} \frac{(1 - r^2) f(e^{it})}{1 + r^2 - 2r \cos(t - \theta)} \, dt.$$

Since $u(x, y)$ is the real part of $f(z)$ and $U(t)$ is the real part of $f(e^{it})$, we can equate the real parts in the latter equation to obtain Equation (11.12), completing the proof Theorem 11.8. $\square$

We now turn our attention to the proof Theorem 11.7.

Proof of Theorem 11.7. The real-valued function

$$P(r, t - \theta) = \frac{1 - r^2}{1 + r^2 - 2r\cos(t - \theta)}$$

is known as the *Poisson kernel*. Expanding the left side of Equation (11.15) in a geometric series gives

$$P(r, t - \theta) = \frac{1}{1 - re^{i(\theta - t)}} + \frac{re^{i(t - \theta)}}{1 - re^{i(t - \theta)}}$$

$$= \sum_{n=0}^{\infty} r^n e^{in(\theta - t)} + \sum_{n=1}^{\infty} r^n e^{in(t - \theta)}$$

$$= 1 + \sum_{n=1}^{\infty} r^n [e^{in(\theta - t)} + e^{in(t - \theta)}] = 1 + 2\sum_{n=1}^{\infty} r^n \cos[n(\theta - t)]$$

$$= 1 + 2\sum_{n=1}^{\infty} r^n (\cos n\theta \cos nt + \sin n\theta \sin nt)$$

$$= 1 + 2\sum_{n=1}^{\infty} r^n \cos n\theta \cos nt + 2\sum_{n=1}^{\infty} r^n \sin n\theta \sin nt.$$

We now use the above result in Equation (11.12) to obtain

$$u(r\cos\theta, r\sin\theta) = \frac{1}{2\pi} \int_{-\pi}^{\pi} P(r, t - \theta) U(t)\, dt$$

$$= \frac{1}{2\pi} \int_{-\pi}^{\pi} U(t)\, dt + \frac{1}{\pi} \int_{-\pi}^{\pi} \sum_{n=1}^{\infty} r^n \cos n\theta \cos nt\, U(t)\, dt$$

$$+ \frac{1}{\pi} \int_{-\pi}^{\pi} \sum_{n=1}^{\infty} r^n \sin n\theta \cos nt\, U(t)\, dt$$

$$= \frac{1}{2\pi} \int_{-\pi}^{\pi} U(t)\, dt + \sum_{n=1}^{\infty} \frac{r^n}{\pi} \cos n\theta \int_{-\pi}^{\pi} \cos nt\, U(t)\, dt$$

$$+ \sum_{n=1}^{\infty} \frac{r^n}{\pi} \sin n\theta \int_{-\pi}^{\pi} \sin nt\, U(t)\, dt$$

$$= \frac{a_0}{2} + \sum_{n=1}^{\infty} a_n r^n \cos n\theta + \sum_{n=1}^{\infty} b_n r^n \sin n\theta.$$

where $\{a_n\}$ and $\{b_n\}$ are the Fourier series coefficients for $U(t)$. This observation establishes the representation for $u(r\cos\theta, r\sin\theta)$ in Equation (11.11). $\square$

Example 11.3. Find the function $u(x, y)$ that is harmonic in the unit disk $|z| < 1$ and takes on the boundary values

$$u(\cos\theta, \sin\theta) = U(\theta) = \frac{\theta}{2}, \quad \text{for} \quad -\pi < \theta < \pi.$$

Solution:

Using Example 10.1, we begin with the Fourier series for $U(\theta)$:

$$U(t) = \sum_{n=1}^{\infty} \frac{(-1)^{n+1}}{n} \sin nt.$$

Using formula 11.11 for the solution of the Dirichlet problem, we obtain

$$u(r\cos\theta, r\sin\theta) = \sum_{n=1}^{\infty} \frac{(-1)^{n+1}}{n} r^n \sin n\theta.$$

Note that the above series representation for $u(r\cos\theta, r\sin\theta)$ takes on the prescribed boundary values at points where $U(\theta)$ is continuous. The boundary function $U(\theta)$ is discontinuous at $z = -1$, which corresponds to $\theta = \pm\pi$; and $U(\theta)$ was not prescribed at these points. Graphs of the approximations $U_7(t)$ and $u_7(x, y) = u_7(r\cos\theta, r\sin\theta)$, which involve the first seven terms in equations are shown in Figure 11.13.

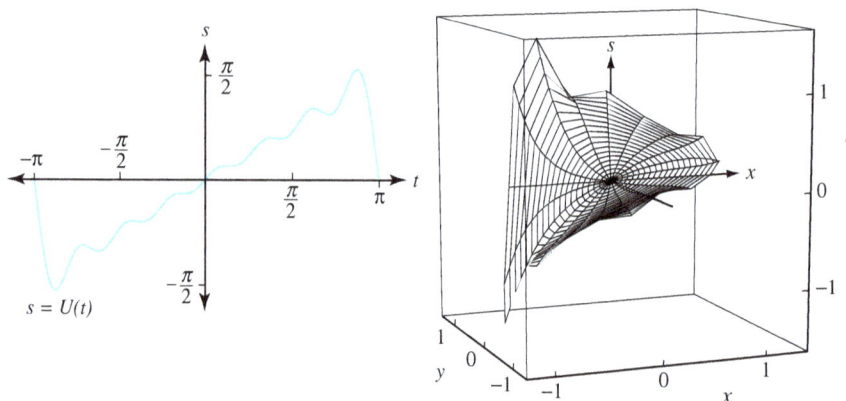

Figure 11.13: Functions $U_7(t)$ and $u_7(r\cos\theta, r\sin\theta)$ for Example 11.3

Exercises for Section 11.2 (Selected answers or hints are on page 457.)

For problems 1–6, find the solution to the given Dirichlet problem in the unit disk D by using the Fourier series representations for the boundary functions that were derived in the examples and exercises of Section 11.1.

1. $U(\theta) = \begin{cases} 1 & \text{for } 0 < \theta < \pi, \\ -1 & \text{for } -\pi < \theta < 0. \end{cases}$

2. $U(\theta) = \begin{cases} \frac{\pi}{2} - \theta & \text{for } 0 \le \theta < \pi, \\ \frac{\pi}{2} + \theta & \text{for } -\pi < \theta < 0. \end{cases}$ (See Figure 11.14)

3. $U(\theta) = \begin{cases} -1 & \text{for } \frac{\pi}{2} < \theta < \pi, \\ 1 & \text{for } -\frac{\pi}{2} < \theta < \frac{\pi}{2}, \\ -1 & \text{for } -\pi < \theta < -\frac{\pi}{2}. \end{cases}$

4. $U(\theta) = \begin{cases} \pi - \theta & \text{for } \frac{\pi}{2} \le \theta \le \pi, \\ \theta & \text{for } -\frac{\pi}{2} \le \theta < \frac{\pi}{2}, \\ -\pi - \theta & \text{for } -\pi < \theta < -\frac{\pi}{2}. \end{cases}$

386

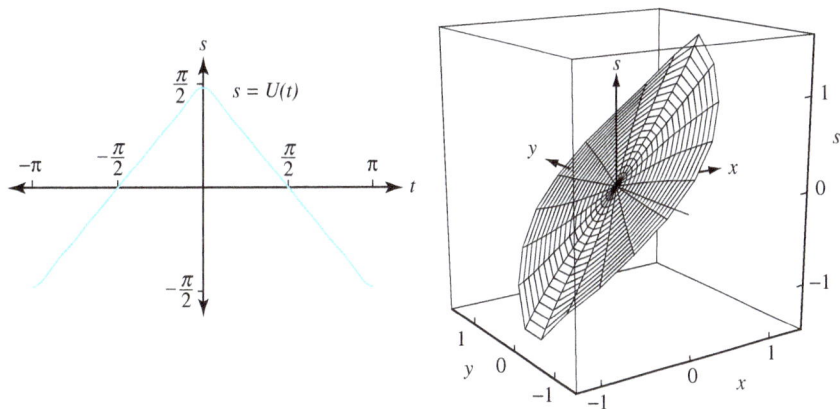

Figure 11.14: Approximations for $U_5(\theta)$ and $u_5(r\cos\theta, r\sin\theta)$ in Exercise 2

5. $U(\theta) = \begin{cases} \pi - \theta & \text{for} \quad \frac{\pi}{2} \leq \theta \leq \pi, \\ \frac{\pi}{2} & \text{for} \quad -\frac{\pi}{2} \leq \theta < \frac{\pi}{2}, \\ \pi + \theta & \text{for} \quad -\pi < \theta < -\frac{\pi}{2}. \end{cases}$ (See Figure 11.15)

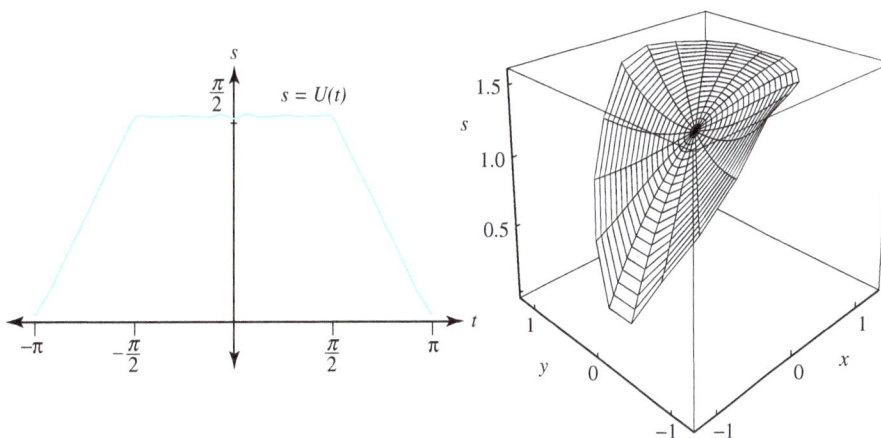

Figure 11.15: Approximations for $U_5(\theta)$ and $u_5(r\cos\theta, r\sin\theta)$ in Exercise 5

6. $U(\theta) = \begin{cases} 1 & \text{for} \quad \frac{\pi}{2} < \theta < \pi, \\ 0 & \text{for} \quad -\frac{\pi}{2} < \theta < \frac{\pi}{2}, \\ -1 & \text{for} \quad -\pi < \theta < -\frac{\pi}{2}. \end{cases}$ (See Figure 11.16)

7. $U(\theta) = \begin{cases} 0 & \text{for} \quad \frac{\pi}{2} \leq \theta \leq \pi, \\ \frac{\pi-\theta}{2} & \text{for} \quad 0 \leq \theta < \frac{\pi}{2}, \\ \frac{\pi+\theta}{2} & \text{for} \quad -\frac{\pi}{2} \leq \theta < 0, \\ 0 & \text{for} \quad -\pi < \theta < -\frac{\pi}{2}. \end{cases}$

8. $U(\theta) = \begin{cases} 0 & \text{for} \quad \frac{\pi}{2} < \theta < \pi, \\ -1 & \text{for} \quad 0 < \theta < \frac{\pi}{2}, \\ 1 & \text{for} \quad -\frac{\pi}{2} < \theta < 0, \\ 0 & \text{for} \quad -\pi < \theta < -\frac{\pi}{2}. \end{cases}$

387

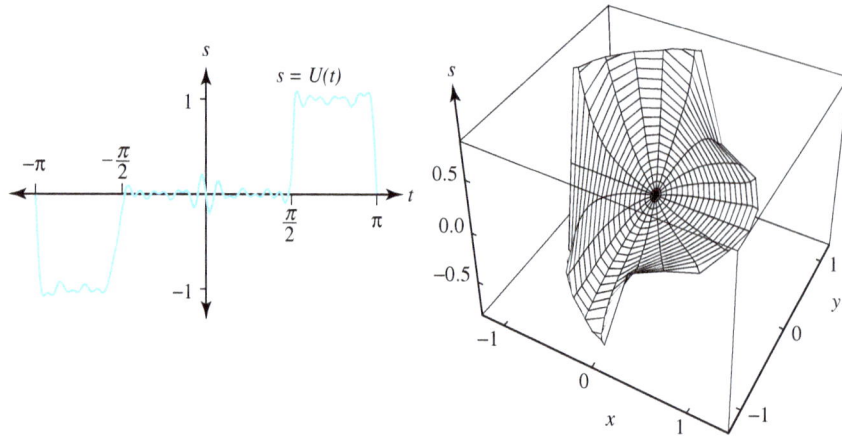

Figure 11.16: Approximations for $U_7(\theta)$ and $u_7(r\cos\theta, r\sin\theta)$ in Exercise 6

9. Write a report on the Dirichlet problem and include some applications.

10. Look up the article on the Poisson integral formula and discuss what you found.

11.3 Vibrations in Mechanical Systems

Consider a spring that resists compression as well as extension, that is suspended vertically from a fixed support, and a body of mass m that is attached at the lower end of the spring. We make the assumption that the mass m is much larger than the mass of the spring so that we can neglect the mass of the spring. If there is no motion then the system is in static equilibrium, as illustrated in Figure 11.17. If the mass is pulled down further and released, then it will undergo an oscillatory motion.

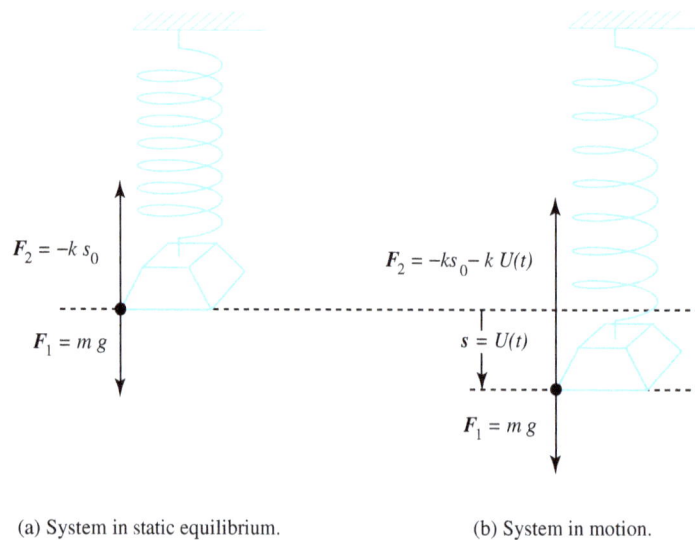

(a) System in static equilibrium.

(b) System in motion.

Figure 11.17: The spring mass system

Suppose there is no friction to slow down the motion of the mass, then we say that the system is *undamped*. We will determine the motion of this mechanical system by considering the forces acting on the mass during the motion. This will lead to a differential equation relating the displacement as a function of time. The most obvious force is that of *gravitational attraction* acting on the mass m and is given by

$$F_1 = mg,$$

where g is the acceleration of gravity. The next force to be considered is the *spring force* acting on the mass and is directed upward if the spring is stretched and downward if it is compressed. It obeys Hooke's law

$$F_2 = ks,$$

where s is the amount the spring is stretched when $s > 0$ and is the amount it is compressed when $s < 0$.

When the system is in static equilibrium and the spring is stretched by the amount s_0, the resultant of the spring force and the gravitational force is zero, which is expressed by the equation

$$mg - ks_0 = 0.$$

Let $s = U(t)$ denote the displacement from static equilibrium with the positive s direction pointed downward as indicated in Figure 10.17(b).The spring force can be written as

$$F_2 = -k[s_0 + U(t)] = -ks_0 - kU(t),$$

and the resultant force F_R is

$$F_R = F_1 + F_2 = mg - ks_0 - kU(t) = -kU(t). \qquad (11.16)$$

The differential equation for motion is obtained by using Newton's second law, which states that the resultant of the forces acting on the mass at any instant satisfies

$$F_R = ma. \qquad (11.17)$$

The distance from equilibrium at time t is measured by $U(t)$, so the acceleration a is given by $a = U''(t)$. Equations (11.16) and (11.17) then give

$$F_R = -kU(t) = mU''(t).$$

Therefore, the undamped mechanical system is governed by the linear differential equation

$$mU''(t) + kU(t) = 0.$$

The general solution to undamped system is known to be

$$U(t) = A\cos\omega t + B\sin\omega t, \quad \text{where} \quad \omega = \sqrt{\frac{k}{m}}.$$

11.3.1 Damped System

If we consider frictional forces that slow down the motion of the mass, then we say that the system is damped. This is visualized by connecting a dashpot to the mass, as indicated in Figure 11.18. For small velocities it is assumed that the frictional force F_3 is proportional to the velocity, that is,

$$F_3 = -cU'(t).$$

The damping constant c must be positive, for if $U'(t) > 0$, then the mass is moving downward and hence F_3 must point upward, which requires that F_3 is negative. The result of the three forces acting on the mass is given by

$$F_1 + F_2 + F_3 = -kU(t) - cU'(t) = mU''(t) = F_R,$$

so the damped mechanical system is governed by the differential equation

$$mU''(t) + cU'(t) + kU(t) = 0.$$

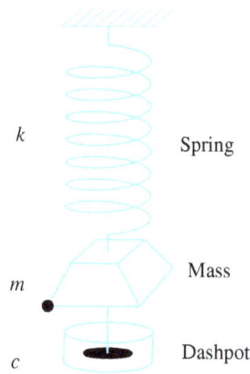

Figure 11.18: The spring mass dashpot system

11.3.2 Forced Vibrations

The vibration discussed earlier are called *free* vibrations because all the forces that affect the motion of the system are internal to the system. We extend our analysis to cover the case in which an external force $F_4 = F(t)$ acts on the mass (see Figure 11.19). Such a force might occur from vibrations of the support to which the top of the spring is attached, or from the effect of a magnetic field on a mass made of iron. As before, we sum the forces F_1, F_2, F_3, and F_4 and set this equal to the resultant force F_R and obtain

$$F_1 + F_2 + F_3 + F_4 = F_R = -KU(t) - cU'(t) + F(t) = mU''(t).$$

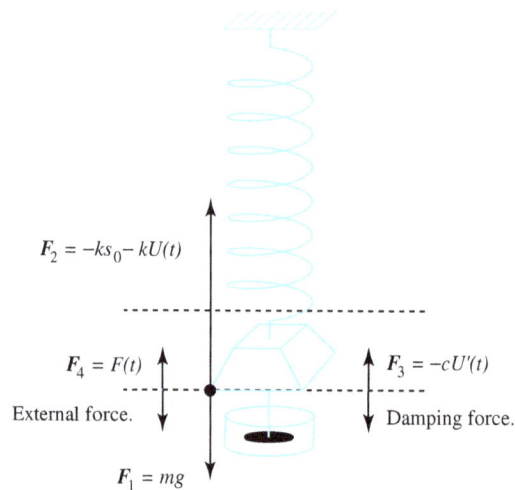

Figure 11.19: The dashpot system with an external force

Thus, the *forced motion* of the mechanical system satisfies the nonhomogeneous linear differential equation

$$mU''(t) + cU'(t) + kU(t) = F(t). \tag{11.18}$$

The function $F(t)$ is called the *input* or *driving force* and the solution $U(t)$ is called the *output* or *response*. Of particular interest are periodic inputs $F(t)$ that can be represented by Fourier series.

For damped mechanical systems that are driven by a periodic input $F(t)$, the general solution involves a *transient part* that vanishes as $t \to +\infty$, and a *steady state part* that is periodic. The transient part of the solution $U_h(t)$ is found by solving the homogeneous differential equation

$$mU_h{}''(t) + cU_h{}'(t) + kU_h(t) = 0.$$

This homogeneous equation has the characteristic equation $m\lambda^2 + c\lambda + k = 0$, and it's roots are $\lambda = \frac{-c \pm \sqrt{c^2 - 4mk}}{2m}$. The coefficients m, c, and k are all positive, and there are three cases to consider.

Case 1. *If $c^2 - 4mk > 0$, the roots are real and distinct, and since $\sqrt{c^2 - 4mk} < c$, it follows that the roots λ_1 and λ_2 are negative real numbers. Thus, for this case, we have*

$$\lim_{t \to +\infty} U_h(t) = \lim_{t \to +\infty} (A_1 e^{\lambda_1 t} + A_2 e^{\lambda_2 t}) = 0.$$

Case 2. *If $c^2 - 4mk = 0$,the roots are real and equal, then $\lambda_1 = \lambda_2 = \lambda$, where λ is a negative real number. Again, for this case we find that*

$$\lim_{t \to +\infty} U_h(t) = \lim_{t \to +\infty} (A_1 e^{\lambda t} + A_2 e^{\lambda t}) = 0.$$

Case 3. *If $c^2 - 4mk < 0$, the roots are complex, then $\lambda = -\alpha \pm \beta i$, where α and β are positive real numbers, and it follows that*

$$\lim_{t \to +\infty} U_h(t) = \lim_{t \to +\infty} (A_1 e^{-\alpha t} \cos \beta t + A_2 e^{-\alpha t} \sin \beta t) = 0.$$

In all three cases, we see that the homogeneous solution $U_h(t)$ decays to 0 as $t \to +\infty$. The steady state solution $U_p(t)$ can be obtained by representing $U_p(t)$ by its Fourier series and substituting $U_p{}''(t)$, $U_p{}'(t)$, and $U_p(t)$ into the nonhomogeneous differential equation and solving the resulting system for the Fourier coefficients of $U_p(t)$. The general solution to Equation (11.18) is then given by

$$U(t) = U_h(t) + U_p(t).$$

Example 11.4. Find the general solution to $U''(t) + 2U'(t) + U(t) = F(t)$, where $F(t)$ is given by the Fourier series $F(t) = \sum\limits_{n=1}^{\infty} \frac{1}{(2n-1)^2} \cos[(2n-1)t]$.

Solution:

First we solve $U_h{}''(t) + 2U_h{}'(t) + U_h(t) = 0$ for the transient solution. The characteristic equation is $\lambda^2 + 2\lambda + 1 = 0$, which has a double root $\lambda = -1$. Hence

$$U_h(t) = A_1 e^{-t} + A_2 t e^{-t}.$$

The steady state solution is obtained by assuming that $U_p(t)$ has the Fourier series representation

$$U_p(t) = \frac{a_0}{2} + \sum_{n=1}^{\infty} a_n \cos nt + \sum_{n=1}^{\infty} b_n \sin nt,$$

and that $U_h{}'(t)$ and $U_h{}''(t)$ can be obtained by termwise differentiation:

$$2U_p{}'(t) = 2\sum_{n=1}^{\infty} nb_n \cos nt - 2\sum_{n=1}^{\infty} na_n \sin nt, \quad \text{and}$$

$$U_p{}''(t) = -\sum_{n=1}^{\infty} n^2 a_n \cos nt - \sum_{n=1}^{\infty} n^2 b_n \sin nt.$$

Substituting these expansions into the differential equation results in

$$F(t) = \frac{a_0}{2} + \sum_{n=1}^{\infty} \left[(1 - n^2)a_n + 2nb_n\right] \cos nt + \sum_{n=1}^{\infty} \left[-2na_n + (1 - n^2)b_n\right] \sin nt.$$

Equating the coefficients with the given series for $F(t)$, we find that $\frac{a_0}{2} = 0$, and that

$$(1 - n^2)a_n + 2nb_n = \begin{cases} \frac{1}{n^2} & \text{when } n \text{ is odd,} \\ 0 & \text{when } n \text{ is even,} \end{cases} \quad -2na_n + (1 - n^2)b_n = 0 \quad \text{for all} \quad n.$$

Solving this linear system for a_n and b_n yields

$$a_n = \begin{cases} \frac{1-n^2}{n^2(1+n^2)^2} & \text{for } n \text{ odd,} \\ 0 & \text{for } n \text{ even,} \end{cases} \quad b_n = \begin{cases} \frac{2n}{n^2(1+n^2)^2} & \text{for } n \text{ odd,} \\ 0 & \text{for } n \text{ even.} \end{cases}$$

The general solution is

$$U(t) = A_1 e^{-t} + A_2 t e^{-t} + \sum_{n=1}^{\infty} \frac{1 - (2n-1)^2}{(2n-1)^2[1 + (2n-1)^2]^2} \cos[(2n-1)t]$$

$$+ \sum_{n=1}^{\infty} \frac{2(2n-1)}{(2n-1)^2[1 + (2n-1)^2]^2} \sin[(2n-1)t].$$

Exercises for Section 11.3 (Selected answers or hints are on page 457.)

For the exercises in this section (1, 2, and 3 given below), solve for $U(t)$ given each of the following Fourier series for $F(t)$ listed below as (a), (b), (c), and (d).

(a) $F(t) = \sum_{n=1}^{\infty} \frac{(-1)^{n+1}}{n} \sin(nt)$, where $F(t) = \frac{t}{2}$ for $-\pi < t < \pi$.

(b) $F(t) = \sum_{n=1}^{\infty} \frac{\sin(\frac{n\pi}{2})}{n} \cos(nt)$, where $F(t) = \begin{cases} -\frac{\pi}{4}, & \text{for} \quad \frac{\pi}{2} < t < \pi, \\ +\frac{\pi}{4}, & \text{for} \quad -\frac{\pi}{2} < t < \frac{\pi}{2}, \\ -\frac{\pi}{4}, & \text{for} \quad -\pi < t < -\frac{\pi}{2}. \end{cases}$

(c) $F(t) = \sum_{n=1}^{\infty} \frac{4\sin(\frac{n\pi}{2})}{n^2\pi} \sin(nt)$, shown in Figure 11.20.

(d) $F(t) = \sum_{n=1}^{\infty} \frac{1-(-1)^n}{2n^2} \cos(nt)$, shown in Figure 11.21.

1. $U''(t) + 2U'(t) + 2U(t) = F(t)$.

2. $U''(t) + 3U'(t) + 2U(t) = F(t)$.

3. $U''(t) + 4U'(t) + 4U(t) = F(t)$.

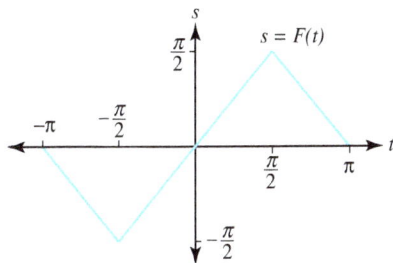

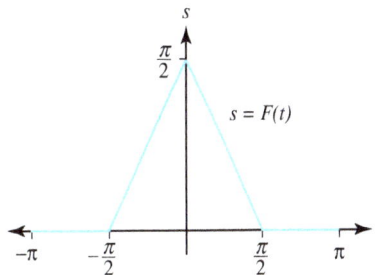

Figure 11.20: Graph of $U(t)$ for the exercises   Figure 11.21: Graph of $F(t)$ for the exercises

11.4   The Fourier Transform

Let $U(t)$ be a real-valued function with period 2π which is piecewise continuous such that $U'(t)$ also exists and is piecewise continuous. Then $U(t)$ has the *complex Fourier series* representation

$$U(t) = \sum_{n=-\infty}^{\infty} c_n e^{int}, \quad \text{where}$$

$$c_n = \frac{1}{2\pi} \int_{-\pi}^{\pi} U(t)e^{-int}\, dt \quad \text{for all} \quad n.$$

The coefficients $\{c_n t\}$ are complex numbers. Previously, we expressed $U(t)$ as a real trigonometric series:

$$U(t) = \frac{a_0}{2} + \sum_{n=1}^{\infty}(a_n \cos nt + b_n \sin nt). \tag{11.19}$$

Hence, a relationship between the coefficients is

$$a_n = c_n + c_{-n} \quad \text{for} \quad n = 0, 1, 2, \ldots, \quad \text{and}$$
$$b_n = i(c_n - c_{-n}) \quad \text{for} \quad n = 1, 2, \ldots,$$

These relations are easy to establish. We start by writing

$$U(t) = c_0 + \sum_{n=1}^{\infty} c_n e^{int} + \sum_{n=1}^{\infty} c_{-n} e^{-int} \tag{11.20}$$

$$= c_0 + \sum_{n=1}^{\infty} c_n(\cos nt + i\sin nt) + \sum_{n=1}^{\infty} c_{-n}(\cos nt - i\sin nt)$$

$$= c_0 + \sum_{n=1}^{\infty}[(c_n + c_{-n})\cos nt + i(c_n - c_{-n})\sin nt].$$

Comparing Equation (11.20) with Equation (11.19), we see that $a_0 = 2c_0$, $a_n = c_n + c_{-n}$, and $b_n = i(c_n - c_{-n})$.

If $U(t)$ and $U'(t)$ are piecewise continuous and have period $2L$, then $U(t)$ has the *complex Fourier series* representation

$$U(t) = \sum_{n=-\infty}^{\infty} c_n e^{i\frac{\pi n t}{L}}, \quad \text{where} \tag{11.21}$$

$$c_n = \frac{1}{2L} \int_{-L}^{L} U(t)e^{-i\frac{\pi n t}{L}} \, dt \quad \text{for all} \quad n. \tag{11.22}$$

We have seen how periodic functions are represented by trigonometric series. Many practical problems involve nonperiodic functions. A representation analogous to Fourier series for a nonperiodic function $U(t)$ is obtained by considering the Fourier series of $U(t)$ for $-L < t < L$ and then taking the limit as $L \to \infty$. The result is known as the Fourier transform of $U(t)$.

Start with a nonperiodic function $U(t)$ and consider the periodic function $U_L(t)$ with period $2L$, where

$$U_L(t) = U(t) \qquad \text{for} \quad -L < t \le L, \quad \text{and}$$
$$U_L(t) = U_L(t + 2L) \qquad \text{for all} \quad t.$$

Then $U_L(t)$ has the complex Fourier series representation

$$U_L(t) = \sum_{n=-\infty}^{\infty} c_n e^{i\frac{\pi n t}{L}}. \tag{11.23}$$

We introduce some terminology to discuss the terms in Equation (11.23), first

$$w_n = \frac{\pi n}{L} \quad \text{is called the } \textbf{frequency.} \tag{11.24}$$

If t denotes time, then the units for w_n are radians per unit time. The set of all possible frequencies is called the *frequency spectrum, i.e.,*

$$\left\{ \ldots, \frac{-3\pi}{L}, \frac{-2\pi}{L}, \frac{-\pi}{L}, \frac{\pi}{L}, \frac{2\pi}{L}, \frac{3\pi}{L}, \ldots \right\}.$$

It is important to note that as L increases, the spectrum becomes finer and approaches a continuous spectrum of frequencies. It is reasonable to expect that the summation in the Fourier series for $U_L(t)$ will give rise to an integral over $[-\infty.\infty]$. This is stated in the following important theorem.

Theorem 11.9 (Fourier Transform). *Let $U(t)$ and $U'(t)$ be piecewise continuous, and*

$$\int_{-\infty}^{\infty} |U(t)| \, dt < M,$$

for some positive constant M. The Fourier transform $F(w)$ of $U(t)$ is defined as

$$F(w) = \frac{1}{2\pi} \int_{-\infty}^{\infty} U(t)e^{-iwt} \, dt. \tag{11.25}$$

At points of continuity, $U(t)$ has the integral representation

$$U(t) = \int_{-\infty}^{\infty} F(w)e^{iwt} \, dw.$$

and at a point $t = a$ of discontinuity of U, the integral converges to $\frac{U(a^-)+U(a^+)}{2}$.

Note: It is common to express the fact that U is transformed into F by using the operator notation

$$\mathrm{F}\big(U(t)\big) = F(w).$$

Proof of Theorem 11.9. Set $\Delta w_n = w_{n+1} - w_n = \frac{\pi}{L}$ and $\frac{1}{2L} = \frac{1}{2\pi}\Delta w_n$. These quantities are used in conjunction with Equations (11.21), (11.22), (11.23) and the frequency in (11.24) to obtain

$$U_L(t) = \sum_{n=-\infty}^{\infty} \left[\frac{1}{2L} \int_{-L}^{L} U(t)e^{-iw_n t}\, dt \right] e^{iw_n t} \tag{11.26}$$

$$= \sum_{n=-\infty}^{\infty} \left[\frac{1}{2\pi} \int_{-L}^{L} U(t)e^{-iw_n t}\, dt \right] e^{iw_n t}\Delta w_n.$$

If we define $F_L(w)$ by

$$F_L(w) = \frac{1}{2\pi} \int_{-L}^{L} U(t)e^{-iwt}\, dt,$$

then Equation (11.26) can be written as

$$U_L(t) = \sum_{n=-\infty}^{\infty} F_L(w_n)e^{iw_n t}\Delta w_n. \tag{11.27}$$

As L gets large, $F_L(w_n)$ approaches $F(w_n)$ and Δw_n tends to zero. Thus the limit on the right-hand side of Equation (11.27) can be viewed as an integral. This substantiates the Fourier integral representation

$$U(t) = \int_{-\infty}^{\infty} F(w)e^{iwt}dw.$$

A more rigorous proof of this fact can be found in advanced texts. $\qquad\square$

The following table depicts some important properties of the Fourier transform.

Linearity	$\text{F}\big(aU_1(t) + bU_2(t)\big) = a\,\text{F}(U_1(t)) + b\,\text{F}\big(U_2(t)\big);$
Symmetry	If $\text{F}\big(U(t)\big) = F(w)$, then $\text{F}\big(F(t)\big) = \frac{1}{2\pi}U(-w);$
Time scaling	$\text{F}(U(at)) = \frac{1}{|a|}F(\frac{w}{a});$
Time shifting	$\text{F}(U(t - t_0)) = e^{-it_0 w}F(w);$
Frequency shifting	$\text{F}\big(e^{-iw_0 t}U(t)\big) = F(w - w_0);$
Time differentiation	$\text{F}\big(U'(t)\big) = iwF(w);$
Frequency differentiation	$\frac{d^n F(w)}{dw^n} = \text{F}\big((-it)^n U(t)\big);$
Moment Theorem	If $M_n = \int_{-\infty}^{\infty} t^n U(t)\, dt$, then $(-i)^n M_n = 2\pi F^{(n)}(0)$.

Table 11.1: Important Properties of the Fourier Transform

Example 11.5. Show that $F\left(e^{-|t|}\right) = \frac{1}{\pi(1+w^2)}$.

Solution:

Using formula 11.25 we obtain

$$
\begin{aligned}
F(w) &= \frac{1}{2\pi} \int_{-\infty}^{\infty} e^{-|t|} e^{-iwt}\, dt \\
&= \frac{1}{2\pi} \int_{-\infty}^{0} e^{(1-iw)t}\, dt + \frac{1}{2\pi} \int_{0}^{\infty} e^{(-1-iw)t}\, dt \\
&= \frac{1}{2\pi(1-iw)} e^{(1-iw)t}\Big|_{t=-\infty}^{t=0} + \frac{1}{2\pi(-1-iw)} e^{(-1-iw)t}\Big|_{t=0}^{t=\infty}. \\
&= \frac{1}{2\pi(1-iw)} + \frac{1}{2\pi(1+iw)} \\
&= \frac{1}{\pi(1+w^2)}.
\end{aligned}
$$

Example 11.6. Show that $F\left(\frac{1}{1+t^2}\right) = \frac{1}{2} e^{-|w|}$.

Solution:

Using the result of Example 11.5 and the symmetry property and the symmetry property, we obtain

$$
F\left(\frac{1}{\pi(1+t^2)}\right) = \frac{1}{2\pi} e^{-|-w|} = \frac{1}{2\pi} e^{-|w|}.
$$

By the linearity property, with each term is multiplied by π, we get

$$
F\left(\frac{1}{1+t^2}\right) = \frac{1}{2} e^{-|w|}.
$$

Exercises for Section 11.4 (Selected answers or hints are on page 458.)

1. Find $\mathfrak{F}\big(U(t)\big)$ for each of the following:

   (a) $U(t) = \begin{cases} 1 & \text{for } |t| < 1, \\ 0 & \text{for } |t| > 1. \end{cases}$

   (b) $U(t) = \begin{cases} 1 - |t| & \text{for } |t| \le 1, \\ 0 & \text{for } |t| > 1. \end{cases}$

2. Let $U(t) = \begin{cases} \sin t & \text{for } |t| \le \pi, \\ 0 & \text{for } |t| > \pi. \end{cases}$ Show that $\mathfrak{F}\big(U(t)\big) = \frac{i \sin \pi w}{\pi(1-w^2)}$.

3. Use the symmetry and linearity properties and the result of Exercise 1a to show that

$$
\mathfrak{F}\left(\frac{\sin t}{t}\right) = \begin{cases} \frac{1}{2} & \text{for } |w| < 1, \\ 0 & \text{for } |w| > 1. \end{cases}
$$

4. Let $U(t) = e^{-\frac{t^2}{2}}$. Show that $\mathfrak{F}\big(U(t)\big) = \frac{1}{\sqrt{2\pi}} e^{-\frac{w^2}{2}}$.

   Hint: Use the integral definition and combine the terms in the exponent, then complete the square and use the fact that $\int_{-\infty}^{\infty} e^{-\frac{r^2}{2}}\, dt = \sqrt{2\pi}$.

396

5. Use the time scaling property and the example in the text to show that

$$\mathfrak{F}(e^{-a|t|}) = \frac{|a|}{\pi(a^2 + w^2)}.$$

6. Use the symmetry and linearity properties and the result of Exercise 2 to show that

$$\mathfrak{F}\left(\frac{i \sin \pi t}{1 - t^2}\right) = \begin{cases} \frac{i \sin w}{2} & \text{for} \quad |w| \le \pi, \\ 0 & \text{for} \quad |w| > \pi. \end{cases}$$

7. Use the time differentiation property and the result of Exercise 4 to show that

$$\mathfrak{F}\left(te^{-\frac{r^2}{2}}\right) = \frac{-iwe^{-\frac{w^2}{2}}}{\sqrt{2\pi}}.$$

8. Use the symmetry and linearity properties and the results of Exercise 1b to show that

$$F\left(\frac{\sin^2 \frac{t}{2}}{t^2}\right) = \begin{cases} \frac{1 - |w|}{4\pi} & \text{for} \quad |w| \le 1, \\ 0 & \text{for} \quad |w| > 1. \end{cases}$$

9. Write a report on the Fourier transform. Discuss some of the ideas you found in the literature that are not mentioned in the text.

11.5   The Laplace Transform

11.5.1   From Fourier Transforms to Laplace Transforms

We have seen that certain real-valued functions $f(t)$ have a Fourier transform and that the integral

$$g(w) = \int_{-\infty}^{\infty} f(t)e^{-iwt}, dt$$

defines the complex function $g(w)$ of the real variable w. If we multiply the integrand $f(t)e^{-iwt}$ by $e^{-\sigma t}$, we create a complex function $G(\sigma + iw)$ of the complex variable $\sigma + iw$:

$$G(\sigma + iw) = \int_{-\infty}^{\infty} f(t)e^{-\sigma t}e^{-iwt}dt = \int_{-\infty}^{\infty} f(t)e^{-(\sigma + iw)t}dt.$$

The function $G(\sigma + iw)$ is called the **two-sided Laplace transform** of $f(t)$, and it exists when the Fourier transform of the function $f(t)e^{-\sigma t}$ exists. From the Fourier transform theory, we can state that a sufficient condition for $G(\sigma + iw)$ to exist is that

$$\int_{-\infty}^{\infty} |f(t)|e^{-\sigma t} \, dt < \infty$$

exists. For a given function $f(t)$, this integral is finite for values of σ that lie in some interval $a < \sigma b$.

The two-sided Laplace transform uses the lower limit of integration, that is, $t = -\infty$, and hence requires a knowledge of the past history of the function $f(t)$, $i.e.$, $t < 0$. For most physical applications, one is interested in the behavior of a system only for $t \ge 0$. Mathematically speaking, the initial conditions $f(0), f'(0), f''(0), \dots$, are a consequence of the past history of the system and are often all that is necessary to know. For this reason, it is useful to define the one-sided Laplace transform of $f(t)$, which is commonly referred to simply as the **Laplace transform** of $f(t)$, which is also defined as an integral:

$$\mathcal{L}\big(f(t)\big) = F(s) = \int_0^\infty f(t)e^{-st}\,dt, \quad \text{where} \quad s = \sigma + i\omega. \tag{11.28}$$

If the defining integral 11.28 for the Laplace transform exists for $s_0 = \sigma_0 + i\omega$, then values of σ with $\sigma > \sigma_0$ imply that $e^{-\sigma t} < e^{-\sigma_0 t}$, and thus

$$\int_0^\infty |f(t)|\,e^{-\sigma t}\,dt < \int_0^\infty |f(t)|\,e^{-\sigma_0 t}\,dt < \infty,$$

and it follows that $F(s)$ exists for $s = \sigma + i\omega$. Therefore, the Laplace transform $\mathcal{L}(f(t))$ is defined for all points s in the right half-plane $\mathrm{Re}(s) > \sigma_0$.

Another way to view the relationship between the Fourier transform and the Laplace transform is to consider the function $U(t)$ given by

$$U(t) = \begin{cases} f(t) & \text{for} \quad t \geq 0, \\ 0 & \text{for} \quad t < 0. \end{cases}$$

Then the Fourier transform theory shows us that

$$U(t) = \frac{1}{2\pi} \int_{-\infty}^\infty \left[\int_{-\infty}^\infty U(t)e^{-i\omega t}\,dt \right] e^{i\omega t}\,d\omega,$$

and since the integrand $U(t)$ is zero for $t < 0$, this equation can be written as

$$f(t) = \frac{1}{2\pi} \int_{-\infty}^\infty \left[\int_0^\infty f(t)e^{-i\omega t}\,dt \right] e^{i\omega t}\,d\omega.$$

Use the change of variable $s = \sigma + i\omega$ and $d\omega = \frac{ds}{i}$, where $\sigma > \sigma_0$ is held fixed, then the new limits of integration are from $s = \sigma - i\omega$ to $s = \sigma + i\omega$. The resulting equation is

$$f(t) = \frac{1}{2\pi} \int_{\sigma-i\infty}^{\sigma+i\infty} \left[\int_0^\infty f(t)e^{-st}\,dt \right] e^{st}\,ds.$$

Therefore, the Laplace transform is

$$\mathcal{L}\big(f(t)\big) = F(s) = \int_0^\infty f(t)e^{-st}\,dt, \quad \text{where} \quad s = \sigma + i\omega,$$

and the inverse Laplace transform is

$$\mathcal{L}^{-1}\big(F(s)\big) = f(t = \frac{1}{2\pi} \int_{\sigma-i\infty}^{\sigma+i\infty} F(s)e^{st}\,ds. \tag{11.29}$$

11.5.2   Properties of the Laplace Transform

Although a function $f(t)$ may be defined for all values of t, its Laplace transform is not influenced by values of $f(t)$, where $t < 0$. The Laplace transform of $f(t)$ is actually defined for the function $U(t)$ given by

$$U(t) = \begin{cases} f(t) & \text{for} \quad t \geq 0, \\ 0 & \text{for} \quad t < 0. \end{cases}$$

A sufficient condition for the existence of the Laplace transform is that $|f(t)|$ does not grow too rapidly as $t \to +\infty$. We say that the function f is of **exponential order** if there exists real constants $M > 0$ and K, such that

$$|f(t)| \le Me^{Kt} \quad \text{holds for all} \quad t \ge 0.$$

All functions in this chapter are assumed to be of exponential order. The next theorem shows that their Laplace transform $F(\sigma + i\tau)$ exists for values of s in a domain that includes the right half-plane $\text{Re}(s) > K$.

Theorem 11.10 (Existence of the Laplace Transform). *If f is of exponential order, then its Laplace transform $L\big(f(t)\big) = F(s)$ is given by*

$$F(s) = \int_0^\infty f(t)e^{-st}\, dt, \quad \text{where} \quad s = \sigma + i\omega.$$

The defining integral for F exists at points $s = \sigma + i\tau$ in the right half plane $\sigma > K$.

Proof. Using $s = \sigma + i\tau$ we see that $F(s)$ can be expressed as

$$F(s) = \int_0^\infty f(t)e^{-\sigma t} \cos \tau t\, dt - i \int_0^\infty f(t)e^{-\sigma t} \sin \tau t\, dt.$$

Then for values of $\sigma > K$, we have

$$\int_0^\infty |f(t)| e^{-\sigma t} |\cos \tau t|\, dt \le M \int_0^\infty e^{(K-\sigma)t}\, dt \le \frac{M}{\sigma - K}, \quad \text{and}$$

$$\int_0^\infty |f(t)| e^{-\sigma t} |\sin \tau t|\, dt \le M \int_0^\infty e^{(K-\sigma)t}\, dt \le \frac{M}{\sigma - K},$$

which imply that the integrals defining the real and imaginary parts of F exist for values of $\text{Re}(s) > K$. $\square$

Note: The domain of definition of the defining integral for the Laplace transform $\mathcal{L}\big(f(t)\big)$ seems to be restricted to a half plane. However, the resulting formula $F(s)$ might have a domain much larger than this half plane. Later we will show that $F(s)$ is an analytic function of the complex variable s. For most applications involving Laplace transforms that we will study, the Laplace transforms are rational functions that have the form $\frac{P(s)}{Q(s)}$, where P and Q are polynomials, and some other important ones will have the form $\frac{e^{as}P(s)}{Q(s)}$.

Theorem 11.11 (Linearity of the Laplace Transform). *Let f and g have Laplace transforms F and G, respectively. If a and b are constants, then*

$$\mathcal{L}\big(af(t) + bg(t)\big) = aF(s) + bG(s).$$

Proof. Let K be chosen so that both F and G are defined for $\text{Re}(s) > K$, then

$$\mathcal{L}\big(af(t) + bg(t)\big) = \int_0^\infty [af(t) + bg(t)]e^{-st}\, dt$$

$$= a \int_0^\infty f(t)e^{-st}dt + b \int_0^\infty g(t)e^{-st}\, dt$$

$$= aF(s) + bG(s).$$

$\square$

Theorem 11.12 (Uniqueness of the Laplace Transform). *Let f and g have Laplace transforms F and G, respectively. If $F(s) \equiv G(s)$, then $f(t) \equiv g(t)$.*

Proof. If σ is sufficiently large, then the integral representation, Equation (11.29), for the inverse Laplace transform can be used to obtain

$$f(t) = \mathcal{L}^{-1}(F(s)) = \frac{1}{2\pi i} \int_{\sigma-i\infty}^{\sigma+i\infty} F(s)e^{st}\, ds = \frac{1}{2\pi i} \int_{s-i\infty}^{s+i\infty} G(s)e^{st}\, ds$$

$$= \mathcal{L}^{-1}i(G(s)) = g(t).$$

$\square$

Example 11.7. Show that the Laplace transform of the step function given by

$$f(t) = \begin{cases} 1 & \text{for } 0 \le t < c, \\ 0 & \text{for } c < t \end{cases} \quad \text{is} \quad \mathcal{L}(f(t)) = \frac{1 - e^{-cs}}{s}.$$

Solution:

Using the integral definition for $\mathcal{L}(f(t))$, we obtain

$$\mathcal{L}(f(t)) = \int_0^\infty f(t)e^{-st}\, dt$$

$$= \int_0^c e^{-st}\, dt + \int_c^\infty e^{-st}(0)\, dt$$

$$= \frac{-e^{-st}}{s} \Big|_{t=0}^{t=c}$$

$$= \frac{1 - e^{-cs}}{s}.$$

Example 11.8. Show that $\mathcal{L}(e^{at}) = \frac{1}{s-a}$, where a is a real constant.

Solution:

We will actually show that the integral defining $\mathcal{L}(e^{at})$ is equal to the formula $F(s) = \frac{1}{s-a}$ for values of s with $\operatorname{Re}(s) > a$, and the extension to other values of s is inferred by our knowledge about the domain of a rational function. Using straightforward integration techniques we find that

$$\mathcal{L}(e^{at}) = \int_0^\infty e^{at}e^{-st}\, dt$$

$$= \lim_{R \to +\infty} \int_0^R e^{(a-s)t}\, dt$$

$$= \lim_{R \to +\infty} \frac{e^{(a-s)R}}{a-s} + \frac{1}{s-a}.$$

Let $s = \sigma + i\tau$ be held fixed, or where $\sigma > a$. Then since $a - \sigma$ is a negative real number we have $\lim_{R \to +\infty} e^{(a-s)R} = 0$ and this result can be used in the latter equation to obtain the desired conclusion.

The property of linearity can be used to find new Laplace transforms from known ones.

Example 11.9. Show that $\mathcal{L}(\sinh at) = \frac{a}{s^2-a^2}$.

Solution:

Since $\sinh at = \frac{1}{2}e^{at} - \frac{1}{2}e^{-at}$, we obtain

$$\mathcal{L}(\sinh at) = \frac{1}{2}\mathcal{L}(e^{at}) - \frac{1}{2}\mathcal{L}(e^{-at}) = \frac{1}{2}\left(\frac{1}{s-a}\right) - \frac{1}{2}\left(\frac{1}{s+a}\right) = \frac{a}{s^2-a^2}.$$

The technique of integration by parts is also helpful in finding new Laplace transforms.

Example 11.10. Show that $\mathcal{L}(t) = \frac{1}{s^2}$.

Solution:

Using integration by parts we obtain

$$\mathcal{L}(t) = \lim_{R \to +\infty} \int_0^R te^{-st}\, dt$$

$$= \lim_{R \to +\infty} \left(\frac{-t}{s}e^{-st} - \frac{1}{s^2}e^{-st}\right)\Bigg|_{t=0}^{t=R}$$

$$= \lim_{R \to +\infty} \left(\frac{-R}{s}e^{-sR} - \frac{1}{s^2}e^{-sR}\right) + 0 + \frac{1}{s^2}.$$

For values of s in the right half plane $\mathrm{Re}(s) > 0$, an argument similar to that in Example 11.8 shows that the last limit approaches zero. This observation establishes the result.

Example 11.11. Show that $\mathcal{L}(\cos bt) = \frac{s}{s^2+b^2}$.

Solution:

A direct approach using the definition is tedious. Let us assume that the complex constants $\pm ib$ are permitted and hence following the Laplace transforms exist:

$$\mathcal{L}(e^{ibt}) = \frac{1}{s-ib}, \quad \text{and} \quad \mathcal{L}(e^{-ibt}) = \frac{1}{s-ib}.$$

Using the linearity of the Laplace transform we obtain

$$\mathcal{L}(\cos bt) = \frac{1}{2}\mathcal{L}(e^{ibt}) + \frac{1}{2}\mathcal{L}(e^{-ibt}) = \frac{1}{2}\frac{1}{s-ib} + \frac{1}{2}\frac{1}{s+ib} = \frac{s}{s^2+b^2}.$$

Inverting the Laplace transform is usually accomplished with the aid of a table of known Laplace transforms and the technique of partial fraction expansion.

Example 11.12. Find $\mathcal{L}^{-1}\left(\frac{3s+6}{s^2+9}\right)$.

Solution:

Using linearity and lines 6 and 7 of Table 11.2, we obtain

$$\mathcal{L}^{-1}\left(\frac{3s+6}{s^2+9}\right) = 3\mathcal{L}^{-1}\left(\frac{s}{s^2+9}\right) + 2\mathcal{L}^{-1}\left(\frac{3}{s^2+9}\right) = 3\cos 3t + 2\sin 3t.$$

Table 11.2 gives the Laplace transforms of some well-known functions, and Table 11.3 highlights some important properties of Laplace transforms.

	$f(t)$	$F(s) = \int_0^\infty f(t)e^{-st}\,dt$
1.	1	$\frac{1}{s}$
2.	t^n	$\frac{n!}{s^{n+1}}$
3.	$U_c(t)$ unit step function	$\frac{e^{-cs}}{s}$
4.	e^{at}	$\frac{1}{s-a}$
5.	$t^n e^{at}$	$\frac{n!}{(s-a)^{n+1}}$
6.	$\cos bt$	$\frac{s}{s^2+b^2}$
7.	$\sin bt$	$\frac{b}{s^2+b^2}$
8.	$e^{at}\cos bt$	$\frac{s-a}{(s-a)^2+b^2}$
9.	$e^{at}\sin bt$	$\frac{b}{(s-a)^2+b^2}$
10.	$t\cos bt$	$\frac{s^2-b^2}{(s^2+b^2)^2}$
11.	$t\sin bt$	$\frac{2bs}{(s^2+b^2)^2}$
12.	$\cosh at$	$\frac{s}{s^2-a^2}$
13.	$\sinh at$	$\frac{a}{s^2-a^2}$

Table 11.2: Well-Known Laplace Transforms

Definition	$\mathcal{L}(\,f(t)) = F(s)$
First derivative	$\mathcal{L}(\,f'(t)) = sF(s) - f(0)$
Second derivative	$\mathcal{L}(\,f''(t)) = s^2 F(s) - sf(0) - f'(0)$
Integral	$\mathcal{L}(\int_0^t f(\tau)\,d\tau) = \frac{F(s)}{s}$
Multiplication by t	$\mathcal{L}(tf(t)) = -F'(s)$
Division by t	$\mathcal{L}(\frac{f(t)}{t}) = \int_s^\infty F(\sigma)\,d\sigma$
s axis Shifting	$\mathcal{L}(e^{at}f(t)) = F(s-a)$
t axis Shifting	$\mathcal{L}(U_a(t)f(t-a)) = e^{-as}F(s) \quad$ for $\quad a > 0$
Convolution	$\mathcal{L}\big(h(t)\big) = F(s)G(s) \quad$ where $\quad h(t) = \int_0^t f(t-\tau)g(\tau)\,d\tau$

Table 11.3: Properties of the Laplace Transform

Exercises for Section 11.5 (Selected answers or hints are on page 459.)

1. Show that $\mathcal{L}(1) = \frac{1}{s}$ by using the integral definition for the Laplace transform. Assume that $\mathrm{Re}(s) > 0$.

2. Show that $\mathcal{L}(t^2) = \frac{2}{s^3}$ by using the integral definition for the Laplace transform. Assume that $\mathrm{Re}(s) > 0$.

3. Find $\mathcal{L}\big(f(t)\big)$ for each of the following.

(a) $U(t) = \begin{cases} t & \text{for } 0 \le t < c, \\ 0 & \text{otherwise.} \end{cases}$

(b) $U(t) = \begin{cases} 1 & \text{for } 1 < t < 2, \\ 0 & \text{otherwise.} \end{cases}$

(c) $U(t) = \begin{cases} e^{at} & \text{for } 0 \le t < 1, \\ 0 & \text{otherwise.} \end{cases}$

(d) $U(t) = \begin{cases} \sin t & \text{for } 0 \le t \le \pi, \\ 0 & \text{otherwise.} \end{cases}$

4. Use the linearity of Laplace transform and Table 11.2 to find:

(a) $\mathcal{L}(3t^2 - 4t + 5)$.

(b) $\mathcal{L}(2\cos 4t)$.

(c) $\mathcal{L}(e^{2t-3})$.

(d) $\mathcal{L}(6e^{-t} + 3\sin 5t)$.

(e) $\mathcal{L}((t+1)^4)$.

(f) $\mathcal{L}(\cosh 2t)$.

5. Use the linearity of the inverse Laplace transform and Table 11.3 to find:

(a) $\mathcal{L}^{-1}\left(\frac{1}{s^2+25}\right)$.

(b) $\mathcal{L}^{-1}\left(\frac{4}{2} - \frac{6}{s^2}\right)$.

(c) $\mathcal{L}^{-1}\left(\frac{1+s^2-s}{s^4}\right)$.

(d) $\mathcal{L}^{-1}\left(\frac{2s+9}{s^2+9}\right)$.

(e) $\mathcal{L}^{-1}\left(\frac{6s}{s^2-4}\right)$.

(f) $\mathcal{L}^{-1}\left(\frac{2s+1}{s(s+1)}\right)$.

6. Write a report on how complex analysis is used in the study of Laplace transforms. Include ideas and examples that are not mentioned in the text.

11.6   Laplace Transforms of Derivatives and Integrals

Theorem 11.13 (Differentiation of $f(t)$). *Let $f(t)$ and $f'(t)$ be continuous for $t \ge 0$, and of exponential order. Then,*

$$\mathcal{L}(f'(t)) = sF(s) - f(0), \quad \text{where} \quad F(s) = \mathcal{L}(f(t)).$$

Proof. Let K be chosen large enough so that both $f(t)$ and $f'(t)$ are of exponential order K. If $\text{Re}(s) > K$, then $\mathcal{L}(f'(t))$ is given by

$$\mathcal{L}(f'(t)) = \int_0^\infty f'(t)e^{-st}\, dt.$$

Next, using integration by parts, we can write the equation this as

$$\mathcal{L}(f'(t)) = \lim_{R \to +\infty} \left[f(t)e^{-st}\right]\Big|_{t=0}^{t=R} + s\int_0^\infty f(t)e^{st}\, dt.$$

Now, $f(t)$ is of exponential order K, and $\mathrm{Re}(s) > K$. It follows that $\lim_{R \to +\infty} f(R)e^{-sR} = 0$, so

$$\mathcal{L}(f'(t)) = -f(0) + s \int_0^\infty f(t)e^{-st}\, dt = sF(s) - f(0).$$

$\square$

An easy consequence of Theorem 11.13 is the following, which we state without proof.

Corollary 11.1. *If $f(t)$, $f'(t)$, and $f''(t)$ are of exponential order, then*

$$\mathcal{L}(f''(t)) = s^2 f(s) - sf(0) - f'(0).$$

Example 11.13. Show that $\mathcal{L}(\cos^2 t) = \frac{s^2+2}{s(s^2+4)}$.

Solution:

Let $f(t) = \cos^2 t$, then $f(0) = 1$ and $f'(t) = -2\sin t \cos t = -\sin 2t$. Using the fact that $\mathcal{L}(-\sin 2t) = -\frac{2}{s^2+4}$, Theorem 11.13 implies that

$$-\frac{2}{s^2+4} = \mathcal{L}(f'(t)) = s\mathcal{L}(\cos^2 t) - 1.$$

from which it follows that $\mathcal{L}(\cos^2 t) = -\frac{2}{s(s^2+4)} + \frac{1}{s} = \frac{s^2+2}{s(s^2+4)}$.

Theorem 11.14. *(Integration of $f(t)$] Let $f(t)$ be continuous for $t \geq 0$, and of exponential order. Further, let $F(s)$ be its Laplace transform, then*

$$\mathcal{L}\left(\int_0^t f(\tau)\, d\tau\right) = \frac{F(s)}{s}.$$

Proof. Let $g(t) = \int_0^t f(\tau)d\tau$, then $g'(t) = f(t)$ and $g(0) = 0$. If we can show that g is of exponential order, then Theorem 11.13 implies that

$$\mathcal{L}(f(t)) = \mathcal{L}(g'(t)) = s\mathcal{L}(g(t)) - 0 = s\mathcal{L}\left(\int_0^t f(\tau)\, d\tau\right).$$

Since $f(t)$ is of exponential order, there are positive values M and K so that

$$|g(t)| \leq \int_0^t f(\tau)\, d\tau \leq M \int_0^t e^{K\tau}\, d\tau = \frac{M}{K}(e^{Kt} - 1) \leq e^{Kt}.$$

Therefore, g is of exponential order. $\square$

Example 11.14. Show that $\mathcal{L}(t^2) = \frac{2}{s^3}$ and $\mathcal{L}(t^3) = \frac{6}{s^4}$.

Solution:

Using Theorem 11.14 and the fact that $\mathcal{L}(2t) = \frac{2}{s^2}$ we obtain

$$\mathcal{L}(t^2) = \mathcal{L}\left(\int_0^t 2\tau\, d\tau\right) = \frac{1}{s}\mathcal{L}(2t) = \frac{1}{s}\frac{2}{s^2} = \frac{2}{s^3}.$$

Now we can use the first result $\mathcal{L}(t^2) = \frac{2}{s^3}$ to establish the second one:

$$\mathcal{L}(t^3) = \mathcal{L}\left(\int_0^t 3\tau^2\, d\tau\right) = \frac{1}{s}\mathcal{L}(3t^2) = \frac{1}{s}\frac{6}{s^3} = \frac{6}{s^4}.$$

One of the main uses of the Laplace transform is its role in the solution of differential equations. The utility of the Laplace transform lies in the fact that the transform of the derivative $f'(t)$ corresponds to multiplication of the transform $F(s)$ by s and then the subtraction of $f(0)$. This permits us to replace the calculus operation of differentiation with simple algebraic operations on transforms.

This idea is used to develop a method for solving linear differential equations with constant coefficients. Consider the initial value problem

$$y''(t) + ay'(t) + by(t) = f(t)$$

with initial conditions $y(0) = y_0$ and $y'(0) = d_0$. The linearity property of the Laplace transform can be used to obtain the equation

$$\mathcal{L}(y''(t)) + a\mathcal{L}(y'(t)) + b\mathcal{L}(y(t)) = \mathcal{L}(f(t)).$$

Let $Y(s) = \mathcal{L}(y(t))$ and $F(s) = \mathcal{L}(f(t))$. Now apply Theorem 11.13 and Corollary 11.1 to get

$$\mathcal{L}(y'(t)) = sY(s) - y(0) \quad \text{and} \quad \mathcal{L}(y''(t)) = s^2Y(s) - sy(0) - y'(0).$$

We can rewrite the last equation in the form

$$s^2Y(s) + asY(s) + bY(s) = F(s) + sy(0) + y'(0) + ay(0). \tag{11.30}$$

The Laplace transform $Y(s)$ of the solution $y(t)$ can be shown to be

$$Y(s) = \frac{F(s) + sy(0) + y'(0) + ay(0)}{s^2 + as + b}. \tag{11.31}$$

For many physical problems involving mechanical systems and electric circuits, the transform $F(s)$ is known, and the inverse of $Y(s)$ can easily be computed. This process is referred to as the operational calculus and has the advantage of changing problems in differential equations into problems in algebra. Then the solution obtained will satisfy the specific initial conditions.

Example 11.15. Solve the initial value problem

$$y''(t) + y(t) = 0 \quad \text{with} \quad y(0) - 2 \quad \text{and} \quad y'(0) = 3.$$

Solution:
Since the right-hand side of the differential equation is $f(t) \equiv 0$ we have $F(s) \equiv 0$. The initial conditions yield $\mathcal{L}(y''(t)) = s^2Y(s) - 2s - 3$ and Equation (11.30) becomes $s^2Y(s) + Y(s) = 2s + 3$. Solving we get $Y(s) = \frac{2s+3}{s^2+1}$ and the solution $y(t)$ is assisted by using Table 11.2 and the computation

$$y(t) = \mathcal{L}^{-1}\left(\frac{2s+3}{s^2+1}\right) = 2\mathcal{L}^{-1}\left(\frac{s}{s^2+1}\right) + 3\mathcal{L}^{-1}\left(\frac{1}{s^2+1}\right) = 2\cos t + 3\sin t.$$

Example 11.16. Solve the initial value problem

$$y''(t) + y'(t) - 2y(t) = 0 \quad \text{with} \quad y(0) = 1 \quad \text{and} \quad y'(0) = 4.$$

Solution:

In the spirit of Example 11.15, we use the initial conditions and Equation (11.31) becomes

$$Y(s) = \frac{s+4+1}{s^2+s-2} = \frac{s+5}{(s-1)(s+2)}.$$

Using partial fraction expansion $Y(s) = \frac{2}{s-1} - \frac{1}{s+2}$ and the solution $y(t)$ is

$$y(t) = \mathcal{L}^{-1}(Y(s)) = 2\mathcal{L}^{-1}\left(\frac{1}{s-1}\right) - \mathcal{L}^{-1}\left(\frac{1}{s+2}\right) = 2e^t - e^{-2t}.$$

Exercises for Section 11.6 (Selected answers or hints are on page 459.)

1. Derive $\mathcal{L}(\sin t)$ from $\mathcal{L}(\cos t)$.

2. Derive $\mathcal{L}(\cosh t)$ from $\mathcal{L}(\sinh t)$.

3. Find $\mathcal{L}(\sin^2 t)$.

4. Show that $\mathcal{L}(te^t) = \frac{1}{(s-1)^2}$. *Hint:* Let $f(t) = te^t$ and $f'(t) = te^t + e^t$.

5. Find $\mathcal{L}^{-1}\left(\frac{1}{s(s-4)}\right)$.

6. Find $\mathcal{L}^{-1}\left(\frac{1}{s(s^2+4)}\right)$.

7. Show that $\mathcal{L}^{-1}\left(\frac{1}{s^2(s+1)}\right) = t - 1 + e^t$.

8. Show that $\mathcal{L}^{-1}\left(\frac{1}{s^2(s^2+1)}\right) = t - \sin t$.

   Solve the initial value problem in the following exercises.

9. $y''(t) + 9y(t) = 0$, with $y(0) = 2$ and $y'(0) = 9$.

10. $y''(t) + y(t) = 1$, with $y(0) = 0$ and $y'(0) = 2$.

11. $y''(t) + 4y(t) = -8$, with $y(0) = 0$ and $y'(0) = 2$.

12. $y'(t) + y(t) = 1$, with $y(0) = 2$.

13. $y'(t) - y(t) = -2$, with $y(0) = 3$.

14. $y''(t) - 4y(t) = 0$, with $y(0) = 1$ and $y'(0) = 2$.

15. $y''(t) - y(t) = 1$, with $y(0) = 0$ and $y'(0) = 2$.

16. $y'(t) + 2y(t) = 3e^t$, with $y(0) = 2$.

17. $y''(t) + y'(t) - 2y(t) = 0$, with $y(0) = 2$ and $y'(0) = -1$.

18. $y''(t) - y'(t) - 2y(t) = 0$, with $y(0) = 2$ and $y'(0) = 1$.

11.7 Shifting Theorems and the Step Function

We have seen how the Laplace transform can be used to solve linear differential equations. Familiar functions that arise in solutions to differential equations are $e^{at} \cos bt$ and $e^{at} \sin bt$. The *first shifting theorem* will show how their transforms are related to those of $\cos bt$ and $\sin bt$ by shifting the variable s in $F(s)$. A companion result, called the *second shifting theorem*, will show how the transform of $f(t - a)$ can be obtained by multiplying $F(s)$ by e^{-as}. Loosely speaking, these results show that multiplication of $f(t)$ by e^{at} corresponds to shifting $F(s-a)$, and shifting $f(t-a)$ corresponds to multiplication of the transform $F(s)$ by e^{as}.

Theorem 11.15 (Shifting the Variable "s"). *If $F(s)$ is the Laplace transform of $f(t)$, then*

$$\mathcal{L}\big(e^{at} f(t)\big) = F(s - a).$$

Proof. Using the integral definition $\mathcal{L}\big(f(t)\big) = F(s) = \int_0^\infty f(t)e^{-st}\,dt$, we see that

$$\mathcal{L}\big(e^{at} f(t)\big) = \int_0^\infty e^{at} f(t)e^{-st}dt = \int_0^\infty f(t)e^{-(s-a)t}\,dt = F(s - a).$$

$\square$

Definition 11.3 (Unit Step Function). *For $a \geq 0$ the unit step function $U_a(t)$ is*

$$U_a(t) = \begin{cases} 0 & \text{for} \quad t < a. \\ 1 & \text{for} \quad t > a. \end{cases}$$

Figure 11.22 depicts the graph of $U_a(t)$.

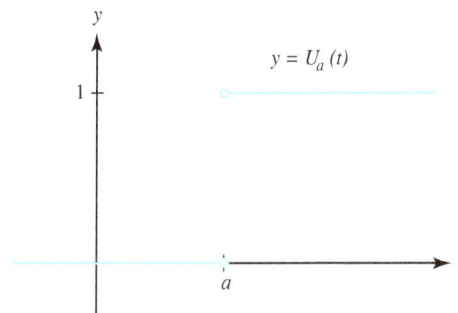

Figure 11.22: Graph of the unit step function $y = U_a(t)$ in Definition 11.3

Theorem 11.16 (Shifting the Variable "t"). *If $F(s)$ is the Laplace transform of $f(t)$ and $a \geq 0$, then*

$$\mathcal{L}\big(U_a(t)f(t - a)\big) = e^{-as}F(s),$$

where $f(t)$ and $U_a(t)(t - a)$ are illustrated in Figure 11.23.

Proof. Using the definition of Laplace transform, we write

$$e^{-as}F(s) = e^{-as}\int_0^\infty f(\tau)e^{-s\tau}\,d\tau = \int_0^\infty f(\tau)e^{-s(a+\tau)}\,d\tau.$$

Using the change of variable $t = a + \tau$ and $dt = d\tau$, we obtain

$$e^{-as}F(s) = \int_a^\infty f(t - a)e^{-st}\,dt.$$

Since $U_a(t) f(t-a) = 0$ for $t < a$, and $U_a(t) f(t-a) = f(t-a)$ for $t > a$, we can write the last equation as

$$e^{-as} F(s) = \int_0^\infty U_a(t) f(t-a) e^{-st}\, dt = \mathcal{L}\big(U_a(t) f(t-a)\big).$$

□

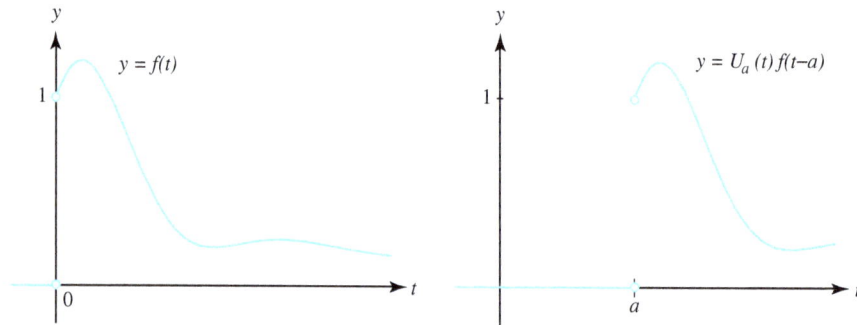

Figure 11.23: Comparison of the functions $f(t)$ and $U_a(t) f(t-a)$

Example 11.17. Show that $\mathcal{L}(t^n e^{at}) = \frac{n!}{(s-a)^{n+1}}$.

Solution:

Let $f(t) = t^n$, then $F(s) = \mathcal{L}(t^n) = \frac{n!}{s^{n+1}}$. Applying Theorem 11.15, we obtain the desired result:

$$\mathcal{L}(t^n e^{at}) = F(s-a) = \frac{n!}{(s-a)^{n+1}}.$$

Example 11.18. Show that $\mathcal{L}(U_c(t)) = \frac{e^{-cs}}{s}$.

Solution:

Set $f(t) = 1$, then $F(s) = \mathcal{L}(1) = \frac{1}{s}$. Now apply Theorem 11.16 and get

$$\mathcal{L}\big(U_c(t)\big) = \mathcal{L}\big(U_c(t) f(t)\big) = \mathcal{L}\big(U_c(t) \cdot 1\big) = e^{-cs} \mathcal{L}(1) = \frac{e^{-cs}}{s}.$$

Example 11.19. Find $\mathcal{L}\big(f(t)\big)$ if $f(t)$ is given in Figure 11.24.

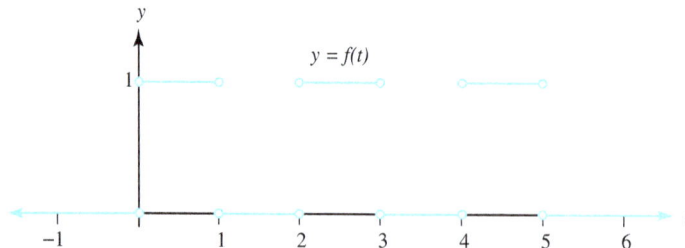

Figure 11.24: The function $y = f(t)$ of Example 11.19

Solution:

We can represent $f(t)$ in terms of step functions:

$$f(t) = 1 - U_1(t) + U_2(t) - U_3(t) + U_4(t) - U_5(t).$$

Using the result of Example 10.18 and linearity, we obtain

$$\mathcal{L}(f(t)) = \frac{1}{s} - \frac{e^{-s}}{s} + \frac{e^{-2s}}{s} - \frac{e^{-3s}}{s} + \frac{e^{-4s}}{s} - \frac{e^{-5s}}{s}.$$

Example 11.20. Use Laplace transforms to solve the initial value problem

$$y''(t) + y(t) = U_\pi(t) \quad \text{with} \quad y(0) = 0 \quad \text{and} \quad y'(0) = 0.$$

Solution:

As usual, let $Y(s)$ denote the Laplace transform of $y(t)$. Then we get

$$s^2 Y(s) + Y(s) = \frac{e^{-\pi s}}{s}.$$

Solving for $Y(s)$, we obtain

$$Y(s) = e^{-\pi s} \frac{1}{s(s^2 + 1)} = \frac{e^{-\pi s}}{s} - \frac{e^{-\pi s} s}{s^2 + 1}.$$

We now use Theorem 11.16 and the facts that $\frac{1}{s}$ and $\frac{s}{s^2+1}$ are the transforms of 1 and $\cos t$, respectively. The solution $y(t)$ computes as follows:

$$y(t) = \mathcal{L}^{-1}\left(\frac{e^{-\pi s}}{s}\right) - \mathcal{L}^{-1}\left(\frac{e^{-\pi s} s}{s^2 + 1}\right) = U_\pi(t) - U_\pi(t)\cos(t - \pi),$$

which can be written in the more familiar form:

$$y(t) = \begin{cases} 0 & \text{for } t < \pi, \\ 1 - \cos t & \text{for } t > \pi. \end{cases}$$

Exercises for Section 11.7 (Selected answers or hints are on page 459.)

1. Find $\mathcal{L}(e^t - te^t)$.

2. Find $\mathcal{L}(e^{-4t} \sin 3t)$.

3. Show that $\mathcal{L}(e^{at} \cos bt) = \frac{s-a}{(s-a)^2 + b^2}$.

4. Show that $\mathcal{L}(e^{at} \sin bt) = \frac{b}{(s-a)^2 + b^2}$.

5. Find $\mathcal{L}^{-1}(F(s))$ for the following:

   (a) $F(s) = \frac{s+2}{s^2 + 4s + 5}$.

   (b) $F(s) = \frac{8}{s^2 - 2s + 5}$.

   (c) $F(s) = \frac{s+3}{(s+2)^2 + 1}$.

   (d) $F(s) = \frac{2s+10}{s^2 + 6s + 25}$.

6. Find $\mathcal{L}(f(t))$ for the following:

   (a) $f(t) = U_2(t)(t - 2)^2$.

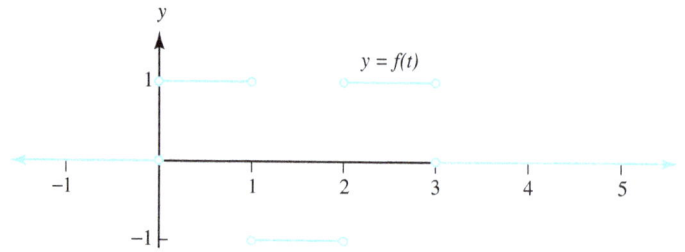

Figure 11.25: The graph $y = f(t)$ (for Problem 6e)

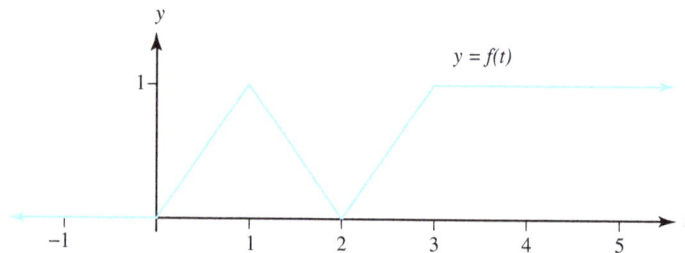

Figure 11.26: The graph $y = f(t)$ (for Problem 6f)

(b) $f(t) = U_1(t)e^{1-t}$.

(c) $f(t) = U_{3\pi}(t)\sin(t - 3\pi)$.

(d) $f(t) = 2U_1(t) - U_2(t) - U_3(t)$

(e) Let $f(t)$ be the function given in Figure 11.25.

(f) Let $f(t)$ be the function given in Figure 11.26.

   Hint: The function is the integral of the one in Exercise 13.

7. Find $\mathcal{L}^{-1}(\frac{e^{-s} + e^{-2s}}{s})$.

8. Find $\mathcal{L}^{-1}(\frac{1 - e^{-s} + e^{-2s}}{s^2})$.

9. Solve the initial value problem for each of the following:

   (a) $y''(t) + 2y'(t) + 2y(t) = 0$, with $y(0) = -1$ and $y'(0) = 1$.

   (b) $y''(t) + 4y'(t) + 5y(t) = 0$, with $y(0) = 1$ and $y'(0) = -2$.

   (c) $2y''(t) + 2y'(t) + y(t) = 0$, with $y(0) = 0$ and $y'(0) = 1$.

   (d) $y''(t) - 2y'(t) + y(t) = 2e^t$, with $y(0) = 0$ and $y'(0) = 0$.

   (e) $y''(t) + 2y'(t) + y(t) = 6te^{-t}$, with $y(0) = 0$ and $y'(0) = 0$.

   (f) $y''(t) + 2y'(t) + y(t) = 2U_1(t)e^{1-t}$, with $y(0) = 0$ and $y'(0) = 0$.

   (g) $y''(t) + y(t) = U_{\pi/2}(t)$, with $y(0) = 0$ and $y'(0) = 1$.

11.8   Multiplication and Division by t

Sometimes the solutions to nonhomogeneous linear differential equations with constant coefficients involve the functions $t\cos bt$, $t\sin bt$, or $t^n e^{at}$ as part of the solution. We now show how the Laplace transforms of $tf(t)$ and $\frac{f(t)}{t}$ are related to the Laplace transform of $f(t)$. The transform of $tf(t)$ will be obtained via differentiation and the transform of $\frac{f(t)}{t}$ will be obtained via integration. To be precise, we state the following theorems.

Theorem 11.17 (Multiplication by "t"). *If $F(s)$ is the Laplace transform of $f(t)$, then*

$$\mathcal{L}\big(tf(t)\big) = -F'(s).$$

Proof. By definition we have $F(s) = \int_0^\infty f(t)e^{-st}\,dt$. Leibniz's rule for partial differentiation under the integral sign permits us to write

$$F'(s) = \frac{\partial}{\partial s} \int_0^\infty f(t)e^{-st}\,dt$$

$$= \int_0^\infty \frac{\partial}{\partial s}[f(t)e^{-st}]\,dt$$

$$= \int_0^\infty [-tf(t)e^{-st}]\,dt$$

$$= -\int_0^\infty tf(t)e^{-st}\,dt$$

$$= -\mathcal{L}\big(tf(t)\big).$$

$\square$

Theorem 11.18 (Division by t). *Let both $f(t)$ and $\frac{f(t)}{t}$ have Laplace transforms and let $F(s)$ denote the transform of $f(t)$. If $\lim\limits_{t\to 0^+} \frac{f(t)}{t}$ exists,*

$$\mathcal{L}\left(\frac{f(t)}{t}\right) = \int_s^\infty F(\sigma)\,d\sigma.$$

Proof. Since $F(\sigma) = \int_0^\infty f(t)e^{-\sigma t}\,dt$, we integrate $F(\sigma)$ from s to ∞:

$$\int_s^\infty F(\sigma)d\sigma = \int_s^\infty \left[\int_0^\infty f(t)e^{-\sigma t}\,dt\right] d\sigma.$$

The order of integration in equation in the double integral is reversed:

$$\int_s^\infty F(\sigma)\,d\sigma = \int_0^\infty \left[\int_s^\infty f(t)e^{-\sigma t}d\sigma\right] dt$$

$$= \int_0^\infty \left[\frac{-f(t)}{t}e^{-\sigma t}\Big|_{\sigma=s}^{\sigma=\infty}\right] dt$$

$$= \int_0^\infty \frac{f(t)}{t}e^{-st}\,dt$$

$$= \mathcal{L}\left(\frac{f(t)}{t}\right).$$

$\square$

Example 11.21. Show that $\mathcal{L}(t\cos bt) = \frac{s^2-b^2}{(s^2+b^2)^2}$.

Solution:

Let $f(t) = \cos bt$, then $F(s) = \mathcal{L}(\cos bt) = \frac{s}{s^2+b^2}$. Hence, we can differentiate $F(s)$ to obtain the desired result:

$$\mathcal{L}(t\cos bt) = -F'(s) = -\frac{s^2+b^2-2s^2}{(s^2+b^2)^2} = \frac{s^2-b^2}{(s^2+b^2)^2}.$$

Example 11.22. Show that $\mathcal{L}(\frac{\sin t}{t}) = \arctan \frac{1}{s}$.

Solution:

Let $f(t) = \sin t$ and $F(s) = \frac{1}{s^2+1}$. Since $\lim\limits_{t \to 0^+} \frac{\sin t}{t} = 1$, we can integrate $F(s)$ to obtain the desired result:

$$\mathcal{L}\left(\frac{\sin t}{t}\right) = \int_s^\infty \frac{d\sigma}{\sigma^2+1} = -\arctan\frac{1}{\sigma}\Big|_{\sigma=s}^{\sigma=\infty} = \arctan\frac{1}{s}.$$

Some types of differential equations involve the terms $ty'(t)$ or $ty''(t)$. Laplace transforms can be used to find the solution if we use the additional substitutions

$$\mathcal{L}(ty'(t)) = -sY'(s) - Y(s), \quad \text{and} \tag{11.32}$$
$$\mathcal{L}(ty''(t)) = -s^2Y'(s) - 2sY(s) + y(0). \tag{11.33}$$

Example 11.23. Use Laplace transforms to solve the initial value problem

$$ty''(t) - ty'(t) - y(t) = 0 \quad \text{with} \quad y(0) = 0.$$

Solution:

Let $Y(s)$ denote the Laplace transform of $y(t)$, then using the substitutions 11.32 and 11.33 results in

$$-s^2Y'(s) - 2sY(s) + 0 + sY'(s) + Y(s) - Y(s) = 0. \tag{11.34}$$

This equation involves $Y'(s)$ and can be written as a first-order linear differential equation:

$$Y'(s) + \left(\frac{2}{s-1}\right)Y(s) = 0. \tag{11.35}$$

The integrating factor ρ for the differential equation is

$$\rho = \exp\left(\int \frac{2}{s-1}\, ds\right) = e^{2\ln(s-1)} = (s-1)^2.$$

Multiplying Equation (11.35) by ρ produces

$$(s-1)^2 Y'(s) + 2(s-1)Y(s) = \frac{d}{ds}\left[(s-1)^2Y(s)\right] = 0.$$

Now we integrate the equation $\frac{d}{ds}\left[(s-1)^2Y(s)\right] = 0$ with respect to s and the results is $(s-1)^2Y(s) = C$, where C is the constant of integration. Hence the solution to Equation (11.34) is

$$Y(s) = \frac{C}{(s-1)^2}.$$

The inverse of the transform $Y(s)$ in equation is the desired solution

$$y(t) = Cte^t.$$

Exercises for Section 11.8 (Selected answers or hints are on page 460.)

1. Find the Laplace transform for each of the following:

(a) $\mathcal{L}(te^{-2t})$.

(b) $\mathcal{L}(t^2 e^{4t})$.

(c) $\mathcal{L}(t \sin 3t)$.

(d) $\mathcal{L}(t^2 \cos 2t)$.

(e) $\mathcal{L}(t \sinh t)$.

(f) $\mathcal{L}(t^2 \cosh t)$.

2. Show that $\mathcal{L}(\frac{e^t - 1}{t}) = \ln \frac{s}{s-1}$.

3. Show that $\mathcal{L}(\frac{1 - \cos t}{t}) = \ln \frac{s^2}{s^2 + 1}$.

4. Find $\mathcal{L}(t \sin bt)$.

5. Find $\mathcal{L}(te^{at} \cos bt)$.

6. Find $\mathcal{L}^{-1}\left(\ln \frac{s^2 + 1}{(s-1)^2}\right)$.

7. Find $\mathcal{L}^{-1}(\ln \frac{s}{s+1})$.

8. Solve the initial value problem for each of the following:

(a) $y''(t) + 2y'(t) + y(t) = 2e^{-t}$, with $y(0) = 0$ and $y'(0) = 1$.

(b) $y''(t) + y(t) = 2 \sin t$, with $y(0) = 0$ and $y'(0) = -1$.

(c) $ty''(t) - ty'(t) - y(t) = 0$, with $y(0) = 0$.

(d) $ty''(t) + (t - 1)y'(t) - 2y(t) = 0$, with $y(0) = 0$.

(e) $ty''(t) + ty'(t) - y(t) = 0$, with $y(0) = 0$.

(f) $ty''(t) + (t - 1)y'(t) + y(t) = 0$, with $y(0) = 0$.

9. Solve the Laguerre equation $ty''(t) + (1 - t)y'(t) + y(t) = 0$, with $y(0) = 1$.

10. Solve the Laguerre equation $ty''(t) + (1 - t)y'(t) + 2y(t) = 0$, with $y(0) = 1$.

11.9 Inverting the Laplace Transform

So far, most of the applications involving the Laplace transform involve a transform (or part of a transform) that is expressed by

$$Y(s) = \frac{P(s)}{Q(s)}. \tag{11.36}$$

where P and Q are polynomials that have no common factors. The inverse of $Y(s)$ is found by using its partial fraction representation and referring to Table 11.2. We now show how the theory of complex variables can be used to systematically find the partial fraction representation. The first result is an extension of Lemma 8.1 to n linear factors. The proof is left for the reader.

Theorem 11.19 (Nonrepeated Linear Factors). *Let $P(s)$ be a polynomial of degree at most $n-1$. If $Q(s)$ has degree n, and has distinct complex roots $a_1, a_2, \ldots, a_n$, then Equation (11.36) has the representation*

$$Y(s) = \frac{P(s)}{(s - a_1)(s - a_2)\cdots(s - a_n)} = \sum_{k=1}^{n} \frac{\text{Res}[Y, a_k]}{s - a_k}. \tag{11.37}$$

413

Theorem 11.20 (A Repeated Linear Factor). *If $P(s)$ and $Q(s)$ are polynomials of degree μ and v, respectively, and $\mu < v + n$ and $Q(a) \neq 0$, then Equation (11.36) has the representation*

$$Y(s) = \frac{P(s)}{(s-a)^n Q(s)} = \sum_{k=1}^{n} \frac{A_k}{(s-a)^k} + R(s), \tag{11.38}$$

where R is the sum of all partial fractions that do not involve factors of the form $(s-a)^j$. Furthermore, the coefficients A_k can be computed with the formula

$$A_k = \frac{1}{(n-k)} \lim_{s \to a} \left[\frac{d^{n-k}}{ds^{n-k}} \left(\frac{P(s)}{Q(s)} \right) \right] \quad for \quad k = 1, 2, \ldots, n. \tag{11.39}$$

Proof. We employ the method of residues. First, multiplying both sides of Equation (11.38) by $(s-a)^n$ gives

$$\frac{P(s)}{Q(s)} = \sum_{j=1}^{n} A_j (s-a)^{n-j} + R(s)(s-a)^n.$$

We can differentiate both sides of this equation $n - k$ times to obtain

$$\frac{d^{n-k}}{ds^{n-k}} \left(\frac{P(s)}{Q(s)} \right) = \sum_{j=1}^{k} A_j \frac{(n-j)!}{(k-j)!} (s-a)^{k-j} + \frac{d^{n-k}}{ds^{n-k}} [R(s)(s-a)^n].$$

In this equation, take the limit as $s \to a$. It is left as an exercise for the reader to fill in the steps to obtain

$$\lim_{s \to a} \frac{d^{n-k}}{ds^{n-k}} \left(\frac{P(s)}{Q(s)} \right) = (n-k)! A_k.$$

which establishes Equation (11.39). $\qquad \square$

Example 11.24. Let $Y(s) = \frac{s^3 - 4s + 1}{s(s-1)^3}$. Find $\mathcal{L}^{-1}(Y(s))$.

Solution:

From Equations (11.37) and (11.38) we can write

$$\frac{s^3 - 4s - 1}{s(s-1)^3} = \frac{A_3}{(s-1)^3} + \frac{A_2}{(s-1)^2} + \frac{A_1}{s-1} + \frac{B_1}{s}.$$

The coefficient B_1 is found by the calculation

$$B_1 = \text{Res}[Y, 0] = \lim_{s \to 0} \frac{s^3 - 4s + 1}{(s-1)^3} = -1.$$

The coefficients A_1, A_2, and A_3 are found by using Theorem 11.20. In this case $a = 1$ and $\frac{P(s)}{Q(s)} = \frac{s^3 - 4s + 1}{s}$, so we get

$$A_3 = \lim_{s \to 1} \left(\frac{P(s)}{Q(s)} \right) = \lim_{s \to 1} \frac{s^3 - 4s + 1}{s} = -2.$$

$$A_2 = \frac{1}{1!} \lim_{s \to 1} \frac{d}{ds} \left(\frac{P(s)}{Q(s)} \right) = \lim_{s \to 1} \left(2s - \frac{1}{s^2} \right) = 1.$$

$$A_1 = \frac{1}{2} \lim_{s \to 1} \frac{d^2}{ds^2} \left(\frac{P(s)}{Q(s)} \right) = \frac{1}{2} \lim_{s \to 1} \left(2 + \frac{2}{s^3} \right) = 2.$$

Hence, the partial fraction representation is

$$Y(s) = \frac{-2}{(s-1)^3} + \frac{1}{(s-1)^2} + \frac{2}{s-1} - \frac{1}{s},$$

and the inverse is

$$y(t) = -t^2 e^t + t e^t + 2 e^t - 1.$$

Theorem 11.21 (Irreducible Quadratic Factors). *Let P and Q be polynomials with real coefficients such that the degree of P is at most 1 larger than the degree of Q. If T does not have a factor of the form $(s-a)^2 + b^2$, then*

$$Y(s) = \frac{P(s)}{Q(s)} = \frac{P(s)}{\left[(s-a)^2 + b^2\right] T(s)} = \frac{2A(s-a) - 2Bb}{(s-a)^2 + b^2} + R(s), \ \ where$$

$$A + iB = \frac{P(a+ib)}{Q'(a+ib)}. \tag{11.40}$$

Proof. Since P, Q, and Q' have real coefficients, it follows that

$$P(a - ib) = \overline{P(a+ib)} \quad \text{and} \quad Q'(a - ib) = \overline{Q'(a+ib)}.$$

The polynomial Q has simple zeros at $s = a \pm ib$, this implies that $Q'(a \pm ib) \neq 0$. Therefore, we obtain

$$\operatorname{Res}[Y, a \pm ib] = \lim_{s \to a \pm ib} \left(\frac{s - (a \pm ib)}{Q(s) - Q(a \pm ib)} P(s) \right) = \frac{P(a \pm ib)}{Q'(a \pm ib)}, \tag{11.41}$$

from which it is easy to see that

$$\operatorname{Res}[Y, a - ib] = \overline{\operatorname{Res}[Y, a - ib]}. \tag{11.42}$$

If we set $A + iB = \operatorname{Res}[Y, a + ib]$ and use Theorem 11.19 and Equations (11.40), (11.41), and (11.42), then we find that

$$Y(s) = \frac{A + iB}{s - a - ib} + \frac{A - iB}{s - a + ib} + R(s).$$

The first two terms on the right side of this equation are now combined to obtain

$$\frac{(A + iB)(s - a + ib) + (A - iB)(s - a - ib)}{(s-a)^2 + b^2} = \frac{2A(s-a) - 2Bb}{(s-a)^2 + b^2},$$

which completes the proof of the theorem. $\qquad \square$

Example 11.25. Let $Y(s) = \frac{5s}{(s^2+4)(s^2+9)}$. Find $\mathcal{L}^{-1}\big(Y(s)\big)$.

Solution:

Here we have $P(s) = 5s$ and $Q(s) = s^4 + 13s^2 + 36$, and the roots of $Q(s)$ occur at $0 \pm 2i$ and $0 \pm 3i$. Computing the residues gives

$$\operatorname{Res}[Y, 2i] = \frac{P(2i)}{Q'(2i)} = \frac{5(2i)}{4(2i)^3 - 26(2i)} = \frac{1}{2}, \quad \text{and}$$

$$\operatorname{Res}[Y, 3i] = \frac{P(3i)}{Q'(3i)} = \frac{5(3i)}{4(3i)^3 - 26(3i)} = -\frac{1}{2}.$$

We find that $A_1 + iB_1 = \frac{1}{2} + 0i$ and $A_2 + iB_2 = -\frac{1}{2} + 0i$, which correspond to $a_1 + ib_1 = 0 + 2i$ and $a_2 + ib_2 = 0 + 3i$, respectively. Thus we obtain

$$Y(s) = \frac{2(\frac{1}{2})(s-0) - (2)(0)(2)}{s^2 + 4} + \frac{2(-\frac{1}{2})(s-0) - (2)(0)(3)}{s^2 + 9} = \frac{s}{s^2 + 4} - \frac{s}{s^2 + 9},$$

and the desired solution is

$$\mathcal{L}^{-1}(Y(s)) = \mathcal{L}^{-1}\left(\frac{s}{s^2 + 4}\right) - \mathcal{L}^{-1}\left(\frac{s}{s^2 + 9}\right) = \cos 2t - \cos 3t.$$

Example 11.26. Find $\mathcal{L}^{-1}(Y(s))$ if $Y(s) = \frac{s^3 + 3s^2 - s + 1}{s(s+1)^2(s^2+1)}$.

Solution:

The partial fraction expression for $Y(s)$ has the form

$$Y(s) = \frac{D}{s} + \frac{C_1}{s+1} + \frac{C_2}{(s+1)^2} + \frac{2A(s-0) - 2B(1)}{(s-0)^2 + 1^2}.$$

Since the linear factor s is nonrepeated, we have

$$D = \text{Res}[Y(s), 0] = \lim_{s \to 0} \frac{s^3 + 3s^2 - s + 1}{(s+1)^2(s^2+1)} = 1.$$

Since the factor $s + 1$ is repeated, we have

$$\begin{aligned}
C_1 &= \text{Res}[Y(s), -1] \\
&= \lim_{s \to -1} \frac{d}{ds}\left(\frac{s^3 + 3s^2 - s + 1}{s(s^2+1)}\right) \\
&= \lim_{s \to -1} \frac{-3s^4 + 4s^3 - 1}{s^2(s+1)^2} \\
&= -2; \\
C_2 &= \text{Res}[(s+1)Y(s), -1] \\
&= \lim_{s \to -1} \frac{s^3 + 3s^2 - s + 1}{s(s^2+1)} \\
&= -2.
\end{aligned}$$

The term $s^2 + 1$ is an irreducible quadratic, with roots $\pm i$, so that

$$A + iB = \text{Res}[Y, i] = \lim_{s \to i} \frac{s^3 + 3s^2 - s + 1}{s(s+1)^2(s+i)} = \frac{1-i}{2},$$

and we obtain $A = \frac{1}{2}$ and $B = -\frac{1}{2}$. Therefore,

$$\begin{aligned}
Y(s) &= \frac{1}{s} + \frac{-2}{s+1} + \frac{-2}{(s+1)^2} + \frac{2\frac{1}{2}(s-0) - 2(-\frac{1}{2})(1)}{(s-0)^2 + 1^2} \\
&= \frac{1}{s} - \frac{2}{s+1} - \frac{2}{(s+1)^2} + \frac{s+1}{s^2+1}.
\end{aligned}$$

Now we use Table 11.2 to get

$$y(t) = 1 - 2e^{-t} - 2te^{-t} + \cos t + \sin t.$$

Example 11.27. Use Laplace transforms to solve the system

$$y'(t) = y(t) - x(t), \quad \text{with} \quad y(0) = 1.$$
$$x'(t) = 5y(t) - 3x(t) \quad \text{with} \quad x(0) = 2.$$

Solution:

Let $Y(s)$ and $X(s)$ denote the Laplace transforms of $y(t)$ and $x(t)$, respectively. If we take the transforms of the two differential equations and get

$$sY(s) - 1 = Y(s) - X(s),$$
$$sX(s) - 2 = 5Y(s) - 3X(s),$$

which can be written as

$$(s-1)Y(s) + X(s) = 1.$$
$$5Y(s) - (s+3)X(s) = -2.$$

Cramer's rule can be used to solve for $Y(s)$ and $X(s)$:

$$Y(s) = \frac{\begin{vmatrix} 1 & 1 \\ -2 & -s-3 \end{vmatrix}}{\begin{vmatrix} s-1 & 1 \\ 5 & -s-3 \end{vmatrix}} = \frac{-s-3+2}{(s-1)(-s-3)-5} = \frac{s+1}{(s+1)^2+1}$$

$$X(s) = \frac{\begin{vmatrix} s-1 & 1 \\ 5 & -2 \end{vmatrix}}{\begin{vmatrix} s-1 & 1 \\ 5 & -s-3 \end{vmatrix}} = \frac{-2s+2-5}{(s-1)(-s-3)-5} = \frac{2(s+1)+1}{(s+1)^2+1}.$$

The desired solution is obtained by computing the inverse transforms:

$$y(t) = e^{-t}\cos t,$$
$$x(t) = e^{-t}(2\cos t + \sin t).$$

According to Equation (11.29) of Section 11.5, the inverse Laplace transform is given by the integral formula

$$f(t) = \mathcal{L}^{-1}\big(F(s)\big) = \frac{1}{2\pi i}\int_{\sigma_0 - i\infty}^{\sigma_0 + i\infty} F(s)e^{st}\,ds,$$

where σ_0 is any suitably chosen large positive constant. This improper integral is a contour integral taken along the vertical line $s = \sigma_0 + i\tau$ in the complex $s = \sigma + i\tau$ plane. We shall show how the residue theory in Chapter 8 is used to evaluate it. Cases where the integrand has either infinitely many poles or has branch points is left for the reader to research in advanced texts. We state the following more elementary result.

Theorem 11.22 (Inverse Laplace Transform). *Let* $F(s) = \frac{P(s)}{Q(s)}$, *where* $P(s)$ *and* $Q(s)$ *are polynomials of degree* m *and* n, *respectively, and* $n > m$. *The inverse Laplace transformation* $F(s)$, *denoted by* $f(t)$, *is given by*

$$f(t) = \mathcal{L}^{-1}\big(F(s)\big) = \sum \text{Res}[F(s)e^{st}, s_k], \tag{11.43}$$

where the sum is taken over all of the residues of the complex function $F(s)e^{st}$ *at its poles* s_k.

417

Proof. Let σ_0 be chosen so that all the poles of $F(s)e^{st}$ lie to the left of the vertical line $s = \sigma_0 + i\tau$. Let Γ_R denote the contour consisting of the vertical line segment between the points $\sigma_0 \pm iR$ and the left semicircle $C_R : s = \sigma_0 + Re^{i\theta}$, where $\frac{\pi}{2} \leq \theta \leq \frac{3\pi}{2}$, as illustrated in Figure 11.27. A slight modification of the proof of Jordan's lemma shows that

$$\lim_{R \to +\infty} \int_{C_R} \frac{P(s)}{Q(s)} e^{st} \, ds = 0.$$

The residue theorem and above limit imply that

$$\mathcal{L}^{-1}(F(s)) = \lim_{R \to +\infty} \frac{1}{2\pi i} \int_{\Gamma_R} \frac{P(s)}{Q(s)} e^{st} \, ds = \sum \text{Res}[F(s)e^{st}, s_k].$$

$\square$

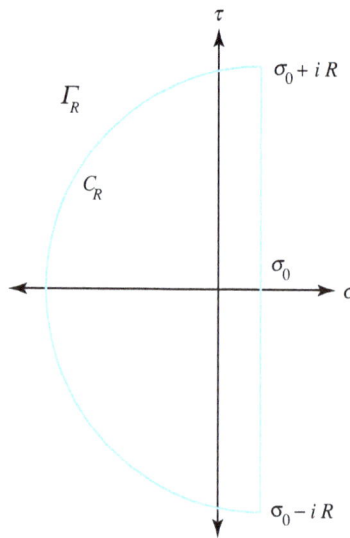

Figure 11.27: The contour Γ_R.

Theorem 11.23 (Heaviside Expansion Theorem). *Let $P(s)$ and $Q(s)$ be polynomials of degree m and n, respectively, where $n > m$. If $Q(s)$ has n distinct simple zeros at the points $s_1, s_2, \ldots, s_n$, then $\frac{P(s)}{Q(s)}$ is the Laplace transform of the function $f(t)$ given by*

$$f(t) = \mathcal{L}^{-1}\left(\frac{P(s)}{Q(s)}\right) = \sum_{k=1}^{n} \frac{P(s_k)}{Q'(s_k)} e^{s_k t}. \tag{11.44}$$

Proof. If $P(s)$ and $Q(s)$ are polynomials and s_k is a simple zero of $Q(s)$, then

$$\text{Res}[F(s)e^{st}, s_k] = \lim_{s \to s_k} \frac{s - s_k}{Q(s) - Q(s_k)} P(s)e^{st} = \frac{P(s_k)}{Q'(s_k)} e^{s_k t}.$$

This outcome allows us to write the residues in Equation (11.43) in the more convenient form given in Equation (11.44). $\square$

Example 11.28. Find the inverse Laplace transform of the function

$$F(s) = \frac{4s + 3}{s^3 + 2s^2 + s + 2}.$$

418

Solution:

Here we have $P(s) = 4s + 3$ and $Q(s) = (s + 2)(s^2 + 1)$, so that Q has simple zeros located at the points $s_1 = -2$, $s_2 = i$, and $s_3 = -i$. Using $Q'(s) = 3s^2 + 4s + 1$, calculation reveals that $\frac{P(-2)}{Q'(-2)} = \frac{-8+3}{12-8+1} = -1$ and $\frac{P(\pm i)}{Q'(\pm i)} = \frac{\pm 4i+3}{-2\pm 4i} = \frac{1}{2} \mp i$. Applying formula 11.44, we see that $f(t)$ is given by

$$f(t) = \frac{P(-2)}{Q'(-2)}e^{-2t} + \frac{P(i)}{Q'(i)}e^{it} + \frac{P(-i)}{Q'(-i)}e^{-it}$$

$$= -e^{-2t} + (\frac{1}{2} - i)e^{it} + (\frac{1}{2} + i)e^{-it}$$

$$= -e^{-2t} + \frac{e^{it} + e^{-it}}{2} + 2\frac{e^{it} - e^{-it}}{2i}$$

$$= -e^{-2t} + \cos t + 2\sin t.$$

Exercises for Section 11.9 (Selected answers or hints are on page 460.)

1. Use partial fractions to find the inverse Laplace transform of

   (a) $Y(s) = \frac{2s+1}{s(s-1)}$.

   (b) $Y(s) = \frac{2s^3 - s^2 + 4s - 6}{s^4}$.

   (c) $Y(s) = \frac{4s^2 - 6s - 12}{s(s+2)(s-2)}$.

   (d) $Y(s) = \frac{s^3 - 5s^2 + 6s - 6}{(s-2)^4}$.

   (e) $Y(s) = \frac{2s^2 + s + 3}{(s+2)(s-1)^2}$.

   (f) $Y(s) = \frac{4-s}{s^2 + 4s + 5}$.

2. Use a contour integral to find the inverse Laplace transform of

   (a) $Y(s) = \frac{1}{s^2 + 4}$.

   (b) $Y(s) = \frac{s+3}{(s-2)(s^2+1)}$.

3. Use the Heaviside Expansion Theorem to find the inverse Laplace transform of

   (a) $Y(s) = \frac{s^3 + s^2 - s + 3}{s^5 - s}$.

   (b) $Y(s) = \frac{s^3 + 2s^2 - s + 2}{s^5 - s}$.

   (c) $Y(s) = \frac{s^3 + 3s^2 - s + 1}{s^5 - s}$.

   (d) $Y(s) = \frac{s^3 + s^2 + s + 3}{s^5 - s}$.

4. Find the inverse Laplace transform of $Y(s) = \frac{s^3 + 2s^2 + 4s + 2}{(s^2 + 1)(s^2 + 4)}$.

5. Solve the following initial value problems.

   (a) $y''(t) + y(t) = 3\sin 2t$, with $y(0) = 0$, and $y'(0) = 3$.

   (b) $y''(t) + 2y'(t) + 5y(t) = 4e^{-t}$, with $y(0) = 1$ and $y'(0) = 1$.

   (c) $y''(t) + 2y'(t) + 2y(t) = 2$, with $y(0) = 1$ and $y'(0) = 1$.

(d) $y''(t) + 4y(t) = 5e^{-t}$, with $y(0) = 2$ and $y'(0) = 1$.

(e) $y''(t) + 2y'(t) + y(t) = t$, with $y(0) = -1$ and $y'(0) = 0$.

(f) $y''(t) + 3y'(t) + 2y(t) = 2t + 5$, with $y(0) = 1$ and $y'(0) = 1$.

6. Solve the following systems of differential equations.

   (a) $x'(t) = 10y(t) - 5x(t)$, $y'(t) = y(t) - x(t)$, with $x(0) = 3$ and $y(0) = 1$.

   (b) $x'(t) = 2y(t) - 3x(t)$, $y'(t) = 2y(t) - 2x(t)$, with $x(0) = 1$ and $y(0) = -1$.

   (c) $x'(t) = 2x(t) + 3y(t)$, $y'(t) = 2x(t) + y(t)$, with $x(0) = 2$ and $y(0) = 3$.

   (d) $x'(t) = 4y(t) - 3x(t)$, $y'(t) = y(t) - x(t)$, with $x(0) = -1$ and $y(0) = 0$.

   (e) $x'(t) = 4y(t) - 3x(t) + 5$, $y'(t) = y(t) - x(t) + 1$, with $x(0) = 0$ and $y(0) = 2$.

   (f) $x'(t) = 8y(t) - 3x(t) + 2$, $y'(t) = y(t) - x(t) - 1$, with $x(0) = 4$ and $y(0) = 2$.

11.10   Convolution

Let $F(s)$ and $G(s)$ denote the transforms of $f(t)$ and $g(t)$, respectively. Then the inverse of the product $F(s)G(s)$ is given by the function $h(t) = (f * g)(t)$ and is called the **convolution** of $f(t)$ and $g(t)$ and can be regarded as a generalized product of $f(t)$ and $g(t)$. Convolutions are helpful in solving integral equations.

Theorem 11.24 (Convolution Theorem). *Let $F(s)$ and $G(s)$ denote the Laplace transforms of $f(t)$ and $g(t)$, respectively. Then the product given by $H(s) = F(s)G(s)$ is the Laplace transformation of the convolution of f and g. It is denoted by $h(t) = (f * g)(t)$, and has the integral representation*

$$h(t) = (f * g)(t) = \int_0^t f(\tau)g(t - \tau)\, d\tau, \quad or \tag{11.45}$$

$$h(t) = (g * f)(t) = \int_0^t g(\tau)f(t - \tau)\, d\tau. \tag{11.46}$$

Proof. The following proof is given for the special case when s is a real number. The general case is covered in advanced texts. Using the dummy variables σ and τ and the integrals defining the transforms, we can express their product as

$$F(s)G(s) = \left[\int_0^\infty f(\sigma)e^{-s\sigma}\, d\sigma \right] \left[\int_0^\infty g(\tau)e^{-s\tau}\, d\tau \right].$$

The product of integrals in this equation can be written as an iterated integral:

$$F(s)G(s) = \int_0^\infty \left[\int_0^\infty f(\sigma)e^{-s(\sigma+\tau)}\, d\sigma \right] g(\tau)\, d\tau.$$

Hold τ fixed, and use the change of variables $t = \sigma + \tau$ and $dt = d\sigma$. Then the inner integral in the equation can be rewritten to obtain

$$F(s)G(s) = \int_0^\infty \left[\int_\tau^\infty f(t - \tau)e^{-st}\, dt \right] g(\tau)\, d\tau = \int_0^\infty \left[\int_\tau^\infty f(t - \tau)g(\tau)e^{-st}dt \right] d\tau.$$

The region of integration for this last iterated integral is the wedge-shaped region in the (t, τ) plane shown in Figure 11.28.

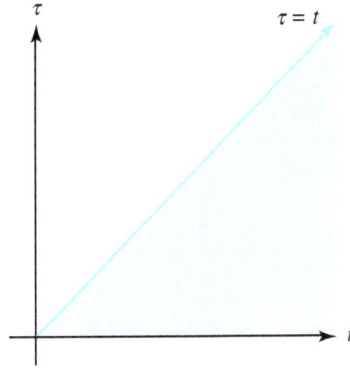

Figure 11.28: The region of integration in the convolution theorem

The order of integration in the integral can be reversed to get:

$$F(s)G(s) = \int_0^\infty \left[\int_0^t f(t - \tau)g(\tau)e^{-st} \, d\tau \right] dt.$$

This equation can be written as

$$F(s)G(s) = \int_0^\infty \left[\int_0^t f(t - \tau)g(\tau) \, d\tau \right] e^{-st} \, dt$$

$$= \mathcal{L}^{-1} \left(\int_0^t f(t - \tau)g(\tau) \, d\tau \right),$$

which establishes Equation (11.46). Since we can interchange the role of the functions $f(t)$ and $g(t)$, Equation (11.45) follows immediately. $\square$

Table 11.4 summarizes some important convolution properties.

Commutativity	$f * g = g * f$
Distributivity	$f * (g + h) = f * g + f * h$
Associativity	$(f * g) * h = f * (g * h)$
Zero	$f * 0 = 0$

Table 11.4: Convolution Properties

Example 11.29. Show that $\mathcal{L}^{-1}\left(\frac{2s}{(s^2+1)^2}\right) = t \sin t.$

***Solution*:**
Let $F(s) = \frac{1}{s^2+1}$, $G(s) = \frac{2s}{s^2+1}$, $f(t) = \sin t$, $g(t) = 2\cos t$, respectively. Applying the convolution theorem we get

$$\mathcal{L}^{-1}\left(\frac{1}{s^2+1}\frac{2s}{s^2+1}\right) = \mathcal{L}^{-1}\big(F(s)G(s)\big)$$

$$= \int_0^t 2\sin(t-\tau)\cos\tau\,d\tau$$

$$= \int_0^t \left[2\sin t\cos^2\tau - 2\cos t\sin\tau\cos\tau\right]d\tau$$

$$= \left[\sin t(\tau + \sin\tau\cos\tau) - \cos t\sin^2\tau\right]\Big|_{\tau=0}^{\tau=t}$$

$$= t\sin t + \sin^2 t\cos t - \cos t\sin^2 t$$

$$= t\sin t.$$

Example 11.30. Use the convolution theorem to solve the integral equation

$$f(t) = 2\cos t - \int_0^t (t-\tau)f(\tau)\,d\tau.$$

Solution:

Letting $F(s) = \mathcal{L}\big(f(t)\big)$ and using $\mathcal{L}(t) = \frac{1}{s^2}$ in the convolution theorem we obtain

$$F(s) = \frac{2s}{s^2+1} - \frac{1}{s^2}F(s).$$

Solving for $F(s)$ we get

$$F(s) = \frac{2s^3}{(s^2+1)^2} = \frac{2s}{s^2+1} - \frac{2s}{(s^2+1)^2},$$

and the solution is

$$f(t) = 2\cos t - t\sin t.$$

Engineers and physicists sometimes consider forces that produce large effects that are applied over a very short time interval. The force acting at the time an earthquake starts is an example. This leads to the idea of a *unit impulse function* $\delta(t)$. Consider a small positive constant a. The function $\delta_a(t)$ is defined by

$$\delta_a(t) = \begin{cases} \frac{1}{a} & \text{for } \ 0 < t < a, \\ 0 & \text{otherwise.} \end{cases}$$

The unit impulse function is obtained by letting the interval width go to zero, *i.e.*,

$$\delta(t) = \lim_{a\to 0}\delta_a(t).$$

Figure 11.29 shows the graph of $\delta_a(t)$ for $a = 10$, 40, and 100. Although $\delta(t)$ is called the *Dirac delta function*, it is not an ordinary function. To be precise it is a distribution, and the theory of distributions permits manipulations of $\delta(t)$ as though it were a function. For our work, we will treat $\delta(t)$ as a function and investigate its properties.

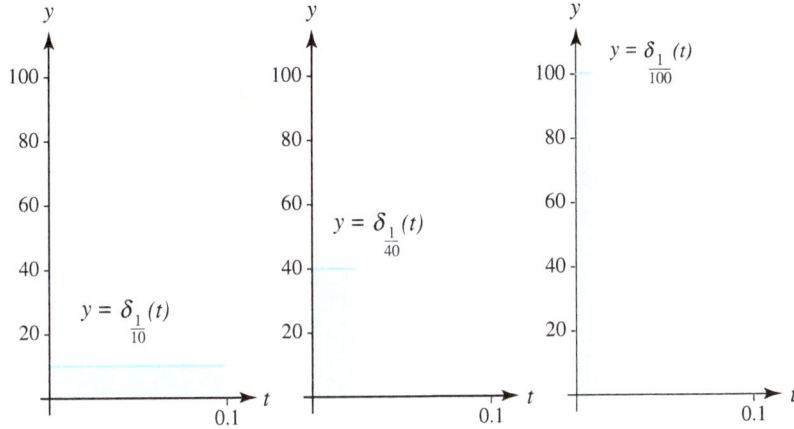

Figure 11.29: Graphs of $y = \delta_a(t)$ for $a = 10, 40,$ and 100

Example 11.31. Show that $\mathcal{L}\big(\delta(t)\big) = 1$.

Solution:

By definition, the Laplace transform of $\delta_a(t)$ is

$$\mathcal{L}\big(\delta_a(t)\big) = \int_0^\infty \delta_a(t) e^{-st}\, dt = \int_0^a \frac{1}{a} e^{-st}\, dt = \frac{1 - e^{-sa}}{sa}.$$

Letting $a \to 0$ in equation, and using L'Hôpital's rule, we obtain

$$\mathcal{L}\big(\delta(t)\big) = \lim_{a \to 0} \mathcal{L}\big(\delta_a(t)\big) = \lim_{a \to 0} \frac{1 - e^{-sa}}{sa} = \lim_{a \to 0} \frac{0 + se^{-sa}}{s} = 1.$$

We now turn our attention to the unit impulse function. First, consider the function $f_a(t)$ obtained by integrating $\delta_a(t)$:

$$f_a(t) = \int_0^t \delta_a(\tau)\, d\tau = \begin{cases} 0 & \text{for} \quad t < 0, \\ \frac{t}{a} & \text{for} \quad 0 \le t \le a, \\ 1 & \text{for} \quad a < t. \end{cases}$$

Then it is easy to see that $U_0(t) = \lim\limits_{a \to 0} f_a(t)$ (see Figure 11.30).

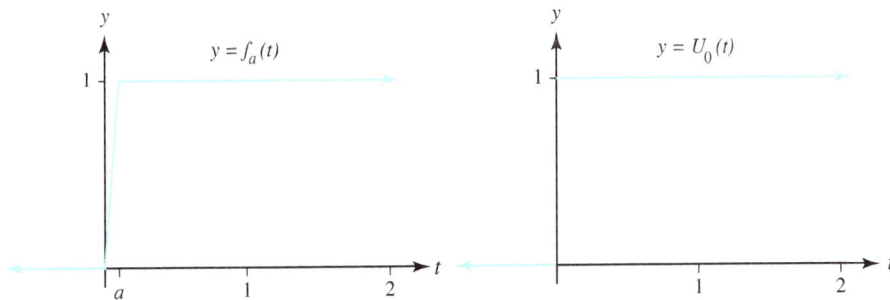

Figure 11.30: The integral of $\delta_a(t)$ is $f_a(t)$, which becomes $U_0(T)$, which becomes $U_0(t)$ when $a \to 0$

The response of a system to the unit impulse function is illustrated in the next example.

Example 11.32. Solve the initial value problem

$$y''(t) + 4y'(t) + 13y(t) = 3\delta(t) \quad \text{with} \quad y(0) = 0 \quad \text{and} \quad y'(0^-) = 0.$$

Solution:

Taking transforms results in $(s^2 + 4s + 13)Y(s) = 3\mathcal{L}(\delta(t)) = 3$, so that

$$Y(s) = \frac{3}{s^2 + 4s + 13} = \frac{3}{(s+2)^2 + 3^2},$$

so the solution is

$$y(t) = e^{-2t} \sin 3t.$$

Note: The condition $y'(0^-) = 0$ is not satisfied by the "solution" $y(t)$. Recall that all solutions using the Laplace transform are to be considered zero for values of $t < 0$. Hence the graph of $y(t)$ is given in Figure 11.31. We see that $y,'(t)$ has a jump discontinuity of magnitude $+3$ at the origin. This happens because either $y(t)$ or $y'(t)$ must have a jump discontinuity at the origin whenever the Dirac delta function occurs as part of the input or driving function.

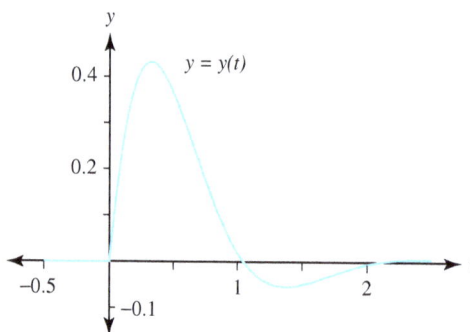

Figure 11.31: The solution $y = y(t)$ to Example 11.32

We now illustrate how the convolution method can be used to solve initial value problems.

Example 11.33 (IVP Convolution Method). Show initial value problem

$$ay''(t) + by'(t) + cy(t) = g(t), \quad \text{with} \quad y(0) = y_0 \quad \text{and} \quad y'(0) = y_1$$

has the unique solution

$$y(t) = u(t) + (h * g)(t),$$

where $u(t)$ is the solution to the homogeneous equation

$$au''(t) + bu'(t) + cu(t) = 0, \quad \text{with} \quad u(0) = y_0 \quad \text{and} \quad u'(0) = y_1,$$

and $h(t)$ is the function whose Laplace transform given by $H(s) = \frac{1}{as^2+bs+c}$.

Solution:

The particular solution is found by solving the equation

$$av''(t) + bv'(t) + cv(t) = g(t), \quad \text{with} \quad v(0) = 0 \quad \text{and} \quad v'(0) = 0.$$

Taking the Laplace transform of both sides of this equation produces

$$as^2 V(s) + bsV(s) + cV(s) = G(s).$$

Solving for $V(s)$ in the latter equation yields $V(s) = \frac{1}{as^2+bs+c}G(s)$. If we set $H(s) = \frac{1}{as^2+bs+c}$, then $V(s) = H(s)G(s)$, and the particular solution is given by the convolution

$$v(t) = (h * g)(t).$$

The general solution is $y(t) = u(t) + v(t) = u(t) + (h*g)(t)$. To verify that the initial conditions are met we compute

$$y(0) = u(0) + v(0) = y_0 + i0 = y_0, \quad \text{and}$$
$$y'(0) = u'(0) + v'(0) = y_1 + 0 = y_1.$$

Example 11.34. Use the convolution method to solve the IVP

$$y''(t) + y(t) = \tan t, \quad \text{with} \quad y(0) = 1 \quad \text{and} \quad y'(0) = 2.$$

Solution:

First solve $u''(t) + u(t) = 0$ with $u(0) = 1$ and $u'(0) = 2$. Taking the Laplace transform yields $s^2 U(s) - s - 2 + U(s) = 0$. Solving for $U(s)$ gives $U(s) = \frac{s+2}{s^2+1}$, and it follows that

$$u(t) = \cos t + 2\sin t.$$

Second, observe that $H(s) = \frac{1}{s^2+1}$ and $h(t) = \sin t$, so that

$$v(t) = (h * g)(t) = \int_0^t \sin(t - s)\tan(s)\,ds$$

$$= \left[\cos(t)\ln\frac{\cos s}{1 + \sin s} - \sin(t - s)\right]\Bigg|_{s=0}^{s=t}$$

$$= \cos(t)\ln\frac{\cos t}{1 + \sin t} + \sin(t).$$

Therefore, the solution is

$$y(t) = u(t) + v(t) = \cos t + 3\sin t + \cos(t)\ln\frac{\cos t}{1 + \sin t}.$$

Exercises for Section 11.10 (Selected answers or hints are on page 461.)

1. Find the indicated convolution for each of the following:

   (a) $t * t$.

   (b) $t * \sin t$.

   (c) $e^t * e^{2t}$.

   (d) $\sin t * \sin 2t$.

2. Use convolution to find $\mathcal{L}^{-1}\big(F(s)\big)$ for each of the following:

   (a) $F(s) = \frac{2}{(s-1)(s-2)}$.

   (b) $F(s) = \frac{6}{s^3}$.

   (c) $F(s) = \frac{1}{s(s^2+1)}$.

(d) $F(s) = \frac{s}{(s^2+1)(s^2+4)}$.

3. Prove the distributive law for convolution: $f * (g + h) = f * g + f * g$.

4. Use the convolution theorem and mathematical induction to show that
$$\mathcal{L}^{-1}\left(\frac{1}{(s-a)^n}\right) = \frac{1}{(n-1)!}t^{n-1}e^{at}.$$

5. Find $\mathcal{L}^{-1}\left(\frac{s}{s-1}\right)$.

6. Find $\mathcal{L}^{-1}\left(\frac{s^2}{s^2+1}\right)$.

7. Use the convolution theorem to solve the initial value problem:
$$y''(t) + y(t) = 2\sin t, \quad \text{with} \quad y(0) = 0 \quad \text{and} \quad y'(0) = 0.$$

8. Use the convolution theorem to show that the solution to the initial value problem $y''(t) + \omega^2 y(t) = f(t)$, with $y(0) = 0$ and $y'(0) = 0$ is
$$y(t) = \frac{1}{\omega}\int_0^t f(\tau)\sin[\omega(t-\tau)]\,d\tau.$$

9. Find $\mathcal{L}\left(\int_0^t e^{-\tau}\cos(t-\tau)\,d\tau\right)$.

10. Find $\mathcal{L}\left(\int_0^t (t-\tau)^2 e^\tau\,d\tau\right)$.

11. Let $F(s) = \mathcal{L}(f(t))$. Use convolution to show that
$$\mathcal{L}^{-1}\left(\frac{F(s)}{s}\right) = \int_0^t f(\tau)\,d\tau.$$

12. Use the convolution theorem to solve the following integral equations.

   (a) $f(t) + 4\int_0^t (t-\tau)f(\tau)\,d\tau = 2$.
   (b) $f(t) = e^t + \int_0^t e^{t-\tau}f(\tau)\,d\tau$.
   (c) $f(t) = 2t + \int_0^t \sin(t-\tau)f(\tau)\,d\tau$.
   (d) $6f(t) = 2t^3 + \int_0^t (t-\tau)^3 f(\tau)\,d\tau$.

13. Solve the initial value problem for each of the following:

   (a) $y''(t) - 2y'(t) + 5y(t) = 2\delta(t)$, with $y(0) = 0$ and $y'(0) = 0$.
   (b) $y''(t) + 2y'(t) + y(t) = \delta(t)$, with $y(0) = 0$ and $y'(0) = 0$.
   (c) $y''(t) + 4y'(t) + 3y(t) = 2\delta(t)$, with $y(0) = 0$ and $y'(0) = 0$.
   (d) $y''(t) + 4y'(t) + 3y(t) = 2\delta(t-1)$, with $y(0) = 0$ and $y'(0) = 0$.

14. Use the IVP convolution method to solve the following initial value problems.

   (a) $y''(t) - 2y'(t) + 5y(t) = 8\exp(-t)$, with $y(0) = 1$ and $y'(0) = 2$.
   (b) $y''(t) + 2y'(t) + y(t) = t^4$, with $y(0) = 1$ and $y'(0) = 2$.
   (c) $y''(t) + 4y'(t) + 3y(t) = 24t^2 e^{-t}$, with $y(0) = 1$ and $y'(0) = 2$.
   (d) $y''(t) + 4y'(t) + 3y(t) = 2te^{-t}$, with $y(0) = 1$ and $y'(0) = 2$.

Answers to Selected Exercises

Section 1.1. The Origin of Complex Numbers
(**Exercise questions are on page** 8)

1. Mimic the argument the text gives in showing $2 + \sqrt{-1} = \sqrt[3]{2 + \sqrt{-121}}$.

3a. The roots are $x_1 = -\frac{1}{3}$, $x_2 = -\frac{1}{3}$, $x_3 = \frac{2}{3}$.

5a. Use Formula (1.3) to get $x = \sqrt[3]{18 + 26\sqrt{-1}} + \sqrt[3]{18 - 26\sqrt{-1}}$. Assume, as Bombelli did that this expression can be put in the form $(u + v\sqrt{-1}) + (u - v\sqrt{-1})$, where u and v are *integers*. Next, imitate the logic in the text that leads to Equations (1.4), (1.5), and (1.6) to get $u(u^2 - 3v^2) + iv(3u^2 - v^2) = 18 + 26i$. The only factors of 18 are 1, 2, 3, 6, 9, and 18, so you can deduce (explain your reasoning) that $u = 3$ and $v = 1$ solve this system. Thus, one solution to $x^3 - 30x - 36 = 0$ is $x = 6$. Divide $x^3 - 30x - 36$ by $x - 6$ and solve the resulting quadratic to get the remaining solutions: $x = -3 \pm \sqrt{3}$.

5c. Proceed as with part a. The solutions are $x = 8$, $x = -4 \pm 2\sqrt{3}$.

Section 1.2. The Algebra of Complex Numbers
(**Exercise questions are on page** 15)

1a. $i^{275} = \left(i^2\right)^{137} i = (-1)^{137} i = -i$.

1c. 0.

1e. $2 + 2i$.

1g. 3.

1i. $-\frac{27}{5} + \frac{11}{5}i$.

3. Let $z = x + iy$ be an arbitrary complex number. Then $z\bar{z} = (x + iy)(x - iy) = x^2 + y^2$, which is obviously a real number.

5a. Since z_1 is a root of the polynomial P, $P(z_1) = 0$. Use Properties (1.12) through (1.14) of Theorem 1.1 to show that $\overline{P(\overline{z_1})} = P(z_1)$, which implies $\overline{P(\overline{z_1})} = 0$. Next show that if $\overline{P(\overline{z_1})} = 0$, then $P(\overline{z_1}) = 0$, confirming that $\overline{z_1}$ is also a root of P.

5c. Find a polynomial for part a, another for part b, and multiply them together.

7. Use the (ordered pair) definition for multiplication to verify that if $z = (x, y)$ is any complex number, then $(x, y)(1, 0) = (x, y)$.

9a. We would want to find a number $\zeta = (a, b)$ such that for any $z = (x, y)$ we have $z * \zeta = z$. Obviously, if $\zeta = (1, 1)$, then according to the definition of $*$ we would have $z * \zeta = (x, y) * (1, 1) = (x, y) = z$. Thus, the multiplicative identity in this case would have to be $\zeta = (1, 1)$.

9b. For any complex number $w = (x, y)$ we would have $(0, a * (x, y) = (0, a)$, which can't possibly equal $(1, 1)$.

11. Let $z_1 = (x_1, y_1)$, $z_2 = (x_2, y_2)$, and $z_3 = (x_3, y_3)$ be arbitrary complex numbers. Then

$$
\begin{aligned}
z_1(z_2 + z_3) &= (x_1, y_1)\big[(x_2, y_2) + (x_3, y_3)\big] \\
&= (x_1, y_1)\big[(x_2 + x_3, y_2 + y_3)\big] \\
&= \big(x_1(x_2 + x_3) - y_1(y_2 + y_3),\ x_1(y_2 + y_3) + (x_2 + x_3)y_1\big) \\
&= \cdots \\
&= (x_1x_2 - y_1y_2,\ x_1y_2 + x_2y_1) + (x_1x_3 - y_1y_3,\ x_1y_3 + x_2y_3) \\
&= z_1z_2 + z_1z_3.
\end{aligned}
$$

Complete the missing steps in $\cdots$ above by using algebraic properties of *real* numbers.

13. $(2 + 3i)^{-1} = \frac{2}{13} - \frac{3}{13}i, \quad (7 - 5i)^{-1} = \frac{7}{74} + \frac{5}{74}i.$

15. Mimic the argument given in the text for multiplication.

Section 1.3 The Geometry of Complex Numbers, Part I
(Exercise questions are on page 20)

1a. $\sqrt{10}$.

1c. 2^{25}.

1e. $(x - 1)^2 + y^2$.

3a. Inside, since $|(\frac{1}{2} + i) - i| = \frac{1}{2}$, which is less than 2.

3c. Outside, since $|(2 + 3i) - i| = |2 + 2i| = \sqrt{8}$, which is greater than 2.

5. Let $z_1 = (x_1, y_1)$ and $z_2 = (x_2, y_2)$. Since neither z_1 nor z_2 equals zero, they are perpendicular iff their dot product is zero. But their dot product is $(x_1, y_1) \cdot (x_2, y_2) = x_1x_2 + y_1y_2$, which is precisely $\mathrm{Re}(z_1\bar{z}_2)$.

7. Let $z = x + iy$. Then

$$
\begin{aligned}
\sqrt{2}|z| \geq |\mathrm{Re}(z)| + |\mathrm{Im}(z)| \quad &\text{iff} \quad \sqrt{2}|z| \geq |x| + |y| \\
&\text{iff} \quad 2|z|^2 \geq |x^2| + 2|x|\,|y| + |y^2| \\
&\text{iff} \quad 2x^2 + 2y^2 \geq |x^2| + 2|x|\,|y| + |y^2| \\
&\text{iff} \quad x^2 - 2|x|\,|y| + y^2 \geq 0 \\
&\text{iff} \quad (|x| - |y|)^2 \geq 0,
\end{aligned}
$$

which is clearly true. A proper argument will start with this last inequality and work backwards to the appropriate conclusion.

9. By the triangle inequality, $|z_1 - z_2| = |z_1 + (-z_2)| \le |z_1| + |-z_2| = |z_1| + |z_2|$.

11. Let $z = (a, b)$. Then $\bar{z} = (a, -b)$, $-z = (-a, -b)$, and $-\bar{z} = (-a, b)$. The line segment from z to $\bar{z}$ is perpendicular to the line segment from $\bar{z}$ to $-z$ since the vector from z to $\bar{z}$ is $\bar{z} - z = (0, -2b)$. The vector from $\bar{z}$ to $-z$ is $(-2a, 0)$, and the dot product of these is clearly zero. A similar argument works for the other line segments. It is also easy to show that the diagonals intersect at the origin, establishing symmetry there.

13. This is an equivalent form of the vector equation between the points $z_1 = (x_1, y_1)$ and $z_2 = (x_2, y_2)$. Explain!

15. By repeated application of Equation (1.25), we have
$$|z_1 z_2 z_3| = |(z_1 z_2) z_3| = |z_1 z_2|\,|z_3| = |z_1|\,|z_2|\,|z_3|.$$

17. $|z - w|^2 = (z - w)\overline{(z - w)} = (z - w)(\bar{z} - \bar{w}) = |z|^2 - \bar{z}w - z\bar{w} + |w|^2$.
$|1 - \bar{z}w|^2 = (1 - \bar{z}w)\overline{(1 - \bar{z}w)} = (1 - \bar{z}w)(1 - z\bar{w}) = 1 - \bar{z}w - z\bar{w} + |z|^2|w|^2$. If $|z| = 1$, $|z - w|^2$ reduces to $1 - \bar{z}w - z\bar{w} + |w|^2$, and $|1 - \bar{z}w|^2$ becomes $1 - \bar{z}w - z\bar{w} + |w|^2$. Thus $|z - w|^2 = |1 - \bar{z}w|^2$, and the conclusion follows. If $|w| = 1$ we get the same result.

19. By Inequality (1.24), we see that $|z_1| - |z_2| \le |z_1 - z_2|$. Also, $|z_2| - |z_1| \le |z_2 - z_1| = |z_1 - z_2|$, so that $|z_1| - |z_2| \ge -|z_1 - z_2|$. Putting these two inequalities together gives $-|z_1 - z_2| \le |z_1| - |z_2| \le |z_1 - z_2|$, from whence the conclusion follows.

21. Let $z_1 = (x_1, y_1)$ and $z_2 = (x_2, y_2)$. Then $\mathrm{Re}(z_1 \bar{z}_2) = x_1 x_2 + y_1 y_2$. Also, $|z_1 \bar{z}_2| = \sqrt{(x_1 x_2 + y_1 y_2)^2 + (-x_1 y_2 + x_2 y_1)^2}$. If either z_1 or z_2 equals 0, then clearly $\mathrm{Re}(z_1 \bar{z}_2) = |z_1 \bar{z}_2|$. If neither equals 0, the two quantities are equal precisely when $-x_1 y_2 + x_2 y_1 = 0$. This occurs when the points z_1 and z_2 lie on a straight line through the origin. Show the details for this last statement.

23. The inequality $\left| \sum_{k=1}^{n} z_k \right| \le \sum_{k=1}^{n} |z_k|$ is clearly true when $n = 1$. Suppose that, for some $j > 1$,

$\left| \sum_{k=1}^{j} z_k \right| \le \sum_{k=1}^{j} |z_k|$. Then, using the triangle inequality and our induction assumption,

$\left| \sum_{k=1}^{j+1} z_k \right| = \left| \left(\sum_{k=1}^{j} z_k \right) + z_{j+1} \right| \le \left| \sum_{k=1}^{j} z_k \right| + |z_{j+1}| \le \left(\sum_{k=1}^{j} |z_k| \right) + |z_{j+1}| = \sum_{k=1}^{j+1} |z_k|$.

25a. By definition, an ellipse is the locus of points the sum of whose distances from two fixed points is constant. Since $|z - z_1|$ gives the distance from the point z to the point z_1, the set $\{z : |z - z_1| + |z - z_2| = K\}$ is precisely those points satisfying that definition.

25c. Letting $z_1 = 2i$, and $z_2 = -2i$, we compute $K = |3 + 2i - 2i| + |3 + 2i + 2i| = 3 + 5 = 8$. With $z = (x, y)$, the equation in Exercise 25a becomes $\sqrt{x^2 + (y - 2)^2} + \sqrt{x^2 + (y + 2)^2} = 8$. Show the details that squaring both sides, simplifying, squaring again, and simplifying again gives $4x^2 + 3y^2 = 48$. In standard form, $x^2 + \frac{3}{4}y^2 = 12$.

Section 1.4 The Geometry of Complex Numbers, Part II
(Exercise questions are on page 27)

1a. $-\frac{\pi}{4}$.

1c. $2\frac{\pi}{3}$.

1e. $-\frac{\pi}{3}$.

1g. $-\frac{\pi}{6}$.

3a. $4(\cos \pi + i \sin \pi) = 4e^{i\pi}$.

3c. $7(\cos \frac{-\pi}{2} + i \sin \frac{-\pi}{2}) = 7e^{-\frac{\pi}{2}}$.

3e. $\frac{1}{2}(\cos \frac{\pi}{2} + i \sin \frac{\pi}{2}) = \frac{1}{2}e^{i\frac{\pi}{2}}$.

3g. $5(\cos \theta + i \sin \theta) = 5e^{i\theta}$, where $\theta = \text{Arctan}(\frac{4}{3})$.

5a. i.

5c. $4 + i4\sqrt{3}$.

5e. $\sqrt{2} - i\sqrt{2}$.

5g. $-e^2$.

7. When $z = \sqrt{3} + i$; $\text{Arg}(iz) = \text{Arg}(z) + \frac{\pi}{2}$; $\text{Arg}(-z) = \text{Arg}(z) - \pi$; $\text{Arg}(-iz) = \text{Arg}(z) - \frac{\pi}{2}$.

9. All z except $z = 0$ and the negative real numbers. Prove this assertion!

11. Let $\theta \in \arg(\frac{1}{z})$. Then $\frac{1}{z} = re^{i\theta}$. Hence, $z = \frac{1}{r}e^{-i\theta}$, so $-\theta \in \arg(z)$, or $\theta \in -\arg(z)$. Thus, $\arg(\frac{1}{z}) \subseteq -\arg(z)$. The proof that $-\arg(z) \subseteq \arg(\frac{1}{z})$ is similar.

13a. Let $0 \neq z = x + iy$. Since $z\bar{z} = x^2 + y^2 > 0$, $\text{Arg}(z\bar{z}) = 0$.

15. From the figure it is clear that $\text{Arg}(z - z_0) = \phi$, and $|z - z_0| = \rho$. The exponential form for $z - z_0$ then gives the desired conclusion.

Section 1.5 The Algebra of Complex Numbers, Part II
(Exercise questions are on page 34)

1a. $-16 - i16\sqrt{3}$.

1c. -64.

3. $\cos 3\theta = \cos^3 \theta - 3\cos\theta \sin^2 \theta$, $\quad \sin 3\theta = 3\cos^2 \theta \sin\theta - \sin^3 \theta$.

5a. $\sqrt{2}\cos(\frac{\pi}{4} + \frac{2k\pi}{3}) + i\sqrt{2}\sin(\frac{\pi}{4} + \frac{2k\pi}{3})$, for $k = 0, 1, 2$.

5c. $2 \pm i$ and $-2 \pm i$.

5e. $2\cos(\frac{\pi}{8} + \frac{k\pi}{2}) + i2\sin(\frac{\pi}{8} + \frac{k\pi}{2})$ for $k = 0, 1, 2, 3$.

7. Since the coefficients are real, roots come in conjugates. Thus, $z - i$, and $z + i$ are factors. Dividing the polynomial by $z^2 + 1$ yields a quadratic, which can be solved with the quadratic formula (Theorem 1.5) to get $2 - i$ and $2 + i$ for the remaining solutions.

9. $2\sqrt{3} + 2i$, $-4i$, $-2\sqrt{3} + 2i$.

11a. Verify that $(1 - z)(1 + z + z^2 + \cdots + z^n) = 1 - z^{n+1}$.

11b. Let $z = e^{i\theta}$. For the left hand side of part (a), use De Moivre's formula. Keep $z = e^{i\theta}$ on the right hand side and multiply numerator and denominator by $e^{-i\frac{\theta}{2}}$. Simplify, and then equate real parts of the left and right hand sides.

13. Use Exercise ([11a](#)) and recall that if z_k is an nth root of unity, then $z_k^n = 1$.

15. The four roots are $\pm 1 \pm i$ (show the details). Use the roots as linear factors in conjugate pairs to get $z^4 + 1 = (z^2 + 2z + 2)(z^2 - 2z + 2)$.

Section 1.6 The Topology of Complex Numbers
(Exercise questions are on page 41)

1a. $z(t) = t + it$ for $0 \leq t \leq 1$.

1c. $z(t) = t + i$ for $0 \leq t \leq 1$.

3a. $z(t) = t + it^2$ for $0 \leq t \leq 2$.

3c. $z(t) = 1 - t + i(1 - t)^2$ for $0 \leq t \leq 1$.

5a. $z(t) = \cos t + i \sin t$ for $-\frac{\pi}{2} \leq t \leq \frac{\pi}{2}$.

7a. $z(t) = \cos t + i \sin t$ for $0 \leq t \leq \frac{\pi}{2}$.

9b. Open: (i), (iv), (v), (vi), and (vii). Connected: (i)–(vi). Domains: (i), (iv), (v), and (vi). Regions: (i)–(vi). Closed regions: (iii). Bounded: (iii), (v), and (vii).

11. Let $R = \max\{|z_1|, |z_2|, \ldots, |z_n|\}$. Clearly, $S \subseteq \overline{D_R(0)}$. Thus, S is bounded.

13. Let $C : z(t) = \big(x(t), y(t)\big)$, $a \leq t \leq b$ be any curve joining -2 and 2. Then $x(a) = -2$, and $x(b) = 2$. By the intermediate value Theorem, there is some $t^* \in (a, b)$ such that $x(t^*) = 0$. But this means $z(t^*) = \big(0, y(t^*)\big)$ is not in the set in question. Explain why!

15a. We prove the contrapositive. Suppose z_0 is accumulation point of S, but that z_0 does not belong to S. By definition of an accumulation point, every deleted neighborhood, $D_\varepsilon^*(z_0)$, contains at least one point of S. Therefore, every (non-deleted) neighborhood $D_\varepsilon(z_0)$ also contains at least one point of S *and* at least one point not in S (namely, z_0). This condition implies that z_0, which does not belong to S, is a boundary point of S. (Show the details for this last assertion). Thus, the set S is not closed.

Section 2.1 Functions and Linear Mappings
(Exercise questions are on page 52)

1a. $6 + \frac{1}{2}i$.

1c. $2i$.

3a. $u(x, y) = x^3 - 3xy^2$; $v(x, y) = 3x^2y - y^3$.

3c. $u(x, y) = \frac{x^2 - y^2}{x^4 + 2x^2y^2 + y^4}$; $v(x, y) = \frac{-2xy}{x^4 + 2x^2y^2 + y^4}$.

5a. 1.

5c. $-\frac{1}{2} + i\frac{\sqrt{3}}{2}$.

5e. -1.

7a. 0.

7c. $\ln\sqrt{2} + i\frac{\pi}{4}$, or $\frac{1}{2}\ln 2 + i\frac{\pi}{4}$.

7e. Yes, because if $f(z_1) = f(z_2)$ (where $z_1 = r_1 e^{i\theta_1}$ and $z_2 = r_2 e^{i\theta_2}$, and θ_1 and θ_2 are the Arguments of z_1 and z_2, respectively), then $\ln r_1 + i\theta_1 = \ln r_2 + i\theta_2$. Equating real and imaginary parts gives $\ln r_1 = \ln r_2$, so $r_1 = r_2$ (because the function $\ln$ is one-to-one). Also, $i\theta_1 = i\theta_2$, so $r_1 e^{i\theta_1} = r_2 e^{i\theta_2}$, i.e., $z_1 = z_2$.

9a. If $\mathbf{E}(z) = \frac{k}{\bar{z} - \bar{z}_0}$, then with rods at the points $z_0 = 0$, $1 - i$, and $1 + i$, each carrying a charge of $\frac{q}{2}$ coulombs per unit length, the total charge at z will be $\frac{k}{\bar{z} - 0} + \frac{k}{\bar{z} - (\overline{1-i})} + \frac{k}{\bar{z} - (\overline{1+i})}$. Combining terms and solving (using the quadratic formula) for when the numerator equals zero (tedious, but good for you!), reveals the total charge to be zero when $z = \frac{2}{3} \pm i\frac{\sqrt{2}}{3}$. Be sure to show the details of your calculations.

11a. Clearly, f is onto, because if $w \in B$, then by definition of B there exists a point $z \in A$ such that $f(z) = w$. Suppose that $f(z_1) = f(z_2)$ for some values z_1 and z_2 in A. Then, because A is a subset of D, z_1 and z_2 both belong to D. But f is one-to-one on D. Therefore, $z_1 = z_2$.

13. The triangle with vertices $-5 - 2i$, -6, and $3 + 2i$.

15a. $w = f(z) = \frac{3+2i}{13}z + \frac{7+9i}{13}$.

15c. $w = f(z) = \frac{i}{5}z + \frac{7+4i}{5}$.

17. Let $f(z = Az + B$ and $g(z) = Cz + E$ be two linear transformations. Then $h(z) = f\big(g(z)\big) = f(Cz + E) = A(Cz + E) + B = ACz + (E + B)$, which is the required form for a linear transformation.

Section 2.2 The Mappings $w = z^n$ and $w = z^{\frac{1}{n}}$
(Exercise questions are on page 59)

1a. Using Equations (2.9) we see that, if $A = \{(x, y) : y = 1\}$, then
$f(A) = \{(u, v) : u = x^2 - 1, \ v = 2x\} = \{(u, v) : u = \frac{v^2}{4} - 1\}$.

1c. The region in the upper half plane $\text{Im}(w) > 0$ that lies between the parabolas $u = 4 - \frac{v^2}{16}$ and $u = \frac{v^2}{4} - 1$.

1e. The point (x, y) in the xy plane is mapped to the point $(u, v) = (x^2 - y^2, 2xy)$. For any x, $u = x^2 - \frac{v^2}{4x^2}$. If $x = 1$ then $u = 1 - \frac{v^2}{4}$. If $x = 2$ then $u = 4 - \frac{v^2}{16}$. Your only remaining task is to show that the strip $\{(x, y) : 1 < x < 2\}$ is mapped between these two parabolas.

1g. The infinite strip $\{(u, v) : 1 < v < 2\}$, which is the region in the uv plane between $v = i$ and $v = 2i$. Show the details in a manner similar to the answer for part a.

3a. The points that lie to the extreme right or left of the branches of the hyperbola $x^2 - y^2 = 4$.

5. The region in the w plane that lies to the right of the parabola $u = 4 - \frac{v^2}{16}$.

7a. The set $\{re^{i\theta} : r > 8 \text{ and } \frac{3\pi}{4} < \theta < \pi\}$.

7c. The set $\{re^{i\theta} : r > 64 \text{ and } \frac{3\pi}{2} < \theta < 2\pi\}$.

9. See also problem 2. The fallacy lies in the assumption implicit in the second equality that $\sqrt{z_1 z_2} = \sqrt{z_1}\sqrt{z_2}$ for all complex numbers z_1 and z_2. Assuming the principal square root is used, then $\sqrt{z_1 z_2} = |z_1 z_2|^{\frac{1}{2}} e^{i\frac{\text{Arg}(z_1 z_2)}{2}}$. This quantity will equal $\sqrt{z_1}\sqrt{z_2} = |z_1|^{\frac{1}{2}} e^{i\frac{\text{Arg}(z_1)}{2}} |z_2|^{\frac{1}{2}} e^{i\frac{\text{Arg}(z_2)}{2}}$ precisely when $\text{Arg}(z_1 z_2) = \text{Arg}(z_1) + \text{Arg}(z_2)$— explain! The latter equality is plainly false when $z_1 = z_2 = -1$. (Again, explain.) To give a very thorough answer to this problem, you should state precisely when the last equality *is* true, and prove your assertion.

11. The right half plane given by $\text{Re}(z) > 0$ is mapped onto the region in the right half plane satisfying $u^2 - v^2 > 0$ and lies to the right of $u^2 - v^2 = 0$. This is the region between the lines $u = v$ and $u = -v$ in the right half of the w plane. A similar analysis can be applied to the case where $b = 0$.

Section 2.3 Limits and Continuity
(Exercise questions are on page 66)

1a. $-3 + 5i$.

1c. $-4i$.

1e. $1 - \frac{3}{2}i$.

3. $\lim\limits_{z \to z_0} (e^x \cos y + ix^2 y) = \lim\limits_{(x,y) \to (x_0, y_0)} (e^x \cos y + ix^2 y)$. Theorem 2.1 now implies the result since the real and imaginary parts of the last expression have limits that imply the desired conclusion. You should show the details for this, of course.

5a. $\lim\limits_{z \to 0} \frac{|z|^2}{z} = \lim\limits_{z \to 0} \frac{z\bar{z}}{z} = \lim\limits_{z \to 0} z = 0$.

7a. i.

7c. 1.

9. No. To see why, approach 0 along the real and imaginary axes respectively.

11a. If $z \to -1$ along the upper semicircle $r = 1$, $0 < \theta \le \pi$, then
$\lim\limits_{z \to -1} f(z) = \lim\limits_{\theta \to \pi} [\cos(\frac{\theta}{2}) + i\sin(\frac{\theta}{2})] = \cos(\frac{\pi}{2}) + i\sin(\frac{\pi}{2}) = i$.

13. The real part is continuous since $\lim\limits_{z \to z_0} xe^y = \lim\limits_{(x,y) \to (x_0, y_0)} xe^y = x_0 e^{y_0}$. A similar argument shows the imaginary part is continuous. Theorem 2.1 then implies that function f is continuous.

15. No. The limit does not exist. Show why.

17. Rewrite f as in Exercise 11, and mimic the argument for part a with an arbitrary negative real number taking the role of -1.

19. Let $\varepsilon > 0$ be given. Since $\lim\limits_{z \to z_0} f(z) = 0$, there is some number δ such that $f(z) \in D_{\frac{\varepsilon}{M}}(0)$ whenever $z \in D_{\delta}^*(z_0)$. Show this implies that if $z \in D_{\delta}^*(0)$, then $|f(z)g(z) - 0| = |f(z)| |g(z)| < \varepsilon$, so that $f(z)g(z) \in D_{\varepsilon}(0)$.

21a. We have remarked that Example 2.16 shows that the function $h(z) = \bar{z}$ is continuous for all z. Since f is continuous for all z, we can apply Theorem 2.4 to the function $f \circ h$ to conclude that $g(z) = f(h(z)) = f(\bar{z})$ is continuous for all z.

23. Make use of standard techniques. For example, to show that $f + g$ is continuous, use Theorem 2.2 applied to the sum of two functions.

Section 2.4 Branches of Functions
(Exercise questions are on page 71)

1a. The sector $\rho > 0$, $\pi/4 < \phi < \pi/2$.

1c. The sector $\rho > 0$, $-\pi/4 < \phi < \pi/4$.

3. Since $f_{2\pi}(z) = r^{\frac{1}{2}} \cos \frac{\theta}{2} + i r^{\frac{1}{2}} \sin \frac{\theta}{2}$, where $2\pi < \theta \leq 4\pi$ (explain!), we see that the point $(r^{\frac{1}{2}} \cos \frac{\theta}{2}, r^{\frac{1}{2}} \sin \frac{\theta}{2})$ will lie in the lower half plane (again, explain). Thus, the range of $f_{2\pi}(z)$ is $\{z : \text{Im}(z) \leq 0, z \neq 0\}$.

5a. $f_1(z) = |z|^{\frac{1}{3}} e^{i \frac{\text{Arg}(z)}{3}}$, so $\left(f_1(z)\right)^3 = \left(|z|^{\frac{1}{3}} e^{i \frac{\text{Arg}(z)}{3}}\right)^3 = |z| e^{i \frac{\text{Arg}(z)}{3}} = z$. This shows that f_1 is indeed a branch of the cube root function.

7. The function $f_{\frac{\pi}{4}}(z) = r^{\frac{1}{3}} e^{i \frac{\theta}{3}}$, where $0 \neq z = r e^{i\theta}$, and $\frac{\pi}{4} < \theta \leq \frac{9\pi}{4}$ does the job. Explain why, and find the range of this function, or of a different function that you concoct.

9a. For $k = 0, 1, 2$ we have $f_k(z) = e^{i \frac{\text{Arg}(z) + 2\pi k}{3}}$ as the three branches of the cube root with domains $D_k = \{z : z \neq 0\}$. As in the text, slit each domain along the negative real axis, and stack D_0, D_1, and D_2 directly above each other. Join the edge of D_0 in the upper half plane to the edge of D_1 in the lower half plane. Join the edge of D_1 in the upper half plane to the edge of D_2 in the lower half plane. Finally, join the edge of D_2 in the upper half plane to the edge of D_0 in the lower half plane. To really impress your teacher, make a sketch or real 3D model of this surface!

Section 2.5 The Reciprocal Transformation $w = \frac{1}{z}$
(Exercise questions are on page 77)

1. The circle $C_{\frac{5}{2}}\left(-\frac{5}{2}i\right) = \{w : |w + \frac{5}{2}i| = \frac{5}{2}\}$.

3. The circle $C_{\frac{1}{6}}\left(-\frac{1}{6}\right) = \{w : |w + \frac{1}{6}| = \frac{1}{6}\}$.

5. The circle $C_{\sqrt{2}}(1 - i) = \{w : |w - 1 + i| = \sqrt{2}\}$.

7. The circle $C_{\frac{4}{5}}\left(\frac{6}{5}\right) = \{w : |w - \frac{6}{5}| = \frac{4}{5}\}$.

9a. Let $\varepsilon > 0$ be given. Choose $R = \frac{1}{\varepsilon}$. Suppose $|z| > R$. Then $\frac{1}{|z|} < \frac{1}{R} = \varepsilon$, so $|f(z) - 0| = |\frac{1}{z}| < \varepsilon$, i.e., $f(z) \in D_\varepsilon(0)$.

11. The exterior of the disk $D_1\left(-\frac{i}{2}\right) = \{(u, v) : u^2 + \left(v + \frac{1}{2}\right)^2 > 1\}$.

13. The disk $D_{\sqrt{2}}(1 - i) = \{(u, v) : (u - 1)^2 + (v + 1)^2 < 2\}$.

15. The intersection of $D_{\frac{1}{2}}\left(\frac{1}{2}\right) = \{(u, v) : (u - \frac{1}{2})^2 + v^2 < \frac{1}{4}\}$ and $D_{\frac{1}{2}}\left(-\frac{i}{2}\right) = \{(u, v) : u^2 + (v + \frac{1}{2})^2 < \frac{1}{4}\}$.

17. The map $w = -1 + \frac{2}{z}$ (with inverse $z = \frac{2}{w+1}$) has The map $w = -1 + \frac{2}{z}$ (with inverse $z = \frac{2}{w+1}$) has

$$|z - 1| < 1 \iff \left| \frac{2}{w+1} - 1 \right| < 1 \iff \left| \frac{2}{u + iv + 1} - 1 \right| < 1 \iff \left| \frac{2(u + 1 - iv)}{(u+1)^2 + v^2} - 1 \right| < 1.$$

The last expression simplifies to $4u[(u + 1)^2 + v^2] > 0$, which occurs iff $u = \operatorname{Re}(w) > 0$. Show the details!

19. Let $\varepsilon > 0$ be given, Choose $R = \frac{2}{\varepsilon} + 1$. Assume $|z| > R = \frac{2}{\varepsilon} + 1$. Then $|z - 1| \geq |z| - 1 > (\frac{2}{\varepsilon} + 1) - 1 = \frac{2}{\varepsilon}$. Therefore $|\frac{2}{z-1}| < \varepsilon$, so $|\frac{z+1}{z-1} - 1| = |\frac{2}{z-1}| < \varepsilon$. To see how to get R, start with $|\frac{z+1}{z-1} - 1| < \varepsilon$, and work backwards.

21. Broadly speaking, $\pm\infty$ are designations for limits in Calculus indicating quantities that get arbitrarily positive or negative. There is no such measure in Complex Analysis. Further, the point ∞ can be given a meaningful definition on the Riemann Sphere. There is no such analogy for $\pm\infty$. Elaborate and give some other comparisons.

Section 3.1 Differentiable and Analytic Functions
(Exercise questions are on page 83)

1a. $f'(z) = 15z^2 - 8z + 7$.

1c. $h'(z) = \frac{3s}{(z+2)^2}$ for $z \neq -2$.

3. Parts (a), (b), (e), (f) are entire, and (c) is entire provided that $g(z) \neq 0$ for all z.

5b. The result is clearly true when $n = 1$. Assume for some $n > 1$ that $P'(z) = a_1 + 2a_2 z + \cdots + na_n z^{n-1}$. Consider $Q(z) = \sum_{k=0}^{n+1} a_k z^k = \sum_{k=0}^{n} a_k z^k + a_{n+1} z^{n+1}$. Since the derivative of the sum of two terms is the sum of the derivatives, we have $Q'(z) = \frac{d}{dz}\left(\sum_{k=0}^{n} a_k z^k \right) + \frac{d}{dz}(a_{n+1} z^{n+1})$. The induction assumption gives the result.

7a. $-4i$.

7c. 3.

7e. -16.

9. $\frac{d}{dz} z^{-n} = \frac{d}{dz}(\frac{1}{z^n})$. Apply the quotient rule: $\frac{d}{dz}(\frac{1}{z^n}) = \frac{(z^n)\frac{d}{dz}(1) - 1\frac{d}{dz}(z^n)}{(z^n)^2}$. Now simplify.

11. Evaluate $\lim_{z \to 0} \frac{f(z) - f(0)}{z - 0} = \lim_{z \to 0} \frac{|z|^2 - 0}{z - 0} = \lim_{z \to 0} \frac{z\bar{z}}{z} = \lim_{z \to 0} \bar{z} = 0$. Follow the hint for the rest.

13. $\frac{f(z_2) - f(z_1)}{z_2 - z_1} = \frac{i^3 - 1^3}{i - 1} = \frac{-i - 1}{i - 1} = i$. The minimum modulus of points on the line $y = 1 - x$ is $\frac{\sqrt{2}}{2}$ (prove this!). But $f'(z) = 3z^2$, and the only solutions to the equation $3z^2 = i$ have moduli equal to $\frac{\sqrt{3}}{3}$ (prove), which is less than $\frac{\sqrt{2}}{2}$ (prove this also).

Section 3.2 The Cauchy-Riemann Equations

(**Exercise questions are on page** 93)

1a. $u(x, y) = -y$, $v(x, y) = x + 4$; $u_x = v_y = 0$, $u_y = -v_x = -1$. The partials are continuous everywhere, so $f'(z) = u_x + iv_x = i$ for all z.

1c. $u_x = v_y = -2(y + 1)$; $u_y = -v_x = -2x$. The partials are continuous everywhere, so $f'(z) = u_x + iv_x = -2(y + 1) + i2x$ for all z.

1e. f is differentiable only at $z = i$, and $f''(i) = 0$.

1g. $u_x = v_y = 2x$, $u_y = 2y$, $v_x = 2y$. The conditions necessary for Theorem (3.4) are satisfied if and only if $y = 0$, and for $z = (x, 0)$, $f'(z) = 2x$.

3. $a = 1$ and $b = 2$.

5. $f'(z) = f''(z) = e^x \cos y + ie^x \sin y$ by Theorem (3.4).

7a. $u_x = -e^y \sin x$, $v_y = e^y \sin x$, $u_y = e^y \cos x$, and $-v_x = -e^y \cos x$. The Cauchy-Riemann equations hold if and only if both $\sin x = 0$ and $\cos x = 0$, which is impossible.

9a. $u_x = \sinh x \sin y = v_y$ and $u_y = \cosh x \sin y = -v_x$. The partials are continuous everywhere, so f is entire.

11a. f is differentiable only at points on the coordinate axes, so f is nowhere analytic.

11c. f is differentiable and analytic inside quadrants I and III.

13. The form of the definition is identical, but the meaning is more subtle in the complex case. For starters, the limit must exist when $z \to z_0$ from any direction in the complex case. The real case is limited to two directions.

15. Since $f = u + iv$ is analytic, u and v must satisfy the Cauchy-Riemann equations. Since f is not constant, this means the functions u and $-v$ do not satisfy the Cauchy-Riemann equations. Explain why this is the case, and then use Theorem (3.3) to conclude that $g = u - iv$ is not analytic.

Section 3.3 Harmonic Functions

(**Exercise questions are on page** 100)

1a. u is harmonic for all values of (x, y).

3. $c = -a$.

5a. $v(x, y) = x^3 - 3xy^2 + c$.

5c. $u(x, y) = -e^y \cos x + c$.

7. By the chain rule, $U_x(x, y) = u_x(x, -y)$, $U_y(x, y) = -u_y(x, -y)$; $U_{xx}(x, y) = u_{xx}(x, -y)$, and $U_{yy}(x, y) = u_{yy}(x, -y)$. Hence, $U_{xx}(x, y) + U_{yy}(x, y) = u_{xx}(x, -y) + u_{yy}(x, -y) = 0$.

9. The function $f = u + iv$ must be analytic, hence so is $f^2 = u^2 - v^2 + i(2uv)$. The result then follows by Theorem 3.8

11. $u_\theta = -rv_r$ implies $u_{\theta\theta} = -rv_{r\theta}$, and $u_{\theta r} = -rv_{rr} - v_r$. Also, $v_\theta = ru_r$ implies $v_{\theta\theta} = ru_{r\theta}$, and $v_{\theta r} = ru_{rr} + u_r$. From these calculations we get
$$r^2 u_{rr} + ru_r + u_{\theta\theta} = (rv_{\theta r} - ru_r) + (ru_r) + (-rv_{r\theta}) = 0.$$

13a. $f(z) = \frac{1}{x+iy} = \frac{x}{x^2+y^2} + i\frac{-v}{x^2+y^2}$.

15. The equipotentials are concentric circles with radii 1, 2, 3, and 4. The streamlines are lines from the origin making an angle of $\frac{k\pi}{8}$ radians for $k = 0, 1, \ldots, 7$.

Section 4.1 Sequences and Series

(Exercise questions are on page 108)

1a. 0.

1c. i.

3. Let $\varepsilon > 0$ be given. Since $\lim\limits_{n\to\infty} z_n = z_0$, there exists N_ε such that if $n > N_\varepsilon$ then $z_n \in D_e(z_0)$, i.e., $|z_n - z_0| < \varepsilon$. But since $|\bar{z}_n - \bar{z}_0| = |z_n - z_0|$, we see that if $n > N_\varepsilon$, then $\bar{z}_n \in D_\varepsilon(\bar{z}_0)$.

5. This is a "telescoping sum" and we have for the nth partial sum $S_n = -\frac{1}{i} + \frac{1}{n+i}$ (show the details for this). Then $\lim\limits_{n\to\infty} S_n = \lim\limits_{n\to\infty} (-\frac{1}{i} + \frac{1}{n+i}) = \lim\limits_{n\to\infty} (i + \frac{n-i}{n^2+1}) = i + 0 = i$.

7. No. In polar form we have $\lim\limits_{n\to\infty} (e^{i\frac{\pi}{4}})^n = \lim\limits_{n\to\infty} e^{i\frac{n\pi}{4}}$. These points oscillate around the eight roots of unity as Example 4.2 indicated.

9. Since $\sum\limits_{n=1}^{\infty} z_n$ converges, $\lim\limits_{n\to\infty} S_n = S$, where S is a complex number. But then $\lim\limits_{n\to\infty} S_{n-1} = S$, so $\lim\limits_{n\to\infty} z_n = \lim\limits_{n\to\infty} (S_n - S_{n-1}) = \lim\limits_{n\to\infty} S_n - \lim\limits_{n\to\infty} S_{n-1} = 0$.

11. $\sum\limits_{n=1}^{\infty} (a + ib)(x_n + iy_n) = \sum\limits_{n=1}^{\infty} [(ax_n - by_n) + i(bx_n + ay_n)]$. By Theorem 4.4, this expression equals $(au - bv) + i(bu + av) = (a + ib)(u + iv)$. Explain why in detail.

13. Duplicate the part of the Theorem that shows $\lim\limits_{n\to\infty} x_n = u$, but replace x_n with y_n and u with v.

15. Following the hint, for $\varepsilon > 0$ there exists numbers N_ε and M_ε such that $n > N_\varepsilon$ implies $|z_n - \zeta_1| < \frac{\varepsilon}{2}$, and $n > M_\varepsilon$ implies $|z_n - \zeta_2| < \frac{\varepsilon}{2}$. Let $L_\varepsilon = \max\{N_\varepsilon, M_\varepsilon\}$. It follows that if $n > L_\varepsilon$, then $|\zeta_1 - \zeta_2| = |\zeta_1 - z_n + z_n - \zeta_2| \leq |\zeta_1 - z_n| + |z_n - \zeta_2| < \varepsilon$.

17. Let $\varepsilon > 0$ and suppose $\lim\limits_{n\to\infty} z_n = 0$. This means there exists N_ε such that $n > N_\varepsilon$ implies $z_n \in D_\varepsilon(0)$, that is, $|z_n - 0| < \varepsilon$. But then $||z_n| - 0| = |z_n - 0| < \varepsilon$, so also we have $|z_n| \in D_\varepsilon(0)$. Therefore, $\lim\limits_{n\to\infty} |z_n| = 0$. The other direction is similar. Show the details.

Section 4.2 Julia and Mandelbrot Sets

(Exercise questions are on page 117)

1a. If $z = r(\cos\theta + i\sin\theta) \neq 0$, show $N(z) = \frac{1}{2}(r - \frac{1}{r})\cos\theta + i\frac{1}{2}(r + \frac{1}{r})\sin\theta$. The result now follows—explain why!

1c. If $z_0 \neq 0$ is real then obviously $z_1 = N(z_1) = \frac{1}{2}(z_0 - \frac{1}{z_0})$ is real. Assume z_n is real for some $n > 1$. Then $z_{n+1} = N(z_n) = \frac{1}{2}(z_n - \frac{1}{z_n})$ is also real, provided $z_n \neq 0$.

3. For $f(z) = az + b$, if our initial guess is z_0, then $z_1 = z_0 - \frac{az_0 + b}{a} = -\frac{b}{a}$. But this is the solution to the equation $f(z) = 0$, so our iteration either stops here or with z_0 (if by chance we had set $z_0 = -\frac{b}{a}$).

5. The Julia set for $f_{-2}(z) = z^2 - 2$ is connected by Theorem 4.9 because the orbit of 0 under f_2 is $\{-2, 2, 2, 2, \ldots\}$, which is a bounded set.

7. Suppose $c \in M$, and let $\{z_k\}$ be the orbit of 0 under f_c. By definition of M, there is some real number N such that $|z_k| < N$ for all k. Let $\{w_k\}$ be the orbit of 0 under $f_{\bar{c}}$. Show by induction that $w_k = \bar{z}_k$ for all k. Once you have that, it is straight forward to conclude that the set $\{w_k\}$ is bounded.

9. There are many examples. The number -2, is in the Mandelbrot set, but its negative, 2, is not. Find another example, and justify your assertion.

11. If we let $c = -\frac{1}{4}\sqrt{3}i$, then $\left|1 - \sqrt{1 - 4c}\right| = \left|1 - \sqrt{1 + \sqrt{3}i}\right| = \left|1 - \left(\frac{\sqrt{2}}{2}\sqrt{3} + \frac{\sqrt{2}}{2}i\right)\right|$ (show the details for this conclusion). But this last quantity equals $\sqrt{3 - \sqrt{6}}$ (explain), which is less than 1 (again, explain).

13. Since $|f'(z_0)| < 1$, we can choose ρ such that $|f'z_0)| < \rho < 1$. Using the same technique as Theorem 4.10, show that if $z \in D_r^*(z_0)$, then $|f(z) - z_0| < \rho|z - z_0|$. In other words, $|z_1 - z_0| < \rho|z - z_0|$, where $z_1 = f(z)$. An easy induction argument now gives that, for all k, $|z_k - z_0| < \rho^k|z - z_0|$, where z_k is the kth iterate of z_0. Since $\rho < 1$, this condition implies $\lim_{k \to \infty} z_k = z_0$. Show the details.

Section 4.3 Geometric Series and Convergence Theorems
(Exercise questions are on page 122)

1. By Theorem 4.12, $\sum_{n=0}^{\infty} \frac{(1+i)^n}{2^n} = \frac{1}{1 - (\frac{1+i}{2})} = 1 + i$ (show the details), since $|\frac{1+i}{2}| < 1$ (show this also).

3. The series converges by the ratio test. Show the details.

5a. Converges in $D_{\frac{\sqrt{2}}{2}}(0)$.

5c. Converges in $D_5(i)$.

7. $|S_n| = |\frac{1}{1-z} - \frac{z^n}{1-z}| \geq |\frac{z^n}{1-z}| - |\frac{1}{1-z}| = |z^n||\frac{1}{1-z}| - |\frac{1}{1-z}|$. Now use the fact that $|z| > 1$ to get the desired conclusion.

9. Mimic the argument most calculus texts give for real series, but replace $|x|$ with $|z|$.

11. If $f(z) = \sum_{n=0}^{\infty} z^{(2^n)}$, then $f(z^2) = \sum_{n=0}^{\infty} (z^2)^{(2^n)} = \sum_{n=0}^{\infty} z^{(2 \cdot 2^n)} = \sum_{n=0}^{\infty} z^{(2^{n+1})} = \sum_{n=1}^{\infty} z^{(2^n)}$. The conclusion follows from this. Explain in detail, especially the second equality for $f(z^2)$.

Section 4.4 Power Series Functions
(Exercise questions are on page 128)

1. The series for $f(z)$ converges absolutely if $\lim\limits_{n\to\infty} |\frac{c_{n+1}}{c_n}||z - \alpha| < 1$. If $\lim\limits_{n\to\infty} |\frac{c_{n+1}}{c_n}| = 0$, the series converges for all z. If $\lim\limits_{n\to\infty} |\frac{c_{n+1}}{c_n}| = \infty$, the series converges only when $z = \alpha$. If $\lim\limits_{n\to\infty} |\frac{c_{n+1}}{c_n}|$ is finite but not zero, then the series converges if $|z - \alpha| < \frac{1}{\lim\limits_{n\to\infty} |\frac{c_{n+1}}{c_n}|} = \rho$.

3a. ∞.

3c. $\frac{3}{5}$.

3e. $\frac{1}{3}$.

3g. $\frac{4}{3}$.

3i. 1.

5. Show that $\limsup\limits_{n\to\infty} |c_n^2|^{\frac{1}{n}} = \left(\limsup\limits_{n\to\infty} |c_n|^{\frac{1}{n}}\right)^2$.

7. The Theorem establishes $f^{(k)}(z) = \sum\limits_{n=k}^{\infty} n(n-1)\cdots(n-k+1)c_n(z-\alpha)^{n-k}$ when $k = 1$.

   Assume the Theorem is true for some $k > 1$, and set $g(z) = \sum\limits_{n=0}^{\infty} b_n(z-\alpha)^n$, where $b_n = (n+k)(n+k-1)\cdots(n+1)c_{n+k}$. In other words, $g(z) = f^{(k)}(z)$ (confirm this). Applying the case when $k = 1$ to the function g gives

$$g'(z) = f^{(k+1)}(z) = \sum_{n=1}^{\infty} n b_n (z-\alpha)^{n-1}$$

$$= \sum_{n=1}^{\infty} n(b_n)(n+k)(n+k-1)\cdots(n+1)c_{n+k}(z-\alpha)^{n-1}$$

$$= \sum_{n=k+1}^{\infty} n(n-1)\cdots(n-k+1)(n-k)c_n(z-\alpha)^{n-(k+1)}$$

   (confirm this also), which is what we needed to establish.

9a. Since $s^n - t^n = (s^{n-1} + s^{n-2}t + s^{n-3}t^2 + \cdots + st^{n-2} + t^{n-1})(s - t)$ (verify!), the conclusion follows from division and the triangle inequality.

11. The series converges for all values of z by the ratio test.

Section 5.1 Elementary Functions
(Exercise questions are on page 135)

1. Recall that $\sum\limits_{n=0}^{\infty} c_n z^n$ is compact notation for $c_0 + \sum\limits_{n=1}^{\infty} c_n z^n$, and that $0! = 1$. Then, by definition, $\exp(0) = \sum\limits_{n=0}^{\infty} \frac{1}{n!} 0^n = \frac{1}{0!} + \sum\limits_{n=1}^{\infty} \frac{1}{n!} 0^n = 1$.

3. Let n be an integer, and set $z = i2n\pi$. Then $e^{i2n\pi} = \cos(2n\pi) + i\sin(2n\pi) = 1$. Conversely, suppose $e^z = e^{x+iy} = 1$. Then $e^x e^{iy} = e^x(\cos y + i\sin y) = 1 + 0i$. This implies $\sin y = 0$.

439

Because e^x is always positive and $e^x \cos y = 1$, $y = 2n\pi$ for some integer n. This result forces $x = 0$, so $z = x + iy = 0 + i2n\pi$, establishing Equation (5.3). Equation (5.4) comes from observing that $e^{z_1} = e^{z_2}$ iff $e^{z_1 - z_2} = 1$, and then appealing to Equation (5.3).

5a. Using properties of the exponential function, $e^z = -4$ iff $z = x + iy$ with $y = (2n+1)\pi$ where n is an integer, and $e^x = 4$. Thus, $x = \ln 4$, and $z = \ln 4 + i(2n+1)\pi$, where n is an integer.

5c. $z = \ln 2 + i(-\frac{\pi}{6} + 2n\pi)$, where n is an integer.

7. The conclusion follows immediately from Equation (5.4).

9a. $\exp(\bar{z}) = \sum\limits_{n=0}^{\infty} \frac{1}{n!}(\bar{z})^n = \lim\limits_{k \to \infty}\left(\sum\limits_{n=0}^{k} \frac{1}{n!}(\bar{z})^n\right) = \lim\limits_{k \to \infty}\left(\overline{\sum\limits_{n=0}^{k} \frac{1}{n!}z^n}\right) = \overline{\left(\lim\limits_{k \to \infty} \sum\limits_{n=0}^{k} \frac{1}{n!}z^n\right)}$,

because the conjugate is a continuous function (explain). This last quantity, of course, equals $\overline{\exp(z)}$.

11a. Method 1: $\lim\limits_{z \to 0} \frac{e^z - 1}{z} = \lim\limits_{z \to 0} \frac{\sum\limits_{n=1}^{\infty} \frac{1}{n!}z^n}{z} = \lim\limits_{z \to 0} \sum\limits_{n=1}^{\infty} \frac{1}{n!}z^{n-1} = 1$. Justify the last equality.

Method 2: Using L'Hôpital's rule (Theorem 3.2), $\lim\limits_{z \to 0} \frac{e^z - 1}{z} = \lim\limits_{z \to 0} \frac{e^z}{1} = 1$.

13a. ie^{iz}.

13c. $(a + ib)e^{(a+ib)z}$.

15. $\sum\limits_{n=0}^{\infty} e^{inz} = \sum\limits_{n=0}^{\infty} (e^{iz})^n$. This is a geometric series. Show that $\text{Im}(z) > 0$ implies $|e^{iz}| = |e^{i(x+iy)}| < 1$, so that the series converges by Theorem 4.12.

17. Show that $e^{x^2 - y^2} \sin 2xy$ is the imaginary part of $\exp(z^2)$, and therefore harmonic by Theorem 3.8.

Section 5.2 The Complex Logarithm
(**Exercise questions are on page** 142)

1a. $2 + i\frac{\pi}{2}$.

1c. $\ln 2 + i\frac{3\pi}{4}$.

1e. $\ln 3 + i(1 + 2n)\pi$, where n is an integer.

1g. $\ln 4 + i(\frac{1}{2} + 2n)\pi$, where n is an integer.

3a. $(\frac{e\sqrt{2}}{2})(1 - i)$.

3c. $1 + i(-\frac{1}{2} + 2n)\pi$, where n is an integer.

5a. Since $\text{Arg}(1 + i) = \frac{\pi}{4}$, the function $f(z) = \log_{\frac{\pi}{4}}(z - 1 - i)$ defined for $z = re^{i\theta} \neq 1 + i$, where $\frac{\pi}{4} < \theta < \frac{9\pi}{4}$, is analytic, and $f'(z) = \frac{1}{z-1-i}$ for those values of z.

5c. We set $f(z) = z\text{Log}(z)$, and deduce that $f'(z) = 1 + \text{Log}(z)$ for $z = re^{i\theta} \neq 0$, where $-\pi < \theta < \pi$.

5e. Note that $z^2 - z + 2 = 0$ when $z = \frac{1 \pm i\sqrt{7}}{2}$. Also, $\mathrm{Arg}(\frac{1 \pm i\sqrt{7}}{2}) = \pm \mathrm{Arctan}(\sqrt{7})$ respectively. For $\alpha = \mathrm{Arctan}(\sqrt{7})$, the function $w = f(z) = \log_\alpha(z^2 - z + 2)$ is differentiable for $z = re^{i\theta} \neq \frac{1 - i\sqrt{7}}{2}$, where $\alpha < \theta < \alpha + 2\pi$. For $\alpha = -\mathrm{Arctan}(\sqrt{7})$, the function $w = f(z) = \log_\alpha(z^2 - z + 2)$ is differentiable for $z = re^{i\theta} \neq \frac{1 + i\sqrt{7}}{2}$, where $\alpha < \theta < \alpha + 2\pi$. Furthermore, in each case $\frac{dw}{dz} = f'(z) = \frac{2z-1}{z^2-z+2}$ for z in the respective domains.

7a. $\ln(x^2 + y^2) = 2\mathrm{Re}\big(\mathrm{Log}(z)\big)$, and $\mathrm{Log}(z)$ is analytic for $\mathrm{Re}(z) > 0$.

9. According to Equation (5.20), $f(z) = \log_{6\pi}(z+2)$ yields
$$f(-5) = \log_{6\pi}(-1) = \ln|-1| + i(7\pi) = 7\pi i.$$

11a. The function $f(z) = \log_0(z+2)$ does the job. Explain why.

11c. The function $f(z) = \log_{-\frac{\pi}{2}}(z+2)$ works. Explain why.

13. There are many possibilities, such as $z_1 = 1$, $z_2 = -1$. Explain.

15. Any branch of the logarithm is defined as an inverse of the exponential. Since there is no value z for which $\exp(z) = 0$, there can be no branch of the logarithm that is defined at the number 0.

Section 5.3 Complex Exponents
(Exercise questions are on page 147)

1a. $\cos(\ln 4) + i\sin(\ln 4)$.

1c. $\cos 1 + i\sin 1$.

3. Note that $0 \cdot \log(z) = \{0 \cdot \zeta : \zeta \in \log(z)\}$. This set amounts to the single element zero. Thus, for $z \neq 0$, $z^0 = \exp(0 \cdot \log z) = \exp(0) = 1$.

5. $2z_{n-1} - 2z_{n-2} = 2(1+i)^{n-1} - 2(1+i)^{n-2} = 2(1+i)^{n-2}[(1+i) - 1]$. This last expression simplifies to $2i(1+i)^{n-2}$. Now, $z_n = (1+i)^n$. Since Log is a one-to-one function, the problem is solved by showing $\mathrm{Log}[(1+i)^n] = \mathrm{Log}[2i(1+i)^{n-2}]$. Use properties of the logarithm to do this.

7. No. $1^{a+ib} = e^{a2\pi n}\cos(b2\pi n) + ie^{a2\pi n}\sin(b2\pi n)$, where n is an integer.

9. The number c must be real, and $|i^c| = 1$.

Section 5.4 Trigonometric and Hyperbolic Functions
(Exercise questions are on page 156)

1. $\frac{d}{dz}\cos z = \frac{d}{dz}\left[\sum_{n=0}^{\infty}(-1)^n \frac{z^{2n}}{(2n)!}\right] = \sum_{n=1}^{\infty}(-1)^n \frac{(2n)z^{2n-1}}{(2n)!}$. Explain why the index n begins at 1 in the last expression. The result follows from simplification and reindexing.

3. $\tan z = \frac{2\cos x \sin x}{2(\cos^2 x \cosh^2 y + \sin^2 x \sinh^2 y)} + i\frac{2\cosh y \sinh y}{2(\cos^2 x \cosh^2 y + \sin^2 x \sinh^2 y)}$. The numerators simplify to $\sin 2x$ and $\sinh 2y$, respectively. Show that the denominator equals $\cos 2x + \cosh 2y$ by using the identities $\cos 2x = \cos^2 x - \sin^2 x$ and $\cosh^2 y - \sinh^2 y = 1$.

5a. This follows immediately from $\sin(z_1 + z_2) = \sin z_1 \cos z_2 + \cos z_1 \sin z_2$.

5c. This follows immediately from $\sinh z = \sinh x \cos y + i \cosh x \sin y$, where we replace $z = x + iy$ with $z = x + iy + i\pi = x + i(y + \pi)$.

5e. This follows immediately from $\sin z = \sin x \cosh y + i \cos x \sinh y$, where we replace $z = x + iy$ with $iz = -y + ix$.

7a. $-\frac{1}{z^2}\cos(\frac{1}{z})$, valid for $z \neq 0$.

7c. $2z \sec z^2 \tan z^2$, valid for $z \neq (k + \frac{1}{2})\pi$, where k is an integer.

7e. $z \cosh z + \sinh z$, valid for all z.

9a. Use the same methods as in Exercise 11a of Section 5.1.

11. By Identity (5.35), $\sin z = 0$, if and only if $\sin x \cosh y + i \cos x \sinh y = 0$. Equate real and imaginary parts to show this occurs iff $x = k\pi$, where k is an integer.

13. Combining 5.38 and 5.39, and letting $z = x + iy$, we get $|\sin z|^2 + |\cos z|^2 = \sin^2 x + \sinh^2 y + \cos^2 x + \sinh^2 y = 1 + 2\sinh^2 y$. This quantity equals 1 iff $y = 0$ (when z is a real number), and is greater than 1 otherwise.

15a. Consider the real part of Identity (5.35), and appeal to Theorem 3.8.

15c. Consider the imaginary part of $\sin(iz)$, and appeal to Theorem 3.8.

17. $Z = 10 + 10i$.

Section 5.5 Inverse Trigonometric and Hyperbolic Functions
(Exercise questions are on page 161)

1a. $(\frac{1}{2} + 2n)\pi \pm i \ln 2$, where n is an integer.

1c. $(\frac{1}{2} + 2n)\pi \pm i \ln(3 + 2\sqrt{2})$, where n is an integer.

1e. $-(\frac{1}{2} + n)\pi + i \ln \sqrt{3}$, where n is an integer.

1g. $i(\frac{1}{2} + 2n)\pi$, where n is an integer.

1i. $\ln(\sqrt{2} + 1) + i(\frac{1}{2} + 2n)\pi$ and $\ln(\sqrt{2} - 1) + i(-\frac{1}{2} + 2n)\pi$, where n is an integer.

1k. $i(\frac{1}{4} + n)\pi$, where n is an integer.

Section 6.1 Complex Integrals
(Exercise questions are on page 166)

1a. $2 - 3i$.

1c. 1.

1e. $\frac{\sqrt{2}\pi}{8} + \frac{\sqrt{2}}{2} - 1 + i(\frac{\sqrt{2}}{2} - \frac{\sqrt{2}\pi}{8})$.

3. Using Equation (6.8) we get $\int_0^\infty e^{-zt}\, dt = \lim_{T\to\infty} \int_0^T e^{-zt}\, dt = \lim_{T\to\infty}(-\frac{1}{z}e^{-zT} + \frac{1}{z}e^{-z(0)}) = \frac{1}{z} + \lim_{T\to\infty}(-\frac{1}{z}e^{-zT})$. Show that $\text{Re}(z) > 0$ implies this last limit equals zero.

5. This follows from Equation (6.8), and the fact that if u and v are differentiable, then f is differentiable, and $\frac{d}{dt}\left[\frac{1}{2}(f(t))^2\right] = f(t)f'(t)$.

Section 6.2 Contours and Contour Integrals

(Exercise questions are on page 178)

1a. $C_1 : z_1(t) = 2e^{it}, 0 \le t \le \frac{\pi}{2}; \quad C_2 : z_2(t) = -t + i(2-t)\, 0 \le t \le 2.$

3a. The approximation simplifies to $-2\sqrt{2} + 2 \approx -0.828427.$

3b. $-\frac{2}{3}.$

5a. $-32i.$

5b. $-8\pi i.$

7a. $0.$

7c. $-2\pi i.$

7e. $i - 2.$

7g. $-4 - i\pi.$

9a. $2\pi i.$

9b. $0.$

11. $-1 + \frac{2i}{3}.$

13. $-2e.$

15. $\exp(1 + i) - 1.$

17. $\sin(1 + i).$

19. The absolute value of the integrand is $\sqrt{x^2 + (1 - x^2)\cos^2\theta}^{\,n}$. This expression simplifies to $\sqrt{x^2 \sin^2\theta + \cos^2\theta}^{\,n}$ (show the details for this assertion). The maximum of this expression occurs when $x = 1$. Now simplify and apply the ML inequality.

Section 6.3 The Cauchy-Goursat Theorem

(Exercise questions are on page 191)

1a. Analytic everywhere except at $z = \pm\frac{i}{\sqrt{2}}$. We break the integral up using partial fractions: $\int_{C_1^+(0)} \frac{z}{2z^2+1}\, dz = \int_{C_1^+(0)} \frac{1/4}{z - \frac{i}{\sqrt{2}}}\, dz + \int_{C_1^+(0)} \frac{1/4}{z + \frac{i}{\sqrt{2}}}\, dz$. Both $\pm\frac{i}{\sqrt{2}}$ lie inside $C_1^+(0)$, so Corollary 6.1 gives $\int_{C_1^+(0)} \frac{1/4}{z - \frac{i}{\sqrt{2}}}\, dz + \int_{C_1^+(0)} \frac{1/4}{z + \frac{i}{\sqrt{2}}}\, dz = \frac{1}{4}(2\pi i) + \frac{1}{4}(2\pi i) = \pi i.$

1c. Analytic everywhere except $z = (n + \frac{1}{2})\pi$, where n is an integer, so $\int_{C_1^+(0)} f(z)\, dz = 0$, since all non analytic points lie outside the circle $C_1(0)$.

3. By the quadratic formula (see Theorem 1.5), $4z^2 - 4z + 5 = 0$ when $z = \frac{1}{2} \pm i$ (verify). Since both these points lie outside $C_1(0)$, the function $(4z^2 - 4z + 5)^{-1}$ is analytic inside $C_1(0)$, so $\int_{C_1^+(0)} (4z^2 - 4z + 5)^{-1}\, dz = 0$ by the Cauchy-Goursat Theorem.

5a. $4\pi i.$

5b. $2\pi i.$

7a. $\frac{\pi i}{4}$.

7b. $-\frac{\pi i}{4}$.

7c. 0.

9. $-\frac{4i}{3}$.

11. 0.

13a. $4\pi i$.

13b. 0.

Section 6.4 The Fundamental Theorems of Integration
(**Exercise questions are on page** 196)

1. $\frac{4}{3} + 3i$.

3. $-e^2 + i$.

5. $-1 + i\frac{\pi+2}{2}$.

7. $-\frac{7}{6} + i\frac{1}{2}$.

9. $-1 - \sinh 1 + \cosh 1$.

11. $\ln\sqrt{2} - \frac{\pi}{4} + i(\ln\sqrt{2} + \frac{\pi}{4} - 1)$.

13. $\text{Log}(1+i) - \text{Log}(2) + \text{Log}(2+i) = -\ln\sqrt{2} + \ln\sqrt{5} + i\left(\frac{\pi}{4} + \text{Arctan}(\frac{1}{2})\right)$.

15. Parametrize C with $z(t) = z_1 + (z_2 - z_1)t$, $0 \le t \le 1$. Then we see that $\int_C 1\, dz = \int_0^1 z'(t)\, dt = \int_0^1 (z_2 - z_1)\, dt = (z_2 - z_1)t \Big|_{t=0}^{t=1} = z_2 - z_1$.

17. $-1 + i$.

19. We know that an antiderivative of the function $fg' + gf'$ is fg by the product rule. Since fg' and gf' are analytic (explain why!), Theorem 6.9 gives us $\int_C [f(z)g'(z) + g(z)f'(z)]\, dz = f(z)g(z)\Big|_{z=z_1}^{z=z_2}$. The conclusion follows from this.

Section 6.5 Integral Representations
(**Exercise questions are on page** 201)

1. $4\pi i$.

3. $-i\frac{\pi}{2}$.

5. $-i\frac{\pi}{3}$.

7. $2\pi i$.

9. $\frac{2\pi i}{(n-1)!}$.

11. $\frac{\pi}{8} - i\frac{\pi}{8}$.

13a. $i\pi \sinh 1$.

13b. $i\pi \sinh 1$.

15a. π.

15b. $-\pi$.

17. 0.

19. Let $f(z) = (z^2 - 1)^n$, which is analytic everywhere. By Cauchy's integral formulas, $P_n(z) = \frac{1}{2^n n!} f^{(n)}(z) = \frac{1}{2^n n!} \left[\frac{n!}{2\pi i} \int_C \frac{f(\xi)}{(\xi - z)^{n+1}} \, d\xi \right]$. The conclusion follows from this. Show the details.

Section 6.6 The Theorems of Morera and Liouville
(Exercise questions are on page 207)

1a. $(z + 1 + i)(z + 1 - i)(z - 1 + i)(z - 1 - i)$.

1c. $(z + i)(z - i)(z - 2 + i)(z - 2 - i)$.

3. We know that the complex cosine is an entire function that is not a constant. By Liouville's Theorem, it is not bounded.

5a. $|f^{(4)}(1)| \le \frac{4!(10)}{2\pi 3^5} 6\pi = \frac{80}{3^3}$. (Explain.)

5b. $|f^{(4)}(0)| \le \frac{4!(10)}{2\pi 2^5} 4\pi = 15$. (Explain.)

7a. If $|f(z)| \ge m$ for all z in D, where $m > 0$, then the function $\frac{1}{f}$ is analytic in D. Apply the maximum modulus Theorem to the function $\frac{1}{f}$ to get your result.

9. Let $f(z) = u(z) + iv(z)$, where v is a harmonic conjugate of u, so that f is analytic in D. The function $F(z) = \exp\left(f(z)\right)$ is also analytic in D, so that $|F|$ does not take on a maximum in D by the maximum modulus Theorem. But $|F(z)| = \exp\left(u(z)\right)$ for all z (show why). This leads to the conclusion since u is a real valued function, and the real valued function exp is an increasing function. Explain this last part in detail.

11. (By contraposition) If f does not have a zero, then $\frac{1}{f}$ is analytic in $D_1(0)$, so its maximum occurs on the boundary. Since f is constant on the boundary, we conclude that both the maximum and the minimum of f are the same, which means f is constant.

Section 7.1 Uniform Convergence
(Exercise questions are on page 214)

1a. By definition, $f(-1) = \frac{1}{1-(-1)} = \frac{1}{2}$. It appears from the graph that the value of the upper function is approximately 1 (certainly larger than $\frac{1}{2}$), so the graph of S_n must be above the graph of f.

1c. From the graph, we approximate $S_n(1) = 5$. As $S_n(x) = \sum_{k=0}^{n-1} x^k$, we deduce that $n = 5$. Explain.

3a. We see that $|\frac{1}{k^2} z^k| \leq \frac{1}{k^2}$ for $z \in \bar{D}_1(0)$. By the Weierstrass M-test, the series $\sum_{k=1}^{\infty} \frac{1}{k^2} z^k$

converges uniformly on $\bar{D}_1(0) = \{z : |z| \leq 1\}$, because the series $\sum_{k=1}^{\infty} \frac{1}{k^2}$ converges.

5. The crucial step in the theorem is the statement, "Moreover, for all $z \in \bar{D}_r(\alpha)$ it is clear that $|c_k(z - \alpha)^k| = |c_k||z - \alpha|^k \leq |c_k| r^k$." If we allowed $r = 1$, we would not be able to claim that $\sum_{k=0}^{\infty} |c_k| r^k$ converges. Explain.

7a. Let us say that $\{f_n\}$ and $\{g_n\}$ converge uniformly on T to f and g respectively. Let $\varepsilon > 0$ be given. The uniform convergence of $\{f_n\}$ means there exists an integer N_ε such that $n \geq N_\varepsilon$ implies $|f_n(z) - f(z)| < \frac{\varepsilon}{2}$ for all $z \in T$. Likewise, there exists an integer M_ε such that $n \geq M_\varepsilon$ implies $|g_n(z) - g(z)| < \frac{\varepsilon}{2}$ for all $z \in T$. If we set $L_\varepsilon = Max\{N_\varepsilon, M_\varepsilon\}$, then for $n \geq L_\varepsilon$, $\left|\big(f_n(z) + g_n(z)\big) - \big(f(z) + g(z)\big)\right| \leq |f_n(z) - f(z)| + |g_n(z) - g(z)| < \frac{\varepsilon}{2} + \frac{\varepsilon}{2} = \varepsilon$ for all $z \in T$.

7b. For all n, let $f_n(x) = x$, and $g_n(x) = \frac{1}{n}$, for all $x \in T$, where T is the real numbers. Then $f_n(x)$ converges uniformly to x, and $g_n(x)$ converges uniformly to 0 (verify). However, even though $f_n(x)g_n(x)$ converges to 0 (explain), the convergence is not uniform (verify). Can you come up with a different example?

9a. For $z \in A$, $|n^{-z}| = |\exp[-(x + iy)\ln n]| = |\exp(-iy)| \,|\exp(-x \ln n)| = n^{-x}$. Since $z \in A$, we know $\text{Re}(z) = x \geq 2$, so $n^{-x} \leq \frac{1}{n^2}$. Thus, with $M_n = \frac{1}{n^2}$, we see that $\zeta(z)$ converges uniformly on A by the Weierstrass M-test.

Section 7.2 Taylor Series Representations
(Exercise questions are on page 221)

1a. $\sinh z = \sum_{n=0}^{\infty} \frac{z^{2n+1}}{(2n+1)!}$ for all z.

1c. $\text{Log}(1 + z) = \sum_{n=1}^{\infty} \frac{(-1)^{n-1}}{n} z^n$ for all $z \in D_1(0)$.

3a. $\frac{1-z}{z-2} = \frac{z-1}{1-(z-1)} = (z-1)\left[\frac{1}{1-(z-1)}\right]$. Expand the expression in brackets by replacing z with $z - 1$ in the geometric series (valid, therefore, for $|z - 1| < 1$), then multiply by the $(z - 1)$ term.

5. $f(z) = \frac{1}{1-z} = \frac{1}{1-i}\left[\frac{1}{1-\frac{z-i}{1-i}}\right]$. Expand the expression in brackets by replacing z with $\frac{z-i}{1-i}$ in the geometric series (valid, therefore, for $|\frac{z-i}{1-i}| < 1$, or $|z - i| < \sqrt{2}$). Explain.

7a. By Taylor's Theorem, $\frac{f^{(n)}(0)}{n!} = (3 + (-1)^n)^n$. Therefore, $\frac{f^{(3)}(0)}{3!} = 8$, so $f^{(3)}(0) = 48$.

9a. Observe that $1 + zf(z) + z^2 f(z) = 1 + \sum_{n=0}^{\infty} c_n z^{n+1} + \sum_{n=0}^{\infty} c_n z^{n+2}$. Reindex and write this as $1 + \sum_{n=1}^{\infty} c_{n-1} z^n + \sum_{n=2}^{\infty} c_{n-2} z^n = 1 + z + \sum_{n=2}^{\infty} (c_{n-1} + c_{n-2}) z^n$. Now use the relation $c_n = c_{n-1} + c_{n-2}$ for $n \geq 2$ to conclude $1 + zf(z) + z^2 f(z) = f(z)$, then solve for $f(z)$.

11. The point z is on the circle $C_\rho(\alpha)$ with center α, so $z \neq \alpha$. Also, z_0 is in the interior of this circle, so again $z \neq z_0$.

13. To verify Identity (7.15), let $h(z) = \beta f(z)$. Clearly $\frac{h^{(n)}(\alpha)}{n!} = \frac{\beta f^{(n)}(\alpha)}{n!} = \beta a_n$. By Taylor's Theorem, $h(z) = \beta f(z) = \sum_{n=0}^{\infty} \beta a_n (z-\alpha)^n$.

15. Use the fact that $f'(z) = [z - (-1+i) + (-1+i)]^{-1}$ and expand $f'(z)$ in powers of $[z - (-1+i)]$. Then apply Corollary 7.2.

17a. By definition, $f(-z) = -f(z)$, so using the chain rule, we see that $f'(z) = \frac{d}{dz}f(z) = -\frac{d}{dz}f(-z) = -f'(-z)(-1) = f'(-z)$. But this means that f' is an even function.

17c. If f is even, then by part b f' is odd, so $f'(0) = -f'(-0) = -f'(0)$. Of course, this implies $f'(0) = 0$. Similarly, from part a f'' is even, so $f'''(0) = 0$. An induction argument gives $f^{(2n-1)}(0) = 0$ for all positive integers n. Show the details.

19a. It is easy to show that $f^{(n)}(0) = n!$ for all positive integers n. Do so via mathematical induction.

19b. The point $z = \frac{1}{2}$ is a removable singularity, since f may be redefined at $\frac{1}{2}$ to be analytic. State what f should equal at that point.

Section 7.3 Laurent Series Representations
(Exercise questions are on page 230)

1. $\frac{1}{z^3 - z^4} = \sum_{n=0}^{\infty} z^{n-3}$ for $0 < |z| < 1$, $\frac{1}{z^3 - z^4} = -\sum_{n=1}^{\infty} \frac{1}{z^{n+3}}$ for $|z| > 1$.

3. $\sum_{n=0}^{\infty} \frac{(-1)^n 2^{2n+1} z^{2n-3}}{(2n+1)!}$ for $|z| > 0$.

5. $\sum_{n=0}^{\infty} \frac{(-1)^n}{(2n+1)! z^{2n+1}}$ valid for $|z| > 0$.

7. $\sum_{n=1}^{\infty} \frac{2 z^{4n-7}}{(4n-2)!}$ valid for $|z| > 0$.

9. $z^{-1}(4-z)^{-2} = \frac{1}{16z} + \sum_{n=0}^{\infty} \frac{(n+2)z^n}{4^{n+3}}$ for $|z| < 4$,

$z^{-1}(4-z)^{-2} = \sum_{n=1}^{\infty} \frac{n(4)^{n-1}}{z^{n+2}}$ for $|z| > 4$.

11. $\text{Log}(\frac{z-a}{z-b}) = \sum_{n=1}^{\infty} \frac{b^n - a^n}{nz^n}$ valid for $|z| > b$. Explain.

13. $\csc z = \frac{1}{z} + \frac{z}{6} + \frac{7z^3}{360} + \cdots$.

15a. This identity is obtained by straightforward substitution, and partial fraction decomposition.

15b. $f(z) = \frac{1}{3} + \frac{1}{3} \sum_{n=1}^{\infty} 2^{-n}(z^n + z^{-n})$.

17. Since $f(z) = \sum_{n=-\infty}^{\infty} a_n z^n$ is valid for $|z| = 1$ (explain), letting $z = e^{i\theta}$ gives $f(e^{i\theta}) = \sum_{n=-\infty}^{\infty} a_n e^{in\theta}$ immediately. By Laurent's Theorem, $a_n = \frac{1}{2\pi i} \int_{c_1^+(0)} \frac{f(z)}{z^{n+1}} dz$ for all integers

n (explain). Parametrizing $C_1^+(0)$ with $z(\theta) = e^{i\phi}$ for $0 \le \phi \le 2\pi$ gives the desired result. Show the details.

19. Since $\sum\limits_{n=1}^{\infty} c_{-n}(z-\alpha)^{-n}$ converges for $|z-\alpha| > r$, the ratio test guarantees that the series converges absolutely for $\{z : |z-\alpha| \ge s\}$, where $s > r$ (show the details). Thus, if $|z-\alpha| = s$, the series $\sum\limits_{n=1}^{\infty} |c_{-n}|s^{-n}$ converges. Since $|c_{-n}(z-\alpha)^{-n}| \le |c_{-n}|s^{-n}$ for all $|z| \ge s$, the Weierstrass M test gives us our conclusion. Explain.

Section 7.4 Singularities, Zeros, and Poles
(Exercise questions are on page 237)

1a. Zeros of order 4 at $\pm i$.

1c. Simple zeros at $-1 \pm i$.

1e. Simple zeros at $\pm i$ and $\pm 3i$.

1g. Simple zeros at $\frac{\sqrt{3}\pm i}{2}$, $\frac{-\sqrt{3}\pm i}{2}$, and $\pm i$.

1i. Zeros of order 2 at $\frac{1 \pm i\sqrt{3}}{2}$ and -1.

1. Simple zeros at $\frac{1\pm i}{\sqrt{2}}$ and $\frac{-1\pm i}{\sqrt{2}}$, and a zero of order 4 at the origin.

2a. Poles of order 3 at $\pm i$, and a pole of order 4 at 1.

2c. Simple poles at $\frac{\sqrt{3}\pm i}{2}$, $\frac{-\sqrt{3}\pm i}{2}$, and $\pm i$.

2e. Simple poles at $\pm\sqrt{3}i$ and $\frac{\pm i}{\sqrt{3}}$.

2g. Simple poles at $z = n\pi$ for $n = \pm 1, \pm 2, \dots$.

2i. Simple poles at $z = n\pi$ for $n = \pm 1, \pm 2, \dots$, and a pole of order 3 at the origin.

2k. Simple poles at $z = 2n\pi$ for $n = 0, \pm 1, \pm 2, \dots$.

3a. Removable singularity at the origin.

3c. Essential singularity at the origin.

3e. Removable singularity at the origin, and a simple pole at -1.

3g. Removable singularity at the origin.

5. By Theorem 7.11, $f(z) = (z - z_0)^k h(z)$, where h is analytic at z_0 and $h(z_0) \ne 0$. We compute

$$f'(z) = k(z-z_0)^{k-1}h(z) + (z-z_0)^k h'(z)$$
$$= (z-z_0)^{k-1}[kh(z) + (z-z_0)h'(z)]$$
$$= (z-z_0)^{k-1}g(z),$$

where $g(z) = kh(z) + (z-z_0)h'(z)$. Explain why $g(z_0) \ne 0$, why g is analytic at z_0, and why Theorem 7.11 now gives the conclusion.

448

7. If it so happens that $m = n$, and the coefficients in the Laurent expansions for f and g about z_0 are negatives of each other, then $f + g$ will have a Taylor series representation at z_0, making z_0 a removable singularity (show the details for this). If $m \neq n$, then it is easy to show that $f + g$ still has a pole. State why, and what the order of the pole is.

9. Appeal to Theorem 7.12 and mimic the argument given in the solution to Problem 5.

11a. Simple poles at $z = \frac{1}{n\pi}$ for $n = \pm 1, \pm 2, \ldots$, and a nonisolated singularity at the origin.

Section 7.5 Applications of Taylor and Laurent Series
(Exercise questions are on page 242)

1a. No. Otherwise $0 = \lim\limits_{n\to\infty} f(\frac{1}{2n}) = f(\lim\limits_{n\to\infty} \frac{1}{2n}) = f(0)$. On the other hand, $1 = \lim\limits_{n\to\infty} f(\frac{1}{2n-1}) = f\left(\lim\limits_{n\to\infty} \frac{1}{2n-1}\right) = f(0)$. Justify and explain.

1b. Yes. There is a simple function with this property. Find it.

1c. No. Use Corollary 7.10 to show that for all z in some disk $D_r(0)$ we have $f(z) = z^3$, and $f(z) = -z^3$, and explain why this is impossible.

3a. Let $z_n = \frac{1}{\pi n}$. Explain.

3b. No, the function f is not analytic at zero (explain why), which is required by the corollary.

5. For $x \neq 0$, $\lim\limits_{x\to 0} |x \sin \frac{1}{x}| \leq \lim\limits_{x\to 0} |x| = 0$. This implies $\lim\limits_{x\to 0} f(x) = 0 = f(0)$. For the complex case, show that there is an essential singularity at 0 and use Theorem 7.17.

Section 8.1 Calculation of Residues
(Exercise questions are on page 250)

1a. 1.

1c. 1.

1e. 1.

1g. 0.

1i. e.

1k. 0.

1m. 4.

3a. $\frac{\pi + i\pi}{8}$.

3c. $(1 - \cos 1)2\pi i$.

3e. $i2\pi \sinh 1$.

3g. $\frac{2\pi i}{3}$.

5a. $-\frac{4\pi i}{25}$.

7a. $-\frac{\pi\sqrt{3}}{24}$.

9a. $\frac{1}{z+1} - \frac{1}{z+2}$.

9c. $\frac{1}{z^2} - \frac{2}{z} + \frac{3}{z+4}$.

9e. $\frac{2}{z-1} + \frac{1}{(z-1)^2} - \frac{2}{(z-1)^3}$.

11. By Theorem 8.2 we have $\text{Res}[g, n] = \lim\limits_{z \to n}(z - n)g(z)$, where n is any integer. Since $g(z) = \pi f(z) \cot \pi z = \pi f(z)\frac{\cos(\pi z)}{\sin(\pi z)}$, and because f, is analytic at n, we use L'Hôpital's rule to get $\lim\limits_{z \to n} \frac{z-n}{\sin(\pi z)} = 1$. Explain how this gives the result.

Section 8.2 Trigonometric Integrals
(Exercise questions are on page 255)

1. $\frac{\pi}{2}$.

3. $\frac{\pi}{2}$.

5. $\frac{\pi}{4}$.

7. $\frac{5\pi}{32}$.

9. $\frac{\pi}{18}$.

11. $\frac{5\pi}{8}$.

13. $\frac{3\pi}{8}$.

15. $\frac{2\pi}{\sqrt{d^2-a^2-b^2}}$.

Section 8.3 Improper Integrals of Rational Functions
(Exercise questions are on page 259)

1. $\frac{\pi}{8}$.

3. 0.

5. $\frac{7\pi}{18}$.

7. $\frac{\pi}{9}$.

9. $\frac{\pi}{9}$.

11. $\frac{2\pi}{3}$.

13. $\frac{2\pi}{3}$.

15. $\frac{\pi}{8a^3}$.

Section 8.4 Improper Integrals of Trigonometric Functions

(Exercise questions are on page 262)

1. $\int_{-\infty}^{\infty} \frac{\cos x}{x^2+9} \, dx = \frac{\pi}{3e^3}$, and $\int_{-\infty}^{\infty} \frac{\sin x}{x^2+9} \, dx = 0$.

3. $\frac{\pi}{4e^2}$.

5. $\frac{\pi}{5}\left(\frac{1}{2e^2} - \frac{1}{3e^3}\right)$.

7. $\frac{\pi \cos 1}{2e^2}$.

9. $\frac{\pi \sin 1}{2e}$.

11. $\frac{\pi \cos 2}{e^2}$.

13. The inequality $\left| \int_{C_R} \frac{\exp(iz)P(z)}{Q(z)} \, dz \right| \le \frac{2\varepsilon}{\pi} \int_0^{\frac{\pi}{2}} e^{\frac{-2R\theta}{\pi}} R \, d\theta < \varepsilon$ in Jordan's lemma would not be possible to get if we replaced $\exp(iz)$ by either the complex sine or cosine. Explain!

Section 8.5 Indented Contour Integrals

(Exercise questions are on page 267)

1. 0.

3. $\frac{\pi}{\sqrt{3}}$.

5. $\frac{\pi}{2}$.

7. π.

9. $\frac{2}{\pi}$.

11. $\pi\left(1 - \frac{1}{e}\right)$.

13. $\pi(1 - \cos 1)$.

15. π.

17. $\frac{2\sqrt{3}}{9}\pi$.

Section 8.6 Integrands with Branch Points

(Exercise questions are on page 271)

1. $\frac{2\pi}{\sqrt{3}}$.

3. $\frac{\pi}{2}$.

5. $\pi \ln 2$.

7. $\frac{\pi^3}{8}$.

9. $\frac{1}{4}\pi \ln 2$.

11c. No. The hypotheses of Theorem 8.7 are not satisfied. Explain why they are not.

13. $\frac{\pi}{\sqrt{2}}$.

15. $\dfrac{\pi\sqrt{2}}{1+\sqrt{3}}$.

17. $\dfrac{\pi}{a\sin \pi a}$.

19. π.

21. $\dfrac{\sqrt{\pi}}{2\sqrt{2}}$.

Section 8.7 The Argument Principle and Rouché's Theorem
(Exercise questions are on page 278)

1a. 1.

1c. 5.

3a. Let $f(z) = 15$. Then $|f(z) + g(z)| = |z^5 + 4z| < 6 < |f(z)|$. As f has no roots in $D_1(0)$, neither does g by Rouché's Theorem.

5a. Let $f(z) = -6z^2$. Then $|f(z)+g(z)| = |z^5+2z+1|$. It is easy to show that $|f(z)+g(z)| < |f(z)|$ for $z \in C_1(0)$. Complete the details.

7a. Let $f(z) = 7$. Then $|f(z) + g(z)| \le 6 < |f(z)|$. Show the details and explain why this gives the conclusion you want.

9. Let $f(z) = z^n$. Then $|f(z) + g(z)| = |h(z)| < 1 = |f(z)|$. Complete the argument.

Section 9.1 Basic Properties of Conformal Mappings
(Exercise questions are on page 285)

1a. All z.

1c. All z except $z = 0, -2$.

1e. All z except $z = 0$.

3. $f'(1) = 1$, $\alpha = \text{Arg}\big(f'(1)\big) = 0$, $|f'(1)| = 1$;

$f'(1 + i) = \frac{1}{2} - \frac{i}{2}$, $\alpha = \text{Arg}\big(f'(1+i)\big) = -\frac{\pi}{4}$, $|f'(1+i) = \frac{\sqrt{2}}{2}$;

$f'(i) = -i$, $\alpha = \text{Arg}\big(f'(i)\big) = -\frac{\pi}{2}$, $|f'(1)| = 1$.

$f'(-1)$ Does not exist.

5. $f'(\frac{\pi}{2} + i) = -i\sinh 1$, $\alpha - \text{Arg}\big(f'(-\frac{\pi}{2} + i)\big) = -\frac{\pi}{2}$, $f'(\frac{\pi}{2} + i) = \sinh 1$;

$f'(-\frac{\pi}{2} + i) = i\sinh 1$, $\alpha = \text{Arg}\big(f'(-\frac{\pi}{2} + i)\big) = \frac{\pi}{2}$, $f'(-\frac{\pi}{2} + i) = \sinh 1$;

$f'(0) = 1$, $\alpha = \text{Arg}\big(f'(0)\big) = 0$, $|f'(0)| = 1$.

7. $|f'(a + ib)| = \frac{1}{|2\sqrt{a+ib}|} = \frac{1}{2(a^2+b^2)^{\frac{1}{4}}} \ne 0$, hence $f(z)$ is conformal at $z = a + ib$. The lines $z_1(t) = a + (b + t)i$ and, $z_2(t) = (a + t) + ib$ intersect orthogonally at the point $z_1(0) = z_2(0) = a + ib$, therefore, their image curves will intersect orthogonally at the point $a + ib$.

9. $|f'(a+ib)| = |\cos(a + ib)| = \sqrt{\cos^2 a \cosh^2 b + \sin^2 a \sinh^2 b} \ne 0$, hence $f(z)$ is conformal at $z = a+ib$. The lines $z_1(t) = a+ti$ and $z_2(t) = a+t$ intersect orthogonally at the point $z_1(0) = z_2(0) = a$. Therefore, their image curves will intersect orthogonally at the point $\sin(a + ib)$.

11. First show that the mapping $W = \bar{Z}$ preserves the magnitude, but reverses the sense of angles at Z_0. Then consider the mapping $w = f(z)$ as a composition.

Section 9.2 Bilinear Transformations
(Exercise questions are on page 292)

1. $z = S^{-1}(w) = \frac{-2w+2}{(1+i)w-(1-i)} = \frac{(1-i)(1-w)}{i+w}$.

3. The disk $|w| < 1$.

5. The region $|w| > 1$.

7. $w = S(z) = \frac{-iz+i}{1+z}$.

9. $w = S(z) = \frac{1-iz}{1+z}$.

11. The disk $|w| < 1$.

13. The portion of the disk $|w| < 1$ that lies in the upper half-plane $\text{Im}(w) > 0$.

15. The region of that lies exterior to both the circles $|w - \frac{1}{2}| = \frac{1}{2}$ and $|w - \frac{3}{2}| = \frac{1}{2}$.

17. The equation $z = \frac{az+ib}{cz+d}$ can be written as $cz^2 + (d-a)z - b = 0$, and a quadratic equation has, at most, two distinct solutions.

Section 9.3 Mappings Involving Elementary Functions
(Exercise questions are on page 298)

1. The portion of the disk $|w| < 1$ that lies in the first quadrant $\{(u,v) : u > 0, v > 0\}$.

3. The horizontal strip $\{w : 0 < \text{Im}(w) < 1\}$.

5. The vertical strip $\{w : 0 < \text{Re}(w) < 1\}$.

7. The region $1 < |w|$.

9. The horizontal strip $\{w : 0 < \text{Im}(w) < \pi\}$.

11. The portion of the upper half-plane $\text{Im}(w) > 0$ that lies in the region $|w| > 1$.

13. $Z = z^2 + 1$, $w = Z^{\frac{1}{2}}$, where the principal brance of the square root is used.

15. The unit disk $|w| < 1$.

Section 9.4 Mapping by Trigonometric Functions
(Exercise questions are on page 304)

1. The portion of the disk $|w| < 1$ that lies in the second quadrant $\text{Re}(w) < 0$, $\text{Im}(w) > 0$.

3. The right branch of the hyperbola $\{(u,v) : u^2 - v^2 = \frac{1}{2}\}$.

5. The region in the first quadrant $\{(u,v) : u > 0, v > 0\}$ that lies inside the ellipse given by $\{(u,v) : \frac{u^2}{\cosh^2 1} + \frac{v^2}{\sinh^2 1} = 1\}$ and to the left of the hyperbola $\{(u,v) : u^2 - v^2 = \frac{1}{2}\}$.

7. (a) $\frac{\pi}{3}$, (b) $-\frac{5\pi}{6}$.

11. The upper half-plane $\text{Im}(w) > 0$.

13. The semi-infinite strip $\{(u, v) : 0 < u < \frac{\pi}{2},\ v > 0\}$.

15. The upper half-plane $\text{Im}(w) > 0$.

Section 10.2 The Dirichlet Problem
(Exercise questions are on page 314)

1   $\phi(x, y) = 15 - 9y$.

3   $\phi(x, y) = 5 + \frac{3}{\ln 2} \ln |z|$.

5   $\phi(x, y) = 4 - \frac{4}{\pi}\text{Arg}(z + 3) + \frac{6}{\pi}\text{Arg}(z + 1) - \frac{3}{\pi}\text{Arg}(z - 2)$, or
    $\phi(x, y) = 4 - \frac{4}{\pi}\text{Arctan}(\frac{y}{x+3}) + \frac{6}{\pi}\text{Arctan}(\frac{y}{x+1}) - \frac{3}{\pi}\text{Arctan}(\frac{y}{x-2})$.

7   $\phi(x, y) = \frac{5}{\pi}\text{Arctan}(\frac{1-x^2-y^2}{2y})$.

9   $\phi(x, y) = 5 - \frac{20}{\pi}\text{Arg}(\frac{i(1-z)}{1+z}) = 5 - \frac{20}{\pi}\text{Arctan}(\frac{1-x^2-y^2}{2y})$.

11   $\phi(x, y) = 3 - \frac{12}{\pi}\text{Arg}(\frac{i-iz^2}{1+z^2}) = 3 - \frac{12}{\pi}\text{Arctan}(\frac{1-(x^2+y^2)^2}{4xy})$.

Section 10.3 Poisson's Integral Formula
(Exercise questions are on page 320)

1. $\phi(x, y) = \frac{y}{2\pi} \ln\left[\frac{(x-1)^2+y^2}{(x+1)^2+y^2}\right] + \frac{x-1}{\pi}\text{Arctan}(\frac{y}{x-1}) - \frac{x+1}{\pi}\text{Arctan}(\frac{y}{x+1}) + 1$

3. Both $e^y \cos x$ and $e^{-y} \cos x$ are harmonic in the upper half plane and satisfy the boundary conditions. Also, $\lim\limits_{y \to \infty} e^{-y} \cos x = 0$. It can be show that the Poisson integral formula defines a bounded function in the upper half plane, therefore the desired solution is $\phi(x, y) = e^{-y} \cos x$.

5. Apply Leibniz's rule $\phi_{xx} + \phi_{yy} = \frac{1}{\pi}\int_{-\infty}^{\infty} U(t)\left[\left(\frac{\partial^2}{\partial x^2} + \frac{\partial^2}{\partial y^2}\right)\frac{y}{(x-t)^2+y^2}\right]dt$. The term in brackets in the integrand is
$\frac{\partial^2}{\partial x^2}\frac{y}{(x-t)^2+y^2} + \frac{\partial^2}{\partial y^2}\frac{y}{(x-t)^2+y^2} = \frac{2(3t^2y-6txy+3x^2y-y^3)}{((x-t)^2+y^2)^3} + \frac{2(-3t^2y+6txy-3x^2y+y^3)}{((x-t)^2+y^2)^3} = 0$. Hence the integrand vanishes and $\phi_{xx}(x, y) + \phi_{yy}(x, y) = 0$, which implies that $\phi(x, y)$ is harmonic.

7. $\phi(-x, y) = \frac{y}{\pi}\int_{-\infty}^{\infty}\frac{U(t)\,dt}{(-x-t)^2+y^2} = \frac{y}{\pi}\int_{-\infty}^{\infty}\frac{U(t)\,dt}{(x+t)^2+y^2} = \frac{y}{\pi}\int_{\infty}^{-\infty}\frac{U(-t)(-1)\,dt}{(x-t)^2+y^2}$

$= \frac{y}{\pi}\int_{-\infty}^{\infty}\frac{U(-t)\,dt}{(x-t)^2+y^2} = -\frac{y}{\pi}\int_{-\infty}^{\infty}\frac{U(t)\,dt}{(x-t)^2+y^2} = -\phi(x, y)$.

Section 10.5 Steady State Temperatures
(Exercise questions are on page 328)

1. For $H(x, y, z) = \frac{1}{\sqrt{x^2+y^2+z^2}}$, we get
$H_{xx} + H_{yy} + H_{zz} = \frac{2x^2-y^2-z^2}{(x^2+y^2+z^2)^{\frac{5}{2}}} + \frac{-x^2+2y^2-z^2}{(x^2+y^2+z^2)^{\frac{5}{2}}} + \frac{-x^2-y^2+2z^2}{(x^2+y^2+z^2)^{\frac{5}{2}}} = 0$, and for
$h(x, y) = \frac{1}{\sqrt{x^2+y^2}}$ we have $h_{xx} + h_{yy} = \frac{2x^2-y^2}{(x^2+y^2)^{\frac{5}{2}}} + \frac{-x^2+2y^2}{(x^2+y^2)^{\frac{5}{2}}} = \frac{1}{(x^2+y^2)^{\frac{3}{2}}} \neq 0$.

3. $T(x, y) = 10 + \frac{10}{\pi}\text{Arctan}(\frac{2xy}{x^2-y^2-1}) - \frac{10}{\pi}\text{Arctan}(\frac{2xy}{x^2-y^2+1})$.

5. $T(x,y) = 100 + \frac{100}{\pi}\text{Arctan}(\sin z + 1) - \frac{100}{\pi}\text{Arctan}(\sin z - 1)$.

7. $T(x,y) = \frac{50}{\ln 2}\ln|z|$.

9. $T(x,y) = 25 + \frac{150}{\pi}\text{Re}(\text{Arcsin}z^2)$.

11. $T(x,y) = \frac{200}{\pi}\text{Re}(\text{Arcsin}\frac{1}{z})$.

13. The isothermals are $\{(x,y) : T(x,y) = k\}$.The equation $100 - \frac{100}{\pi}\arctan(\frac{1-x^2-y^2}{2y}) = k$ can be manipulated to yield $c = \tan\frac{\pi}{100}(100-k) = \frac{1-x^2-y^2}{2y}$ which is better recognized as the circle $x^2 + (y+c)^2 = 1 + c^2$.

15. $T(x,y) = 40 + 20\left[\text{Im}(\text{Arcsin}(z))\right]$.

Section 10.6 Two-Dimensional Electrostatics

(Exercise questions are on page 336)

1. $\phi(x,y) = 100 + \frac{100}{\ln 2}\ln|t|$.

3. $\phi(x,y) = 150 - \frac{200x}{x^2+y^2}$.

5. $\phi(x,y) = 50 + \frac{200}{\pi}\text{Re}(\text{Arcsin}z)$.

7. (a) $w = S(z) = \frac{2z-6}{z+3}$, (b) $\phi(x,y) = 200 - \frac{200}{\ln 2}\ln\left|\frac{2z-6}{z+3}\right|$.

Section 10.7 Two-Dimensional Fluid Flow

(Exercise questions are on page 346)

1. (a) $V(r,\theta) = A\overline{\left(1 - \frac{1}{e^{2i\theta}}\right)} = A\overline{(1 - e^{-2i\theta})} = A(1 - \cos 2\theta - i\sin 2\theta)$;

(c) $z = 1$ and $z = -1$.

3. (a) Speed $= A|\bar{z}|$. The minimum speed is $A|1-i| = A\sqrt{2}$.

(b) The maximum pressure in the channel occurs at the point $1 + i$.

5a. $\Psi(r,\theta) = Ar^{\frac{3}{2}}\sin\frac{3\theta}{2}$.

Section 10.8 The Joukowski Airfoil

(Exercise questions are on page 354)

1. The condition $z + \frac{1}{z} = w$ implies that $z^2 + 1 = zw$. Rewrite as $z^2 - zw + 1 = 0$, and then use the quadratic formula.

3. (a) $x^2 + (y-a)^2 = 1 + a^2$

(b) Use the inverse $x + iy = \frac{1-u^2-v^2}{(1-u)^2+v^2} + i\frac{2v}{(1-u)^2+v^2}$ and substitute for x and y in part (a) to obtain the equation $\frac{4(u-av)}{(1-u)^2+v^2} = 0$, which yields the line $v = \frac{1}{a}u$,

(c) the slope is $\text{Arctan}(\frac{1}{a}) = \frac{\pi}{2} - \text{Arctan}(a)$.

Section 10.9 The Schwarz-Christoffel Transformation
(Exercise questions are on page 360)

1. $f'(z) = A(z-a)^{-\frac{\pi-k\pi}{\pi}} = A(z-a)^{k-1}$. Integrate and get $f(z) = \frac{A}{k}(z-a)^k$, then choose $A = k$.

3. $f'(z) = A(z+1)^{\frac{1}{2}}(z-1)^{\frac{1}{2}} = A\left[\frac{z}{(z^2-1)^{\frac{1}{2}}} + \frac{1}{(z^2-1)^{\frac{1}{2}}}\right]$. Integrate with the boundary conditions $f(-1) = 0$ and $f(1) = -1$ to get $w = f(z) = \frac{1}{\pi}\left[(z^2-1)^{\frac{1}{2}} + \text{Log}(z + (z^2-1)^{\frac{1}{2}})\right] - i$.

5. $f'(z) = A(z+1)^{-1}z(z-1)^{-1}$, and $w = f(z) = \text{Log}(z^2-1)^{\frac{1}{2}}$.

7. $f'(z) = A(z+1)^1 z^{-1} = A(1 + \frac{1}{z})$. Integrate to get $f(z) = z + \text{Log}(z)$.

9. Select $x_1 = -\frac{1-\alpha}{\alpha}$ and $x_2 = x_3 = 1$, then form $f'(z) = A(z + \frac{1-\alpha}{\alpha})^{-\alpha}(z)(z-1)^{\alpha-1}$.
Computation then reveals that $A = (\frac{1-\alpha}{\alpha})^{\alpha-1}$, which is used to construct the desired function $w = f(z) = \int A(z + \frac{1-\alpha}{\alpha})^{-\alpha}(z)(z-1)^{\alpha-1}\,dz = (z-1)^\alpha (1 + \frac{\alpha z}{1-\alpha})^{1-\alpha}$.

11. $f'(z) = Az^{-\frac{1}{2}}(z-1) = A(z^{\frac{1}{2}} - z^{-\frac{1}{2}})$. Integrate and get $f(z) = -\frac{i}{2}z^{\frac{1}{2}}(z-3)$.

Section 10.10 Image of a Fluid Flow
(Exercise questions are on page 364)

1. $f'(z) = A(z+1)^{-\frac{1}{2}}z(z-1)^{-\frac{1}{2}} = A\frac{z}{(z^2-1)^{\frac{1}{2}}}$. Integrate with the boundary conditions $f(-1) = 0$ and $f(0) = i$ to get $w = f(z) = (z^2-1)^{\frac{1}{2}}$.

3. $w = f(z) = (z-1)^\alpha \left(1 + \frac{\alpha z}{1-\alpha}\right)^{1-\alpha}$.

5. $w = f(z) = -1 + \int_{-1}^z \frac{(\xi-1)^{\frac{1}{4}}}{\xi^{\frac{1}{4}}}\,d\xi$.

$w = f(z) =$
$i + \frac{1}{\pi}\left[4(z-1)^{\frac{1}{4}}z^{\frac{3}{4}} - 2\text{Arctan}(1 - \frac{1}{z})^{\frac{1}{4}} + \text{Log}(1 - (1 - \frac{1}{z})^{\frac{1}{4}}) - \text{Log}(1 + (1 - \frac{1}{z})^{\frac{1}{4}})\right]$.

Section 10.11 Sources and Sinks
(Exercise questions are on page 372)

1. $F_1(w) = \log(\frac{w-1}{w+1})$ is the complex potential for a source at $w_1 = 1$ and sink at $w_2 = -1$. The function $w = S(z) = z^2$ maps $z_1 = 1$ and $z_2 = i$ onto w_1 and w_2, respectively. Therefore, the composition $F_2(z) = F_1(S(z)) = F_1(z^2) = \log(\frac{z^2-1}{z^2+1})$ is the desired complex potential.

3. $F(z) = \log(\sin z)$.

5. $F(z) = \log(z^2 - 1)$.

7. $w = 2(z+1)^{\frac{1}{2}} + \text{Log}\left(\frac{1-(z+1)^{\frac{1}{2}}}{1+(z+1)^{\frac{1}{2}}}\right) + i\pi$.

9. $w = \frac{1}{\pi}\text{Arcsin}(z) + \frac{i}{\pi}\text{Arcsin}(\frac{1}{z}) + \frac{1+i}{2}$.

Section 11.1 Fourier Series
(Exercise questions are on page 381)

1. $U(t) = \frac{4}{\pi} \sum\limits_{n=1}^{\infty} \frac{1}{2n-1} \sin[(2n-1)t]$.

3. $V'(t) = \frac{4}{\pi} \sum\limits_{n=1}^{\infty} \frac{1}{(2n-1)^2} \frac{d}{dt} \cos[(2n-1)t] = -\frac{4}{\pi} \sum\limits_{n=1}^{\infty} \frac{1}{2n-1} \sin[(2n-1)t] = -U(t)$.

5. $\frac{\pi}{2} = V(0) = \frac{4}{\pi} \sum\limits_{n=1}^{\infty} \frac{1}{(2n-1)^2} \cos(0) = \frac{4}{\pi} \sum\limits_{n=1}^{\infty} \frac{1}{(2n-1)^2}$. Now solve for $\sum\limits_{n=1}^{\infty} \frac{1}{(2n-1)^2}$.

7. $U(t) = \frac{4}{\pi} \sum\limits_{n=1}^{\infty} \frac{(-1)^{n-1}}{(2n-1)^2} \sin[(2n-1)t]$.

9. $U(t) = \frac{2}{\pi} \sum\limits_{n=1}^{\infty} \frac{1}{2n-1} \sin[(2n-1)t] - \frac{4}{\pi} \sum\limits_{n=1}^{\infty} \frac{1}{2(2n-1)} \sin[2(2n-1)t]$, where $a_n = 0$ for all n, and $b_{4n} = 0$ for all n.

11. $U(t) = \frac{2}{\pi} \sum\limits_{n=1}^{\infty} \frac{1}{2n-1} \sin[(2n-1)t] + \frac{4}{\pi} \sum\limits_{n=1}^{\infty} \frac{1}{2(2n-1)} \sin[2(2n-1)t]$, where $a_n = 0$ for all n, and $b_{4n} = 0$ for all n.

Section 11.2 The Dirichlet Problem for the Unit Disk
(Exercise questions are on page 386)

1. $u(r\cos\theta, r\sin\theta) = \frac{4}{\pi} \sum\limits_{n=1}^{\infty} \frac{1}{2n-1} r^{2n-1} \sin[(2n-1)\theta]$.

3. $u(r\cos\theta, r\sin\theta) = \frac{4}{\pi} \sum\limits_{n=1}^{\infty} \frac{(-1)^{n-1}}{2n-1} r^{2n-1} \cos[(2n-1)\theta]$.

5. $u(r\cos\theta, r\sin\theta) = \frac{3\pi}{8} + \frac{2}{\pi} \sum\limits_{n=1}^{\infty} \frac{1}{(2n-1)^2} r^{2n-1} \cos[(2n-1)\theta]$

$\qquad -\frac{4}{\pi} \sum\limits_{n=1}^{\infty} \frac{1}{2^2(2n-1)^2} r^{4n-2} \cos[2(2n-1)\theta]$.

7. $u(r\cos\theta, r\sin\theta) = \frac{\pi}{8} + \frac{2}{\pi} \sum\limits_{n=1}^{\infty} \frac{1}{(2n-1)^2} r^{2n-1} \cos[(2n-1)\theta]$

$\qquad +\frac{4}{\pi} \sum\limits_{n=1}^{\infty} \frac{1}{2^2(2n-1)^2} r^{4n-2} \cos[2(2n-1)\theta]$.

Section 11.3 Vibrations in Mechanical Systems
(Exercise questions are on page 392)

1a. $U_h(t) = c_1 e^{-t} \sin(t) + c_2 e^{-t} \cos(t)$,

$U_p(t) = \sum\limits_{n=1}^{\infty} \frac{2(-1)^n}{4+n^4} \cos(nt) + \sum\limits_{n=1}^{\infty} \frac{(-1)^n(n^2-1)}{n(4+n^2)} \sin(nt)$,

$U(t) = c_1 e^{-t} \sin(t) + c_2 e^{-t} \cos(t) + \sum\limits_{n=1}^{\infty} \frac{2(-1)^n}{4+n^4} \cos(nt) + \sum\limits_{n=1}^{\infty} \frac{(-1)^n(n^2-2)}{n(4+n^2)} \sin(nt)$.

1c. $U_h(t) = c_1 e^{-t} \sin(t) + c_2 e^{-t} \cos(t)$,

$$U_p(t) = -\sum_{n=1}^{\infty} \frac{8 \sin(\frac{n\pi}{2})}{n(4+n^2)} \cos(nt) + \sum_{n=1}^{\infty} \frac{4 \sin(\frac{n\pi}{2})(2-n^2)}{n^2(4+n^4)\pi} \sin(nt),$$

$$U(t) = c_1 e^{-t} \sin(t) + c_2 e^{-t} \cos(t) + \sum_{n=1}^{\infty} \frac{2(-1)^n}{4+n^4} \cos(nt) + \sum_{n=1}^{\infty} \frac{(-1)^n(n^2-2)}{n(4+n^2)} \sin(nt).$$

1c. Alternative Answer

$$U_p(t) = -\sum_{n=1}^{\infty} \frac{8(-1)^n}{(2n-1)(16n^4-32n^3+24n^20-8n+5)\pi} \cos\left((2n-1)t\right)$$

$$+ \sum_{n=1}^{\infty} \frac{4(-1)^n(4n^2-4n-1)}{(2n-1)^2(16n^4-32n^3+24n^20-8n+5)\pi} \sin\left((2n-1)t\right),$$

$$U(t) = c_1 e^{-t} \sin(t) + c_2 e^{-t} \cos(t) + \sum_{n=1}^{\infty} \frac{8(-1)^n}{(2n-1)16n^4-32n^3+24n^20-8n+5)\pi} \cos\left((2n-1)t\right)$$

$$+ \sum_{n=1}^{\infty} \frac{4(-1)^n(4n^2-4n-1)}{(2n-1)^2(16n^4-32n^3+24n^20-8n+5)\pi} \sin\left((2n-1)t\right).$$

3a. $U_h(t) = c_1 e^{-2t} + c_2 t e^{-2t}$,

$$U_p(t) = -\sum_{n=1}^{\infty} \frac{4(-1)^n}{(4+n^2)^2} \cos(nt) + \sum_{n=1}^{\infty} \frac{(-1)^n(n^2-4)^2}{n(4+n^2)^2} \sin(nt),$$

$$U(t) = c_1 e^{-2t} + c_2 t e^{-2t} - \sum_{n=1}^{\infty} \frac{4(-1)^n}{(4+n^2)^2} \cos(nt) + \sum_{n=1}^{\infty} \frac{(-1)^n(n^2-4)}{n(4+n^2)^2} \sin(nt).$$

3c. $U_h(t) = c_1 e^{-2t} + c_2 t e^{-2t}$,

$$U_p(t) = -\sum_{n=1}^{\infty} \frac{16 \sin(\frac{n\pi}{2})}{n(4+n^2)^2\pi} \cos(nt) + \sum_{n=1}^{\infty} \frac{4 \sin(\frac{n\pi}{2}(4-n^2))}{n^2(4+n^2)^2\pi} \sin(nt),$$

$$U(t) = c_1 e^{-2t} + c_2 t e^{-2t} - \sum_{n=1}^{\infty} \frac{16 \sin(\frac{n\pi}{2})}{n(4+n^2)^2\pi} \cos(nt) + \sum_{n=1}^{\infty} \frac{4 \sin(\frac{n\pi}{2})(4-n^2)}{n^2(4+n^2)^2\pi} \sin(nt).$$

3c. Alternative Answer

$$U_p(t) = -\sum_{n=1}^{\infty} \frac{16(-1)^n}{2n-1)(4n^2-4n+5)^2\pi} \cos\left((2n-1)t\right) + \sum_{n=1}^{\infty} \frac{4(-1)^n(4n^2-4n-3)}{4n^2-4n+5)^2\pi} \sin\left((2n-1)t\right),$$

$$U(t) = c_1 e^{-2t} + c_2 t e^{-2t} + \sum_{n=1}^{\infty} \frac{16(-1)^n}{2n-1)(4n^2-4n+5)^2\pi} \cos\left((2n-1)t\right)$$

$$+ \sum_{n=1}^{\infty} \frac{4(-1)^n(4n^2-4n-3)}{4n^2-4n+5)^2\pi} \sin\left((2n-1)t\right).$$

Section 11.4 The Fourier Transform
(Exercise questions are on page 396)

1a. $\mathfrak{F}\big(U(t)\big) = \frac{\sin(w)}{\pi w}$.

1c. $\mathfrak{F}\big(U(t)\big) = \frac{1-\cos(w)}{\pi w^2} = \frac{2\sin^2(\frac{w}{2})}{\pi w^2}$.

Section 11.5 The Laplace Transform

(Exercise questions are on page 402)

1. Use $s = \sigma + i\tau$ and the integral

$\int e^{-(\sigma+i\tau)t}\, dt = \frac{e^{-\sigma t}[-\sigma\cos(\tau t)+\tau\sin(\tau t)]}{\sigma^2+\tau^2} + i\frac{e^{-\sigma t}[\tau\cos(\tau t)+\sigma\sin(\tau t)]}{\sigma^2+\tau^2} = u(t) + iv(t)$. Supply the details showing that $\lim\limits_{t\to+\infty} u(t) = 0$ and $\lim\limits_{t\to+\infty} v(t) = 0$. From that conclusion it follows that $\mathcal{L}(1) = \int_0^\infty e^{-(\sigma+i\tau)t}\, dt = 0 + 0i - \frac{-1}{\sigma+i\tau} = \frac{1}{s}$.

3a. $\mathcal{L}\big(f(t)\big) = \frac{1}{s^2} - \frac{ce^{-cs}}{s} - \frac{e^{-cs}}{s^2}$.

3c. $\mathcal{L}\big(f(t)\big) = \frac{1}{s-a} - \frac{e^{a-s}}{s-a}$.

4a. $\mathcal{L}(3t^2 - 4t + 5) = \frac{6}{s^3} - \frac{4}{s^2} + \frac{5}{s}$.

4c. $\mathcal{L}(e^{2t-3}) = \frac{e^{-3}}{s-2}$.

4e. $\mathcal{L}\big((t+1)^4\big) = \frac{24}{s^5} + \frac{24}{s^4} + \frac{12}{s^3} + \frac{4}{s^2} + \frac{1}{s}$.

5a. $\mathcal{L}^{-1}\big(\frac{1}{s^2+25}\big) = \frac{1}{5}\sin 5t$.

5c. $\mathcal{L}^{-1}\big(\frac{1+s^2-s^3}{s^4}\big) = -1 + t + \frac{t^3}{6}$.

5e. $\mathcal{L}^{-1}\big(\frac{6s}{s^2-4}\big) = 3e^{-2t} + 3e^{2t} = 6\cosh 2t$.

Section 11.6 Laplace Transforms of Derivatives and Integrals

(Exercise questions are on page 406)

1. $\mathcal{L}(\sin t) = \frac{1}{s^2+1}$.

3. $\mathcal{L}(\sin^2 t) = \frac{2}{s(s^2+4)}$.

5. $\mathcal{L}^{-1}\big(\frac{1}{s(s-4)}\big) = -\frac{1}{4} + \frac{1}{4}e^{4t}$.

7. $\mathcal{L}^{-1}\big(\frac{1}{s^2(s+1)}\big) = t - 1 + e^{-t}$.

9. $y(t) = 2\cos 3t + 3\sin 3t$.

11. $y(t) = -2 + 2\cos 2t + \sin 2t$.

13. $y(t) = 2 + e^t$.

15. $y(t) = -1 - \frac{1}{2}e^{-t} + \frac{3}{2}e^t = -1 + \sinh t + e^t$.

17. $y(t) = e^{-2t} + e^t$.

Section 11.7 Shifting Theorems and the Step Function

(Exercise questions are on page 409)

1. $\mathcal{L}(e^t - te^t) = \frac{-1}{(s-1)^2} + \frac{1}{s-1}$.

3. $\mathcal{L}(e^{at}\cos bt) = \frac{s-a}{(s-a)^2+b^2}$.

5a. $f(t) = \mathcal{L}^{-1}\big(\frac{s+2}{s^2+4s+5}\big) = e^{-2t}\cos t$.

459

5c. $f(t) = \mathcal{L}^{-1}\left(\frac{s+3}{(s+2)^2+1}\right) = e^{-2t}\cos t + e^{-2t}\sin t$.

7a. $\mathcal{L}\left(U_2(t)(t-2)^2\right) = \frac{2e^{-2s}}{s^3}$.

7c. $\mathcal{L}\left(U_{3\pi}(t)\sin(t-3\pi)\right) = \frac{e^{-3\pi s}}{s^2+1}$.

7e. $\mathcal{L}^{-1}\left(\frac{e^{-s}+e^{-2s}}{s}\right) = U_1(t) + U_2(t)$.

8. $\mathcal{L}\left(f(t)\right) = \frac{1}{s}(1 - 2e^{-s} + 2e^{-2s} - e^{-3s})$.

9a. $y(t) = -e^{-t}\cos t$.

9c. $y(t) = 2e^{-\frac{t}{2}}\sin\frac{t}{2}$.

9e. $y(t) = t^3 e^{-t}$.

9g. $y(t) = \left[1 - \delta(t - \frac{\pi}{2})\right]\left(\sin(t)\right) + (1 - \sin t)U_{\frac{\pi}{2}}(t)$.

Section 11.8 Multiplication and Division by t
(Exercise questions are on page 412)

1a. $\mathcal{L}(te^{-2t}) = \frac{1}{(s+2)^2}$.

1c. $\mathcal{L}(t\sin 3t) = \frac{6s}{(s^2+9)^2}$.

1e. $\mathcal{L}(t\sinh t) = \frac{2s}{(s^2-1)^2}$.

3. $\mathcal{L}\left(\frac{e^t-1}{t}\right) = \ln\frac{s}{s-1}$.

5. $\mathcal{L}(t\sin bt) = \frac{2bs}{(s^2+b^2)^2}$.

7. $\mathcal{L}^{-1}\left(\ln\frac{s^2+1}{(s-1)^2}\right) = \frac{2(e^t-\cos t)}{t}$.

8a. $y(t) = te^{-t} + t^2 e^{-t}$.

8c. $y(t) = Cte^t$.

8e. $y(t) = Ct$.

9. $y(t) = 1 - t$.

Section 11.9 Inverting the Laplace Transform
(Exercise questions are on page 419)

1a. $\mathcal{L}^{-1}\left(\frac{2s+1}{s(s-1)}\right) = -1 + 3e^t$.

1c. $\mathcal{L}^{-1}\left(\frac{4s^2-6s-12}{s(s+2)(s-2)}\right) = 3 + 2e^{-2t} - e^{2t}$.

1e. $\mathcal{L}^{-1}\left(\frac{2s^2+s+3}{(s+2)(s-1)^2}\right) = e^{-2t} + e^t + 2te^t$.

2a. $\mathcal{L}^{-1}\left(\frac{1}{s^2+4}\right) = \frac{1}{2}\sin 2t$.

3a. $\mathcal{L}^{-1}\left(\frac{s^3+s^2-s+3}{s^5-s}\right) = -3+e^t+e^{-t}+\cos t+\sin t = -3+2\cosh t+\cos t+\sin t.$

3c. $\mathcal{L}^{-1}\left(\frac{s^3+3s^2-s+1}{s^5-s}\right) = -1+e^t+e^{-t}-\cos t+\sin t = -1+2\cosh t-\cos t+\sin t.$

4. $\mathcal{L}^{-1}\left(\frac{s^3+2s^2+4s+2}{(s^2+1)(s^2+4)}\right) = \cos t+\sin 2t.$

5a. $y(t) = e^{-t}+e^{-t}\sin 2t.$

5c. $y(t) = e^{-t}+\cos 2t+\sin 2t.$

5e. $y(t) = 1+t.$

6a. $x(t) = 2e^{-2t}-e^t$, and $y(t) = e^{-2t}-2e^t$.

6c. $x(t) = -e^{-t}+2te^{-t}$, and $y(t) = te^{-t}$.

6e. $x(t) = -2+6e^{-t}\cos 2t+6e^{-t}\sin 2t$, and $y(t) = -1+3e^{-t}\cos 2t$.

Section 11.10 Convolution
(Exercise questions are on page 425)

1a. $f(t) = t$, $g(t) = t$, and $(f*g)(t) = \int_0^t f(\tau)g(t-\tau)\,d\tau = \int_0^t \tau(t-\tau)\,d\tau = \frac{t^3}{6}$.

1c. $f(t) = e^t$, $g(t) = e^{2t}$, and $(g*f)(t) = \int_0^t e^{2\tau}e^{t-\tau}\,d\tau = -e^t+e^{2t}$.

2a. $f(t) = \mathcal{L}^{-1}\left(\frac{2}{(s-1)(s-2)}\right) = -2e^t+2e^{2t}$.

2c. $f(t) = \mathcal{L}^{-1}\left(\frac{1}{s(s^2+1)}\right) = 1-\cos t$.

3. Prove the distributive law for convolution: $f*(g+h) = f*g+f*h$.

Proof $f*(g+h) = \int_0^t f(\tau)(g+h)(t-\tau)\,d\tau = \int_0^t f(\tau)g(t-\tau)\,d\tau + \int_0^t f(\tau)h(t-\tau)d\tau = f*g+f*g.$

5. $f(t) = \mathcal{L}^{-1}\left(\frac{s}{s-1}\right) = e^t+\delta(t)$.

7. $y(t) = -t\cos t+\sin t$.

9. $\mathcal{L}\left(\int_0^t e^{-\tau}\cos(t-\tau)\,d\tau\right) = \mathcal{L}(e^t)\mathcal{L}(\cos(t)) = \frac{s}{(s+1)(s^2+1)}$.

11. Given $F(s) = \mathcal{L}(f(t))$ and $G(s) = \mathcal{L}(1) = \frac{1}{s}$ and $g(t) = 1$, we have

$\mathcal{L}^{-1}\left(\frac{F(s)}{s}\right) = F(s)G(s) = (f*g)(t) = \int_0^t f(\tau)g(t-\tau)\,d\tau = \int_0^t f(\tau)\,d\tau.$

12a. $F(s) = \frac{1}{s-2}$, and $f(t) = e^{2t}$.

12c. $F(s) = \frac{2}{s^4-1}$, and $f(t) = \sinh t-\sin t$.

13a. $y(t) = te^{-t}$.

13c. $y(t) = (-e^{3-3t}+e^{1-t})U_1(t)$.

14a. $y(t) = -21te^{-t}-119e^{-t}+120-96t+36t^2-8t^3+t^4$.

461

www.ingramcontent.com/pod-product-compliance
Lightning Source LLC
Chambersburg PA
CBHW051748200326
41597CB00025B/4487